FUNDAMENTALS OF PHYSICAL ACOUSTICS

FUNDAMENTALS OF PHYSICAL ACOUSTICS

DAVID T. BLACKSTOCK
University of Texas
Austin, Texas

A WILEY-INTERSCIENCE PUBLICATION
JOHN WILEY & SONS, INC.
New York · Chichester · Weinheim · Brisbane · Singapore · Toronto

This book is printed on acid-free paper.

Published simultaneously in Canada.

Library of Congress Cataloging-in-Publication Data:

Blackstock, David T.
Fundamentals of physical acoustics/by David T. Blackstock
p. cm
"A Wiley-Interscience publication."
Includes index.
ISBN 0-471-31979-1 (acid-free paper)
1. Sound. I. Title.

QC225.15.B55 2000
534 21–dc21 99-042186

Printed in the United States of America.

10 9 8 7 6

To my wife
Marjorie

CONTENTS

PREFACE

This book is an introduction to physical acoustics. It has grown out of a full-year course given at the University of Texas at Austin for beginning graduate students in engineering and physics. Although the organization described below is in terms of a two-semester presentation, the material may easily be spread over three quarters for universities that operate on the quarter system.

The somewhat unconventional format of the book has a pedagogical basis. Except for Chapter 2, the first half of the book (the first semester) is mainly restricted to plane waves (for special needs some very simple spherical-wave theory is introduced). Most topics requiring the use of nonplanar waves are postponed to the last half of the book (the second semester). By concentrating on plane waves in the first semester, the instructor may introduce the student to a wide variety of acoustical concepts—for example, propagation, reflection and transmission, refraction, normal modes (rectangular geometry), horn theory, and absorption and dispersion—while keeping mathematical complexity to a minimum. Having had a good exposure to the field, the student is then well prepared for the second semester, which covers topics requiring more advanced mathematics, such as spherical and cylindrical waves, nonrectangular wave-guides, directional radiation, and diffraction. An instructor who wishes to follow this plan should use Chapters 1 and 3–9 for the first semester and Chapters 2 and 10–15 for the second semester.

The role (and position) of Chapter 2 also requires explanation. Two separate derivations of the wave equation are presented in the book. Again the motivation is pedagogical. The author feels strongly that since the wave equation is fundamental to the study of physical acoustics at the graduate level, its derivation should be rigorous. However, an advanced derivation is not likely to be meaningful to the beginning student. Two different derivations are therefore provided. The first time, in Chapter 1, the derivation is simple, restricted to plane waves in lossless fluids. By the beginning of the second semester, however, the student is better prepared (and motivated) for a rigorous treatment. On the other hand, engineers and scientists using the book as a reference do not expect to find a rigorous derivation of the wave equation in the middle of the book. As a compromise, therefore, the rigorous derivation appears in Chapter 2. However, the instructor who follows the recommended course sequence should postpone coverage of Chapter 2 until after Chapter 9.

If only one semester is available for the course, the instructor may wish to concentrate on the plane wave chapters, omit Chapter 2, and skip enough of the more advanced topics in Chapters 5–9 to make room for sections of choice from Chapters 10–15. This approach might also fit a two-quarter sequence at universities that operate on the quarter system. Finally, since undergraduates have successfully taken the course at UT Austin, instructors may find the book suitable for a senior-level course.

Although heavily influenced in my acoustical career by my graduate days at Harvard (1956–1960), particularly by the course work under F. V. ("Ted") Hunt, I feel fortunate that my teaching of basic acoustics did not begin until 14 years later. Had teaching begun immediately, the course would undoubtedly have been based on Hunt's model. While that would not have been a bad plan, the experiences of the 14 (and subsequent) years led me to develop my own model. For example, when my academic career began, at the University of Rochester in 1963, Hugh Flynn was already teaching the first-year acoustics course. I therefore developed an undergraduate course on wave motion for electrical engineering students. Certain parts of the present book have their roots in that course. Other parts have been heavily influenced by the research that my students and I have carried out. My belief in the usefulness of time-domain methods (see, e.g., Chapters 1, 3, and 14) stems from my research in nonlinear acoustics, where the time domain offers the easiest approach for certain problems.

The contributions of many persons over the lengthy period of development of this book are gratefully acknowledged. Among early typists of the manuscript were Jane Andrus, Pat Kleinert, Debbie Craig, and Carlie Tilly. Above all was the dedication of Becky Ellis, who never gave up typing the seemingly endless series of modifications I handed her and always found a way to do "all those impossible things" I asked for. Whatever success the book may enjoy as a text is due to the many students who for more than two decades used the material in handout form. Their questions and comments made me realize when a presentation was unclear, and their support, encouragement, and pestering kept me going. They were also the guinea pigs for the homework problems, many of which started out as test questions, given at the end of each chapter. A special word of thanks goes to Fred Cotaras and Jim Ten Cate, who urged and largely implemented the switch from typewriter text to word processor software, in particular, LaTeX. I thank Laura Brewer and Won Ohm for drawing the figures for Chapter 1, Jim Ten Cate for all figures in Chapter 7 and some in Chapter 13, and Robin Cleveland for Fig. 5.11. I owe a special debt to my colleagues Mark Hamilton and Elmer Hixson at the University of Texas, Chris Morfey at the University of Southampton, and Wayne Wright at Kalamazoo College for suggestions, criticism, and encouragement. Allan Pierce several times provided very good advice, and his book was a constant source of help. Finally, writing this book would have been exceedingly difficult without the extended support of Applied Research Laboratories. I am particularly grateful to Chester McKinney, who was always encouraging during the early stages of development of the book, and to Michael Pestorius, who was very supportive during the final stages.

1

INTRODUCTION

A. WHAT IS A WAVE?

Wave motion is such an intuitive concept that it is hard to define. It is easy to give a simple definition of a wave but difficult to give one for which exceptions or counterexamples cannot be found. Let us begin, therefore, by making some simple observations about wave motion.

1. A wave is a disturbance or deviation from a pre-existing condition. The motion of the disturbance constitutes a transfer of information from one point in space to another. Frequently the pre-existing condition is complete quiet, such as static equilibrium. For instance, a ripple on a quiet pond is a disturbance we regard as a wave. An ocean tidal wave (tsunami), however, rides on top of a surface that is already disturbed by swells, breakers, and ripples. In this case the disturbed surface is the normal or pre-existing condition, and the tidal wave is the deviation from it.

2. Time plays a vital role. Static displacement of a string or rubber membrane is indeed a disturbance, but it is not a wave. Moreover, a wave travels with finite, not infinite, speed. If one end of a perfectly rigid rod is hit by a hammer, all points along the rod feel the blow instantaneously, and the rod moves as a unit. No wave motion has occurred, only rigid body translation. In fact, of course, no rod is perfectly rigid. The rod's elasticity provides a means for the impulse to travel at finite speed so that the other end of the rod feels the blow after a short delay. The real mechanics therefore does include wave motion.

3. All mechanical waves travel in a material medium. Sound, for example, needs a fluid, such as air or water, or a solid. It cannot travel through a vacuum. Seismic waves travel through the earth. A wave on a violin string requires the presence of the string in order to exist. Some waves, however, require no medium. For example, electromagnetic waves, which include radio waves, light, and x rays, have no difficulty traveling through outer space.

4. Gross movement of the medium, although it may occur, is not a necessary part of the process of wave motion. A ripple on the surface of a stream may travel upstream or downstream, but the flow of the stream is not a requisite for the propagation even though it may modify the apparent propagation speed. Consider two examples.

(a) Wind blows a floating leaf across the surface of a pond. The leaf is a moving disturbance, and information is transferred. The process does not constitute wave motion, however, because the leaf's transference from one point to another is due solely to the wind, which is a gross movement of the medium.

(b) A ripple set up by the wind travels across the same pond. Now we have a wave because the disturbance, the ripple, is not bodily transported by either the wind or the water.

5. An idealization of many types of wave motion is embodied mathematically in what is called the wave equation,

$$c^2 \nabla^2 u - u_{tt} = 0, \tag{A–1}$$

where u is a physical property associated with the disturbance or signal, the operator ∇^2 is defined by

$$\nabla^2(\ ) = (\ )_{xx} + (\ )_{yy} + (\ )_{zz},$$

c is a constant representing the speed at which the wave travels, and x, y, z and t are rectangular spatial coordinates and time, respectively. Independent variables used as subscripts denote partial differentiation, for example, u_{tt} means $\partial^2 u/\partial t^2$. Is it sufficient to base our definition of a wave on Eq. A–1? The answer is no. "A wave is something that satisfies the wave equation" is too narrow a definition. For example, although Eq. A–1 does describe the behavior of sound waves in a lossless fluid, it is an oversimplification to say that a disturbance in the fluid is a sound wave only if it satisfies this equation. If the viscosity of the fluid is taken into account, Eq. A–1 must be replaced by a more complicated equation (see Chap. 9), which by strict mathematical definition is not a wave equation. Even though solutions of the more complicated equation do not satisfy "the" wave equation, Eq. A–1, they are nevertheless still thought of as waves: damped waves. In other words, solutions of Eq. A–1 generally represent waves, but not all waves obey this equation.[1]

6. The disturbance need not be oscillatory in order to be a wave. We frequently think of waves as having sinusoidal shape, but spikes, rectangular pulses, and noise are all perfectly valid waveforms for waves.

Summary. A wave is the movement of a disturbance or piece of information from one point to another in a medium (except that electromagnetic waves do not require a medium). The movement takes place at finite speed. The disturbance moves with respect to the medium. The shape or form of the disturbance is arbitrary.

Before leaving the general discussion of wave motion, we note the distinction between longitudinal and transverse waves. In a longitudinal wave the disturbance moves parallel to the direction of propagation; in a transverse wave the disturbance moves perpendicular to the direction of propagation. Sound waves and compressional elastic waves (called P waves in geophysics) are longitudinal. Waves on a string or a membrane, shear waves (called S waves in geophysics), and electromagnetic waves are transverse.

Although we have found that the wave equation, Eq. A–1, does not pertain to all wave behavior, it is sufficiently general to warrant close study. Indeed it is the backbone of acoustical theory. Some elementary solutions of the wave equation and their properties are discussed in the next section.

B. PLANE WAVES: SOME BASIC SOLUTIONS

Most of the first part of this text is devoted to plane sound waves. The basic attributes of planar wave motion, not necessarily restricted to sound, are developed and discussed in this section.

By dropping the y and z dependence in Eq. A–1, we obtain the equation for plane waves that travel along the x axis,[2]

$$c^2 u_{xx} - u_{tt} = 0. \tag{B–1}$$

[1] One may even think of examples in which a mathematical solution of the wave equation is not really a wave. Consider a tape recording system. Suppose a record head puts a signal $u(t)$ on a magnetic tape that is moving with velocity v, and a receive head a distance $x = vt$ away picks up the signal. Although the received signal $u(t - x/v)$ may be shown to satisfy the wave equation (with $c = v$), the translation of the signal is due solely to bodily transport by the moving tape. The process is therefore not wave motion.

[2] In the present, general discussion of the wave equation, u is any appropriate property of the waves. When we specialize to sound waves, u stands for the x component of the particle velocity.

This is one of the classical partial differential equations and has many applications in engineering and the sciences. Two other archetype equations are the diffusion equation $ku_{xx} - u_t = 0$, also called the heat conduction equation, and Laplace's equation $u_{xx} + u_{yy} = 0$, which is important in such diverse fields as electrostatics, incompressible flow, and steady heat conduction. Although these two equations look deceptively similar to the wave equation, their solutions are grossly different.[3] In this section basic solutions of the wave equation are developed and analyzed. They are then applied to two general classes of problems, free waves and forced waves.

1. General Solution of the Wave Equation

Although formal methods for finding the general solution of Eq. B–1 are known, a more intuitive and direct approach is sufficient for our purposes. After developing one general solution, we easily find a second one that is independent of the first. The sum of the two is the most general solution.

a. First General Solution

Because past experience tells us that many waves have sinusoidal shape, we might guess a solution of form $u = A \sin ax$. Substitution shows, however, that such a simple function does not satisfy Eq. B–1. In fact, it is clear from Eq. B–1 that the solution must be a function of time as well as distance. Since the simplest combination of x and t is linear, try $u = A \sin(ax + bt)$. Test this solution by substituting it in Eq. B–1:

$$c^2 u_{xx} - u_{tt} = -c^2 a^2 A \sin(ax + bt) + b^2 A \sin(ax + bt). \qquad \text{(B–2)}$$

The right-hand side can equal zero, as required, only if $b = \pm ac$. Taking the minus sign for the moment, we have

$$u = A \sin a(x - ct)$$

as a function that satisfies the wave equation. But is it the only function that works? How about a cosine in place of the sine function,

$$u = A \cos a(x - ct)?$$

This answer is also found to satisfy Eq. B–1 and is therefore a solution. If sines and cosines work, what about the complex exponential function,

$$u = Ae^{\pm ja(x-ct)}?$$

[3]For a discussion of the mathematical properties of hyperbolic (e.g., the wave equation), parabolic (e.g., the diffusion equation), and elliptic (e.g., Laplace's equation) partial differential equations, see, for example, Ref. 1.

It too is found to work. But the sine, cosine, and complex exponential functions are closely related and oscillatory. Suppose we try a nonoscillatory function, such as

$$u = A \ln a(x - ct).$$

This function also satisfies Eq. B–1, as do $A \exp[a(x - ct)^2]$ and $A\sqrt{x - ct}$. Since it begins to appear that the argument $x - ct$ is more important than the particular mathematical function, let us try the simple algebraic term

$$u = A(x - ct)^n.$$

Not only does this function solve Eq. B–1, it does so regardless of the value chosen for n. Because of the linearity of the wave equation, we may form an infinite sum of terms of this sort, in this case taking n to be an integer,

$$u = \sum_{n=-\infty}^{\infty} a_n(x - ct)^n,$$

and still have a solution. Since the summation on the right-hand side is a Laurent series and represents a function $u = f(x - ct)$, we are led to try

$$u = f(x - ct) \tag{B–3}$$

directly, where f is an arbitrary function. To check Eq. B–3, first calculate the derivatives

$$\begin{aligned}
u_x &= f'(x - ct)\frac{\partial}{\partial x}(x - ct) = f'(x - ct), \\
u_{xx} &= f''(x - ct)\frac{\partial}{\partial x}(x - ct) = f''(x - ct), \\
u_t &= f'(x - ct)\frac{\partial}{\partial t}(x - ct) = -cf'(x - ct), \\
u_{tt} &= -cf''(x - ct)\frac{\partial}{\partial t}(x - ct) = c^2 f''(x - ct),
\end{aligned}$$

where f' means the derivative of f with respect to its argument. Substitution in Eq. B–1,

$$c^2 u_{xx} - u_{tt} = c^2 f''(x - ct) - c^2 f''(x - ct) \equiv 0,$$

proves that Eq. B–3 is indeed a solution. Note that no assumptions have been made about the smoothness or regularity[4] of f. Thus f may be a very smooth

[4]The power series solution that led us to try Eq. B–3 might seem to imply something about smoothness of the function $f(x - ct)$. In fact, however, smoothness is not a necessary requirement. Our proof that $f(x - ct)$ is a solution does not require that $f(x - ct)$ be expressed as a power series.

function, such as a sine or cosine, or it may have abrupt changes or even infinite discontinuities. For example, the unit step function $H(x-ct)$, the delta function $\delta(x-ct)$, and the tangent function $\tan(x-ct)$ are all valid solutions of the wave equation.

b. Interpretation

The meaning of the solution Eq. B–3 may be made clear by taking an example. Let f be a simple linear, or ramp, function of unit length, that is,

$$u = A(x-ct)\,, \quad 0 < x - ct < 1\,, \tag{B–4}$$

and for simplicity let the amplitude A be unity. Sketches of u versus x at three specific times are given in Fig. 1.1. For example, at $t=0$ (bottom sketch) u is zero everywhere except in the range $0 < x < 1$, where it is the straight-line segment $u = x$. Note that the sketches in Fig. 1.1 are stacked as a geologist would order them, that is, with the oldest on the bottom. The reason for this will become clear presently. The most important conclusion to be drawn from the figure is that the wave is steadily translated, unchanged in shape, along the x axis. Since adding τ to t and $c\tau$ to x leaves the argument $x - ct$ unchanged, the signal u simply replicates itself at a new place. In a sense this operation defines propagation. A given point on a wave is translated unchanged with speed c.[5]

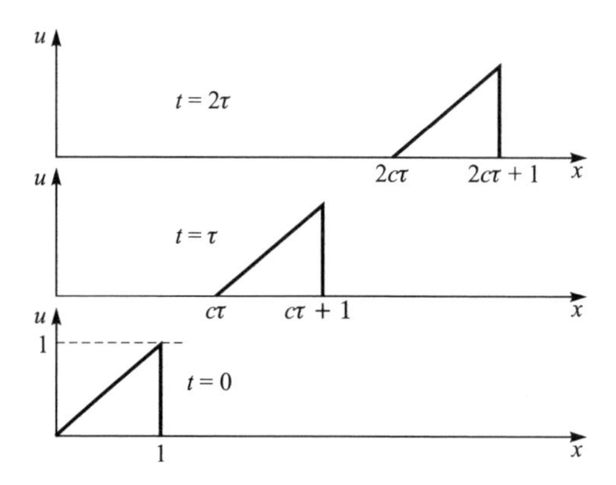

Figure 1.1 Sketches showing the waveform in space when the solution is a section of a ramp function.

[5]A mathematical proof is as follows: Differentiation of $u = f(x-ct)$ yields $du = f'dx - cf'dt$, which, if $u = \text{const}$, reduces to

$$(dx/dt)_{u=\text{const}} = c.$$

Each value of $x - ct$ defines a phase front of the wave. Moreover, each phase front is associated with a particular value of u. In the case of the ramp wave, for example, $x - ct = 1$ signifies the head of the wave, $x - ct = 0$ the tail of the wave, and $x - ct = \frac{1}{2}$ the midpoint. The values of u associated with these points are $1, 0$, and $\frac{1}{2}$, respectively.

Notice that the limitation $0 < x - ct < 1$ in Eq. B–4 is a very important part of the description of the wave. Without the limitation the disturbance would have no beginning or end; it would stretch from $-\infty$ to $+\infty$. Of course, some waves are modeled as being of infinite extent, for example, a very long sinusoidal wave train. If the wave is a pulse, however, its proper description must include some form of limitation. See Problems 1B–2 and 1B–3.

c. *Characteristics Diagram (x, t Plane)*

A simple way to keep track of waves is to use a characteristics diagram, a graphical display in which the axes are the independent variables t and x. As an example, consider the ramp wave just discussed. In particular, follow a given point on the wave from sketch to sketch in Fig. 1.1. By drawing a line connecting the tails in the three sketches, we obtain the path followed by the tail in the x, t plane. This is the line $t = x/c$ in Fig. 1.2. Similarly, a line through the heads defines the path along which the head of the wave travels, line $t = (x - 1)/c$ in Fig. 1.2. The paths along which various phase fronts travel, such as the head and the tail, are the loci of the phase fronts and are called characteristics. Two important properties of the characteristics should be noted: (1) their slope is the inverse of the propagation speed and (2) each characteristic is associated with a specific value of the dependent variable u.

In a sense the characteristics diagram contains the entire solution of the problem in graphical form. Once the main characteristics have been drawn, it is a very simple matter to construct waveforms. See Fig. 1.3. To obtain the spatial waveform at a particular time, say t_0, draw a horizontal line $t = t_0$ across the field of characteristics, as shown in Fig. 1.3a. The intersections of the characteristics with this line provide the values of u from which the spatial waveform, Fig. 1.3b, may be constructed. To obtain the time waveform at a

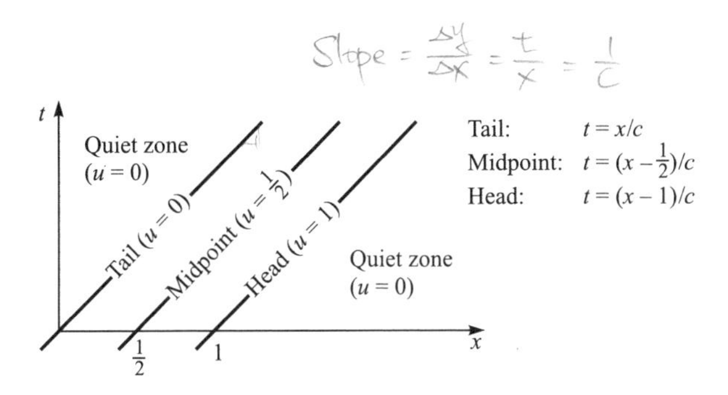

Figure 1.2 Characteristics diagram.

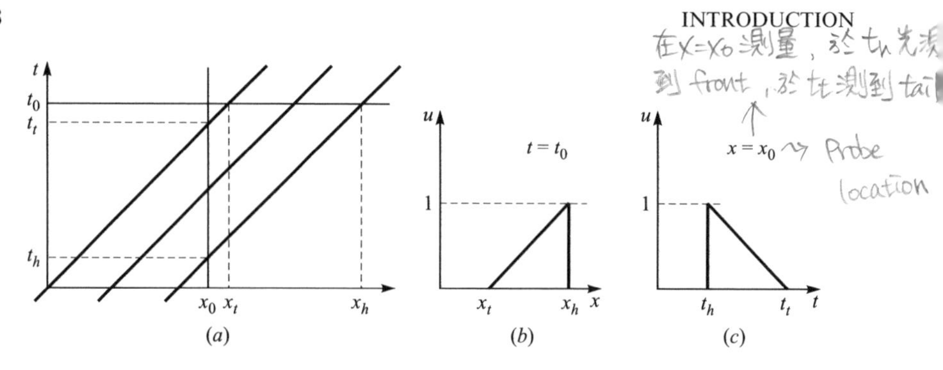

Figure 1.3 Construction of waveforms from the characteristics diagram.

given point in space, say x_0, draw a vertical line at $x = x_0$. From its intersections with the characteristics, construct the time waveform, as shown in Fig. 1.3*c*.

d. Second General Solution

A theorem for linear ordinary differential equations is that the most general solution of an *n*th-order linear differential equation has *n* arbitrary constants. In the case of an *n*th-order linear partial differential equation, the most general solution contains *n* arbitrary functions. Because the wave equation is second order, that is, the highest derivative appearing is the second derivative, the solution requires two arbitrary functions. We already have one, $f(x - ct)$. A second is needed. What is it?

Use a little previous knowledge. The solution $f(x - ct)$ has been found to represent a wave traveling in the direction of increasing x, that is, it is an *outgoing wave*, or forward traveling wave. But we know waves can travel both ways. How should an *incoming wave*, or backward traveling wave, which travels in the direction of decreasing x, be represented? Recall the discussion following Eq. B–2. Two solutions were found possible, $b = -ac$ and $b = +ac$. Choice of the minus sign led to an expression for the forward traveling wave. Choice of the plus sign leads to

$$u = g(x + ct), \tag{B–5}$$

where g, like f, is an arbitrary function. Substitution in Eq. B–1 shows that g indeed satisfies the wave equation.

To demonstrate that g represents a backward traveling wave, start with the same ramp shape used in the previous example. Construction of the shapes at $t = 0$, τ, 2τ yields Fig. 1.4. Notice that the foot of the ramp is now the head of the wave. Figure 1.5 gives the characteristics diagram. It is clear that g represents a backward traveling wave. Note that the characteristics have slope $-1/c$.

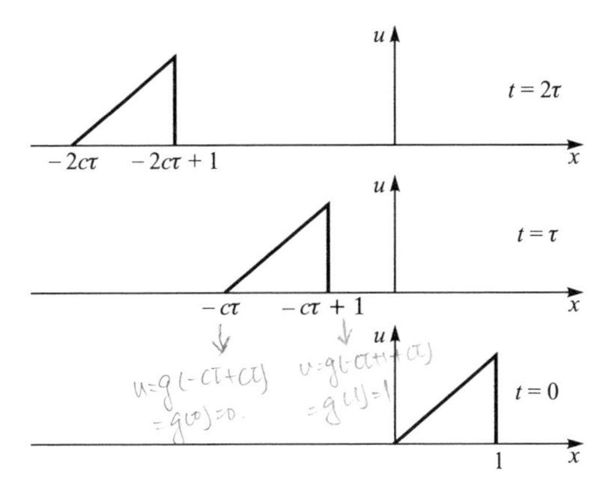

Figure 1.4 Backward traveling, or incoming, wave.

e. Most General Solution

Because the wave equation is linear, superposition may be used. The most general solution is therefore a combination of Eqs. B–3 and B–5:

$$u = f(x - ct) + g(x + ct). \tag{B–6}$$

An alternative form, just as general and more convenient in certain cases, is

$$u = F\left(t - \frac{x}{c}\right) + G\left(t + \frac{x}{c}\right). \tag{B–7}$$

Again, the first term represents an outgoing wave, and the second term an incoming wave.

Given the general solution, how is it applied to solve a particular problem? Particular problems arise when initial conditions and boundary conditions are

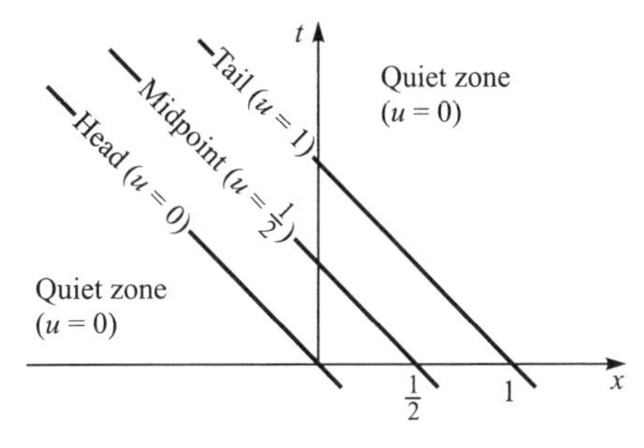

Figure 1.5 Characteristics diagram for a backward traveling wave.

given. These conditions determine the specific functions f and g (or F and G) that fit the particular problem. Two general classes may be identified: (1) free-wave problems and (2) forced-wave problems. A free wave is produced when an initial disturbance (a spatial variation) in the medium is suddenly released at a given time. The sound from a burst balloon is an example of a free wave. A forced wave results from prescribed motion (a time variation) at a particular point or points in space. An example is the sound produced by a loudspeaker. In the next two sections free waves and forced waves are discussed. Wave motion on a string is used to illustrate the ideas. The wave equation still has the form of Eq. B–1 but is written in terms of the string displacement ξ, that is,

$$c^2\xi_{xx} - \xi_{tt} = 0.$$

2. Free Waves

Equation B–6 is the solution convenient for free waves. The application of initial conditions to determine f and g is illustrated in the following problem.

a. Illustrative Problem

A string is deformed into rectangular shape[6] and released from rest. Mathematically, the initial conditions are

$$\xi(x, 0) = \text{rect}(x/2a), \tag{B–8a}$$

$$\xi_t(x, 0) = 0, \tag{B–8b}$$

where ξ is the string displacement and $\text{rect}[(x - x_0)/w]$ is the rectangle function: a unit rise of width w centered at $x = x_0$.[7] See Fig. 1.6. Notice that two initial conditions are needed because the highest time derivative present in the wave equation is the second. Starting with the general solution

$$\xi = f(x - ct) + g(x + ct), \tag{B–9}$$

[6]The infinitely sharp rise and fall at the ends of the rectangular deformation are of course physically unrealizable and moreover violate the derivation given below (Sec. C.2) of the wave equation for a string. Treated as the limit of a smooth rectangle with finite rise and fall, however, the ideal rectangle is useful for visualizing wave behavior.

[7]The rectangle function may also be expressed in terms of the unit step function

$$\text{rect}\left(\frac{x - x_0}{w}\right) = H\left(x - x_0 + \frac{1}{2}w\right) - H\left(x - x_0 - \frac{1}{2}w\right).$$

The unit step function $H(x - x_0)$ is defined to be unity for $x > x_0$, zero for $x < x_0$.

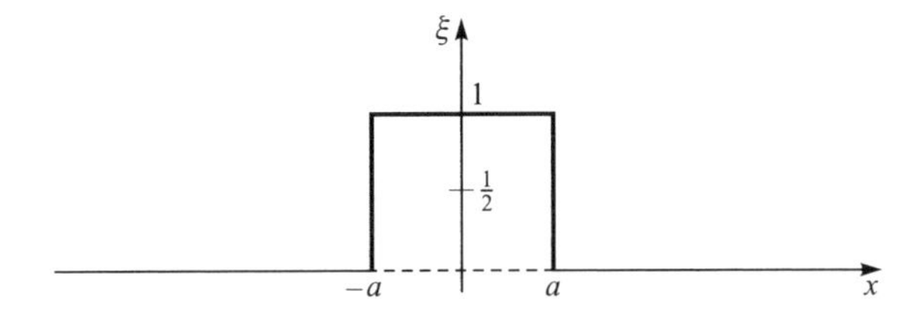

Figure 1.6 Initial displacement of a string.

evaluate it at $t = 0$ and apply the first initial condition, Eq. B–8a:

$$f(x) + g(x) = \mathrm{rect}(x/2a). \tag{B–10}$$

Next find ξ_t,

$$\xi_t = -cf'(x - ct) + cg'(x + ct);$$

evaluate it at $t = 0$, applying the second initial condition, Eq. B–8b,

$$-f'(x) + g'(x) = 0,$$

and integrate,

$$-f(x) + g(x) = C, \tag{B–11}$$

where C is a constant. Subtraction of Eq. B–11 from Eq. B–10 yields the function f, while addition yields g:

$$f(x) = \frac{1}{2}\mathrm{rect}\left(\frac{x}{2a}\right) - \frac{C}{2}, \qquad g(x) = \frac{1}{2}\mathrm{rect}\left(\frac{x}{2a}\right) + \frac{C}{2}.$$

After suitable translation (x is replaced by $x - ct$ as the argument of f but replaced by $x + ct$ as the argument of g), these results are substituted in Eq. B–9,

$$\xi = \frac{1}{2}\,\mathrm{rect}\left(\frac{x - ct}{2a}\right) + \frac{1}{2}\,\mathrm{rect}\left(\frac{x + ct}{2a}\right). \tag{B–12}$$

A sequence of waveforms and the characteristics diagram are shown in Fig. 1.7. The rectangular disturbance splits into two equal pulses, each having half the amplitude of the original disturbance but the same shape. As shown in the next section, an equal split always occurs if the disturbance has zero initial velocity.

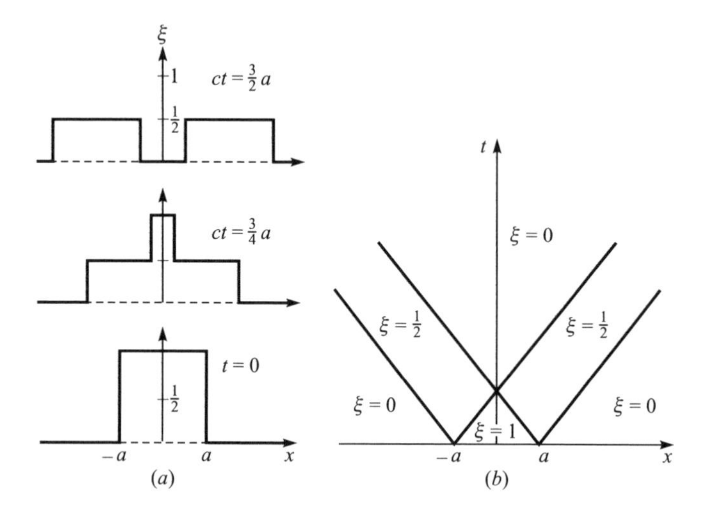

Figure 1.7 String initially deformed into a rectangle function and released from rest. (*a*) Waveforms. (*b*) Characteristics diagram.

b. General Case: d'Alembert's Solution

Equation B–6 may be made to satisfy general initial conditions, such as

$$u(x, 0) = P(x), \tag{B–13a}$$

$$u_t(x, 0) = Q(x). \tag{B–13b}$$

Following the procedure used in the illustrative problem yields

$$P(x) = f(x) + g(x) \tag{B–14}$$

and

$$Q(x) = -c[f'(x) - g'(x)].$$

Integration of this equation from an arbitrary reference point x_0 to x yields

$$\frac{1}{c}\int_{x_0}^{x} Q(y)\, dy = -f(x) + g(x) - K, \tag{B–15}$$

where K is the constant $g(x_0) - f(x_0)$. Subtraction and addition of Eqs. B–14 and B–15 lead, respectively, to

$$f(x) = \frac{1}{2} P(x) - \frac{1}{2c} \int_{x_0}^{x} Q(y)\, dy - \frac{K}{2},$$

$$g(x) = \frac{1}{2} P(x) + \frac{1}{2c} \int_{x_0}^{x} Q(y)\, dy + \frac{K}{2}.$$

When these expressions are substituted in Eq. B–6, the constant terms cancel and the two integrals may be combined. The final result is

$$u = \frac{1}{2} \left\{ P(x-ct) + P(x+ct) + \frac{1}{c} \int_{x-ct}^{x+ct} Q(y)\, dy \right\}. \tag{B–16}$$

Equation B–16 is called d'Alembert's solution of the wave equation, even though it was first derived by Euler (1748).[8] Notice that if $Q = 0$, that is, if $u_t = 0$ when $t = 0$, the forward and backward traveling waves have the same shape as the initial disturbance but half the amplitude. The initial disturbance therefore splits into two equal waves that travel in opposite directions.

Example 1.1. The struck string. An undisturbed string is struck from below by a hammer of width $2a$, as shown in Fig. 1.8. The velocity imparted is v_0. The initial conditions are

$$\xi(x, 0) = 0,$$
$$\xi_t(x, 0) = v_0[H(x+a) - H(x-a)].$$

D'Alembert's solution for this case is

$$\begin{aligned}\xi &= \frac{v_0}{2c} \int_{x-ct}^{x+ct} [H(y+a) - H(y-a)]\, dy \\ &= \frac{v_0}{2c} [R(x+ct+a) - R(x+ct-a) - R(x-ct+a) + R(x-ct-a)],\end{aligned}$$

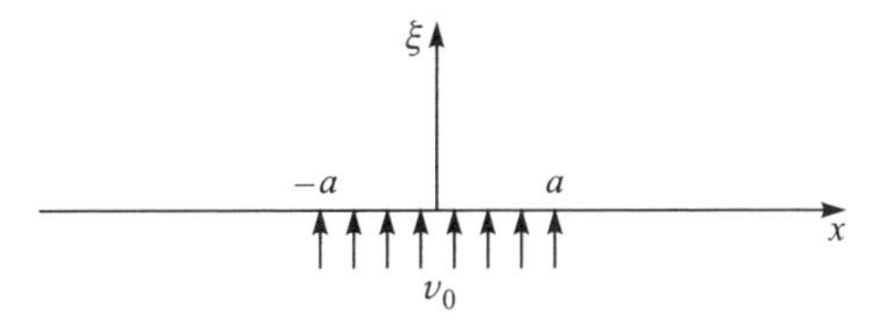

Figure 1.8 Initial conditions for the struck string problem.

[8]Wheeler and Crummett (Ref. 8) give a fascinating account of the derivation and solution of the wave equation for a string. The principal players (and combatants) were d'Alembert, Euler, and Daniel Bernoulli; later contributions come from Lagrange and Fourier. The arguments among them add depth and perspective to our discussion.

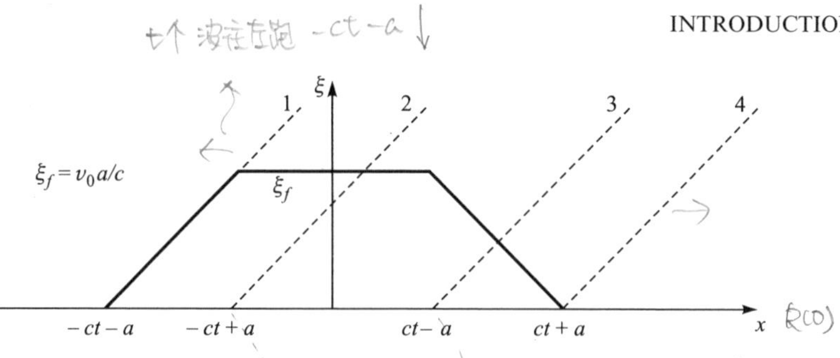

Figure 1.9 Shape of struck string at time $t > a/c$.

where R is the ramp function $[R(x) = xH(x)]$. The first and second terms represent the head and tail, respectively, of the backward traveling wave, the third and fourth terms the tail and head, respectively, of the forward traveling wave. Each term is depicted (and labeled) by a dashed line in Fig. 1.9. According to the solution, lines 2 and 3 are to be subtracted from the sum of lines 1 and 4. The solid curve in Fig. 1.9 results. The center of the string moves upward with velocity v_0 for a short time and then comes to rest with a permanent displacement $\xi_f = v_0 a/c$ (see Problem 1C–3). Does the eventual displacement of the entire string by this amount imply the expenditure of an infinite amount of energy? No. Since the effect of gravity has not been included, the string has no more potential energy in the final state $(\xi = \xi_f)$ than in the initial state $(\xi = 0)$. See Problem 1C–4 for a calculation of the energy associated with the motion of the struck string.

3. Forced Waves

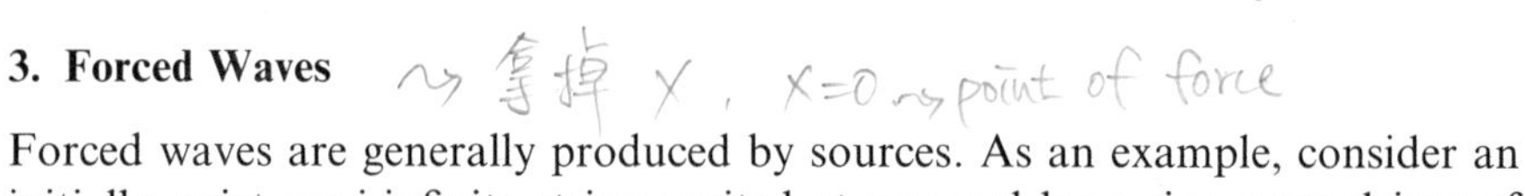

Forced waves are generally produced by sources. As an example, consider an initially quiet semi-infinite string excited at one end by a sine wave driver of angular frequency ω (period $T = 2\pi/\omega$). The driver is turned on for only two cycles. See Fig. 1.10. The boundary conditions are as follows:

1. $\xi(0, t) = \sin \omega t\,[H(t) - H(t - 2T)]$ (source condition).
2. No incoming waves exist (*radiation condition*).[9]

[9]The physical basis of the radiation boundary condition is the lack, on a uniform, semi-infinite string, of any agency to cause reflection. The string has no knots in it, for example. The possibility of incoming waves from $x = \infty$ is ruled out either by fiat, i.e., "no source exists at $x = \infty$," or by noting that if there were a source at $x = \infty$, no wave emitted could reach the region of interest in finite time.

Figure 1.10 Example of a forced wave: string driven at one end by two cycles of a sine wave.

The initial conditions are $\xi = \xi_t = 0$ everywhere at $t = 0$. Start with the general solution in the form of Eq. B–7,

$$\xi = F\left(t - \frac{x}{c}\right) + G\left(t + \frac{x}{c}\right). \tag{B–17}$$

Why use this form rather than Eq. B–6? It is simply a matter of convenience. At $x = 0$, where the displacement is prescribed, Eq. B–17 reduces to the very simple expression $\xi = F(t) + G(t)$. If Eq. B–6 has been used, evaluation at the boundary would have led to the more awkward expression $\xi = f(-ct) + g(ct)$.

The functions F and G are determined by applying the boundary conditions. First, because the radiation condition precludes the G function, Eq. B–17 reduces to

$$\xi = F\left(t - \frac{x}{c}\right).$$

Evaluate this expression at $x = 0$ and apply the source condition:

$$F(t) = \sin \omega t[H(t) - H(t - 2T)].$$

Substituting $t - x/c$ for t, we obtain the final solution

$$\xi = \sin \omega\left(t - \frac{x}{c}\right)\left[H\left(t - \frac{x}{c}\right) - H\left(t - \frac{x}{c} - 2T\right)\right]. \tag{B–18}$$

A typical spatial waveform is shown in Fig. 1.11. Also of interest is the time waveform picked up, say, by a probe, which can be positioned at any point x. Let the probe output (a time signal) be displayed on an oscilloscope. In this way a sequence of time waveforms at probe distances $x = 0$, x_1, and x_2 may be obtained, as shown in Fig. 1.12. The quantity τ is the delay time, that is, the time it takes the signal to travel from the source to the probe.

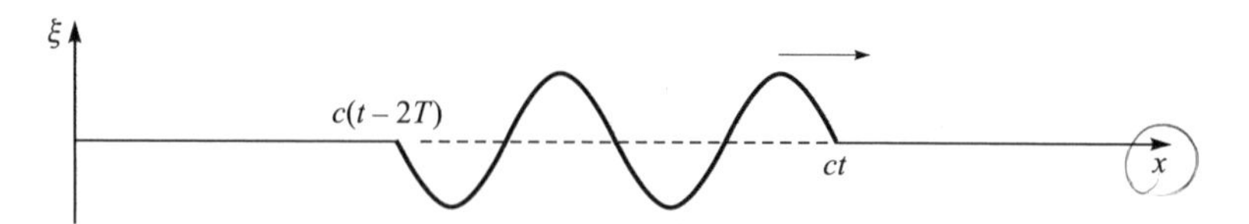

Figure 1.11 Spatial waveform at time $t > 2T$.

The characteristics diagram for this problem is shown in Fig. 1.13. As usual, all the characteristics have the same slope $1/c$. Superposed on the time axis is the source waveform. Above the tail characteristic and below the head characteristic are quiet zones.

4. Relation between Derivatives for a Progressive Wave

A progressive wave is a wave traveling in one direction, for example, an outgoing wave or an incoming wave, as opposed to a standing wave, which is made up of a combination of outgoing and incoming waves. For an outgoing wave

$$u = f(x - ct)$$

the spatial derivative is

$$u_x = f'$$

and the time derivative is

$$u_t = -cf'.$$

In other words, the relation between the two derivatives is $u_t = -cu_x$, or

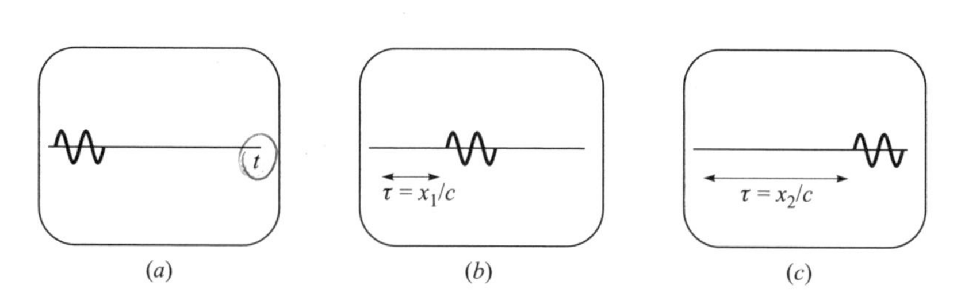

Figure 1.12 Time waveforms for a forced wave at various distances: (*a*) $x = 0$; (*b*) $x = x_1$; (*c*) $x = x_2$.

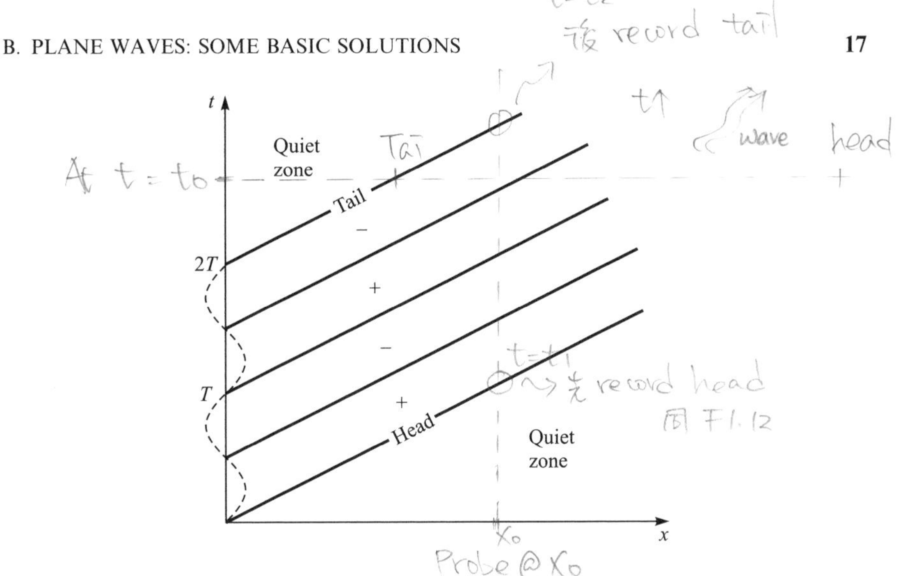

Figure 1.13 Characteristics diagram for a forced wave.

$$u_t + cu_x = 0. \tag{B–19}$$

This expression may be thought of as a first-order wave equation. Its one and only solution is $u = f(x - ct)$; in particular, $g(x + ct)$ is *not* a solution. A graphical construction shows why the x and t derivatives have opposite signs. Consider an outgoing pulse such as the triangular one in Fig. 1.14*a*. It is seen that the x and t derivatives at any given point, such as the midpoint A, are of opposite sign.

In the case of incoming waves, let us deduce the u_x, u_t relation by referring to Fig. 1.14*b*. Since the x and t derivatives at point A now have the same sign,

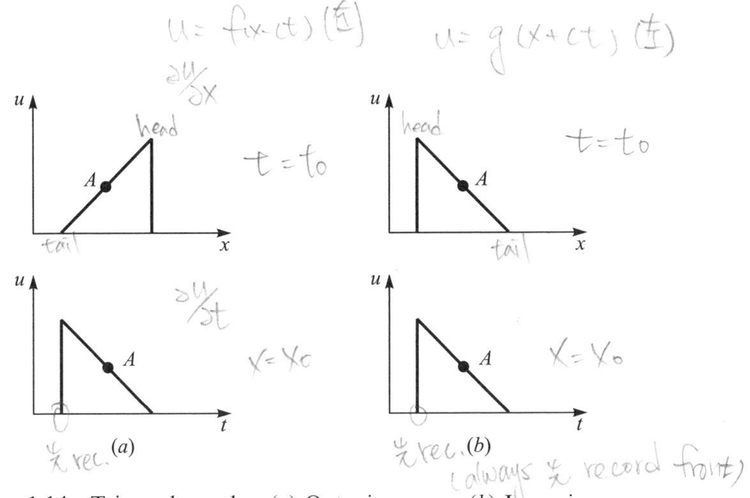

Figure 1.14 Triangular pulse. (*a*) Outgoing wave. (*b*) Incoming wave.

one suspects that the relation is $u_t = +cu_x$, that is, that the first-order wave equation for incoming waves is

$$u_t - cu_x = 0. \tag{B–20}$$

This guess may be confirmed by showing that $g(x + ct)$ is a solution. Note that $f(x - ct)$ is *not* a solution of Eq. B–20.

Problem 1B–6 provides an illustration of the usefulness of the relation between the space and time derivatives.

C. DERIVATION OF WAVE EQUATIONS. IMPEDANCE

How is the wave equation derived? Three common physical systems that carry waves—the electrical transmission line, the flexible string, and the compressible fluid—are considered in this section. The transmission line is taken up first because of its simplicity: it provides the archetype derivation of the wave equation. Next comes the flexible string. Although the derivation does not follow the same steps used for the transmission line, we get a chance to see the process of linearization. Mathematical statements of physical laws underlying wave motion are frequently nonlinear and must be linearized if Eq. B–1 is to be obtained. Finally, the acoustic wave equation is developed. The scope of the derivation is limited to plane waves in lossless fluids in order that the forest not be missed on account of the trees. The wave equation obtained is sufficient for most parts of Chaps. 3–8. A more complete derivation—for three-dimensional waves in more realistic fluids—is given in Chap. 2.

Also contained in this section is an introduction to the concept of impedance.

1. Electrical Transmission Line

a. Lossless Line

A long electrical line, say two parallel wires or a coaxial cable, has a certain amount of inherent capacitance and inductance. Let C be the capacitance per unit length of line, L the inductance per unit length, e the voltage, and i the current. The line may be modeled by a string of incremental series inductances $L\ \Delta x$ and shunt capacitances $C\ \Delta x$ as shown in Fig. 1.15. Consider the voltage drop across the right-hand inductance, which carries the loop current $i(x + \Delta x)$:

$$e(x) - e(x + \Delta x) = L\ \Delta x\ \frac{\partial i(x + \Delta x)}{\partial t},$$

or

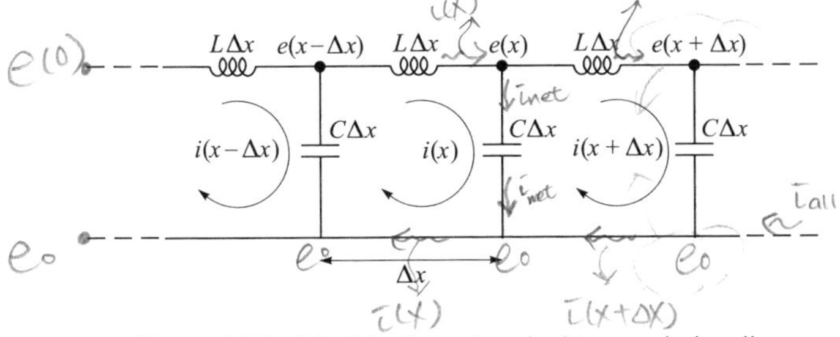

Figure 1.15 Model of an electrical transmission line.

$$\frac{e(x+\Delta x) - e(x)}{\Delta x} + L\,\frac{\partial i(x+\Delta x)}{\partial t} = 0.$$

In the limit as $\Delta x \to 0$ this equation becomes

$$e_x + Li_t = 0. \tag{C–1}$$

Similarly, the net loop current $i(x) - i(x+\Delta x)$ flowing through the middle capacitor is related to the voltage $e(x)$ by

$$e(x) = \frac{1}{C\,\Delta x}\int [i(x) - i(x+\Delta x)]\,dt.$$

Differentiate with respect to time and take the limit as $\Delta x \to 0$:

$$i_x + Ce_t = 0. \tag{C–2}$$

Equations C–1 and C–2 are typical of physical laws that lead to wave motion. Two dependent variables, often called "field variables," in this case current and voltage, are coupled by two first-order partial differential equations. To obtain a wave equation, eliminate one of the variables, say the current i. Differentiate Eq. C–1 with respect to x and Eq. C–2 with respect to t, multiply the latter by L, and subtract. The result is

$$\frac{1}{LC}\,e_{xx} - e_{tt} = 0, \tag{C–3}$$

a wave equation in which the propagation speed is

$$c = \frac{1}{\sqrt{LC}}.$$

An alternative approach is to obtain a wave equation in i by eliminating e,

$$\frac{1}{LC} i_{xx} - i_{tt} = 0.$$

This is a convenient place to introduce the concept of characteristic impedance. Consider an outgoing progressive wave described, say, by $e = f(x - ct)$. Substitution in Eq. C–2 leads to

$$i_x = -Ce_t = Ccf'(x - ct),$$

or, after integration,[10]

$$i = Ccf(x - ct) = \sqrt{C/L}\, e.$$

The ratio of voltage to current is the impedance Z. When, as in the present case, only an outgoing wave is present, the ratio is called the *characteristic impedance* and the special symbol Z_0 is used,

$$Z_0 = \left(\frac{e}{i}\right)_{\text{outgoing}} = \sqrt{\frac{L}{C}}. \tag{C–4}$$

Two important conclusions may be drawn. First, the two field variables e and i have the same waveform. Second, the characteristic impedance is a function only of the properties of the transmission line; in particular, Z_0 does not depend on the shape of the waveform. An assumption commonly made in deriving Eq. C–4 is that the wave is sinusoidal. Although this assumption is useful for more complicated transmission lines, it is by demonstration not necessary for the simple LC line considered here. Impedance is discussed in more detail in Sec. E below.

If the progressive wave is incoming, i.e., $e = g(x + ct)$, one can show that $e/i = -\sqrt{L/C}$. The relation between the field variables is therefore $e = -Z_0 i$. Can you explain why the voltage and current are in phase for an outgoing wave but out of phase for an incoming wave? *Hint*: Whereas voltage is a scalar quantity, current is a vector.

b. Lossy Line

Real electrical lines have losses and leakage, which may be modeled by adding a resistance $R\,\Delta x$ in series with the inductance and a conductance $G\,\Delta x$ in parallel with the capacitance. See Fig. 1.16. The wave equation for this line turns out to be

$$\frac{1}{LC} e_{xx} - e_{tt} = \left(\frac{G}{C} + \frac{R}{L}\right) e_t + \frac{RG}{LC} e, \tag{C–5}$$

[10] The constant of integration is set equal to zero by assuming that ahead of the wave the line is quiet ($e = 0$ and $i = 0$).

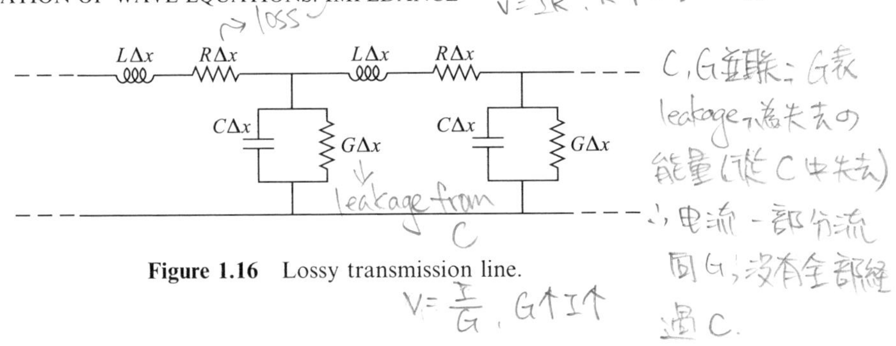

Figure 1.16 Lossy transmission line.

which is called the telegrapher's equation. For the case $G = 0$ this equation has application in acoustics as a model of sound propagation in air-filled porous materials, such as fiberglass (see Problems 2–9, 2–10, 2–19, 2–20). Although the equation has no simple general solutions such as $f(x - ct)$ and $g(x + ct)$, time-harmonic solutions, e.g.,

$$e = e_0 e^{j(\omega t - kx)},$$

where ω is the angular frequency and k is the wave number, may be obtained. One then finds that the characteristic impedance Z_0 depends not only on the inherent electrical properties of the line but also on frequency.

Another interesting lossy line is one in which the resistance is in parallel with the inductance. As Fig. 1.17 shows, the current $i(x)$ is broken up into two parts, $i_1(x)$ through the inductance and $i_2(x)$ through the resistance. The voltage drop across the inductance is

$$e(x - \Delta x) - e(x) = L\,\Delta x\,\frac{\partial i_1}{\partial t},$$

or, in the limit as $\Delta x \to 0$,

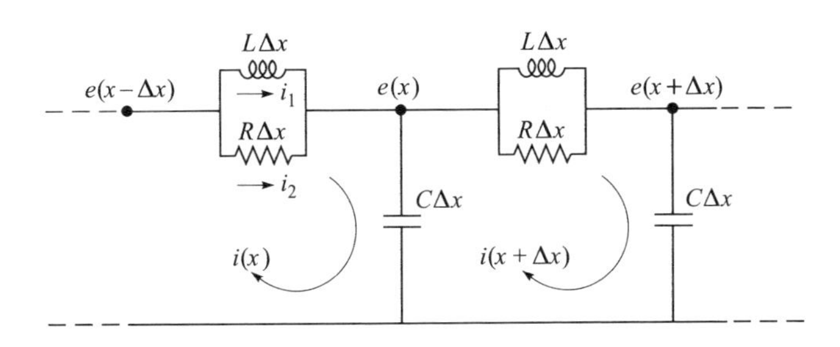

Figure 1.17 Lossy transmission line with inductance and resistance in parallel.

$$\frac{\partial i_1}{\partial t} = -\frac{1}{L}\frac{\partial e}{\partial x}. \qquad \text{(C–6)}$$

Since the voltage drop across the resistance is also $e(x - \Delta x) - e(x)$, the relation between e and i_2 is

$$i_2 = -\frac{1}{R}\frac{\partial e}{\partial x}. \qquad \text{(C–7)}$$

Adding Eq. C–6 to the derivative of Eq. C–7, and noting that $i = i_1 + i_2$, we obtain

$$i_t + \frac{1}{L}e_x + \frac{1}{R}e_{xt} = 0.$$

When this equation is combined with the equation for the current flow through the shunt capacitance (Eq. C–2), the result is

$$(L/R)e_{xxt} + e_{xx} - LCe_{tt} = 0. \qquad \text{(C–8)}$$

This equation is an exact analog of the wave equation for sound in a viscous medium (see Chap. 9). Electrical transmission lines are sometimes used to do analog experiments of acoustical wave motion; see, for example, Ref. 7.

2. Waves on a String

A flexible string having its equilibrium position along the x axis is shown in Fig. 1.18. Let ξ be the transverse displacement of the string from equilibrium, $\mathcal{T}$ be the tension in the string, and s be the arc length along the string. We focus our attention on the displacement ξ of a small element of the string, which at

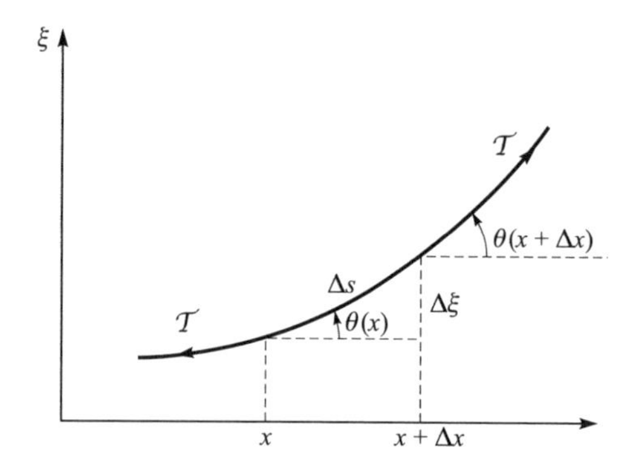

Figure 1.18 Displacement of and forces on a flexible string.

the instant shown in the figure has arc length Δs. The element moves vertically as described by Newton's second law, which relates the mass $\times$ acceleration of the element to the forces acting on it. The forces are the vertical components of the tension at either end of the element. When due consideration is given to various geometrical relations, it turns out that Newton's law reduces to the wave equation.

a. Derivation of the Wave Equation

The derivation begins with assumptions. The tension $\mathcal{T}$ is assumed to be the same everywhere in the string. Elastic forces are neglected, except that the string must be elastic enough to be extensible. Later, in order to linearize, we shall assume that the angle θ associated with the displacement $\Delta\xi$ (see Fig. 1.18) is small. It is not necessary to start off with the small-angle assumption, however.

First, geometrical considerations are used to obtain a relation between ξ and θ. Use of the trigonometric relations (see Fig. 1.18)

$$\frac{\partial x}{\partial s} = \cos\theta, \qquad \frac{\partial \xi}{\partial s} = \sin\theta, \qquad \frac{\partial \xi}{\partial x} = \tan\theta$$

leads to

$$\cos\theta = \left(1 + \xi_x^2\right)^{-1/2}, \qquad \sin\theta = \xi_x\left(1 + \xi_x^2\right)^{-1/2}.$$

Take the derivative of $\cos\theta$ with respect to x,

$$\sin\theta\ \theta_x = \xi_x \xi_{xx}(1 + \xi_x^2)^{-3/2},$$

substitute for $\sin\theta$ from the previous equation, and obtain

$$\theta_x = \cos^2\theta\ \xi_{xx}. \tag{C–9}$$

We are now ready to apply Newton's second law to the string element. The mass Δm of the element, most easily found from its value in the undisturbed state, is $\Delta m = \rho_\ell\, \Delta x$, where ρ_ℓ is the linear density, that is, the mass per unit length of the string. The vertical forces acting on the element are the two transverse components of the tension at the ends of the element. Newton's law is therefore

$$\rho_\ell\, \Delta x\, \xi_{tt} = \mathcal{T}\sin(\theta + \Delta\theta) - \mathcal{T}\sin\theta. \tag{C–10}$$

Dividing by Δx, we obtain

$$\rho_\ell \xi_{tt} = \frac{\mathcal{T}\sin(\theta + \Delta\theta) - \mathcal{T}\sin\theta}{\Delta x},$$

or, in the limit as $\Delta x \to 0$,

$$\rho_\ell \xi_{tt} = \mathcal{T}(\sin\theta)_x = \mathcal{T}\cos\theta\,\theta_x. \tag{C–11}$$

Substitute for θ_x from Eq. C–9:

$$\xi_{tt} - (\mathcal{T}/\rho_\ell)\cos^3\theta\;\xi_{xx} = 0. \tag{C–12}$$

It is now time to linearize. If the small-angle assumption $|\theta| \ll 1$ is invoked, $\cos\theta = 1$, and the classical wave equation for a string is obtained.

$$\xi_{tt} - c^2\xi_{xx} = 0, \tag{C–13}$$

where $c = \sqrt{\mathcal{T}/\rho_\ell}$ is the speed of propagation. Notice that in this case the derivation leads directly to the wave equation. No intermediate stage involving two first-order equations is reached.

b. Two First-Order Equations

A pair of first-order equations analogous to those for the electrical transmission line may, however, be found. Linearized versions of Eqs. C–11 and C–9 are, respectively,

$$\rho_\ell \xi_{tt} - \mathcal{T}\theta_x = 0, \qquad \theta_x - \xi_{xx} = 0.$$

If the second equation is integrated with respect to x and differentiated with respect to time, and if the string velocity $v = \xi_t$ is introduced,[11] we obtain

$$\rho_\ell v_t - \mathcal{T}\theta_x = 0, \tag{C–14a}$$

$$\theta_t - v_x = 0. \tag{C–14b}$$

Now introduce the transverse force f exerted by one element of the string on its neighbor,

$$f = -\mathcal{T}\sin\theta \doteq -\mathcal{T}\theta. \tag{C–15}$$

When this relation, which is similar to $f = -kx$ for a spring, is used to eliminate θ from Eqs. C–14, the result is

[11] The symbol v is used for velocity here because it is transverse to the axis x of the string. In the case of sound waves (next section) the velocity is longitudinal (along the x axis) and is denoted by the symbol u.

$$\rho_\ell v_t + f_x = 0, \tag{C–16a}$$

$$f_t + \mathcal{T} v_x = 0. \tag{C–16b}$$

These two equations are analogous to the two field equations, Eqs. C–1 and C–2, respectively, for the transmission line. The force f is analogous to voltage (both exert a push), and the string velocity v is analogous to current (both represent a flow).[12] Of course, one of the variables, say f, may be eliminated and the usual wave equation obtained, $c^2 v_{xx} - v_{tt} = 0$.

c. Characteristic Impedance

A solution representing an outgoing wave is $f = \phi(x - ct)$. Substitute this solution in Eq. C–16b,

$$\mathcal{T} v_x = -f_t = c\phi'(x - ct),$$

and integrate with respect to x:

$$\mathcal{T} v = c\phi = cf.$$

To continue the analogy to the electrical transmission line, we define a characteristic impedance Z_0 as the ratio of the force f to the velocity v for an outgoing wave,

$$Z_0 \equiv (f/v)_{\text{outgoing}} = \mathcal{T}/c = \sqrt{\mathcal{T}\rho_\ell}. \tag{C–17}$$

Again it is found that the characteristic impedance is determined only by properties of the medium; Z_0 does not depend on the form or shape of the traveling wave.

d. Boundary Conditions

If the end of a string, say at $x = \ell$, is fixed, the boundary condition is

$$\xi(\ell, t) = 0 \quad \text{(fixed end)}. \tag{C–18}$$

If the end is free, the end element of the string experiences no restoring force f (because no element exists at $\ell + \Delta x$ to pull it up or down). By virtue of Eq. C–15, therefore, $\theta(\ell, t) = 0$, or, since $\xi_x = \tan\theta \doteq \theta$,

$$\xi_x(\ell, t) = 0 \quad \text{(free end)}. \tag{C–19}$$

[12] The analogy used in this book is called the *impedance* analogy. In the *mobility* analogy, also in common use, current is taken to be analogous to force (and pressure) while voltage is likened to velocity. For a discussion of the impedance and mobility analogies, see, for example, Ref. 6.

In summary, a fixed end is a point of zero displacement, a free end a point of zero slope. Much use is made of these conditions in Chap. 6.

e. Energy

A string element of length dx has mass $dm = \rho_\ell \, dx$ and velocity ξ_t. Its kinetic energy dE_{kinetic} is therefore

$$dE_{\text{kinetic}} = \tfrac{1}{2}\rho_\ell \xi_t^2 \, dx. \tag{C–20}$$

By virtue of being stretched an amount

$$\begin{aligned} ds - dx &= \left(\frac{1}{\cos\theta} - 1\right)dx \\ &= \left(\sqrt{1+\xi_x^2} - 1\right)dx \\ &\doteq \tfrac{1}{2}\xi_x^2 \, dx, \end{aligned}$$

the element has stored energy dE_{stored}:

$$dE_{\text{stored}} = \mathcal{T}\left(\tfrac{1}{2}\,\xi_x^2 \, dx\right) = \tfrac{1}{2}\,\rho_\ell c^2 \xi_x^2 dx. \tag{C–21}$$

The total energy $dE = dE_{\text{kinetic}} + dE_{\text{stored}}$ of the element is therefore

$$dE = \tfrac{1}{2}\,\rho_\ell c^2(\xi_x^2 + c^{-2}\xi_t^2)\,dx. \tag{C–22}$$

Integration over the length of the string gives the total string energy

$$E = \frac{1}{2}\rho_\ell c^2 \int (\xi_x^2 + c^{-2}\xi_t^2)\,dx. \tag{C–23}$$

If the wave motion is progressive, regardless of whether it is outgoing ($\xi_t = -c\xi_x$) or incoming ($\xi_t = c\xi_x$), Eq. C–23 simplifies to

$$E = \rho_\ell c^2 \int \xi_x^2 \, dx. \tag{C–24}$$

An example of the use of Eqs. C–23 and C–24 is given in Problem 1C–4.

f. Forced Vibration

What if the string is driven by a time-varying external force? An example is the motion of a telephone line excited by gusts of wind blowing over the line. Let $f_\ell(x, t)$ denote the force per unit length exerted by the external force, where x has been included in the argument of f_ℓ in case we wish to consider problems in which the force varies from point to point along the string. Newton's law for the forced string is Eq. C–10 with the force term added,

$$\rho_\ell \, \Delta x \, \xi_{tt} = \mathcal{T} \sin(\theta + \Delta\theta) - \mathcal{T} \sin\theta + f_\ell \, \Delta x. \tag{C–25}$$

Carrying through the steps leading from Eq. C–10 to Eq. C–13 yields the following *inhomogeneous* wave equation:

$$\xi_{xx} - \frac{1}{c^2}\xi_{tt} = -\frac{1}{\mathcal{T}} f_\ell(x, t). \tag{C–26}$$

See Sec. 11B.3.b for the analogous equation for forced motion of a circular membrane and an example.

3. Sound Waves

The wave equation for sound is derived from the conservation equations for fluids. Because our scope at this stage is limited to lossless plane waves, only the one-dimensional conservation equations for nondissipative fluids are needed. The effect of body forces, such as gravity, is also neglected. A more complete and general derivation is given in Chap. 2. Before discussing the conservation equations, we first review some concepts about continuous media, in particular, what is meant by fluid particle and fluid velocity. The coordinate systems used to describe fluid motion are also discussed.

Although liquids and gases are composed of molecules, our common experience is that fluids behave as continuous media. When we speak of a fluid particle, therefore, we do not mean a molecule of the fluid but rather a large enough collection of molecules that the average of their random motions is zero. To fix ideas, consider a *gedanken* experiment in which the density ρ of a gas at rest is measured. Weigh a box of gas, subtract the weight of the box, and divide by g to get the mass M of the gas. Finally, divide M by the box volume V to obtain the density. Now cut the box in half and repeat the measurement. The density is of course the same: half the volume of gas has half the mass. Repeated halvings of the box are found to have no effect on the density—up to a point. When the box becomes small enough that it contains only a "small number" of molecules, cutting the box in half may not divide the molecules, which are in constant motion, into two equal groups. In other words, the probability of finding half the molecules in each half of the box decreases as box size decreases. Further halvings produce ever more erratic measures of the density. The results of our *gedanken* experiment are shown in Fig. 1.19, a plot of density against length L of a side of the box. We see that density of a fluid particle is a useful concept provided the particle is not so small that behavior of individual molecules is important. The same is true of other macroscopic properties, such as pressure, temperature, and fluid velocity. For a gas the mean free path ℓ of the molecules may be used to estimate the minimum particle size. The mean free path is the average distance a molecule travels between collisions with other molecules. To be conservative, assume that the continuum model of a fluid breaks down when the smallest length important in a problem is of the

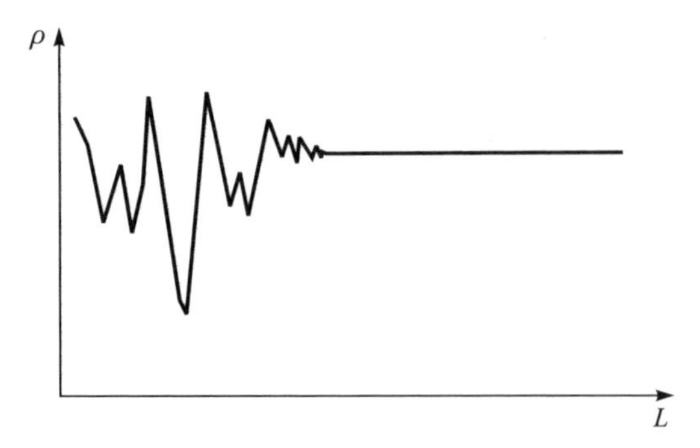

Figure 1.19 Dependence of density on size of a fluid sample.

same order of magnitude as the mean free path. The value of ℓ for air at STP[13] is about 0.1 μm ($= 10^{-7}$ m), and the number of molecules contained in a volume $\ell^3 = 0.001\mu\text{m}^3$ is about 25,000. A fluid particle is thus a very large collection of molecules. Although the argument here is specifically for gases, the same general conclusions hold for liquids.

In the same way that a fluid particle is distinguished from an individual molecule of the fluid, fluid velocity means velocity of a fluid particle, not molecular velocity.

Two coordinate systems for describing fluid motion are in common use. In the *Lagrangian*, or *material*, coordinate system, fluid properties, such as pressure, density, and particle velocity, are expressed in terms of their variation from fluid particle to fluid particle, regardless of the fact that the particles are in motion. The coordinates are therefore the fluid particles themselves. In the other system, called *Eulerian* or *spatial*, coordinates are fixed in space; fluid properties are described at fixed spatial points without regard to which particles are present at the time of the description. Two examples may make the difference between the two coordinate systems clear. Particle velocity may be measured by using tracer particles, such as smoke in air or ink or dye in water. The velocity is determined by measuring the displacement of a tracer particle over a very short time interval, for example, by taking a short time exposure photograph of the flow field. Since the tracer particle follows a fluid particle, the measurement yields the velocity in Lagrangian coordinates. It would, however, be very difficult to measure pressure in this way. A pressure sensor, such as a microphone (for air) or hydrophone (for water), cannot easily be moved to follow a fluid particle. Instead, the sensor is usually fixed in space and, as such, provides an Eulerian measure of the pressure. The same is true of most devices used to measure temperature. Because most acoustical measurements are

[13]STP stands for standard temperature and pressure. The values used in this book are 20°C and 1 atm $= 1.013 \times 10^5$ Pa, respectively. The pascal, Pa, is equal to 1 N/m^2.

Eulerian in nature, Eulerian coordinates are used in the derivation of the conservation equations presented here. In linear acoustics, with which most of this text is concerned, the distinction between the two coordinate systems is not significant. Particle motions are so small that the wave equation has the same mathematical form in both Lagrangian and Eulerian coordinates. In nonlinear acoustics, however, the particle motions are large enough that the distinction is in general not trivial; see, for example, Ref. 4.

a. One-Dimensional Continuity Equation (Conservation of Mass)
Consider the flow of a compressible fluid through a duct of constant but otherwise arbitrary cross section (area S). Figure 1.20 shows a duct of circular cross section. The segment between x and $x + \Delta x$ is the control volume (CV), albeit an infinitesimal one, for this case. Mass flows into and out of the CV, and we wish to know the rate at which the mass inside changes. First list the assumptions:

1. The CV is fixed in space (the coordinate system is thus Eulerian).
2. The flow is one-dimensional, depending only on x and t.

As a consequence of the second assumption, the particle velocity u and density (indeed all flow properties) are the same over a cross section. The quantity $\rho u S$, which has units of mass per unit time, represents the mass flow of the fluid. A statement of conservation of mass for the control volume is as follows:

$$\begin{array}{c}\text{Time rate of increase}\\ \text{of mass inside the CV}\end{array} = \begin{array}{c}\text{Net mass flow into the CV}\\ \text{through the surfaces of the CV,}\end{array}$$

or, in mathematical terms,

$$\frac{\partial}{\partial t}(S\rho\,\Delta x) = \rho u S|_x - \rho u S|_{x+\Delta x}. \qquad \text{(C–27)}$$

On the left-hand side, ρ may be taken to be the average density in the control volume. When, in a moment, the limit $\Delta x \to 0$ is taken, ρ will become a true

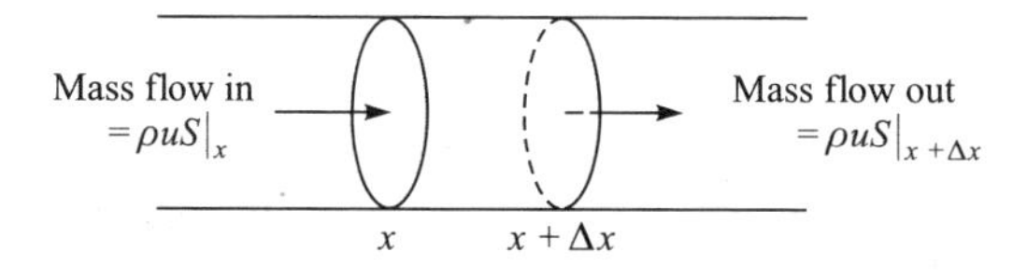

Figure 1.20 Mass flow through the surfaces of a cylindrical control volume.

point function. Since S is constant and Δx is not a function of time, Eq. C–27 may be rearranged as follows:

$$\frac{\partial \rho}{\partial t} = \frac{\rho u|_x - \rho u|_{x+\Delta x}}{\Delta x}.$$

In the limit as $\Delta x \to 0$, the right-hand side becomes $-\partial(\rho u)/\partial x$, and we obtain

$$\rho_t + (\rho u)_x = 0. \qquad \text{(C–28)}$$

This is the equation of continuity.

An alternative form is found by expanding the second term in Eq. C–28,

$$\rho_t + u\rho_x + \rho u_x = 0, \qquad \text{(C–29)}$$

and denoting the first two terms by $D\rho/Dt$:

$$\frac{D\rho}{Dt} + \rho u_x = 0. \qquad \text{(C–30)}$$

The operator D/Dt is called the total or material derivative. Operating on any quantity q,

$$\frac{Dq}{Dt} = \frac{\partial q}{\partial t} + u\frac{\partial q}{\partial x} \qquad \text{(C–31)}$$

means the total time rate of change of property q for a fluid particle. The first term $\partial q/\partial t$ is the time rate of change of q the fluid particle would experience if it were not moving. The second term is the additional rate of change caused by the particle's movement into a new region of the fluid, where conditions may be different. Note that because Dq/Dt means the time rate of change of q as we follow a fluid particle, the equivalent derivative in Lagrangian coordinates is the ordinary partial derivative $\partial q/\partial t$.[14]

b. Conservation of Momentum in One Dimension

The same control volume is used to derive a momentum balance relation; see Fig. 1.21. The momentum per unit volume is ρu, and the momentum flux

[14]An alternative way to obtain Eq. C–31 is to use the chain rule from advanced calculus. If a quantity q is a function of two variables x and t, the total derivative with respect to t is

$$\frac{dq}{dt} = q_t + q_x\frac{dx}{dt}.$$

In application of this result to fluid flow, x is distance and t is time, and dx/dt thus represents the fluid velocity u. The material derivative D/Dt is therefore simply the total derivative d/dt. The notation D/Dt is traditional in hydrodynamics.

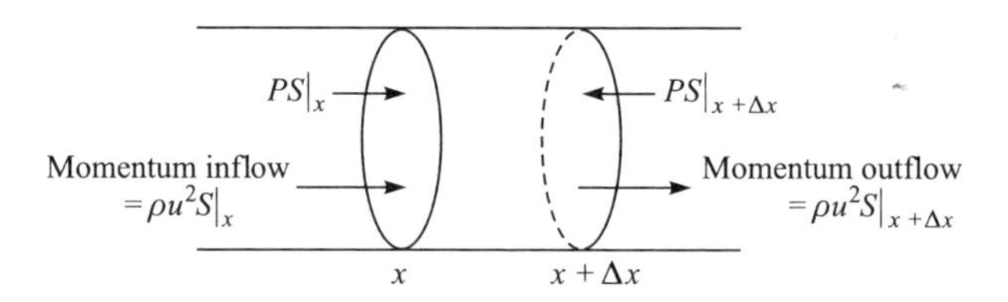

Figure 1.21 Momentum flow and forces acting for a cylindrical control volume.

(momentum per unit area per unit time) is $\rho u^2 S$. Newton's second law ($F = ma$), with account taken of the momentum inflow across the boundaries of the CV, is

Time rate of increase of momentum inside the CV	=	Net momentum inflow across boundaries	+	Sum of the forces acting on the CV.

Forces are generally of two kinds, body forces and surface forces. To the assumptions listed for the derivation of continuity, we add these two about forces:

3. Body forces are not significant. The most common body force is gravity; its effect is considered in Chaps. 2 and 8. Gravity generally affects sound propagation only over large distances.
4. The fluid is inviscid. In other words the only significant surface force is that due to the pressure P.

Given these assumptions, the only forces acting on the control volume are $PS|_x$ at the left end and $PS|_{x+\Delta x}$ at the right end.[15] The mathematical version of the verbal statement of momentum conservation is therefore

$$\frac{\partial}{\partial t}(\rho u S\ \Delta x) = \rho u^2 S|_x - \rho u^2 S|_{x+\Delta x} + PS|_x - PS|_{x+\Delta x}.$$

Divide through by $S\ \Delta x$ and take the limit as $\Delta x \to 0$:

$$(\rho u)_t + (\rho u^2)_x + P_x = 0. \tag{C-32}$$

[15]The symbol P is used in this book for two different quantities. Here P stands for the total pressure, which is the sum of the ambient pressure p_0 and the excess or acoustic pressure p. In Chap. 4 (and a few other places) P is used to mean the (complex) amplitude of a time-harmonic acoustic pressure variation, i.e., $p = P(x, y, z)e^{j\omega t}$. The usage meant should be clear from the context.

This is one form of the momentum equation. An alternative form is found by expanding the first two terms:

$$\rho_t u + \rho u_t + (\rho u)_x u + \rho u u_x + P_x = 0.$$

Because the first and third terms cancel by virtue of the continuity equation, Eq. C–28, the final result is

$$\rho(u_t + u u_x) + P_x = 0, \qquad \text{(C–33)}$$

which is simpler than Eq. C–32. Since the terms in parentheses are equal to Du/Dt, still another alternative form is

$$\rho \frac{Du}{Dt} + P_x = 0. \qquad \text{(C–34)}$$

Had the longitudinal force due to viscosity been included in our derivation, we should have obtained, in place of Eq. C–33,

$$\rho(u_t + u u_x) + P_x = (\lambda + 2\mu)u_{xx}, \qquad \text{(C–35)}$$

where μ is the shear viscosity coefficient and λ is the dilatational viscosity coefficient. The viscosity term is discussed in some detail in Chap. 2, and the attenuation of sound due to viscosity is covered in Chap. 9.

In preparation for the next section, we state a fifth assumption, which is somewhat related to the fourth:

5. The flow is lossless.

c. *Equation of State*

In general an energy conservation equation should be considered as well as a thermodynamic equation of state. Because of assumption 5, however, a separate equation for energy is not needed. The energy equation and losses are taken up in Chaps. 2 and 9.

An equation of state is simply a relation between thermodynamic variables. A well-known example is the perfect gas law, one form of which is

$$P = R\rho T, \qquad \text{(C–36)}$$

where T is the absolute temperature and R is the universal gas constant divided by the molecular weight of the gas (for air $R = 287.08$ J/kg K). An equation of state particularly useful in acoustics is one that relates pressure to density and entropy,

$$P = P(\rho, s), \qquad \text{(C–37)}$$

where s is the entropy per unit mass. When losses are negligible (assumption 5), as is often the case in acoustics, the entropy remains constant. The pressure is then a function of density alone:

$$P = P(\rho). \tag{C–38}$$

Equation C–38 is a general form of the isentropic equation of state.

For gases the most frequently used isentropic equation of state is the so-called adiabatic gas law

$$(P/p_0) = (\rho/\rho_0)^{\gamma}, \tag{C–39}$$

where γ is the ratio of specific heats (for air $\gamma = 1.4$) and p_0 and ρ_0 are the static values of P and ρ, respectively. In Chap. 2 we show that Eq. C–39 follows from the perfect gas law, Eq. C–36, when the entropy is constant.

In the case of liquids Eq. C–39 is not applicable. A more general approach is needed. For any fluid (liquid or gas) one may express the general isentropic equation of state (Eq. C–38) as a Taylor series in the condensation $(\rho - \rho_0)/\rho_0$,

$$P = p_0 + A\,\frac{\rho - \rho_0}{\rho_0} + \frac{B}{2!}\left(\frac{\rho - \rho_0}{\rho_0}\right)^2 + \frac{C}{3!}\left(\frac{\rho - \rho_0}{\rho_0}\right)^3 + \cdots. \tag{C–40}$$

The coefficients $A, B, C, \ldots$ are determined from experiments or from other analyses. Since the condensation is usually very small, the most important of the coefficients is A. To evaluate A, we introduce the sound speed c, which is defined by

$$\begin{aligned} c^2 &\equiv \left.\frac{\partial P}{\partial \rho}\right|_{s\,=\,\mathrm{const}} \\ &= \frac{dP}{d\rho} \quad \text{for an isentropic process.} \end{aligned} \tag{C–41}$$

The sound speed may be thought of as a new thermodynamic variable, since it is derived from other thermodynamic quantities. Differentiation of Eq. C–40 yields

$$c^2 = \frac{A}{\rho_0} + \frac{B}{\rho_0}\frac{\rho - \rho_0}{\rho_0} + \frac{C}{2!\rho_0}\left(\frac{\rho - \rho_0}{\rho_0}\right)^2 + \cdots.$$

In the limit of vanishing condensation ($\rho \to \rho_0$), c^2 becomes a constant, which we denote c_0^2. The coefficient A is thus equal to $\rho_0 c_0^2$, which is the adiabatic bulk modulus of the fluid for small departures from equilibrium (the inverse, $1/\rho_0 c_0^2$, is sometimes called the compressibility of the fluid). Indeed, sound speed is often defined in terms of the adiabatic bulk modulus, namely, $c = \sqrt{A/\rho_0}$.

The distinction between c and c_0 should be noted. As Fig. 1.22 shows, c^2 is the slope of the P, ρ curve (for an isentropic process) at any point, whereas c_0^2 is the slope at the static operating point p_0, ρ_0. The quantity c_0 is found in tables of sound speed, such as the one given in Appendix A. Strictly speaking, c_0 should be called the *small-signal sound speed* because it is equal to c only in the limit as $\rho \to \rho_0$. It is traditional, however, to shorten the term to *sound speed.*

Equation C–40 may be simplified by eliminating the static components. Let p stand for the *acoustic* or *excess pressure*

$$p \equiv P - p_0, \tag{C–42}$$

and let $\delta\rho$ stand for the excess density,

$$\delta\rho \equiv \rho - \rho_0. \tag{C–43}$$

The isentropic equation of state then takes on the convenient form

$$p = c_0^2 \delta\rho \left[1 + \frac{B}{2!A} \frac{\delta\rho}{\rho_0} + \frac{C}{3!A} \left(\frac{\delta\rho}{\rho_0} \right)^2 \cdots \right]. \tag{C–44}$$

Values of B/A, the coefficient of the first nonlinear term, have been tabulated for a number of liquids; see, for example, Ref. 3.

Example 1.2. Adiabatic gases. Let us evaluate the coefficients A, B/A, C/A, $\cdots$ for gases. Start with the adiabatic gas law, Eq. C–39, and expand it by using the binomial theorem:

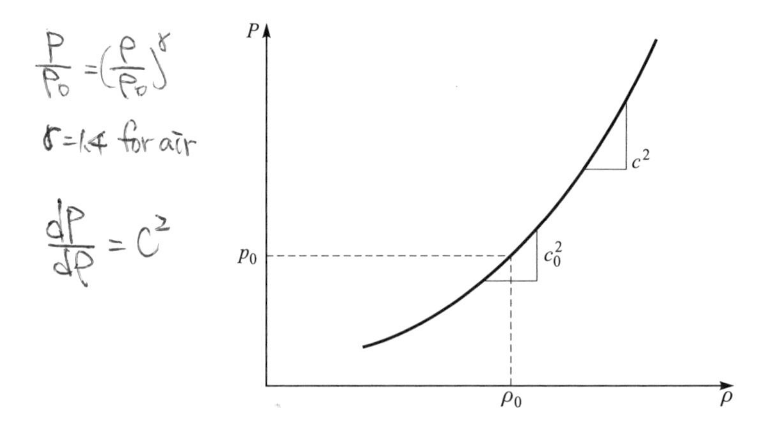

Figure 1.22 Square of the sound speed is the slope of the pressure-density curve.

$$1 + \frac{p}{p_0} = \left(1 + \frac{\delta\rho}{\rho_0}\right)^\gamma$$
$$= 1 + \frac{\gamma\delta\rho}{\rho_0} + \frac{\gamma(\gamma-1)}{2!}\left(\frac{\delta\rho}{\rho_0}\right)^2 + \frac{\gamma(\gamma-1)(\gamma-2)}{3!}\left(\frac{\delta\rho}{\rho_0}\right)^3 + \cdots,$$

or

$$p = \frac{\gamma p_0}{\rho_0}\delta\rho\left[1 + \frac{(\gamma-1)}{2!}\frac{\delta\rho}{\rho_0} + \frac{(\gamma-1)(\gamma-2)}{3!}\left(\frac{\delta\rho}{\rho_0}\right)^2 + \cdots\right].$$

Compare this equation with Eq. C–44 and identify

$$c_0 = \sqrt{\gamma p_0/\rho_0}, \tag{C–45}$$
$$B/A = \gamma - 1, \tag{C–46}$$
$$C/A = (\gamma - 1)(\gamma - 2), \tag{C–47}$$

etc. The most important of these relations is Eq. C–45, which shows how the small-signal sound speed for a given gas depends on the static pressure and density of the gas. For example, very light gases like helium and hydrogen have a high sound speed. Sound travels more slowly in heavier gases, such as air or carbon dioxide. An alternative expression for the sound speed, found by applying Eq. C–36, is

$$c_0 = \sqrt{\gamma R T_0}, \tag{C–48}$$

where T_0 is the static temperature. Here is another important result: the sound speed in a gas varies as the square root of the absolute temperature.

d. Historical Note

The value of c_0 given by Eq. C–45 or C–48 is sometimes called the Laplacian sound speed because the formula first appeared (1816) in a derivation given by Pierre Simon Laplace (1749–1827). More than a century earlier a derivation by Sir Isaac Newton (1642–1727) in Book II of the *Principia* (1687) had yielded (in modern notation) the following formula for sound speed:

$$b_0 = \sqrt{p_0/\rho_0}, \tag{C–49}$$

where the symbol b_0 indicates Newton's prediction. When the speed of sound was measured in air, however, the Newtonian value was found to be about 16% low. The discrepancy became one of the unexplained mysteries of physics. Many ad hoc explanations were presented, including ones by Newton. Although Laplace's analysis (1816) correctly solved the mystery, his explanation did not gain immediate acceptance, and the ad hoc theories continued into the middle of the nineteenth century. In modern terminology Newton's formula is the speed of sound for an isothermal gas, whereas Laplace's formula applies to an adiabatic gas.[16] Since sound normally propagates with negligible flow of heat, Laplace's formula is the appropriate one. Newton is sometimes criticized for assuming that sound is an isothermal process when in fact it is an adiabatic process. The criticism is unjust. The concepts of isothermal and adiabatic processes were not known in Newton's day (Ref. 5).

e. The Linear Wave Equation

In the absence of sound the fluid is assumed to be quiet, that is, the values of pressure, density, and particle velocity are taken to be $P = p_0$, $\rho = \rho_0$, and $u = 0$, respectively. Moreover, we assume the quiet fluid to be homogeneous, i.e., $\rho_0 \neq \rho_0(x)$ and $p_0 \neq p_0(x)$. Some of these restrictions are relaxed in Chaps. 2 and 8.

Although Eqs. C–28, C–33, and C–44 are nonlinear, they may be made linear by introducing the so-called small-signal approximation. Widely used in acoustics, this approximation is valid even for the loudest sounds most of us experience. Because most sound waves disturb the status quo in the fluid only a tiny bit, the variables associated with sound—excess pressure p, excess density $\delta\rho$, and particle velocity u—may be assumed to be small quantities of first order. What this means is that each is small compared to an appropriate static quantity of the same dimensions. We start by assuming

$$|\delta\rho| \ll \rho_0. \tag{C–50}$$

If this inequality is true, then, as Eq. C–44 shows, the restriction on the acoustic pressure is[17]

$$|p| \ll \rho_0 c_0^2. \tag{C–51}$$

Similarly, it can be shown that the restriction on the particle velocity is

$$|u| \ll c_0. \tag{C–52}$$

These three relations, taken together or separately (since one of them implies the other two), are what is meant by the small-signal assumption.

[16]To see why Eq. C–49 implies an isothermal process, notice that if the temperature is constant, Eq. C–36 reduces to Boyle's law $P = K\rho$ (K is a constant), a linear P, ρ relation. In turn, Eq. C–39 shows that a linear P, ρ relation requires γ to be 1. For $\gamma = 1$, Eq. C–45 reduces to Newton's formula.

[17]Notice that, however natural it might seem, the static pressure p_0 is not the measure against which $|p|$ is assumed small. Although in gases p_0 and $\rho_0 c_0^2$ are comparable (see Eq. C–45), in liquids they are not. For example, in water $\rho_0 c_0^2$ is about 20,000 atm. An underwater sound wave having an amplitude of 1 atm is thus truly a small signal. The fact that its amplitude is not small relative to the static pressure is irrelevant (except for the possibility of cavitation; see Problem 1E–7).

The small-signal assumption is now used to linearize the conservation equations. Each term in Eqs. C–28, C–33, and C–44 is assigned an order based on the number of small quantities multiplied together. First consider the continuity equation. Given Eq. C–43, Eq. C–29 may be expressed as

$$\underline{\delta\rho_t} + \underline{\underline{u\delta\rho_x}} + \underline{\rho_0 u_x} + \underline{\underline{\delta\rho u_x}} = 0, \tag{C–53}$$

(recall that ρ_0 is a constant). The first and third terms have been underlined once to show that they are first-order terms: each contains just a single small quantity. The twice-underlined terms are second order because they are products of small quantities. If the first-order terms are small, as assumed, the second-order terms must be so much smaller that by comparison they are negligible. Dropping them, we obtain

$$\delta\rho_t + \rho_0 u_x = 0. \tag{C–54}$$

This is the linear form of the continuity equation for a homogeneous medium at rest.[18] The momentum equation may be linearized similarly. Expanded, Eq. C–33 becomes

$$\underline{\rho_0 u_t} + \underline{\underline{\rho_0 u u_x}} + \underline{\underline{\delta\rho u_t}} + \underline{\underline{\underline{\delta\rho u u_x}}} + \underline{p_x} = 0.$$

In this case a third-order term (triply underlined) is present in addition to two second-order terms. When all but the first-order terms are dropped (for the same reasons given above), the result is

$$\rho_0 u_t + p_x = 0. \tag{C–55}$$

The isentropic equation of state may be linearized by inspection; Eq. C–44 becomes

$$p = c_0^2 \delta\rho. \tag{C–56}$$

The reader may be concerned that second-order terms have been dropped without regard to relative magnitudes of derivatives. Consider Eq. C–53, for example. What if $\delta\rho$ were to vary rapidly with distance but slowly with time? The second term would then not necessarily be small compared to the first. The comparison "rapidly with distance but slowly with time" is, however, vague because distance and time have different units. The rigorous way to investigate this question is to write the equations in terms of dimensionless variables so that all derivatives are comparable. Furthermore, dimensional coefficients,

[18] More general cases are considered in Chap. 2.

such as ρ_0, should be suppressed. To this end we introduce the following dimensionless variables:

$$\delta\rho^\dagger = \delta\rho/\rho_0,\ u^\dagger = u/c_0,\ \tau = t/t_c,\ z = x/c_0 t_c,$$

where t_c is a characteristic time. (Although t_c need not be specified in this case, a convenient choice is often $t_c = 1/\omega$, where ω is angular frequency.) Equation C–53 now becomes

$$\delta\rho_\tau^\dagger + u^\dagger \delta\rho_z^\dagger + u_z^\dagger + \delta\rho^\dagger u_z^\dagger = 0.$$

If, as already assumed, $|\delta\rho^\dagger| \ll 1$ and $|u^\dagger| \ll 1$, then the quadratic terms are much smaller than the linear terms, regardless of the differentiation. A more physical way to arrive at the same result is to recognize that since acoustic waves travel with speed c_0, changes with respect to space and time must be such that the magnitude of $\delta\rho_t$ must be approximately equal to $|c_0 \delta\rho_x|$ (recall Sec. B.4). In magnitude, therefore, the first two terms in Eq. C–53 are approximately $|c_0\delta\rho_x|$ and $|u\delta\rho_x|$, respectively. Since the derivatives are now the same in both terms, and since by the small-signal assumption $|u| \ll c_0$, the second term is clearly negligible compared to the first.

The final step is to combine the three linearized equations and obtain a (linear) wave equation. If u and p are to be adopted as the primary field variables (the usual convention), the first step is to eliminate the excess density between Eqs. C–54 and C–56:

$$p_t + \rho_0 c_0^2 u_x = 0. \qquad \text{(C–57)}$$

Equations C–55 and C–57 are analogous to the pair of transmission line equations, Eqs. C–1 and C–2, respectively. Pressure and voltage are paired as analogs, as are particle velocity and current. (What are the electrical component analogs of ρ_0 and $\rho_0 c_0^2$?) The pressure may be eliminated by subtracting the time derivative of Eq. C–55 from the spatial derivative of Eq. C–57. The final result is the linear wave equation

$$c_0^2 u_{xx} - u_{tt} = 0. \qquad \text{(C–58)}$$

The reader should confirm that elimination of u rather than p leads to

$$c_0^2 p_{xx} - p_{tt} = 0. \qquad \text{(C–59)}$$

f. Characteristic Impedance

To find the characteristic impedance for an outward traveling sound wave, start with the traveling wave solution,

$$u = f(x - c_0 t),$$

and use either the momentum equation (Eq. C–55) or the continuity equation (in the form of Eq. C–57) to calculate the pressure. Choosing the former, we find

$$p_x = -\rho_0 u_t = \rho_0 c_0 f'(x - c_0 t),$$

or, on integration,

$$p = \rho_0 c_0 f(x - c_0 t) = \rho_0 c_0 u.$$

The characteristic impedance is therefore

$$Z_0 \equiv \left(\frac{p}{u}\right)_{\text{outgoing}} = \rho_0 c_0. \qquad \text{(C–60)}$$

Can you show that the relation for an incoming wave is $p = -Z_0 u$?

Values of Z_0 for a number of substances are given in Appendix A. Table 1.1 gives the values of c_0 and Z_0 at STP (20°C, 1 atm in this text) for the most common acoustical fluids. The unit of impedance is the *rayl*, named after Lord Rayleigh. The unit is MKS rayl in the MKS system; divide by 10 to convert to CGS rayls (CGS system).

D. SPHERICAL AND CYLINDRICAL SOUND WAVES OF ONE DIMENSION

Although the first part of this book is primarily about sound waves that are plane, some elementary results for one-dimensional spherical and cylindrical waves are given here for occasional use later. A detailed treatment of spherical and cylindrical waves is postponed until Chaps. 10 and 11, respectively.

Table 1.1 Sound Speed and Characteristic Impedance of Air and Water

Fluid ($p_0 = 1$ atm)	c_0 (m/s)	Z_0 (MKS rayls)
Air at 20°C	343	415
Water at 20°C	1481	1.48×10^6
Seawater at 13°C	1500	1.54×10^6

1. Three-Dimensional Wave Equation

In vector notation the three-dimensional continuity and momentum equations for lossless fluids are (see Chap. 2)

$$\rho_t + \nabla\cdot(\rho\mathbf{u}) = 0 \tag{D–1}$$

and

$$\rho[\mathbf{u}_t + (\mathbf{u}\cdot\nabla)\mathbf{u}] + \nabla P = 0, \tag{D–2}$$

respectively. Their corresponding linear forms are

$$\delta\rho_t + \rho_0\nabla\cdot\mathbf{u} = 0 \tag{D–3}$$

and

$$\rho_0\mathbf{u}_t + \nabla p = 0. \tag{D–4}$$

To these two equations add the linearized isentropic equation of state, Eq. C–56.

For nonplanar waves it is frequently useful to introduce a new field variable, the velocity potential ϕ defined by

$$\mathbf{u} = \nabla\phi. \tag{D–5}$$

Substitution of this expression in the momentum equation, Eq. D–4, makes it possible to integrate the latter once. The result is[19]

$$p = -\rho_0\phi_t. \tag{D–6}$$

Combination of Eqs. D–3, D–5, D–6, and C–56 then yields

$$\nabla^2\phi - \frac{1}{c_0^2}\phi_{tt} = 0. \tag{D–7}$$

Once this equation is solved for ϕ, the pressure and particle velocity may be found from Eqs. D–6 and D–5, respectively.

The three one-dimensional forms of Eq. D–7 are as follows:

[19] A "constant of integration," say $\rho_0 K'(t)$, could in principle be added to the right-hand side of Eq. D–6. The constant may be gotten rid of by defining a new potential $\phi_{\text{new}} = \phi - K(t)$. Equations D–5 and D–6 then follow, with ϕ replaced by ϕ_{new}. Assume that this step has already been taken. Physical arguments may also be used to show that Eq. D–6 is the desired integral of the momentum equation. If the fluid has been quiet ($p = 0$) at some time in the past, one must take $\rho_0 K'(t) = 0$.

$$\text{Plane waves:} \quad \phi_{xx} - \frac{1}{c_0^2}\,\phi_{tt} = 0. \tag{D–8}$$

$$\text{Cylindrical waves:} \quad \phi_{rr} + \frac{1}{r}\,\phi_r - \frac{1}{c_0^2}\,\phi_{tt} = 0. \tag{D–9}$$

$$\text{Spherical waves:} \quad \phi_{rr} + \frac{2}{r}\,\phi_r - \frac{1}{c_0^2}\,\phi_{tt} = 0. \tag{D–10}$$

A general equation that covers all three cases is

$$\phi_{rr} + \frac{a}{r}\,\phi_r - \frac{1}{c_0^2}\,\phi_{tt} = 0, \tag{D–11}$$

where $a = 0$, 1, and 2 for plane, cylindrical, and spherical waves, respectively.

2. Solutions for One-Dimensional Waves

As we already know, the general solution of the plane-wave equation is

$$\phi = f(x - c_0 t) + g(x + c_0 t).$$

An equally general solution exists for spherical waves. Without approximation, Eq. D–10 may be rearranged and put in the following form:

$$(r\phi)_{rr} - \frac{1}{c_0^2}\,(r\phi)_{tt} = 0. \tag{D–12}$$

Because this equation has the same form as that for plane waves, its solution has the same form, namely,

$$r\phi = f(r - c_0 t) + g(r + c_0 t),$$

or

$$\phi = \frac{f(r - c_0 t)}{r} + \frac{g(r + c_0 t)}{r}. \tag{D–13}$$

The first term represents an outgoing, spherically diverging wave. The amplitude diminishes as $1/r$.[20] The second term represents an incoming, spherically converging wave. As r decreases, the wave amplitude increases. Although spherical waves are not discussed in detail until Chap. 10, simple examples are given in the next section and in Chap. 3.

[20] The drop in amplitude does not imply dissipation. Rather it results from the spreading of the wave over an ever larger area. "Geometrical spreading" is a term used to mean the diminution of a wave due solely to the way a wave diverges as it propagates.

No similarly general solution can be found for cylindrical waves. The mathematical magic needed to convert Eq. D–9 to plane-wave form does not exist. Because using the combination $r\phi$ does the trick for spherical waves, physical intuition suggests trying the combination $\sqrt{r}\phi$ for cylindrical waves. Alas, the transformation does not quite work. Equation D–10 rearranged with $\sqrt{r}\phi$ as the dependent variable becomes

$$(\sqrt{r}\phi)_{rr} - \frac{1}{c_0^2}(\sqrt{r}\phi)_{tt} + \frac{(\sqrt{r}\phi)}{4r^2} = 0.$$

Because of the presence of the last term, this equation fails to have the plane-wave form. For large values of r, however, the last term is very small. By neglecting it, we obtain the following approximate solution:

$$\phi = \frac{f(r - c_0 t)}{\sqrt{r}} + \frac{g(r + c_0 t)}{\sqrt{r}}, \tag{D–14}$$

which shows that at great distances the amplitude of a cylindrical wave diminishes as $1/\sqrt{r}$. It is emphasized, however, that this solution is valid only in the farfield, that is, far from the origin. In Chap. 11, which is devoted to cylindrical waves, special exact solutions of Eq. D–9 are obtained by using Bessel functions.

3. Sound from a Pulsating Sphere

Assume that a pulsating sphere of radius r_0 is surrounded by an infinite homogeneous fluid containing no other sound sources. The sphere pulsates with angular frequency ω in such a way that the velocity potential ϕ varies sinusoidally at the surface of the sphere. Determine the wave motion in the fluid ($r > r_0$). In other words, adapt the general solution Eq. D–13 so as to satisfy the following conditions:

(a) The signal at $r = r_0$ is $\phi = A \sin \omega t$ (source condition).

(b) The fluid ($r > r_0$) contains no reflecting objects or sources (radiation condition).

Because of condition (b), no incoming wave can be generated. The function g in Eq. D–13 must therefore be zero. The outgoing wave part of the solution is conveniently expressed as

$$\phi = \frac{F(t - r/c_0)}{r}.$$

Application of condition (a) yields

$$A \sin \omega t = \frac{F(t - r_0/c_0)}{r_0},$$

which may be solved for $F(t)$,

$$F(t) = r_0 A \sin \omega(t + r_0/c_0).$$

Replacing t with the desired argument $t - r/c_0$ leads to the final form of the solution,

$$\phi = \frac{Ar_0}{r} \sin[\omega t - k(r - r_0)], \tag{D–15}$$

where $k = \omega/c_0 = 2\pi/\lambda$ is the wave number and λ is the wavelength.

Next find p, u, and the impedance. It is expedient at this point to switch to complex exponentials. Equation D–15 may be written

$$\phi = \frac{Ar_0}{r} \operatorname{Im} e^{j[\omega t - k(r - r_0)]}, \tag{D–16}$$

where Im means "imaginary part of" (similarly, Re means "real part of"). Use of Eqs. D–5 and D–6 yields

$$u = -jk\left(1 + \frac{1}{jkr}\right)\phi \tag{D–17}$$

and

$$p = -j\omega\rho_0\phi, \tag{D–18}$$

respectively. (The final answer is found by taking the imaginary part of each expression, since in going from Eq. D–15 to Eq. D–16, we represented the original function as the imaginary part of a complex exponential.) The impedance for an outgoing, time-harmonic, spherical wave is therefore

$$Z \equiv \frac{p}{u} = \frac{\rho_0 c_0}{1 + 1/jkr}. \tag{D–19}$$

A great deal may be learned by analyzing Eq. D–19. First of all, this is the first time we have encountered a complex impedance. A complex impedance signifies a phase difference (other than 180°) between the pressure and particle velocity signals. The switch from trigonometric functions to complex exponentials was in fact done partly to make it easy to represent the phase difference. To focus on the phase, express the impedance in polar form,

$$Z = \rho_0 c_0 \frac{kr}{\sqrt{1 + k^2 r^2}} e^{j \cot^{-1} kr}. \qquad \text{(D–20)}$$

At great distances ($kr \gg 1$), the phase angle $\cot^{-1} kr$ is zero, and Z reduces to the plane-wave value $\rho_0 c_0$. The reason is that the wavefronts become nearly plane (the curvature over a wavelength is very small). At the opposite extreme, at small distances ($kr \ll 1$), the phase angle approaches 90°. More about this in a moment.

Another observation is that the impedance is frequency dependent. This fact is a clue to dealing with more complicated source signals, such as transients or noise: break up the signal into its Fourier components and analyze each separately.

Of particular interest is the impedance presented by the medium to the source. Let Z_s stand for the value of Z evaluated at the source ($r = r_0$),

$$Z_s = \frac{\rho_0 c_0}{1 + 1/jkr_0}. \qquad \text{(D–21)}$$

Detailed analysis of this quantity, which is sometimes called the load or radiation impedance (since it represents the acoustical load imposed on the pulsating sphere by the medium), is left to Problem 1D–2. It is easy to see, however, that if $kr_0 \gg 1$ (the source circumference large compared to a wavelength), the source impedance is simply the plane-wave value $\rho_0 c_0$. Because the source "sees" a pure resistance, it is relatively easy to radiate power into the fluid (see Sec. E.3 below). On the other hand, if $kr_0 \ll 1$, Z_s becomes completely imaginary,

$$Z_s = j\omega \rho_0 r_0.$$

Because the source now sees a pure reactive load, it cannot radiate any power into the medium. Consequently, for good radiation efficiency the source should not be small compared to a wavelength. What does this imply about the size of a woofer (a loudspeaker used for low frequency radiation)? A tweeter (a loudspeaker used for high frequency radiation)?

E. SIGNALS, IMPEDANCE, INTENSITY AND POWER, AND LEVELS

This section contains information about a variety of concepts and quantities useful in acoustics.

1. Time and Frequency Domains

Acoustical signals are so varied that many ways of describing them have developed. For example, a signal may be called transient or steady state, broadband or narrowband, monochromatic or multifrequency, deterministic or random, intense or weak. Here we say a little about representing signals in the time and frequency domains.[21]

A common approach in the presentation of acoustics is to concentrate on sinusoidal waves. The special properties of the sine, cosine, and complex exponential functions may be used to treat problems and derivations in a very simple manner; recall, for example, the pulsating sphere problem solved and analyzed in the previous section. Even when the special properties are not needed, however, the tendency is not to give up the assumption that the wave is time harmonic. The justification, often left unsaid, is that on account of the linearity of the wave equation, any more complicated wave may be represented as a sum of sinusoidal waves. In F. V. Hunt's words, "Blessed be Fourier."[22] The implication is that sound waves are best represented in the frequency domain.

However, exclusive reliance on the frequency domain is not recommended. First, one may come to believe that all problems and derivations necessarily start with sine waves. Second, if the signal is broadband, as it is for a pulse or train of pulses, its frequency domain representation (a Fourier series) may be so complicated that physical interpretation of results may be difficult. Finally some problems are simply easier to solve in the time domain. Notice that the entire introductory analysis of wave motion, Sec. B above, was carried out in the time domain. Had we chosen to use the frequency domain, the analysis would have been much more complicated and the physical effects not so apparent. On the other hand, it is true that many problems are much more easily treated in the frequency domain. The trick is to know when to use one domain and when the other. One goal of this book is for the reader to develop a facility for both domains.

The Fourier transform offers a formal means of entering the frequency domain. Given a time domain signal $u(x, t)$, the frequency domain version is

$$\bar{u}(x, \omega) = \mathcal{F}[u(x, t)] = \int_{-\infty}^{\infty} u(x, t) e^{-j\omega t}\, dt, \tag{E–1}$$

where $\mathcal{F}$ stands for the Fourier transform and is defined by the integration operation indicated. Conversely, to return to the time domain, use the inverse transform,

$$u(x, t) = \mathcal{F}^{-1}[\bar{u}(x, \omega)] = \frac{1}{2\pi} \int_{-\infty}^{\infty} \bar{u}(x, \omega) e^{jt\omega}\, d\omega. \tag{E–2}$$

[21] Signal processing, although a very important part of acoustics, is not dealt with in this book. For treatments of signal processing with applications to acoustics, see, for example, Ref. 2.
[22] A statement Hunt was fond of writing on the blackboard in his introductory class in acoustics at Harvard in the 1950s and 1960s.

A more direct approach often used is simply to assume a time-harmonic signal, e.g.,[23]

$$u(x, t) = U(x)e^{j\omega t}. \tag{E–3}$$

This method is used extensively, for example, in Chap. 4. The particle velocity amplitude $U(x)$, which is essentially the same as the transform $\bar{u}(x, \omega)$, (1) is a function of frequency as well as space and (2) may be complex. Whether to use Fourier transforms or just assume a time-harmonic signal depends on circumstances. If the wave at the source is known to be a steady-state, single-frequency signal, the time-harmonic assumption is ideal. The description one obtains for the sound field will likely be in its most useful form and require no further processing. If on the other hand the source signal is not monochromatic (and if direct time domain techniques are not useful), the Fourier transform may be preferable.

The wave equation itself, Eq. D–7, is easily transformed to the frequency domain. Either by using Fourier transforms or by assuming a time-harmonic signal $\phi = \Phi(x, y, z)e^{j\omega t}$, one obtains

$$\nabla^2\Phi + k^2\Phi = 0. \tag{E–4}$$

Equation E–4, called the *Helmholtz equation*, is frequently used as the starting point in analyses of wave motion.

2. Impedance

Impedance is often described as the ratio of a "push" variable q_p (such as voltage or pressure) to a corresponding "flow" variable q_f (such as current or particle velocity). Also to be kept in mind is the fact that impedance is a frequency domain concept. To see why, try using time domain versions of q_p and q_f:

$$Z = \frac{q_p(x, y, z; t)}{q_f(x, y, z; t)}. \tag{E–5}$$

As it stands, this ratio is not very useful because it is in general time dependent. However, let q_p be a time-harmonic function, i.e., $q_p = Q_p(x, y, z)e^{j\omega t}$. Provided the system relating q_f to q_p is linear, q_f must also be time harmonic; denote it $q_f = Q_f(x, y, z)e^{j\omega t}$. In this case the time dependence cancels,

[23] Dependent variables are generally denoted by lowercase letters in this text. For time-harmonic signals, the corresponding capital letter is frequently used to stand for the (complex) amplitude of the signal. Equation E–3 illustrates the practice. The complex amplitude is often called a *phasor*.

$$Z = \left(\frac{q_p}{q_f}\right)_{\text{time harmonic}} = \frac{Q_p}{Q_f}. \tag{E–6}$$

The ratio is now a much more useful quantity. Note that because Q_p and Q_f are in general complex, so too is Z. In certain instances it is not necessary to assume time-harmonic signals because the time dependence cancels regardless of the waveform. For example, in our treatment of lossless transmission lines, strings, and fluids carrying plane sound waves, we were able to calculate the characteristic impedance without assuming time-harmonic waves. Our method worked because the impedance in these cases is real and frequency independent. If the impedance depends on frequency, as it does, for example, for spherical waves (Eq. D–20) or for standing waves in a tube (Chap. 4), it is necessary to begin with time-harmonic signals.

Three different impedances are in common use in acoustics (in each case time-harmonic signals are assumed).

(a) *Acoustic impedance* at a given surface is the ratio of the sound pressure averaged over that surface p_{av} to the volume velocity q of the fluid through the surface,

$$Z_{ac} = p_{av}/q. \tag{E–7}$$

(b) *Specific acoustic impedance* is the ratio of the pressure at a point to the particle velocity at the point,

$$Z_{sp\ ac} = p/u. \tag{E–8}$$

(c) *Mechanical impedance* is the ratio of the force f acting on a specified area to the particle velocity through or of that area,

$$Z_{mech} = f/u. \tag{E–9}$$

For any wave whose pressure is constant over a specified area S, e.g., a plane wave in a duct of cross-sectional area S, we have $p_{av} = p$, $q = Su$, and $f = pS$. In this case Eqs. E–7 and E–9 reduce to

$$Z_{ac} = p/Su = Z_{sp\ ac}/S \tag{E–10}$$

and

$$Z_{mech} = pS/u = Z_{sp\ ac}S, \tag{E–11}$$

respectively. Of the three impedances—represented by Eqs. E–8, E–10, and E–11—the one most commonly encountered is the specific acoustic impedance.

In this book it is often referred to as "impedance" for short and the subscript "sp ac" is omitted.

We repeat for emphasis that, for traveling plane waves, pressure and particle velocity are related to each other as follows:

$$\text{Forward traveling waves: } p = Z_0 u, \tag{E–12a}$$

$$\text{Backward traveling waves: } p = -Z_0 u, \tag{E–12b}$$

where $Z_0 = \rho_0 c_0$ is the characteristic (specific acoustic) impedance of the medium. Why in physical terms are the pressure and particle velocity in phase for a forward traveling wave but out of phase for a backward traveling wave? *Hint*: Consider the propagation of a positive pressure step. If the pressure step travels in the forward direction, in which direction is fluid forced to flow, that is, what is the sign of the accompanying particle velocity step? Now reanswer this question for a pressure step traveling in the backward direction.

3. Intensity and Sound Power

The propagation of an acoustic wave is accompanied by a flow of energy in the direction the wave is traveling. Sound intensity I in a specified direction is defined as the time average of energy flow (energy flow means energy per unit time, or power) through a unit area, where the normal to the area points in the specified direction. Recall that mechanical power is equal to the dot product of the force and the velocity, $\mathbf{f} \cdot \mathbf{u}$. The acoustical adaptation of this expression is

$$p\mathbf{u} \cdot \Delta\mathbf{S} = \mathbf{i} \cdot \Delta\mathbf{S},$$

where $\Delta\mathbf{S}$ is the increment of the area, and $\mathbf{i} \equiv p\mathbf{u}$ is the instantaneous energy flow per unit area. The time average of $\mathbf{i}$ is the intensity

$$\mathbf{I} = \frac{1}{t_{av}} \int_0^{t_{av}} p\mathbf{u}\, dt. \tag{E–13}$$

The choice of integration, or averaging, time t_{av} depends on the type of waveform being considered. For periodic waves t_{av} is the period. For transients t_{av} is the duration of the transient signal. For nonperiodic waves, such as noise, t_{av} must be a "long time," i.e.,

$$\mathbf{I} = \lim_{t_{av} \to \infty} \frac{1}{t_{av}} \int_0^{t_{av}} p\mathbf{u}\, dt. \tag{E–14}$$

Note the directional character of the intensity: no power flows in a direction perpendicular to the particle velocity. The vector nature of the intensity is often suppressed by writing Eq. E–13 as a scalar relation,

$$I = \frac{1}{t_{\text{av}}} \int_0^{t_{\text{av}}} pu \, dt, \tag{E–15}$$

the implication being that I is in the direction of propagation, since that is usually also the direction of the particle velocity vector. In the MKS system the unit of intensity is watts per square meter, in the CGS system watts per square centimeter.

Special formulas for the intensity are now developed for the following two somewhat overlapping classes of problems:

(1) Progressive waves of arbitrary waveform in lossless fluids.
(2) Time-harmonic waves, not necessarily progressive, in arbitrary fluids.

For class (1) first let the progressive wave be traveling in the forward direction. Substitution of the characteristic impedance relation, Eq. E–12a, in Eq. E–15 yields

$$I = \frac{1}{\rho_0 c_0} \frac{1}{t_{\text{av}}} \int_0^{t_{\text{av}}} p^2 \, dt = \frac{p_{\text{rms}}^2}{\rho_0 c_0}, \tag{E–16}$$

where p_{rms} is the root-mean-square pressure

$$p_{\text{rms}} = \sqrt{\frac{1}{t_{\text{av}}} \int_0^{t_{\text{av}}} p^2 \, dt}. \tag{E–17}$$

(For example, if the pressure is a sinusoidal signal of amplitude A, the rms value is $p_{\text{rms}} = A/\sqrt{2}$.) Equation E–16 is quite general in that it holds for periodic waves, transients, and noise. Moreover, as shown in Sec. 10D.2, it also applies to (pure radial) outgoing spherical waves, even though, as shown in Sec. D.3 above, Z is complex for spherical waves.

Does Eq. E–16 also apply to backward traveling plane waves? Not without an important modification. Using Eq. E–12b, one finds $I = -p_{\text{rms}}^2/\rho_0 c_0$. What is the significance of the minus sign? *Hint*: In which direction does the energy flow in this case?

To deal with problems of class (2) (time-harmonic fields), let $p = Pe^{j\omega t}$ and $u = Ue^{j\omega t}$, where P and U are the complex amplitudes. In particular, let $U = |U|e^{j\beta}$. Special care must now be taken in evaluating Eq. E–15 because forming the product pu is a nonlinear operation. Since the actual values of pressure and particle velocity are Re p and Re u (or Im p and Im u), respectively, we must use these values in forming the product. The impedance may be expressed in either cartesian form (real and imaginary components) or polar form (magnitude and phase):

$$Z = \frac{p}{u} = \frac{P}{U} = \mathrm{Re}(Z) + j\mathrm{Im}(Z) = |Z|e^{j\theta},$$

where θ is the phase angle between P and U. Equation E–15 reduces in this case ($t_{\mathrm{av}} = 2\pi/\omega$) to

$$\begin{aligned} I &= \frac{\omega}{2\pi}\int_0^{2\pi/\omega} \mathrm{Re}(Zu)\mathrm{Re}(u)\,dt \\ &= \frac{\omega}{2\pi}\int_0^{2\pi/\omega} |Z||U|^2\cos(\omega t + \beta + \theta)\cos(\omega t + \beta)\,dt \\ &= \frac{1}{2}|Z|\cos\theta|U|^2. \end{aligned}$$

Recognizing that $|Z|\cos\theta = \mathrm{Re}(Z)$ and that $\frac{1}{2}|U|^2 = u_{\mathrm{rms}}^2$, we obtain

$$I = \mathrm{Re}(Z)u_{\mathrm{rms}}^2. \tag{E–18a}$$

This important and useful result is the analog of the electrical circuits law that the power is equal to the resistance times the square of the rms current. Equivalent results, the derivations of which are left to the reader, are

$$I = \frac{p_{\mathrm{rms}}^2}{|Z|}\cos\theta, \tag{E–18b}$$

$$I = \tfrac{1}{2}|P||U|\cos\theta, \tag{E–18c}$$

and

$$I = \tfrac{1}{2}\mathrm{Re}(PU^*) = \tfrac{1}{2}\mathrm{Re}(P^*U) = \tfrac{1}{4}(PU^* + P^*U), \tag{E–18d}$$

where the asterisk means complex conjugate.

Next consider sound power. The power W passing through a surface S is simply the integral of the intensity over that surface,

$$W = \int_S \mathbf{I}\cdot d\mathbf{S}'. \tag{E–19}$$

Some special cases are of interest. First, for a duct in which the intensity is uniform over the cross-sectional area S (and $\mathbf{I}$ and $d\mathbf{S}'$ are parallel), the integration is trival and gives

$$W = IS. \tag{E–20}$$

An obvious application of this result is to plane waves in tubes of constant cross section. It is also useful, however, in the case of horns, where the cross-sectional area is assumed to vary slowly with distance; see Chap. 7. Notice that Eq. E–20 is independent of the direction of the plane wave; power is a scalar quantity. As a second special case, consider an omnidirectional, outgoing spherical wave and let the surface S be a sphere of radius r with the source at the center. Because I is constant over the surface, Eq. E–19 reduces to

$$W = 4\pi r^2 I. \tag{E–21}$$

If the medium is lossless, the power generated by the source is undiminished after the wave has been transmitted any distance r. The intensity therefore varies inversely as the square of the distance. When losses exist, however, the acoustic power is not conserved; some is transformed into heat. Intensity then decreases more rapidly than $1/r^2$.

A very useful formula for the power radiated by an acoustical source may be obtained by letting the surface area be that of the source itself, S_s. Assume that the source velocity is normal to S_s and is time harmonic, $u_s = u_0 e^{j\omega t}$, where the amplitude u_0 is real. Substitution of Eq. E–18a into Eq. E–19 yields

$$W = \frac{1}{2}\int_{S_s} R_s u_0^2 \, dS'. \tag{E–22}$$

The quantity $R_s = \text{Re } Z_s$, called the *radiation resistance*, is the real part of the impedance presented by the medium to the source. For example, find the power emitted by the pulsating sphere considered in Sec. D.3. The value of R_s is found from Eq. D–21. Because the radiation is uniform over the surface of the sphere, Eq. E–22 is easily integrated and reduces to

$$W = 2\pi r_0^2 u_0^2 \frac{\rho_0 c_0 k^2 r_0^2}{1 + k^2 r_0^2}. \tag{E–23}$$

Notice the strong dependence of W on kr_0. In particular, Eq. E–23 quantitatively confirms the statement made at the end of Sec. D that it is difficult to get much power out of a small source (more precisely, a source for which $kr_0 \ll 1$).

4. Sound Pressure Level and Other Levels

The range of amplitudes and intensities of acoustical signals that may be observed and measured is very large. For example, the acoustic pressure of intense rocket engine noise can be as much as 9 orders of magnitude greater than the pressure of the weakest sound detectable by the human ear (threshold of hearing). Because a range of 9 decades in amplitude (18 decades in intensity) is difficult to comprehend, a logarithmic scale is used in acoustics to compress the range. The logarithmic measure devised is called a level, e.g., sound pressure level and intensity level. Although levels are unitless, they are expressed in decibels (dB).

a. Sound Pressure Level

The most widely used level is the *sound pressure level* (SPL),[24] which is a measure of the rms pressure of a sound:

$$\text{SPL} \equiv 20 \log_{10}(p_{\text{rms}}/p_{\text{ref}}) \quad \text{dB}, \tag{E–24}$$

where p_{ref} is a reference pressure. For air the reference pressure is 20 μPa (0.0002 μbar in the CGS system; 1 μbar = 1 dyne/cm^2), which corresponds approximately to the threshold of hearing for young persons in the frequency range of greatest sensitivity (1–4 kHz). The threshold of hearing is therefore approximately 0 dB for frequencies at which the ear is most sensitive. As an example, find the SPL for ordinary speech, for which p_{rms} is about 0.1 Pa. Using Eq. E–24, we obtain SPL = 74 dB. For comparison the SPL close to a jet aircraft at takeoff may be of order 120 dB. What is the SPL of the rocket engine noise mentioned in the previous paragraph?

For water the standard reference pressure is 1 μPa. Formerly a reference of 1 μbar ($= 10^5$ μPa) was used. One must be careful when reading the literature of underwater sound to determine the author's reference. For example, 100 dB re 1 μbar is equal to 200 dB re 1 μPa. Good practice when reporting a level, whether for air or water, is always to state the reference.

b. Intensity Level

The definition of *intensity level* (IL) is

$$\text{IL} = 10 \log_{10}(I/I_{\text{ref}}) \quad \text{dB}. \tag{E–25}$$

The reference intensity in air is 10^{-12} W/m^2. The relation between SPL and IL for traveling plane and spherical waves may be calculated as follows. Since for traveling waves Eq. E–16 holds, the expression for IL may be written

$$\begin{aligned} \text{IL} &= 10 \log_{10} \frac{p_{\text{rms}}^2}{\rho_0 c_0 I_{\text{ref}}} \frac{p_{\text{ref}}^2}{p_{\text{ref}}^2} \\ &= 20 \log_{10} \frac{p_{\text{rms}}}{p_{\text{ref}}} + 10 \log_{10} \frac{p_{\text{ref}}^2}{\rho_0 c_0 I_{\text{ref}}}. \end{aligned}$$

[24]Standard current practice in air acoustics is to use the symbol L_p to represent sound pressure level, where L stands for level and the subscript p signifies pressure. Similarly, L_i is used for intensity level and L_w for power level. These symbols are not, however, widely used in underwater acoustics. Since this text is not restricted to airborne sound, we stick to the older convention in which SPL stands for sound pressure level, IL for intensity level, and PWL for power level.

The first term is, of course, SPL. At STP the second term is very small. In particular, we find

$$\mathrm{IL} = \mathrm{SPL} - 0.16 \quad \mathrm{dB}. \tag{E–26}$$

For practical purposes, therefore, SPL and IL are numerically the same for *progressive* waves in air at STP.

The limitation of Eq. E–26 to progressive waves may be made clear by example. In a perfect standing wave field, like that created by a piston vibrating in a closed-end tube, as much energy is reflected from the closed end as travels forward from the piston. Because the net intensity is everywhere zero, IL $= -\infty$ everywhere. The sound pressure on the other hand varies from maxima (SPL large) at pressure antinodes to nulls (SPL $= -\infty$) at pressure nodes.

No common intensity reference is standard for water.

c. Sound Power Level

The definition of sound power level (PWL) is

$$\mathrm{PWL} = 10 \log_{10}(W/W_{\mathrm{ref}}), \tag{E–27}$$

where $W_{\mathrm{ref}} = 10^{-12}$ W. For reference the power level of a very soft whisper may be as low as 30 dB ($W = 10^{-9}$ W!), while that for the Saturn rocket is about 195 dB ($W = 30$ MW). Note that PWL is a measure of the total acoustical energy per unit time emitted by a source. It cannot be converted into SPL until additional information is provided, in particular, directionality of the source and distance from the source.

We close Sec. E.4 with some sample calculations for an N wave (Fig. 1.23), so called because its time waveform resembles the letter N. Waves of this shape are relatively common. For example, the sonic boom is an N wave; so is the wave produced by a bursting balloon (see Chap. 3) or by an electric spark in air. The wave from an explosion in air is somewhat N shaped after it has traveled a long distance.

Example 1.3. N wave. A plane forward traveling N wave in air has a pressure amplitude $A = 2$ Pa and a duration $2T = 1$ ms.

(a) **SPL**. To find the SPL, we must first calculate the rms pressure by means of Eq. E–17 (since the wave is not a sinusoid, the special result $p_{\mathrm{rms}} = A/\sqrt{2}$ does *not* apply). In this case the averaging time is $2T$, and the expression for p is

$$p = -A\,\frac{t - (t_0 + T)}{T}, \qquad t_0 < t < t_0 + 2T.$$

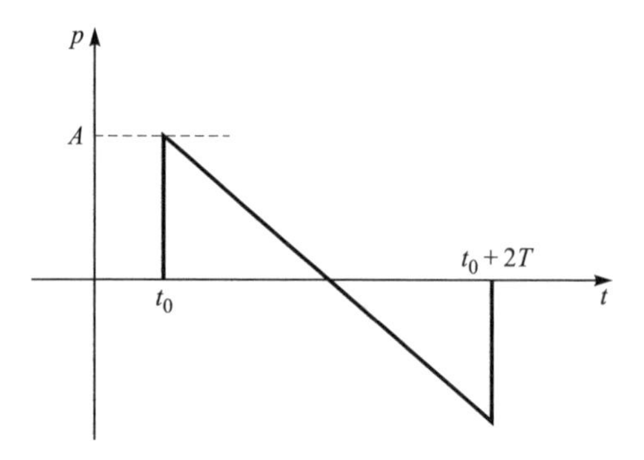

Figure 1.23 Traveling N wave.

Substitution into Eq. E–17 (the calculation may be simplified by shifting the time axis so that its origin coincides with the center of the N wave) yields

$$p_{\text{rms}} = A/\sqrt{3} = 1.15\,\text{Pa in this case.}$$

Equation E–24 is then used to obtain SPL = 95.2 dB. Notice that these results are independent of the duration $2T$ of the N wave.

(b) **Intensity and IL**. Use Eq. E–16 to obtain

$$I = \frac{(2/\sqrt{3})^2}{415} = 3.21 \times 10^{-3}\ \text{W/m}^2.$$

Either Eq. E–25 or E–26 yields

$$\text{IL} = 95.1\,\text{dB}.$$

(c) **Power and PWL**. Suppose that, instead of being plane, the N-shaped pulse is a uniform hemispherical wave having been produced by an explosion on the ground 10 m away. Find the sound power and sound power level produced by the explosion. Because the wave is known to be uniform, Eq. E–21 may be used, except that account must be taken of the fact that the sound is radiated into a hemisphere instead of a full sphere. We obtain

$$W = 2\pi(10^2)(3.21 \times 10^{-3})\ \text{W} = 2.02\ \text{W}.$$

The power level is therefore

$$\text{PWL} = 10\log_{10} 2.02 \times 10^{12}\ \text{dB} = 123\,\text{dB}.$$

What is the sound pressure level 100 m away from the explosion?

REFERENCES

1. W. F. Ames, *Numerical Methods for Partial Differential Equations*, 2nd ed. (Academic Press, New York, 1977).
2. J. S. Bendat and A. G. Piersol, *Engineering Applications of Correlation and Spectral Analysis* (Wiley–Interscience, New York, 1980).
3. R. T. Beyer, "Nonlinear acoustics (experimental)," in D. E. Gray, Ed., *American Institute of Physics Handbook*, 3rd ed. (McGraw-Hill Book Co., New York, 1972), Chap. 3o.
4. D. T. Blackstock, "Lagrangian one-dimensional equations for hydrodynamics for a viscous, thermally conducting fluid," *J. Acoust. Soc. Am.* **33**, 1245–1246 (1961).
5. F. V. Hunt, *Origins in Acoustics* (Yale University Press, New Haven, 1978), pp. 148–154.
6. R. B. Lindsay, *Mechanical Radiation* (McGraw-Hill Book Co., New York, 1960), pp. 243–248.
7. T. L. Szabo, "Lumped-element transmission-line analog of sound in a viscous medium," *J. Acoust. Soc. Am.* **45**, 124–130 (1969).
8. G. F. Wheeler and W. P. Crummett, "The vibrating string controversy," *Am. J. Phys.* **55**, 33–37 (1987).

PROBLEMS

1A–1. In the popular story of how Sir Isaac Newton formulated the law of gravity, Sir Isaac was sitting under an apple tree daydreaming. An apple fell out of the tree and struck him on the head. The incident caused him to conceive the law of gravitational attraction between bodies. Although probably apocryphal, if the story were true, would the sequence of events (not including Sir Isaac's thought processes) illustrate wave motion? Defend your answer.

1A–2. Two Native Americans communicate with each other from hilltop to hilltop using smoke signals. Does this form of communication constitute wave motion? Explain your answer.

1A–3. A line of traffic is backed up behind a traffic light. The automobiles are at rest, all brakes are off, and the gears are in neutral. The rearmost car is suddenly bumped from behind by a truck that was unable to stop in time. The rearmost car bumps into the next car, and so on down the line until the first car is bumped.

(a) Does the "chain reaction" collision process constitute wave motion? Defend your answer.

(b) If you answer "yes" to (a), what physical attributes of the line of cars determine the wave speed? Under what conditions is the wave speed constant? (You may assume elastic collisions, i.e., perfect bumpers.)

1B–1. Laplace's equation is $u_{xx} + u_{yy} = 0$. Let $f(z)$ represent a function of a complex variable $z = x + jy$.

(a) Show that $u = f(z)$ is a solution of Laplace's equation.

(b) Is $f(z^*)$, where z^* is the complex conjugate of z, also a solution?

1B–2. Suppose an infinite string is constrained so that a section in the middle is deformed into one cycle of a sine function as shown in the figure, i.e., $\xi = \sin x$ for $|x| \leq \pi$. The string is quiet up until time $t = 0$, at which point the constraint is released.

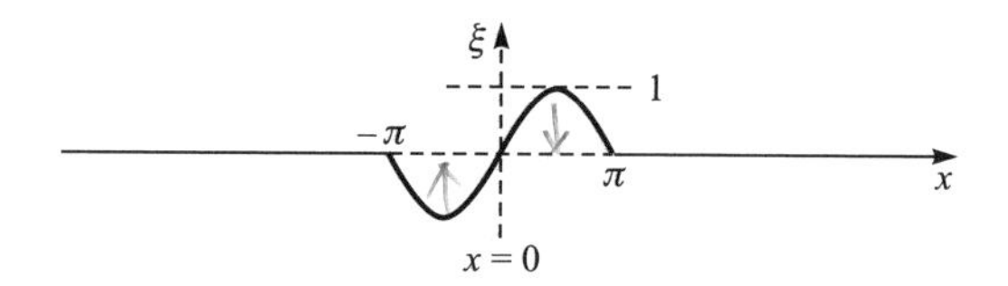

(a) Give the analytical solution of the problem, i.e., find $\xi(x, t)$ for $t > 0$.

(b) Sketch the string displacement $\xi(x)$ for a sequence of times t.

(c) Sketch the characteristics diagram (x, t diagram) for the problem. Give detail, including important points, lines, and regions, and the values of ξ for them if appropriate.

1B–3. An infinite string has zero velocity and the deformation shown in the figure, i.e., $\xi = \cos x$, $-\pi/2 < x < \pi/2$. At time $t = 0$ the constraint is released.

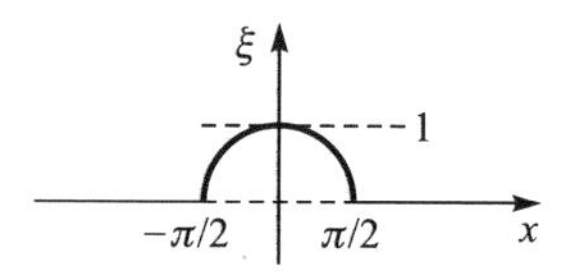

(a) Find the mathematical solution for $t > 0$.

(b) Sketch the waveform on the string $\xi(x)$ for a few representative values of ct, including one for $0 < ct < \pi/2$.

(c) Sketch the characteristics diagram (x, t diagram) for the problem. Give detail, including important points, lines, and regions, and the values of ξ for them if appropriate.

1B–4. At the end $x = 0$ of a semi-infinite string the displacement varies as $\xi(0, t) = \xi_0(1 - e^{-at})H(t)$, where $H(t)$ is the unit step function and a is a real positive constant.

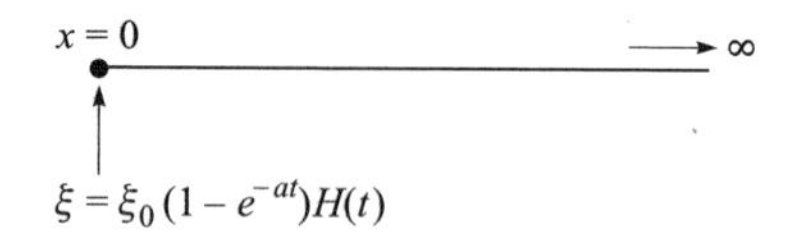

(a) Starting with the general solution $\xi = F(t - x/c) + G(t + x/c)$, evaluate F and G so as to obtain the solution that satisfies the conditions given.

(b) Sketch the string shape for a few representative times.

(c) Draw the characteristics diagram for the problem. Indicate any quiet zone(s) and show the value of ξ on a few sample characteristics, including the one for the head of the wave.

1B–5. This problem is similar to the previous one, but the string is now infinite in both directions. The displacement at the center of the string ($x = 0$) is the same, $\xi(0, t) = \xi_0(1 - e^{-at})H(t)$.

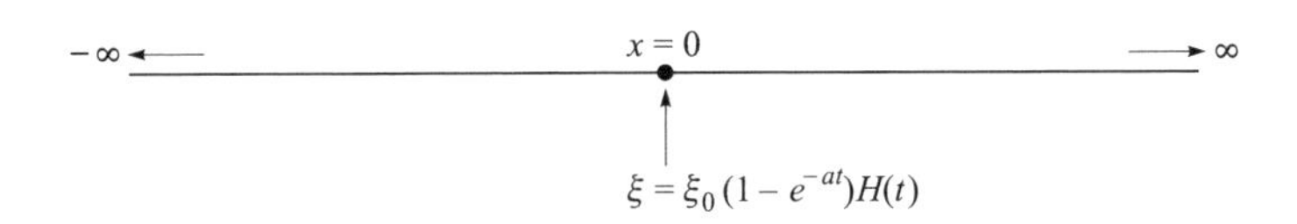

(a) Start with the general solution $\xi = F(t - x/c) + G(t + x/c)$, and find the specific solution for this problem.

(b) Sketch the string shape for a few representative times.

(c) Draw the characteristics diagram. Show detail, such as that specified in part (c) of the previous problem.

1B–6. The initial displacement of an infinite string is like that shown in Problem 1B–3. The resultant wave motion is found to be only a forward traveling wave.

(a) Find the initial velocity distribution $\xi_t(x, 0)$.

(b) What is the amplitude of the wave?

(c) Sketch the characteristics diagram for the problem. Indicate the value of the string displacement ξ in all parts of the diagram.

1C–1. Derive the wave equation for a lossy transmission line, an increment of which is shown in the sketch. Here L is the series inductance per unit length, C is the shunt capacitance per unit length, and R is the series resistance per unit length.

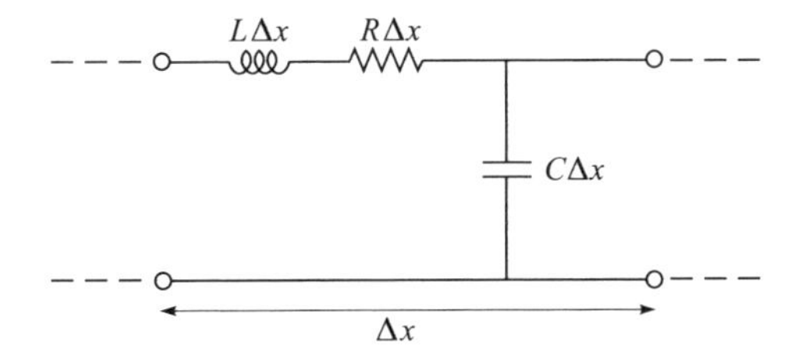

1C–2. An annealed copper wire under a tension of 10 N is to be used in an engineering design. It is required that the wave speed of flexural waves on the wire be at least 65 m/s.

(a) What restriction does this put on the gauge (AWG; see standard handbooks for wire gauge tables) of the wire? Choose the largest diameter wire (smallest gauge number) in the tables that meets this restriction.

(b) What is the characteristic impedance for flexural waves on the wire you have selected?

1C–3. Consider the struck string problem analyzed briefly in Sec. B.2.

(a) Draw the characteristics diagram for the problem.

(b) Sketch the spatial waveform for times for which $ct = a/3$, a, $2a$, and $4a$.

(c) Show that the permanent displacement of the string after the wave passes is $\xi_f = v_0 a/c$.

1C–4. For the struck string problem (Sec. B.2.b) calculate the total energy of the wave motion and show that it is equal to the energy the string has when it is first struck ($\rho_\ell a\, v_0^2$). Make the calculation for three cases:

(a) At time $t = 0+$ (when the string has just been struck).

(b) At time $t > a/c$ (when the forward and backward traveling waves are resolved).

(c) At time $0 < t < a/c$ (when the two waves are not resolved).

What conclusion do you draw from your three answers?

1C–5. Derive the wave equation for a string for which the linear density ρ_ℓ and the tension $\mathcal{T}$ vary along the lingth of the string, i.e., $\rho_\ell = \rho_\ell(x)$ and $\mathcal{T} = \mathcal{T}(x)$.

$$\text{ANSWER:}\quad \xi_{xx} + \frac{\mathcal{T}_x}{\mathcal{T}}\xi_x - \frac{1}{c^2}\xi_{tt} = 0\ .$$

Is c constant for this case? Explain.

1C–6. By using the definition of the sound speed, $c^2 = dP/d\rho$, show that for an adiabatic gas, i.e., $P/p_0 = (\rho/\rho_0)^\gamma$, the sound speed formula is

$$c^2 = \gamma P/\rho = \gamma RT.$$

For small signals reduce this result to

$$c_0^2 = \gamma p_0/\rho_0 = \gamma RT_0.$$

1C–7. Carry out the derivation of the small-signal acoustic wave equation for an isothermal gas, that is, a gas that obeys Boyle's law,

$$P/p_0 = \rho/\rho_0.$$

(Note that this is the form the perfect gas law $P = R\rho T$ takes when T is constant.) Start with the exact conservation laws, plus Boyle's law, and linearize them.

ANSWER: $b_0^2 u_{xx} - u_{tt} = 0$, where $b_0^2 = p_0/\rho_0$.

1C–8. The linearized equation of state for a relaxing fluid is

$$\tau(p_t - c_\infty^2 \delta\rho_t) + (p - c_0^2 \delta\rho) = 0,$$

where τ is a constant called the relaxation time and c_∞ and c_0 are two different sound speeds (normally quite close to each other). Derive the equation for small-signal waves in such a fluid.

(a) Starting with the exact continuity and momentum equations, linearize them.

(b) Combine the results with the equation of state given above and obtain the desired wave equation.

ANSWER: $\tau(u_{tt} - c_\infty^2 u_{xx})_t + (u_{tt} - c_0^2 u_{xx}) = 0.$

1C–9. Consider a perfect gas in which the static pressure p_0 stays constant as the absolute temperature T_0 changes. How does the characteristic impedance vary with T_0?

1D–1. The subject of this problem is the (vector) wave equation in $\mathbf{u}$ for small signals. Start with the linearized three-dimensional conservation equations and the equation of state.

(a) By eliminating p and $\delta\rho$, derive the wave equation in $\mathbf{u}$.

(b) Specialize to one-dimensional radial waves in both cylindrical and spherical coordinates, that is, let $\mathbf{u} = \mathbf{i}_r u$, where $\mathbf{i}_r$ is a unit vector in the radial (r) direction. The result should be the wave equation in u analogous to the one in ϕ, Eq. D–11.

(c) Is the wave equation in u [part (b)] simpler than, the same as, or more complicated than the corresponding wave equation in ϕ? Discuss.

1D–2. The expression for the radiation impedance of a pulsating sphere is given by Eq. D–21.

(a) Separate the impedance into real (resistive) and imaginary (reactive) parts. Write your result in the cartesian form $Z_s = \rho_0 c_0 (r_s + jx_s)$. On the same graph sketch each of the components r_s and x_s as a function of kr_0 out to $kr_0 = 10$. Be sure to plot enough values for kr_0 near the origin to obtain the proper behavior of the functions as $kr_0 \to 0$.

(b) What is the source diameter in terms of wavelengths at which Re (Z_s) reaches 90% of the plane-wave value $\rho_0 c_0$?

(c) Taking the value in part (b) as a criterion, what is the minimum diameter of a source to radiate 1000-Hz waves effectively (1) in air, (2) in fresh water?

1D–3. A simple acoustical model of the ocean is a fluid bounded by two parallel planes (the surface and the bottom), which are separated by a distance D (the ocean depth). A receiver is located a distance r from a source of spherical waves (both source and receiver are in the water). Assume that all sound striking the surface and the bottom is reflected back into the water. It is found that the sound pressure diminishes with r as

$$p \propto 1/r$$

near the source ($r \ll D$) but as

$$p \propto 1/\sqrt{r}$$

very far from the source ($r \gg D$). Explain this phenomenon.

1E–1. Find p_{rms} for the waveforms shown in sketches (a) and (b).

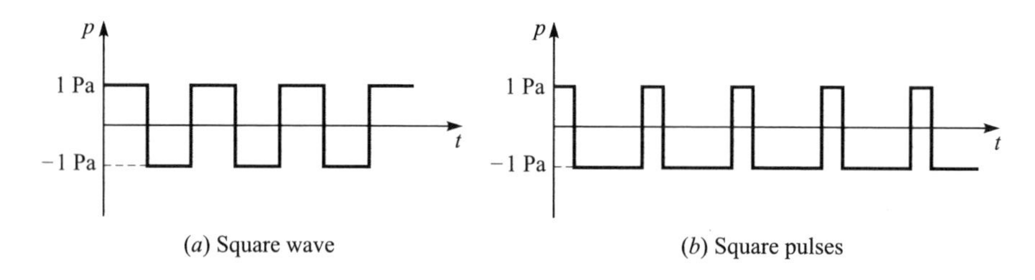

(a) Square wave (b) Square pulses

1E–2. Find the intensity in air of an outgoing plane wave having the waveform of the square wave in Problem 1E–1, sketch (a). Repeat for an incoming square wave. Is there any difference? Discuss, on physical grounds.

1E–3. A sawtooth wave in air has a pressure amplitude of 1 dyne/cm^2. Find the sound pressure level.

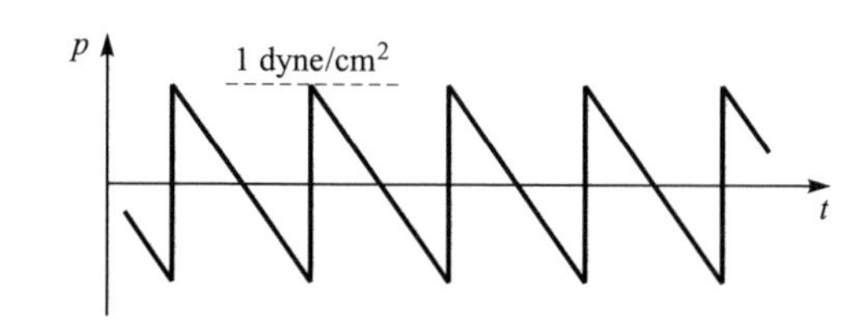

1E–4. A 5-cm-diameter piston driven by a shaker (vibrator) fits snugly in the end of a long tube that is filled with air at STP. Vibrating sinusoidally at a frequency of 100 Hz, the piston produces a plane wave that travels down the tube. An accelerometer mounted on the piston measures an acceleration amplitude of 4 m/s^2. Find the following quantities:

(a) The rms particle velocity of the sound wave

(b) Sound pressure level

(c) Intensity

(d) Acoustic power generated by the piston

1E–5. Find the value of the specific acoustic impedance at Denver, Colorado. Take the altitude of Denver to be 1 mile. Data from the U.S. Standard Atmosphere table may be used to make your calculation. If the long tube described in the previous problem were located at Denver, what would the sound pressure level in the tube be?

1E–6. The SPL of ordinary speech is approximately 74 dB (re 20 μPa). Assume the speech is a plane progressive wave.

(a) Find the rms particle velocity.

(b) Given the same particle velocity in seawater, what is the SPL (re 1 μPa)?

1E–7. A large underwater sound transducer in a freshwater lake generates very intense plane waves of sinusoidal waveform in the water. If during the negative half cycle the total pressure, that is, ambient plus acoustic pressure, becomes zero, cavitation (the production of bubbles) may occur. Recall that the ambient pressure at a depth h in the water is $p_0 + \rho_0 gh$, where ρ_0 is the density of the water, g is the acceleration of gravity, and p_0 is the atmospheric pressure above the water surface. The transducer is operated at a depth of 10 m.

(a) What is the maximum acoustic intensity that may be achieved without running the risk of cavitation?

(b) What SPL (re 1 μPa) corresponds to this intensity?

Note: The actual cavitation threshold depends on other factors besides ambient pressure, for example, frequency and dissolved gas content of the water. Ignore the other factors in working this problem.

1E–8. For certain acoustical signals the usual measures such as p_{rms}, SPL, and I do not give a very useful description of the wave. For example, the wave produced by an underwater explosion is frequently modeled as a sharp rise in pressure (a shock front) followed by an exponentially decaying tail. If for simplicity a plane wave is assumed, the waveform is given by

$$p(x, t) = Ae^{-(t-x/c_0)/\tau}, \quad t - x/c_0 > 0,$$

where A is the amplitude of the shock and τ is the decay time of the tail. Note that an observer at point x does not receive any signal at all until time $t = x/c_0$. Moreover, the duration of the signal is infinite.

(a) Attempt to determine p_{rms}, SPL, and I for this wave. You should find the calculation difficult because the quantities specified depend on the averaging time t_{av}. For example, if $t_{\mathrm{av}} = \infty$, $p_{\mathrm{rms}} = 0$, but if t_{av} is taken to be finite, $p_{\mathrm{rms}} \neq 0$. Parts (b) and (c) are about other measures that may be used to characterize the wave.

(b) Devise a "peak SPL," i.e., a sound pressure level based on the peak pressure rather than the rms pressure.

(c) Find the total energy flux (energy per unit area) of the wave,

$$E = \int_0^\infty pu\, dt.$$

In particular, determine whether this quantity is as ambiguous as the intensity asked for in part (a).

The following problems, denoted G for "general," pertain to two or more sections of the chapter.

1G–1. A sphere of radius $r_0 = 1$ cm pulsates sinusoidally at a frequency of 25 kHz in 20°C fresh water. The sound pressure level at a radial distance $r = 1$ m is 140 dB (re 1 μPa).

(a) Find the SPL and the particle velocity amplitude u_0 at $r = 100$ m.

(b) Find the particle velocity amplitude at the surface of the pulsating sphere.

1G–2. Consider a spherically spreading wave of angular frequency ω. Show that the formula for the intensity at any distance r is

$$I = p_{\mathrm{rms}}^2/\rho_0 c_0.$$

Is the restriction to a wave of "angular frequency ω" really necessary to the applicability of this result? Explain.

1G–3. A 5-kHz siren located on the ground is to produce a sound pressure level of 74 dB at a distance 1 mile away. The siren is omnidirectional, that is, it radiates equally in all directions above the ground. How much acoustic power is required of the siren?

1G–4. A metal plate placed at the interface ($x = 0$) between air ($x > 0$) and fresh water ($x < 0$) vibrates with velocity $u = u_0 \sin \omega t H(t)$. The amplitude u_0 is 0.1 m/s, and the temperature is 20°C.

(a) Find the SPL of the wave radiated into the air.

(b) Find the SPL of the wave radiated into the water.

(c) For the waterborne wave, which is traveling in the negative x direction, sketch the spatial waveform for both the particle velocity and the pressure at time $t = 3T$, where T is the period of the wave.

1G–5. An infinite, air-filled pipe has a membrane in the center. The acoustic pressure on the left is $p = A/2$, that on the right is $p = -A/2$. The air on both sides of the membrane is completely quiet up until time $t = 0$. At $t = 0$ the membrane is ruptured.

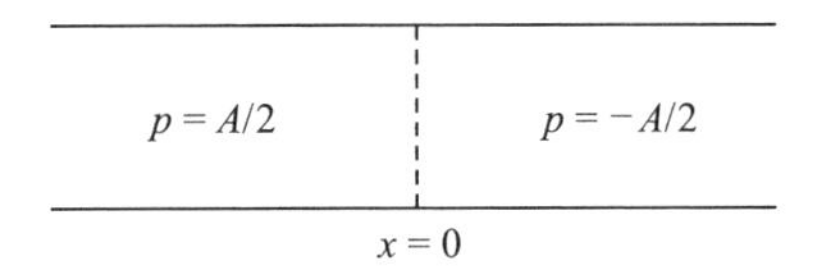

(a) Find the solution for the pressure at subsequent times.

(b) Sketch the spatial distribution of pressure at representative times t_1, t_2, and t_3, where $t_3 > t_2 > t_1 > 0$.

(c) Sketch the characteristics diagram for this problem. Identify the values of p and u in the various sections of the x, t plane.

(d) Sketch the spatial distribution of particle velocity u at time t_1.

1G–6. At time $t = 0$ the spatial pressure waveform of a traveling plane wave in seawater (13°C) is as shown in the sketch, i.e., the wave is two wavelengths of a sine wave.

(a) Find the following quantities:

(i) Frequency, if the wave train were of infinite extent

(ii) Time duration of the signal

(iii) SPL

(iv) Particle velocity amplitude u_0

(v) Intensity

(b) Suppose the wave is an outgoing traveling wave. Sketch the time waveform of the particle velocity signal observed at point x_1.

(c) Given x_2 as the point of observation, reanswer part (b) for an incoming traveling wave.

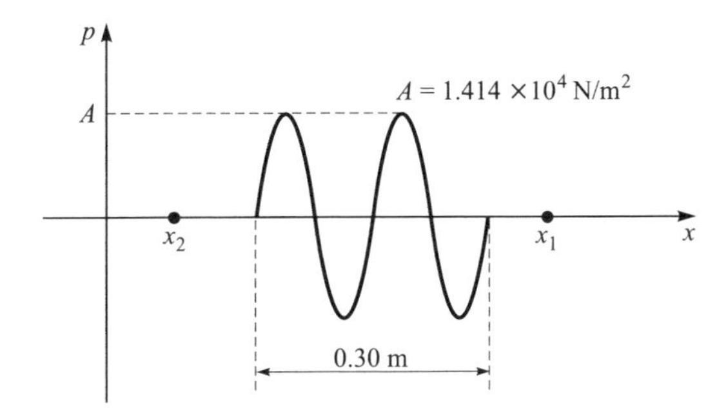

1G–7. The spatial distribution of pressure p and particle velocity u for a plane-wave field in air are measured at time $t = 0$. Two different sets of properties are to be considered.

A. Let the measured properties be

$$p(x, 0) = A \sin kx \operatorname{rect}(x/\lambda),$$
$$u(x, 0) = 0,$$

where k and λ are related by $k = 2\pi/\lambda$.

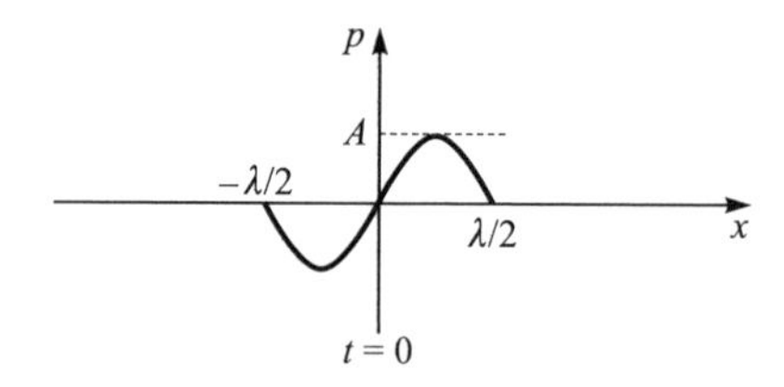

(a) Sketch the time waveform of the pressure $p(t)$ at

(i) $x = \lambda$,

(ii) $x = -\lambda$.

(b) Given $A = \sqrt{2}$ Pa, find the sound pressure level at the same two places.

B. Let the measured properties be

$$p(x, 0) = A \sin kx \operatorname{rect}(x/\lambda),$$
$$u(x, 0) = (A/\rho_0 c_0) \sin kx \operatorname{rect}(x/\lambda).$$

Reanswer (a) and (b).

2

DETAILED DEVELOPMENT OF THE ACOUSTICAL WAVE EQUATION

The derivation of the acoustical wave equation in Chap. 1 is limited to plane waves in lossless fluids. In the present chapter the scope is broadened to cover three-dimensional disturbances in dissipative fluids. Account is also taken of body forces, such as gravity. Section A is devoted to the three-dimensional conservation laws for fluids. Including body forces and dissipation requires two substantive changes to the elementary approach used in Chap. 1. First, terms representing viscous forces and body forces must be added to the momentum equation. Second, a third conservation equation, conservation of energy, must be added in order to account for dissipation due to viscosity and heat transfer. Before linearizing the conservation laws and merging them to form a linear wave equation, we pause in Sec. B to consider briefly the effect of the nonlinear terms. Dropping the dissipation terms, we combine the conservation equations for an adiabatic gas to form a nonlinear wave equation. For plane, progressive waves an exact solution of this equation is possible and is given. In Sec. C the equations of motion are linearized and combined. Wave equations are derived for a variety of different conditions, e.g., lossless fluid at rest, lossless fluid in motion, and viscous thermally conducting fluid.

A. CONSERVATION EQUATIONS AND CONSTITUTIVE RELATION

1. Equation of Continuity

The one-dimensional equation of continuity was derived in Sec. 1C.3.a. A differential element of fluid bounded by a rigid duct of constant cross section was considered. Here we generalize to three-dimensional flow. Two different approaches are used. The first, a direct generalization of the method used in Chap. 1, yields the differential form of the continuity equation. The second approach is more elegant. It leads to an integral form of the continuity equation, which is then converted by application of the divergence theorem to the differential form.

Consider the flow of fluid through the cubical volume element $\Delta V = \Delta x\ \Delta y\ \Delta z$, as shown in Fig. 2.1. Let the vector fluid velocity be given in rectangular coordinates by

$$\mathbf{u} = \mathbf{i}u + \mathbf{j}v + \mathbf{k}w, \qquad \text{(A–1)}$$

and assume that the volume element contains no mass sources or sinks. A verbal statement of conservation of mass for the fluid in the volume element is

Time rate of increase of mass inside ΔV	=	Net mass inflow through surface of ΔV

To find the mass flow through the six faces of the volume element, notice that u is the only velocity component that can carry fluid through the two x faces of the cube. Similarly, v and w are the only components that effect mass flow through the y and z faces, respectively. The mathematical rendition of the verbal statement is therefore

$$\frac{\partial}{\partial t}(\rho\,\Delta x\,\Delta y\,\Delta z) = \Delta y\,\Delta z(\rho u|_x - \rho u|_{x+\Delta x}) + \Delta z\,\Delta x(\rho v|_y - \rho v|_{y+\Delta y}) + \Delta x\,\Delta y(\rho w|_z - \rho w|_{z+\Delta z}).$$

Divide both sides of the equation by $\Delta x\,\Delta y\,\Delta z$, take the limit as Δx, Δy, and $\Delta z \to 0$, and put all the terms on the left-hand side:

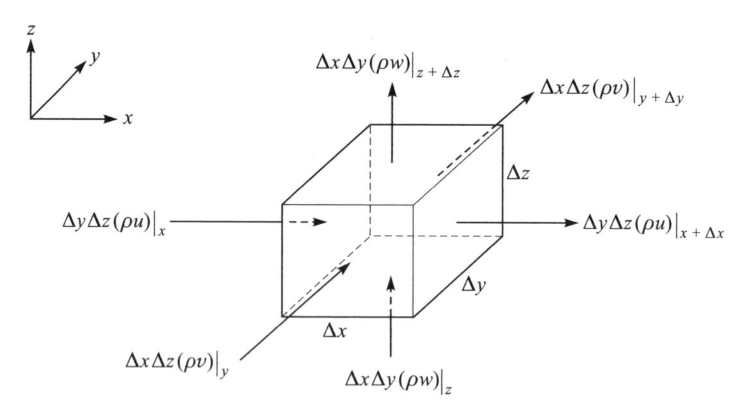

Figure 2.1 Mass flow through a cubical volume element.

$$\frac{\partial \rho}{\partial t} + \frac{\partial(\rho u)}{\partial x} + \frac{\partial(\rho v)}{\partial y} + \frac{\partial(\rho w)}{\partial z} = 0, \tag{A–2}$$

or in vector notation

$$\frac{\partial \rho}{\partial t} + \nabla\cdot(\rho \mathbf{u}) = 0. \tag{A–3}$$

This is one version of the equation of continuity. Another common version is found by expanding the divergence term and making use of the general definition of the material derivative

$$\frac{D(\cdot)}{Dt} \equiv \frac{\partial(\cdot)}{\partial t} + \mathbf{u}\cdot\nabla(\cdot) = \frac{\partial(\cdot)}{\partial t} + u\frac{\partial(\cdot)}{\partial x} + v\frac{\partial(\cdot)}{\partial y} + w\frac{\partial(\cdot)}{\partial z} \tag{A–4}$$

(Eq. A–4 is a generalization of Eq. 1C–31). We obtain

$$\frac{D\rho}{Dt} + \rho\nabla\cdot\mathbf{u} = 0. \tag{A–5}$$

The continuity equation for a region in which mass sources are present is the subject of Problem 2–1.

Next let us develop the integral form of the continuity equation. Consider an arbitrary control volume V fixed in space and bounded by a surface S (Fig. 2.2). In this case the verbal statement is

Time rate of increase of mass inside V	=	Net mass *inflow* through surface S

The left-hand side of the verbal statement is just $\partial/\partial t \int_V \rho \, dV$. Calculation of the right-hand side is a little more subtle. The vector $d\mathbf{S}$ at an arbitrary point on the surface has magnitude dS (the element of area) and direction *outward* normal to the surface. Because the fluid velocity $\mathbf{u}$ is not necessarily normal to the surface S at that point, the mass inflow through dS is given by the dot product $-\rho\mathbf{u}\cdot d\mathbf{S}$. The minus sign appears because $\mathbf{u}$ and $d\mathbf{S}$ are in opposite

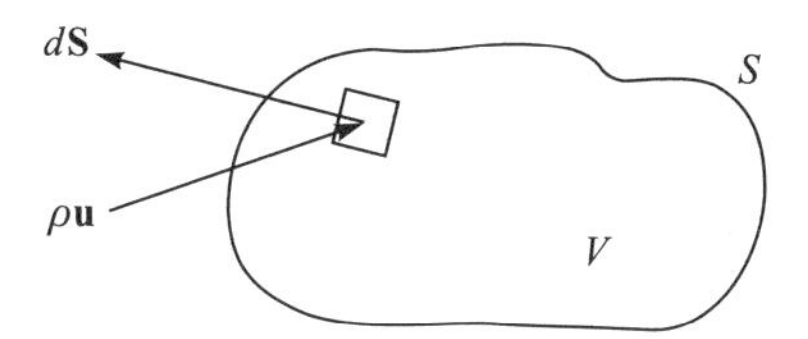

Mass *inflow* through $dS = \rho\mathbf{u}\cdot(-d\mathbf{S})$

Figure 2.2 Mass flow through an arbitrary control volume V.

directions. That is, *inflow* is found by taking the dot product of $\mathbf{u}$ with $(-d\mathbf{S})$. Integrating over the entire surface, we obtain the right-hand side of the verbal statement. The result is, after that term is moved to the left side,

$$\frac{\partial}{\partial t}\int_V \rho \, dV + \int_S \rho\mathbf{u}\cdot d\mathbf{S} = 0. \tag{A–6}$$

This is the integral form of the continuity equation.

To convert Eq. A–6 to the differential form, Eq. A–3, first note that the operator $\partial/\partial t$ may be taken inside the volume integral because the volume V is fixed in space, that is, $V \neq V(t)$. Next the Gauss divergence theorem,

$$\int_S \mathbf{A}\cdot d\mathbf{S} = \int_V \nabla\cdot\mathbf{A} \, dV, \tag{A–7}$$

where $\mathbf{A}$ is any vector, may be used to convert the surface integral in Eq. A–6 to a volume integral. The result is

$$\int_V \left[\frac{\partial \rho}{\partial t} + \nabla\cdot(\rho\mathbf{u})\right] dV = 0.$$

Since the volume V is arbitrary, vanishing of the integral requires that the integrand itself must vanish,

$$\frac{\partial \rho}{\partial t} + \nabla\cdot\rho\mathbf{u} = 0,$$

which is Eq. A–3.

Example 2.1. Mass flow through a plane, small-signal step wave. The integral form of the conservation equations is widely used in fluid mechanics to relate flow into and out of a specified region, or control volume, without solving for the detailed flow within the region. In this example we consider the propagation of a weak, plane pressure step in an otherwise quiet fluid, as shown in Fig. 2.3*a*. Since the step is a small signal, its propagation speed is assumed to be c_0. The fluid properties behind the step are $u = u_1$, $\rho = \rho_1$, and $P = P_1$; ahead of the step (quiet fluid) the values are $u = 0$, $\rho = \rho_0$, and $P = p_0$. If a fixed control volume were drawn about the step in Fig. 2.3*a*, the flow through the front and back sides, as well as the mass inside, would be a function of time. Consequently, Eq. A–6 would be hard to evaluate. Evaluation is made easy, however, by adopting a coordinate system moving with the step, as shown in Fig. 2.3*b* (c_0 is subtracted from every velocity in the flow). A control volume CV is drawn (the dotted box) to enclose the step. Since in this coordinate system the flow is steady, the time derivative term in Eq. A–6 vanishes (the mass within CV is constant), and only the surface integral needs to be evalu-

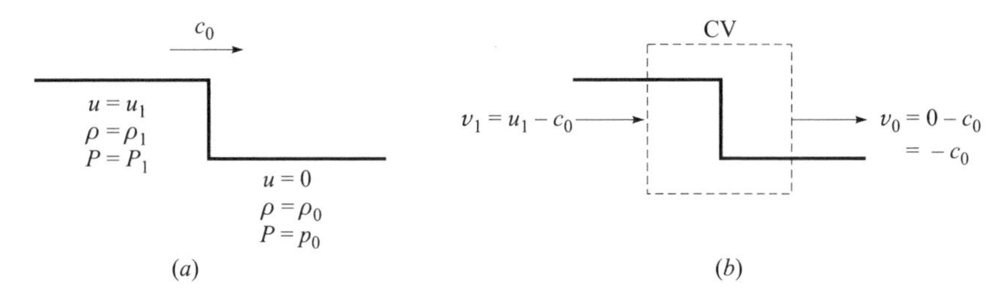

Figure 2.3 Weak pressure step traveling into a quiet fluid. (*a*) Step passes by the fixed observer. (*b*) Observer moves with the step and therefore perceives the step to be at rest.

ated. Integration over the left-hand face yields $-\rho_1 v_1 S_1$, where $v_1 = u_1 - c_0$ is the fluid velocity (in the moving coordinate system) at the face. The minus sign appears because **u** and $d\mathbf{S}$ are oppositely directed. Integration over the right-hand face yields $\rho_0 v_0 S_0$. The other faces of the CV contribute nothing to the integral because on them **u** is normal to $d\mathbf{S}$. Since $S_1 = S_0$, Eq. A–6 reduces to

$$\rho_1 v_1 = \rho_0 v_0, \tag{A–8}$$

i.e., the mass flow into the CV equals the mass flow out. Now return to the original coordinate system ($v_1 = u_1 - c_0$ and $v_0 = -c_0$) and let $\rho_1 = \rho_0 + \delta\rho_1$, where, since the step is a small signal, $\delta\rho_1 \ll \rho_0$. Equation A–8 becomes

$$(\rho_0 + \delta\rho_1)(u_1 - c_0) = -\rho_0 c_0,$$

or, if the quadratic term $u_1 \delta\rho_1$ is dropped because the wave is a small signal,

$$u_1 = \frac{c_0}{\rho_0}\delta\rho_1. \tag{A–9}$$

This example illustrates the power and simplicity of the control volume method. The result was obtained without any information about the actual waveform of the disturbance inside the control volume.

2. Momentum Equation

We also derive the momentum equation by two methods. First, the differential volume element is used and the body force is taken to be gravity. Second, the integral form of the momentum equation is developed for a fluid subject to arbitrary body forces. In both treatments the only surface force considered is that due to pressure. Viscous forces are added later.

a. Inviscid Fluid Subject to Body Forces

Consider the momentum balance equation for fluid in a spatial volume $\Delta V = \Delta x\,\Delta y\,\Delta z$ in a gravitational field. See Fig. 2.4. For simplicity, consider a plan view of the cube at the surface $y = 0$ (right-hand sketch in Fig. 2.4). Momentum is carried by the moving fluid as it flows through ΔV. The net momentum inflow in the x direction through the left, right, bottom, and top faces is as follows:

$$\text{Left and right faces:} \qquad (\rho u)u\,\Delta y\,\Delta z|_x - (\rho u)u\,\Delta y\,\Delta z|_{x+\Delta x},$$
$$\text{Bottom and top faces:} \qquad (\rho u)w\,\Delta x\,\Delta y|_z - (\rho u)w\,\Delta x\,\Delta y|_{z+\Delta z}.$$

The front and back faces of the cube (see perspective view) also participate; the net inflow there is

$$\text{Front and back faces:} \qquad (\rho u)v\,\Delta x\,\Delta z|_y - (\rho u)v\,\Delta x\,\Delta z|_{y+\Delta y}.$$

The forces acting on the cube are the gravitational body force $-\rho g\Delta x\Delta y\Delta z$, which acts in the z direction only, and the surface forces exerted by the fluid surrounding the cubical volume element. If the fluid is inviscid, the only surface force is that due to pressure. In the x direction the net force due to pressure is

$$P\,\Delta y\,\Delta z|_x - P\,\Delta y\,\Delta z|_{x+\Delta x}.$$

Conservation of x momentum, expressed verbally as

$$\begin{matrix}\text{Time rate of change of} \\ x \text{ momentum inside } \Delta V\end{matrix} = \begin{matrix}\text{Sum of exterior} \\ \text{forces acting in} \\ x \text{ direction}\end{matrix} + \begin{matrix}x \text{ momentum} \\ \text{inflow through} \\ \text{surface of } \Delta V\end{matrix}$$

takes the following mathematical form:

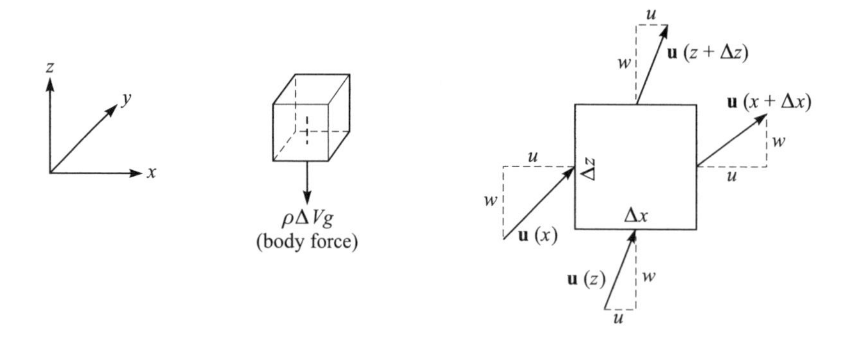

Figure 2.4 Momentum flow through x and z faces of a cubical volume element.

$$\begin{aligned}\frac{\partial}{\partial t}(\rho u\,\Delta x\,\Delta y\,\Delta z) = {} & P\,\Delta y\,\Delta z|_x - P\,\Delta y\,\Delta z|_{x+\Delta x} \\ & + \rho u^2\,\Delta y\,\Delta z|_x - \rho u^2\,\Delta y\,\Delta z|_{x+\Delta x} \\ & + \rho uv\,\Delta x\,\Delta z|_y - \rho uv\,\Delta x\,\Delta z|_{y+\Delta y} \\ & + \rho uw\,\Delta x\,\Delta y|_z - \rho uw\,\Delta x\,\Delta y|_{z+\Delta z}.\end{aligned}$$

Dividing through by $\Delta x\,\Delta y\,\Delta z$ and taking the limit as Δx, Δy, and Δz approach zero, we obtain, after moving all the terms to the left-hand side,

$$x \text{ momentum:}\quad (\rho u)_t + (\rho u^2)_x + (\rho uv)_y + (\rho uw)_z + P_x = 0. \tag{A–10}$$

(The reader is reminded that subscripts t, x, y, and z used in this way mean partial derivatives.) Expansion yields

$$\underline{\rho_t u} + \rho u_t + \underline{(\rho u)_x u} + \rho u u_x + \underline{(\rho v)_y u} + \rho v u_y + \underline{(\rho w)_z u} + \rho w u_z + P_x = 0.$$

The underlined terms drop out by application of the continuity equation, Eq. A–3 [their sum is $u(\partial\rho/\partial t + \nabla\cdot\rho\mathbf{u})$]. What remains is

$$\rho[u_t + uu_x + vu_y + wu_z] + P_x = 0,$$

or

$$x \text{ momentum:}\quad \rho\frac{Du}{Dt} + P_x = 0. \tag{A–11a}$$

Similarly, for the y and z momenta we find

$$y \text{ momentum:}\quad \rho\frac{Dv}{Dt} + P_y = 0, \tag{A–11b}$$

$$z \text{ momentum:}\quad \rho\frac{Dw}{Dt} + P_z = -\rho g. \tag{A–11c}$$

The gravitational body force term appearing in the last equation is negative because we have taken our coordinate system with the z axis pointing upward. If the three component equations are multiplied by unit vectors **i**, **j**, and **k**, respectively, and added, the vector form of the momentum equation is obtained:

$$\rho\frac{D\mathbf{u}}{Dt} + \nabla P = -\rho g\mathbf{k}. \tag{A–12}$$

Although the gravitational body force term is frequently neglected in acoustics, it needs to be taken into account in certain cases. Gravitation has two effects, one indirect and one direct, on acoustic propagation. The indirect effect is stratification of the medium, which is described by the hydrostatic version ($\mathbf{u} = 0$) of the momentum equation, $\partial p_0/\partial z = -\rho_0 g$, where the subscript zero denotes static value. Stratification causes a variation of sound speed with height (in the atmosphere) or depth (in the ocean), which in turn causes refraction of sound. The direct effect is a change in amplitude due to a change in acoustic impedance as a wave propagates upward or downward in the stratified medium. A substantial vertical distance (of the order of 5 km in the atmosphere) must be traversed before this effect is very noticeable, however. For short-range propagation, therefore, the gravitational term may be ignored completely. The direct and indirect effects of gravitation are taken up in Chap. 8.

The integral form of the momentum equation is now developed. See Fig. 2.5. The quantity $(\rho\mathbf{u})\mathbf{u}\cdot(-d\mathbf{S})$ is the momentum per unit time carried by the moving fluid into the control volume V through the element of area dS. The total flow of momentum into V is found by integrating over the surface S. The statement of "Newton's law" for the fluid in V is therefore

$$\begin{matrix}\text{Time rate of} \\ \text{increase of momentum} \\ \text{inside } V\end{matrix} = \begin{matrix}\text{Sum of exterior} \\ \text{forces acting}\end{matrix} + \begin{matrix}\text{Net momentum } \textit{inflow} \\ \text{through } S \text{ carried by} \\ \text{the fluid}\end{matrix}$$

The exterior forces are usually of two kinds, surface forces $\mathbf{F}_s$, such as those due to pressure, and body forces (not necessarily restricted to gravity):

$$\text{Exterior forces} = \mathbf{F}_s + \int_V \mathbf{B}\rho\, dV,$$

where $\mathbf{B}$ is the body force per unit mass. In the case of gravity, for example, $\mathbf{B} = -g\mathbf{k}$. For an inviscid fluid the surface forces are due entirely to the pressure, that is,

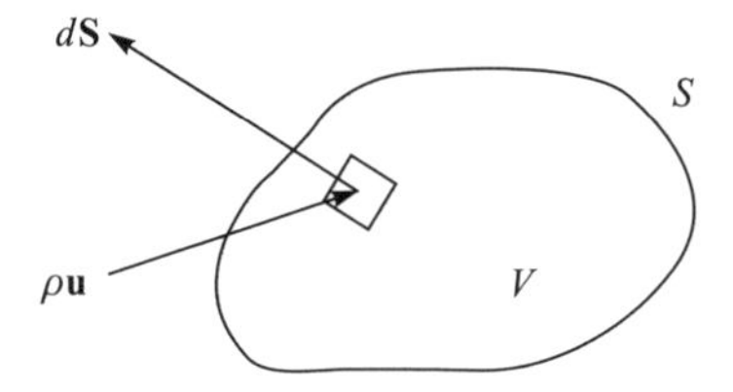

Momentum *inflow* through $d\mathbf{S} = \rho\mathbf{u}[\mathbf{u}\cdot(-d\mathbf{S})]$

Figure 2.5 Momentum flow through an arbitrary control volume V.

$$\mathbf{F}_s = -\int_s P\,d\mathbf{S},$$

where the minus sign arises because of the direction of $d\mathbf{S}$ (outward pointing normal). The mathematical statement of momentum balance for an inviscid fluid is therefore

$$\frac{\partial}{\partial t}\int_V \rho\mathbf{u}\,dV = \int_V \mathbf{B}\rho\,dV - \int_S P\,d\mathbf{S} - \int_S (\rho\mathbf{u})\mathbf{u}\cdot d\mathbf{S}. \tag{A–13}$$

This is the integral form of the momentum equation.

To derive the differential form, we must first convert the two surface integrals into volume integrals. Again the divergence theorem must be used, but a form more general than Eq. A–7 is required. In particular, we have (Ref. 1)

$$\int_S P\,d\mathbf{S} = \int_V \nabla P\,dV, \tag{A–14a}$$

$$\int_S (\rho\mathbf{u})\mathbf{u}\cdot d\mathbf{S} = \int_V \nabla\cdot[(\rho\mathbf{u})(\mathbf{u})]\,dV. \tag{A–14b}$$

Substitute these two relations in Eq. A–13, bring the time derivative operation inside the integral, and collect terms:

$$\int_V \{(\rho\mathbf{u})_t + \nabla\cdot[(\rho\mathbf{u})\mathbf{u}] + \nabla P - \rho\mathbf{B}\}\,dV = 0.$$

Because the volume V is arbitrary, the integrand itself must vanish, i.e.,

$$(\rho\mathbf{u})_t + \nabla\cdot[(\rho\mathbf{u})\mathbf{u}] + \nabla P = \rho\mathbf{B}. \tag{A–15}$$

This is one version of the differential form of the momentum equation. A simpler version may be found by expanding the first two terms,

$$\underline{\rho_t\mathbf{u}} + \rho\mathbf{u}_t + \underline{[\nabla\cdot(\rho\mathbf{u})]\mathbf{u}} + (\rho\mathbf{u}\cdot\nabla)\mathbf{u}.$$

The underlined terms vanish by virtue of the continuity equation, while the other two terms combine to form $\rho D\mathbf{u}/Dt$. Equation A–15 therefore reduces to

$$\rho\frac{D\mathbf{u}}{Dt} + \nabla P = \rho\mathbf{B}. \tag{A–16}$$

When the body force is gravity, Eq. A–16 reduces to Eq. A–12.

Example 2.2. Momentum flow through a plane, small-signal step wave. As an example of the direct use of the momentum equation in integral form, we reconsider the problem of the weak pressure step (Example 2.1, Fig. 2.3). The control volume approach is used to evaluate Eq. A–13 and then to find the relation between the pressure jump $P_1 - p_0$ and the particle velocity jump u_1. Again using the CV for the coordinate system moving with the wave, we note that because the flow is steady, the time derivative term vanishes. If the wave motion is horizontal, in the x direction, say, and the body force is due to gravity, the x component of Eq. A–13 has no contribution from the body force integral. Evaluation of the x components of the remaining two (surface) integrals yields

$$P_1 + \rho_1 v_1^2 = p_0 + \rho_0 v_0^2. \tag{A–17}$$

Denoting $P_1 - p_0$ by the excess pressure p_1 and making use of Eq. A–8, we obtain $p_1 = \rho_0 v_0(v_0 - v_1)$, or in terms of the original coordinate system,

$$p_1 = \rho_0 c_0 u_1. \tag{A–18}$$

This is a novel way of obtaining the relation between the pressure and particle velocity for a progressive plane wave.

b. Viscous Fluid (Body Force Omitted)

Tensor notation is normally used in the derivation of the momentum equation for a viscous fluid because the effect of viscosity is most generally described in terms of stress and rate of strain tensors. Unfortunately, we do not have space here to do justice to the subject. Several excellent references are available (Refs. 5, 8). We merely state the result that when viscosity is included, the momentum equation is (for simplicity, body forces are omitted)

$$\rho \frac{D\mathbf{u}}{Dt} + \nabla P = (\lambda + 2\mu)\nabla(\nabla \cdot \mathbf{u}) - \mu \nabla \times \nabla \times \mathbf{u}, \tag{A–19}$$

where μ is the shear viscosity coefficient and λ is the dilatational viscosity coefficient. Sometimes the right-hand side of Eq. A–19 is expressed as

$$(\lambda + \mu)\nabla(\nabla \cdot \mathbf{u}) + \mu \nabla^2 \mathbf{u},$$

which form may be obtained by using the vector identity

$$\nabla \times \nabla \times \mathbf{u} = \nabla(\nabla \cdot \mathbf{u}) - \nabla^2 \mathbf{u}. \tag{A–20}$$

The combination of the viscosity coefficients $\lambda + 2\mu$ appearing in the momentum equation requires some discussion. The shear viscosity coefficient μ is easily measured by direct means. It is very difficult, however, to measure λ

directly. One way to avoid the difficulty is to use "Stokes's assumption." Plausible reasoning led Stokes to conclude that λ should equal $-\frac{2}{3}\mu$. If Stokes's assumption is valid, the coefficient of the first term on the right-hand side of Eq. A–19 becomes

$$\lambda + 2\mu = \tfrac{4}{3}\mu,$$

and indeed $\frac{4}{3}\mu$ is the factor often given in textbooks. However, measurements of sound absorption for many fluids indicate that although Stokes's assumption is correct for certain very simple fluids such as the noble gases (helium, argon, neon, etc.),[1] it is not generally valid for all fluids. An alternative approach, which sometimes allows the sound absorption data to be explained, is to replace Stokes's assumption with $\lambda = -\frac{2}{3}\mu + \mu_B$, where μ_B is called the bulk viscosity coefficient. The combination $\lambda + 2\mu$ then becomes

$$\lambda + 2\mu = \tfrac{4}{3}\mu + \mu_B = \mu\tilde{V}, \tag{A–21}$$

where

$$\tilde{V} = \frac{4}{3} + \frac{\mu_B}{\mu} \tag{A–22}$$

is called the viscosity number. Using this notation, we rewrite the momentum equation as

$$\rho\frac{D\mathbf{u}}{Dt} + \nabla P = \tilde{V}\mu\nabla(\nabla\cdot\mathbf{u}) - \mu\nabla\times\nabla\times\mathbf{u}. \tag{A–23}$$

Of course nothing new has been added. We have simply replaced one difficult-to-measure coefficient λ with another, μ_B (or $\tilde{V}$).

An alternative form of Eq. A–23 is useful for the next section. By means of the vector identity $(\mathbf{u}\cdot\nabla)\mathbf{u} = \frac{1}{2}\nabla(|\mathbf{u}|^2) - \mathbf{u}\times\nabla\times\mathbf{u}$, one may express the fluid acceleration as

$$\frac{D\mathbf{u}}{Dt} = \mathbf{u}_t + (\mathbf{u}\cdot\nabla)\mathbf{u} = \mathbf{u}_t + \frac{1}{2}\nabla|\mathbf{u}|^2 - \mathbf{u}\times\nabla\times\mathbf{u}.$$

Substitution in Eq. A–23 then yields

$$\rho[\mathbf{u}_t + \tfrac{1}{2}\nabla|\mathbf{u}|^2 - \mathbf{u}\times\nabla\times\mathbf{u}] + \nabla P = \tilde{V}\mu\nabla(\nabla\cdot\mathbf{u}) - \mu\nabla\times\nabla\times\mathbf{u}. \tag{A–24}$$

[1]Calculations based on the kinetic theory of gases also validate Stokes's assumption for monatomic gases.

c. *Irrotational and Rotational Flows*

The concept of rotational and irrotational flow is conveniently presented at this point (Refs. 5–8). The velocity vector may be split into irrotational and rotational parts by introducing the scalar potential ϕ and the vector potential $\mathbf{\Psi}$ as follows:

$$\mathbf{u} = \nabla\phi + \nabla \times \mathbf{\Psi}. \tag{A–25}$$

The interpretation assigned to the two terms becomes evident if we take the divergence of $\mathbf{u}$ and note that the divergence of a curl is zero:

$$\nabla\cdot\mathbf{u} = \nabla^2\phi + \nabla\cdot\nabla \times \mathbf{\Psi} = \nabla^2\phi.$$

Moreover, since the curl of a gradient is zero, taking the curl of $\mathbf{u}$ yields

$$\nabla \times \mathbf{u} = \nabla \times \nabla\phi + \nabla \times \nabla \times \mathbf{\Psi} = \nabla \times \nabla \times \mathbf{\Psi}.$$

The quantity $\nabla \times \mathbf{u}$ is called the vorticity. If the vorticity is zero, the flow is irrotational, and we may express the velocity simply as the gradient of a scalar, i.e., $\mathbf{u} = \nabla\phi$. In most common acoustical phenomena the flow is irrotational. Rotational flows are important very close to boundaries (where fluid shear is important), in wakes, and in separated flows—in short wherever shear viscosity effects are significant. Although very important in many branches of fluid mechanics, rotational flow is not frequently encountered in acoustics.[2]

It is instructive to see how the momentum equation simplifies when the flow is either irrotational or rotational. First, if the flow is irrotational, $\mathbf{u}$ may be expressed as the gradient of a scalar,

$$\mathbf{u} = \nabla\phi,$$

and the $\nabla \times \mathbf{u}$ terms in Eq. A–24 vanish. Consequently Eq. A–24 reduces to

$$\rho\left[\nabla\phi_t + \tfrac{1}{2}\nabla|\nabla\phi|^2\right] + \nabla P = \tilde{V}\mu\ \nabla(\nabla^2\phi). \tag{A–26}$$

If ρ were a constant rather than a variable, this equation could be integrated once with respect to the gradient operator ∇ (see Problem 2–11 for an exact integration when the fluid is inviscid). For small-signal acoustics, Eq. A–26 is linearized,

$$\rho_0\nabla\phi_t + \nabla P - \tilde{V}\mu\nabla(\nabla^2\phi) = 0,$$

and then easily integrated:

[2]Acoustical applications include flow noise, some aspects of aeroacoustics, and absorption of sound due to propagation over a surface (see Sec. 9D).

$$\rho_0\phi_t + P - \tilde{V}\mu\nabla^2\phi = p_0.$$

(The constant of integration is the static pressure p_0; any other constant of integration has been absorbed in the potential function ϕ.) Rearranged, this equation gives the relation between ϕ and the acoustic pressure p,

$$p \equiv P - p_0 = -\rho_0\phi_t + \tilde{V}\mu\,\nabla^2\phi. \tag{A–27}$$

The inviscid form of this relation is Eq. 1D–6, derived in Sec. 1D.1.

If the flow is rotational rather than irrotational, the divergence of $\mathbf{u}$ vanishes, and the momentum equation reduces to

$$\rho(\mathbf{u}_t - \mathbf{u}\times\nabla\times\mathbf{u}) + \nabla P + \mu\nabla\times\nabla\times\mathbf{u} = 0. \tag{A–28}$$

This equation is at the heart of boundary-layer theory but does not play a very prominent role in acoustics.

3. Energy Equation

The energy equation is derived in three main steps. First, an integral form of the equation is obtained by using the control volume approach. Second, the divergence theorem is used to convert the integral form to differential form. Finally, the differential form is greatly simplified by applying the continuity and momentum equations. The use of tensor notation would greatly simplify the derivation. By omitting viscous effects, however, we avoid the need for tensors and yet are able to include all other phenomena of interest. At the end, the viscous energy dissipation term is added ad hoc.

The usual control volume V bounded by a surface S is shown in Fig. 2.6. Increments of energy associated with transport by the fluid, heat transfer, and work done are also shown. Our convention is that the work done *on* the system by the surroundings is positive. A good start on a basic statement of conservation of energy is the first law of thermodynamics: the energy of a system

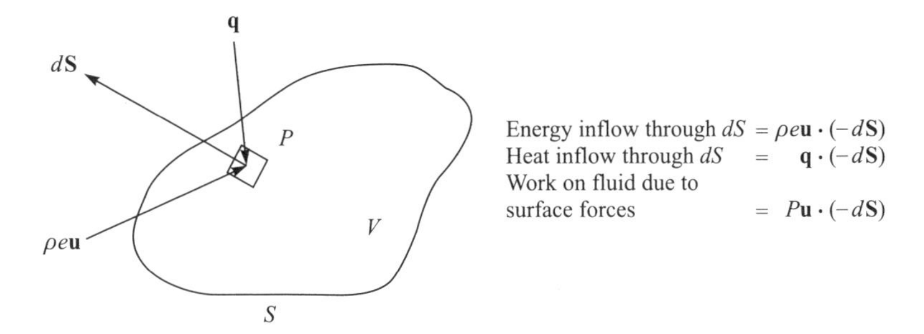

Figure 2.6 Energy flow through an arbitrary control volume V.

increases accordingly as heat is added to it and work is done on it. When account is taken of the energy transported by the flowing fluid, the statement is as follows:

Time rate of increase of energy inside V	=	Net energy *inflow* through surface S	+	Work/unit time done on the fluid in V by surface forces	+	Heat added by transfer across surface S,

or, in mathematical terms,

$$\frac{\partial}{\partial t}\int_V \rho e\, dV = -\int_S \rho e\mathbf{u}\cdot d\mathbf{S} - \int_S P\mathbf{u}\cdot d\mathbf{S} - \int_S \mathbf{q}\cdot d\mathbf{S}. \tag{A–29}$$

Here e is total energy per unit mass and $\mathbf{q}$ is the heat flux vector (energy per unit time per unit area). The first term on the right-hand side is the net energy carried into the control volume by the flow. The minus sign is needed because $d\mathbf{S}$ is directed outward and we are calculating inflow, not outflow, of energy. The second term represents the work per unit time (or power) done by the surface forces, in this case just the pressure. The third term represents the energy added in the form of heat. The first step has now been completed.

The second step, conversion of Eq. A–29 to differential form, is accomplished by using the divergence theorem to transform the surface integrals to volume integrals. The result is, after rearrangement,

$$(\rho e)_t + \nabla\cdot(\rho e\mathbf{u}) + \nabla\cdot(P\mathbf{u}) = -\nabla\cdot\mathbf{q}. \tag{A–30}$$

The third and final step consists of simplifying Eq. A–30 by using the continuity and momentum equations. Expansion of the first two terms on the left-hand side of Eq. A–30 yields

$$\underline{e\rho_t} + \rho e_t + \underline{e\nabla\cdot(\rho\mathbf{u})} + \rho\mathbf{u}\cdot\nabla e = \rho\frac{De}{Dt}$$

because the underlined terms vanish by application of the continuity equation. Next, the total energy e is broken down into its three components (all per unit mass): internal energy ϵ, kinetic energy $\frac{1}{2}u^2 = \frac{1}{2}\mathbf{u}\cdot\mathbf{u}$, and potential energy gz (we assume here that the body force is gravity),

$$e = \epsilon + \tfrac{1}{2}\mathbf{u}\cdot\mathbf{u} + gz.$$

Equation A–30 becomes

$$\rho\frac{D\epsilon}{Dt} + \underline{\rho\mathbf{u}\cdot\frac{D\mathbf{u}}{Dt}} + \underline{\rho g\mathbf{u}\cdot\mathbf{k}} + \underline{\mathbf{u}\cdot\nabla P} + P\,\nabla\cdot\mathbf{u} = -\nabla\cdot\mathbf{q}.$$

Again the underlined terms vanish, this time by application of the momentum equation, Eq. A–12. We are left with

$$\rho \frac{D\epsilon}{Dt} + P\nabla\cdot\mathbf{u} = -\nabla\cdot\mathbf{q}. \tag{A–31}$$

To develop the energy equation further requires that (1) an equation of state be specified, (2) the heat transfer mechanism be identified, and (3) the energy dissipation due to viscosity be included. Item 1 is dealt with in the next section. The most common assumption for item 2 is that the heat flow is due to conduction, which is described by Fourier's law, $\mathbf{q} = -\kappa\,\nabla T$, where T is the (absolute) temperature and κ is the thermal conduction coefficient. In this case Eq. A–31 becomes

$$\rho \frac{D\epsilon}{Dt} + P\nabla\cdot\mathbf{u} = \kappa\,\nabla^2 T. \tag{A–32}$$

As for item 3, a suggestion of the role played by viscosity may be found by considering the one-dimensional form of Eq. A–32,

$$\rho \frac{D\epsilon}{Dt} + Pu_x = \kappa T_{xx}. \tag{A–33}$$

For a viscous fluid subject to no body forces, the momentum equation in one (cartesian) dimension is (see Eq. A–23)

$$\rho \frac{Du}{Dt} = -\left(P - \tilde{V}\mu u_x\right)_x.$$

The form of this equation suggests that the true surface force per unit area (for planar flow) is not just P but rather $P - (\tilde{V}\mu)u_x$. If this correction is made in Eq. A–33, one obtains

$$\rho \frac{D\epsilon}{Dt} + Pu_x = (\tilde{V}\mu)u_x^2 + \kappa T_{xx}. \tag{A–34}$$

Indeed this is the correct one-dimensional energy equation for a thermoviscous fluid. The terms on the right-hand side are the energy dissipation terms. They represent the energy lost because of viscous dissipation and heat transfer, respectively. Note that because the viscous dissipation term is intrinsically nonlinear, it drops out when the small-signal approximation is made. This does not mean, however, that viscous effects are negligible for small-signal waves; the (linear) viscosity terms in the momentum equation remain.

The three-dimensional form of the viscous energy dissipation function is much more complicated than the term that appears in Eq. A–34. The final form of the energy equation, given here without proof, is

$$\rho\frac{D\epsilon}{Dt} + P\nabla\cdot\mathbf{u} = \Phi^{(\mathrm{visc})} + \kappa\nabla^2 T, \qquad \text{(A–35)}$$

where $\Phi^{(\mathrm{visc})}$ is the three-dimensional viscous energy dissipation function (a nonlinear quantity). For reference, the expression for $\Phi^{(\mathrm{visc})}$ in cartesian tensor form is

$$\Phi^{(\mathrm{visc})} = 2\mu d_{ij} d_{ji} + \lambda d_{kk} d_{ii}, \qquad \text{(A–36)}$$

where $d_{ij} = \frac{1}{2}(u_{i,j} + u_{j,i})$ is the rate of deformation tensor. A vector form of $\Phi^{(\mathrm{visc})}$ has been given by Hunt (Ref. 4).

Example 2.3. Energy balance for a plane, small-signal step wave. The problem of a weak pressure step (Examples 2.1 and 2.2, Fig. 2.3) is briefly considered once again, this time to illustrate the direct evaluation of the integral form of the energy equation. Start with Eq. A–29. As usual, the time derivative term vanishes. Evaluation of the heat flow integral yields $-(q_1 S_1 - q_0 S_0)$. This quantity is zero. Why? *Hint*: What is the temperature gradient at faces 1 and 0 of the CV? The remaining two (surface) integrals may be combined and expanded,

$$\int_S \left(e + \frac{P}{\rho}\right)\rho\mathbf{u}\cdot d\mathbf{S} = \int_S \left(\frac{1}{2}u^2 + \epsilon + \frac{P}{\rho}\right)\rho\mathbf{u}\cdot d\mathbf{S} = \int_S \left(\frac{1}{2}u^2 + h\right)\rho\mathbf{u}\cdot d\mathbf{S} = 0,$$

where $h = \epsilon + P/\rho$ is the enthalpy per unit mass. The integral may easily be evaluated. Numerous substitutions, including various perfect gas relations (see next section) and the results of previous calculations for the step wave (Eqs. A–8 and A–18), then lead to an expression for the wave speed c_0 in terms of p_0 and ρ_0. The exercise is a good warmup for Problem 2–4.

4. Equation of State (Constitutive Relation)

In the derivation of the acoustic wave equation given in Sec. 1C.3 no energy equation was needed because the fluid was assumed to be lossless, i.e., isentropic, for which case the equation of state, 1C–37, reduces to a simple pressure-density relation

$$P = P(\rho). \qquad \text{(1C–38)}$$

This equation, the continuity equation, and the momentum equation for an inviscid fluid then constitute a closed set in the three variables ρ, P, and $\mathbf{u}$.

In the present analysis, in which the fluid's viscosity and thermal conduction are taken into account, we have already introduced five variables—$\mathbf{u}$, ρ, P, T, and ϵ—and have only three equations (continuity, momentum, and energy) to connect them. A fourth equation is the equation of state, which must clearly be more general than the modest pressure-density relation used in Chap. 1. For simplicity we use the perfect gas law (later, when we specialize to the small-signal equations, arbitrary fluids are considered). A fifth equation is provided by relations from thermodynamics. When the fourth and fifth equations are combined with Eq. A–35, we obtain an energy equation for a thermoviscous perfect gas in terms of the three variables **u**, ρ, and P. This energy equation and the continuity and momentum equations then constitute a closed set.[3] Later in this section we show that when viscosity and heat conduction are negligible, the energy equation (for a perfect gas) reduces to the adiabatic gas law.

The perfect gas law is

$$P = R\rho T, \tag{1C–36}$$

where R is the gas constant. For a perfect gas the internal energy ϵ depends only on the temperature, in particular,

$$d\epsilon = C_v dT, \tag{A–37}$$

where C_v is the specific heat at constant volume (this is the fifth equation mentioned in the previous paragraph). Another useful thermodynamic relation is

$$R = C_p - C_v, \tag{A–38}$$

or

$$R = (\gamma - 1)C_v, \tag{A–39}$$

where C_p is the specific heat at constant pressure and γ is the ratio of specific heats. Substitution of Eqs. A–37, A–38, A–39 in Eq. A–35 yields

$$\rho \frac{D}{Dt}\left(\frac{P}{\rho}\right) + (\gamma - 1)P\,\nabla\cdot\mathbf{u} = (\gamma - 1)(\Phi^{(\text{visc})} + \kappa\,\nabla^2 T). \tag{A–40}$$

This is one form of the energy equation for a thermoviscous perfect gas.

We now develop two other forms. First the continuity equation is used to substitute $-(1/\rho)\,D\rho/Dt)$ for $\nabla\cdot\mathbf{u}$. Expansion of the derivatives on the left-hand side of Eq. A–40 leads to cancellation of two terms. If the equation is then multiplied by $\rho^{-\gamma}$, the terms may be rearranged in the following form:

$$\frac{D}{Dt}\left(\frac{P}{\rho^{\gamma}}\right) = \rho^{-\gamma}\left[(\gamma - 1)\Phi^{(\text{visc})} + \frac{\kappa}{C_v}\nabla^2\left(\frac{P}{\rho}\right)\right]. \tag{A–41}$$

[3]An alternative, four-equation set, in which T is the fourth variable, is given in Sec. A.6 below.

This equation helps explain where the so-called adiabatic gas law comes from: If the gas is inviscid and thermally nonconducting, the right-hand side vanishes. Integration then gives (for a homogeneous gas) $P/\rho^\gamma = \text{const}$, or

$$P/p_0 = (\rho/\rho_0)^\gamma. \tag{1C–39}$$

"Adiabatic," which means only that no heat flows, is thus a misnomer; Eq. 1C–39 would better be called the "isentropic gas law." Another, somewhat more useful form of the energy equation for a perfect gas may be found by expanding the first term of Eq. A–40 and using the perfect gas law to express the temperature in terms of the pressure and density,

$$\frac{DP}{Dt} + \gamma P \nabla\cdot\mathbf{u} = (\gamma - 1)\Phi^{(\text{visc})} + \frac{\kappa}{C_v}\nabla^2\left(\frac{P}{\rho}\right). \tag{A–42}$$

5. Entropy Equation

An entropy equation equivalent in informational content to the energy equation is sometimes useful. If the continuity equation is used to replace $\nabla\cdot\mathbf{u}$ in the second term on the left-hand side of Eq. A–35 by $\rho D(1/\rho)/Dt$, the result is

$$\rho\left[\frac{D\epsilon}{Dt} + P\,\frac{D(1/\rho)}{Dt}\right] = \Phi^{(\text{visc})} + \kappa\nabla^2 T.$$

The quantity in brackets is equal to $T\,Ds/Dt$, where s is the entropy per unit mass.[4] We thus arrive at the entropy equation

$$\rho T\frac{Ds}{Dt} = \Phi^{(\text{visc})} + \kappa\nabla^2 T. \tag{A–43}$$

An important special case occurs when no energy is dissipated. The entropy equation then reduces to

$$\frac{Ds}{Dt} = 0, \tag{A–44}$$

which means that the entropy of a fluid particle remains constant with time. In most cases we shall study, the entropy is the same for every particle. Equation A–44 then reduces to

$$s = \text{const}. \tag{A–45}$$

[4]Section 5.12 in Panton's book (Ref. 5) or Sec. 2.3 in Thompson's book (Ref. 8).

Instances do arise, however, in which Eq. A–44 must be used as is. For example, consider lossless, vertical propagation of sound through a stratified fluid, such as the atmosphere or the ocean. Because the static value of the entropy s_0 varies with altitude (or depth), the proper characterization of losslessness in this case is Eq. A–44, not A–45. This problem is touched on in Sec. C.3 below and explored in more detail in Chap. 8.

6. Summary and Discussion

Considerable time and space has been given here to a detailed derivation of the conservation equations for fluids. Part of the motivation has been to develop in the reader an appreciation of the physics underlying the ordinary wave equation $c_0^2 \nabla^2 \phi - \phi_{tt} = 0$, which, given the complexity of the exact conservation equations, is surprisingly simple. Another purpose has been to provide a starting point for forays into more complicated areas of acoustics. That is, if the ordinary wave equation is not sufficient to describe a particular acoustical phenomenon, how does one go about developing a more suitable wave equation? It is hoped that the material in the preceding sections may provide clues for answering this question.[5] For convenience the four fundamental equations—continuity, momentum, energy, and state—are restated here (in differential form) for the case of a perfect gas not subject to body forces:

$$\frac{D\rho}{Dt} + \rho \nabla\cdot\mathbf{u} = 0, \tag{A–46}$$

$$\rho \frac{D\mathbf{u}}{Dt} + \nabla P = (\lambda + 2\mu)\nabla(\nabla\cdot\mathbf{u}) - \mu \nabla \times \nabla \times \mathbf{u}, \tag{A–47}$$

$$\rho C_v \frac{DT}{Dt} + P\,\nabla\cdot\mathbf{u} = \Phi^{(\text{visc})} + \kappa \nabla^2 T, \tag{A–48}$$

$$P = R\rho T. \tag{A–49}$$

The restriction to perfect gases is made not because liquids are unimportant. On the contrary, liquids are of great practical significance. The perfect gas law is used here because of its analytic simplicity.

The full equations, A–46 through A–49, are exceedingly difficult to deal with. Progress can be made, however, by considering special cases in which certain terms drop out. Fortunately, many of the special cases correspond closely to real flows of practical interest. Some examples:

1. Incompressible flow. $D\rho/Dt = 0$. Note that in this case continuity reduces to $\nabla\cdot\mathbf{u} = 0$, which means that several terms in the remaining

[5]For a more extensive essay on the conservation equations, see Hunt's paper (Ref. 4). For example, Hunt considers fluids in which the viscosity and heat conduction coefficients are not constant. For a treatment of fluids that are not perfect gases, see Ref. 6.

equations vanish. Incompressible flows are of interest in such diverse fields as hydraulics, oceanography, civil engineering, low-speed aerodynamics, and blood flow.

2. Time-independent or steady flow. Considerable simplification occurs when the time derivative terms vanish. Steady flows are of interest in aerodynamics, hydraulics, pipe flow, and so on.
3. Lossless flow. When λ, μ, $\kappa = 0$, quite a number of terms vanish. Several important fluid motions, including sound propagation, are lossless or nearly so.
4. Small-signal flow. Linearization of the equations brings a most welcome simplification. Most sound waves in normal experience are small-signal disturbances.

Because sound is a fluid phenomenon that involves both compressibility and time variation, examples 1 and 2 in the list above are rarely of interest in acoustics. Examples 3 and 4, on the other hand, are often relevant. Indeed most of the remainder of the text is devoted to the classical wave equation $\phi_{tt} - c_0^2 \nabla^2 \phi = 0$, to which the four equations reduce when the flow is both lossless and small signal.

Before considering lossless, small-signal waves, however, let us briefly investigate a somewhat more general wave motion, namely, that of finite-amplitude disturbances in a lossless gas. A finite-amplitude wave is one whose amplitude is not infinitesimally small. When the amplitude is finite, the nonlinear terms in Eqs. A–46, A–47, A–48, and A–49 cannot be disregarded.

B. NONLINEAR WAVE EQUATION

Nonlinear acoustics is the field of acoustics devoted to very intense sound, specifically to waves of amplitude high enough that the small-signal assumption is violated. Retaining the nonlinear terms makes the conservation equations much harder to solve. On the other hand, even limited progress may be very rewarding. Nonlinearity produces a rich variety of phenomena that are foreign to classical linear acoustics. In the material presented below we first derive the three-dimensional nonlinear wave equation for a lossless perfect gas. Next, a solution is obtained for the special case of plane progressive wave motion generated by arbitrary excitation at the source. Finally, the solution is analyzed for the case in which the source excitation is a pure tone.

1. Introduction

When the nonlinear terms in the conservation equations are not dropped, great mathematical difficulty is encountered. Analytical advances are usually made by considering special cases. Here we assume that the fluid is a lossless perfect gas. This assumption does not affect the continuity equation, Eq. A–46, but Eq. A–47 simplifies to

$$\text{Momentum:} \qquad \rho \frac{D\mathbf{u}}{Dt} + \nabla P = 0, \tag{B–1}$$

and, as shown at the end of Sec. A.4 above, the energy and state equations reduce to the adiabatic gas law,

$$\text{Equation of state:} \qquad \frac{P}{p_0} = \left(\frac{\rho}{\rho_0}\right)^{\gamma}. \tag{B–2}$$

It is now useful to introduce the sound speed c, a thermodynamic quantity defined and discussed in Sec. 1C.3.c; see Fig. 1.22 and Eq. 1C–41, repeated here for convenience,[6]

$$c^2 = \frac{dP}{d\rho}. \tag{1C–41}$$

Taking the indicated derivative of Eq. B–2, we obtain

$$c^2 = \frac{\gamma P}{\rho} = \gamma RT. \tag{B–3}$$

Note that since P, ρ, and T are total, not static, values, the sound speed c in this case is in general not a constant. Equations B–2 and B–3 may be combined to express P and ρ in terms of c:

$$P = p_0 \left(\frac{c}{c_0}\right)^{2\gamma/(\gamma-1)}, \tag{B–4a}$$

$$\rho = \rho_0 \left(\frac{c}{c_0}\right)^{2/(\gamma-1)}. \tag{B–4b}$$

These relations may be used to eliminate P and ρ from the continuity and momentum equations. One obtains

$$\frac{Dc}{Dt} + \frac{\gamma - 1}{2} c \nabla \cdot \mathbf{u} = 0, \quad \text{(continuity)}, \tag{B–5}$$

[6]The derivative is really taken with the entropy held constant. But since the entropy is constant in the absence of losses ($\lambda = \mu = \kappa = 0$), it is appropriate here to use a total derivative rather than a partial derivative.

$$\frac{D\mathbf{u}}{Dt} + \frac{2}{\gamma - 1} c \nabla c = 0, \quad \text{(momentum)}, \tag{B–6}$$

respectively.

The sound speed may now be eliminated, and Eqs. B–5 and B–6 combined, by introducing the velocity potential ϕ. We consider the flow to be irrotational so that $\mathbf{u}$ may be expressed as the gradient of a scalar, $\mathbf{u} = \nabla\phi$. Equation B–6 becomes

$$\nabla\phi_t + \frac{1}{2}\nabla[\nabla\phi\cdot\nabla\phi] + \frac{1}{\gamma - 1}\nabla c^2 = 0,$$

which may be integrated once to yield

$$\phi_t + \frac{1}{2}(\nabla\phi\cdot\nabla\phi) + \frac{1}{\gamma - 1}\, c^2 = \frac{1}{\gamma - 1}\, c_0^2$$

(the constant of integration is found by evaluating the left-hand side at a quiet point in the medium, $\mathbf{u} = 0$, $c = c_0$). Solve for c^2,

$$c^2 = c_0^2 - (\gamma - 1)\left[\phi_t + \tfrac{1}{2}\,\nabla\phi\cdot\nabla\phi\right], \tag{B–7}$$

and use this expression to eliminate c from Eq. B–5. The result is, after a lot of algebra,

$$c_0^2\,\nabla^2\phi - \phi_{tt} = \left[(\nabla\phi)^2\right]_t + (\gamma - 1)\phi_t\,\nabla^2\phi + \nabla\phi\cdot\left[\frac{\gamma - 1}{2}\,\nabla\phi\,\underline{\nabla^2\phi} + \frac{1}{2}\,\nabla(\nabla\phi)^2\right]. \tag{B–8}$$

It is interesting to note that this equation also holds for a medium moving uniformly with speed u_0 in the x direction except that the factor underlined, $\underline{\nabla^2\phi}$, must be replaced by $\nabla^2\phi - u_0^2$.

Equation B–8 is the exact wave equation for sound waves of arbitrary amplitude in a lossless gas. How simple by comparison is the ordinary linear wave equation! To obtain the latter, we must drop all the nonlinear terms, which are on the right-hand side of Eq. B–8. In the next section a solution of the nonlinear wave equation is found for the special case of plane progressive waves.

2. Plane Progressive Waves of Finite Amplitude

A relatively simple, exact solution of the nonlinear wave equation may be obtained when the waves are plane progressive. Rather than attack Eq. B–8 directly, however, we begin with the two antecedent equations, Eqs. B–5 and B–6, which for plane waves reduce to

$$c_t + uc_x + \frac{\gamma - 1}{2} cu_x = 0, \tag{B–9}$$

$$u_t + uu_x + \frac{2}{\gamma - 1} cc_x = 0. \tag{B–10}$$

This is a second-order system. A first integral may be found by a somewhat unorthodox method: we seek a relation $c = c(u)$ that makes both equations the same. If the relation exists, then $c_t = c'u_t$ and $c_x = c'u_x$, where c' means dc/du. If these expressions are substituted in Eqs. B–9 and B–10, one finds that the two equations become the same if

$$\frac{\gamma - 1}{2}\frac{c}{c'} = \frac{2}{\gamma - 1} cc',$$

or

$$\frac{dc}{du} = \pm\frac{\gamma - 1}{2}. \tag{B–11}$$

The sign determines the direction of the progressive wave motion. Choosing the plus sign, which pertains to forward traveling progressive waves, integrating once, and evaluating the constant of integration at a place where the fluid is completely quiet ($u = 0$, $c = c_0$), we obtain

$$c = c_0 + \frac{\gamma - 1}{2} u. \tag{B–12}$$

As can readily be proved, substitution of this relation in either Eq. B–9 or B–10 yields

$$u_t + (c_0 + \beta u)u_x = 0, \tag{B–13}$$

where

$$\beta = \tfrac{1}{2}(\gamma + 1). \tag{B–14}$$

Equations B–9 and B–10 have thus been reduced to a single first-order equation, Eq. B–13. Physically, what we have done is to find a wave equation that is valid for forward traveling waves only. Selection of the minus sign in Eq. B–11 would have led to the equation valid for backward traveling waves, namely Eq. B–13 with c_0 replaced by $-c_0$. See Problem 2–12 for an approach that leads to a more general form of Eq. B–13.

Equation B–13 may be solved rather easily. First notice that the linearized version (term βuu_x neglected) has the following familiar solution:

$$u = f\left(t - \frac{x}{c_0}\right), \tag{B–15}$$

which represents a wave traveling in the $+x$-direction with speed c_0. Since the sum $c_0 + \beta u$ appears in Eq. B–13 in place of c_0, perhaps $c_0 + \beta u$ represents the true propagation speed dx/dt. If so, then

$$u = f\left(t - \frac{x}{c_0 + \beta u}\right) \tag{B–16}$$

should be the solution of Eq. B–13. Indeed Eq. B–16 is the solution, as direct substitution in Eq. B–13 shows (see Problem 2–13).

Now analyze the meaning of Eq. B–16. Because $|u|$ is generally much smaller than c_0, the difference between the two propagation speeds, c_0 and $c_0 + \beta u$, would seem to be negligible. However, small departures of $c_0 + \beta u$ from c_0 have a cumulative effect. The peak of a wave (where u has its greatest value) travels fastest, the trough (where u is least) slowest. Consequently, the peak tends to catch up with the trough, as sketches (*a*), (*b*), and (*c*) in Fig. 2.7 show. The multivalued waveform shown in (*d*) is physically unrealizable. Actual multivaluedness is prevented by the effects of viscosity and heat conduction, which have been neglected in our treatment here. Dissipation becomes very important whenever any segment of the waveform becomes very steep, as in (*c*). Very steep segments are actually shock waves. The effect of nonlinearity is therefore generally to cause intense sound waves to distort and form shocks.

Two physical effects cause the actual propagation speed to be $c_0 + \beta u$ instead of just c_0. First, convection causes the propagation velocity to be

$$\frac{dx}{dt} = c + u, \tag{B–17}$$

that is, the sound wave is carried along by the fluid particles it sets in motion. Second, the sound speed c is itself not constant. The peak of the wave is a place where the gas, being compressed, is a little warmer. Since $c \propto \sqrt{T}$ (see Eq. B–3), the sound speed is greatest at a peak. Conversely, at a trough, where the gas is a little colder because it has been expanded, c has its minimum value. It turns out that the mathematical manifestation of this effect is Eq. B–12. Note that the nonconstancy of c depends on γ. In particular, nonconstancy disappears

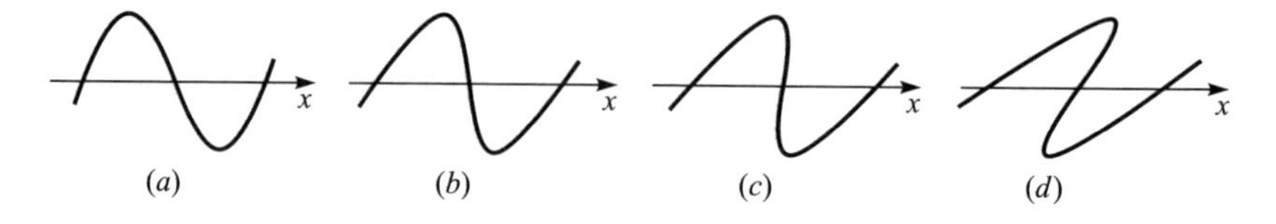

Figure 2.7 Cumulative distortion of a plane progressive wave.

($c = c_0$) when $\gamma = 1$, a value that implies a linear pressure-density relation (see Eq. B–2). Nonconstancy of c is thus associated with nonlinearity of the P-ρ relation.

Combination of Eqs. B–17 and B–12 yields

$$\frac{dx}{dt} = c_0 + \beta u, \tag{B–18}$$

which is the propagation speed implied by Eq. B–16. The quantity β is sometimes referred to as the coefficient of nonlinearity because, as Eq. B–16 shows, if β were zero, the small-signal solution Eq. B–15 would hold. A similar analysis can be performed for liquids. The main difference is that the formula for β becomes

$$\beta = 1 + \frac{B}{2A}, \tag{B–19}$$

where B and A are the coefficients found in Eq. 1C–40 and discussed at the end of Sec. 1C.3.c. For water, for example, $B/A \doteq 5$, which value leads to $\beta_{\text{water}} = 3.5$. For air ($\gamma = 1.4$) the value of the coefficient is $\beta_{\text{air}} = 1.2$. In a sense, therefore, water is more nonlinear acoustically than air.

3. Second-Harmonic Distortion

The deformation of the wave discussed in the previous section can be represented in terms of the generation and growth of harmonic distortion components in the propagating wave. Consider the wave motion generated in a lossless tube by sinusoidal vibration of a piston at one end of the tube. See Fig. 2.8. Starting with the general solution equation, Eq. B–16, we evaluate f by setting $x = 0$ and applying the boundary condition

$$u(0, t) = u_0 \sin \omega t.$$

$$\therefore f(t) = u_0 \sin \omega t.$$

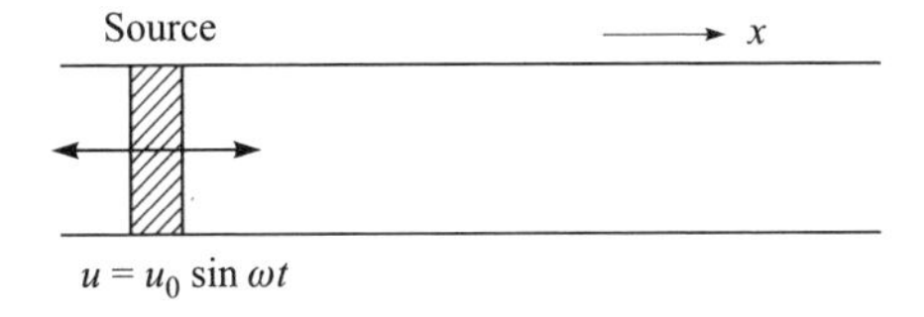

Figure 2.8 Finite-amplitude wave generated by vibration of a piston in a tube.

Substituting $t - x/(c_0 + \beta u)$ for t gives the solution for values of $x > 0$:

$$u = u_0 \sin \omega[t - x/(c_0 + \beta u)]. \tag{B–20}$$

Although Eq. B–20 is the solution in principle, its form, $u = u(x, t, u)$, is inconvenient. One way to obtain an explicit solution $u(x, t)$ is to expand u as a Taylor series. The expansion is facilitated by first doing some rearranging and simplifying. A preliminary expansion valid for $|\beta u| < c_0$ is

$$\frac{x}{c_0 + \beta u} = \frac{x}{c_0}\left[1 - \beta\frac{u}{c_0} + \left(\beta\frac{u}{c_0}\right)^2 \cdots\right].$$

If $|\beta u/c_0|$ is small compared with unity (what is the value of $\beta u_0/c_0$ for a sound pressure level of 140 dB in air?), the expansion may be terminated after two terms. Equation B–20 then becomes

$$u = u_0 \sin[y + \beta kxu/c_0], \tag{B–21}$$

where $y = \omega t - kx$ and $k = \omega/c_0$. We now expand the sine function in Taylor series about the point y:

$$u = u_0\left[\sin y + \left(\beta kx\frac{u}{c_0}\right)\underbrace{(\sin y)'}_{\cos y} + \frac{1}{2!}\left(\beta kx\frac{u}{c_0}\right)^2\underbrace{(\sin y)''}_{-\sin y} + \cdots\right].$$

Moreover, wherever u appears on the right-hand side, it is to be expressed by its Taylor series expansion. Applied repeatedly, this procedure generates the following power series in the small parameter $\epsilon \equiv u_0/c_0$ (here ϵ has nothing to do with the same symbol used earlier to mean internal energy per unit mass):

$$u = u_0\{\sin y + \epsilon[\beta kx \sin y \cos y] + \epsilon^2(\beta kx)^2[\sin y \cos^2 y - \tfrac{1}{2}\sin^3 y] + \cdots\}.$$

The smaller the value of ϵ, the more rapidly the series converges. Stopping with the first two terms and noting that $\sin y \cos y = \frac{1}{2}\sin 2y$, we obtain

$$u = u_0\left[\sin(\omega t - kx) + \frac{1}{2}\beta\epsilon kx \sin 2(\omega t - kx) + \cdots\right]. \tag{B–22}$$

Thus in addition to the expected fundamental component $u_0 \sin(\omega t - kx)$, the sound also contains a second-harmonic signal, the amplitude of which grows linearly with the propagation distance x. Experiments confirm this prediction (see, for example, Ref. 2). The growth of second-harmonic distortion with distance is a manifestation of the cumulative distortion of the wave shown in Fig. 2.7.

If the ϵ^2 term is included in Eq. B–22, it will be found to contain a third-harmonic component and also a (new) fundamental component. Both amplitudes vary as x^2. Because the new fundamental turns out to be out of phase with the original fundamental, the total fundamental decreases with distance as the wave propagates. In other words, harmonic distortion components are not generated "for free." They develop at the expense of the fundamental.

4. Sum- and Difference-Frequency (Intermodulation) Distortion

When the piston is excited at two angular frequencies ω_1 and ω_2, called primary frequencies, the distortion components include signals at the sum frequency $(\omega_2 + \omega_1)$ and difference frequency $(\omega_2 - \omega_1)$ as well as at the second and higher harmonics of the two primaries. The analysis for this case is left as a problem (Problem 2–14). Generation of the difference frequency signal has a very interesting application, called the *parametric acoustic array*, in the area of beamforming (Ref. 11; see also Ref. 3).

C. SMALL-SIGNAL WAVE EQUATION

We now linearize the conservation equations. The justification is that most acoustical disturbances in our experience are small enough that the nonlinear terms are not important. The general linearization procedure is discussed in detail in Sec. 1C.3.e for the case of plane waves in a lossless quiet fluid at rest. Hence only brief arguments are necessary in this section. Although linearization always leads to a small-signal wave equation, the exact form the equation takes depends on the assumptions about the nature of the wave motion and the medium. Six different cases are described below. In the first three the fluid is assumed lossless. The last three are for various dissipative fluids. Body forces are neglected in all but the third case.

1. Lossless Medium at Rest

The simplest and most common acoustical problem occurs when body forces are not significant and the medium may be characterized as inviscid and thermally nonconducting. In this case the continuity and momentum equations reduce to

$$\rho_t + \mathbf{u}\cdot\nabla\rho + \rho\nabla\cdot\mathbf{u} = 0, \tag{C–1a}$$

$$\rho[\mathbf{u}_t + (\mathbf{u}\cdot\nabla)\mathbf{u}] + \nabla P = 0, \tag{C–1b}$$

which are known as Euler's equations. Similarly, the equation of state is the isentropic one, Eq. 1C–40, or, in more convenient form (see Eq. 1C–44),

$$P - p_0 = c_0^2(\rho - \rho_0)\left[1 + \frac{B}{2!A}\left(\frac{\rho - \rho_0}{\rho_0}\right) + \frac{C}{3!A}\left(\frac{\rho - \rho_0}{\rho_0}\right)^2 + \cdots\right]. \qquad \text{(C–1c)}$$

As noted earlier, the absence of losses makes an energy equation unnecessary. When no acoustical disturbance is present, these three equations are satisfied by

$$\rho = \rho_0, \qquad P = p_0, \qquad \mathbf{u} = 0. \qquad \text{(C–2)}$$

Equation C–2 is the "zero-order solution," which simply describes a quiet medium at rest.

Now suppose a small departure from quiet conditions occurs, expressed by writing

$$\begin{aligned} \rho &= \rho_0 + \delta\rho, & |\delta\rho| &\ll \rho_0, \\ P &= p_0 + p, & |p| &\ll \rho_0 c_0^2, \\ \mathbf{u} &= 0 + \mathbf{u}, & |\mathbf{u}| &\ll c_0, \end{aligned} \qquad \text{(C–3)}$$

where the inequalities at the right mean that $\delta\rho$, p, and $\mathbf{u}$ are taken to be "small quantities of first order." Substitute these expressions into the continuity, momentum, and state equations:

$$\delta\rho_t + \underline{\mathbf{u}\cdot\nabla\delta\rho} + \rho_0\,\nabla\cdot\mathbf{u} + \underline{\delta\rho\ \nabla\cdot\mathbf{u}} = 0,$$

$$\rho_0\mathbf{u}_t + \underline{\rho_0(\mathbf{u}\cdot\nabla)\mathbf{u}} + \underline{\delta\rho\,\mathbf{u}_t} + \underline{\delta\rho(\mathbf{u}\cdot\nabla)\mathbf{u}} + \nabla p = 0,$$

$$p = c_0^2\delta\rho + \underline{\frac{B}{2A}\,\rho_0 c_0^2\left(\frac{\delta\rho}{\rho_0}\right)^2} + \cdots.$$

The underlined terms are of second or higher order because they contain products of first-order quantities. If the first-order terms are small, the terms of second and higher order are so exceedingly small that they may be neglected. The result is

$$\delta\rho_t + \rho_0\nabla\cdot\mathbf{u} = 0, \qquad \text{(C–4a)}$$

$$\rho_0\mathbf{u}_t + \nabla p = 0, \qquad \text{(C–4b)}$$

$$p = c_0^2\delta\rho, \qquad \text{(C–4c)}$$

and the linearization has been accomplished.

Next, Eqs. C–4 must be combined. To reduce the set to a single equation in p, first use Eq. C–4c to eliminate $\delta\rho$ from Eq. C–4a:

$$p_t + \rho_0 c_0^2 \nabla \cdot \mathbf{u} = 0.$$

If the time derivative of this equation is subtracted from the divergence of Eq. C–4b, the result is the classical wave equation,

$$p_{tt} - c_0^2 \nabla^2 p = 0. \tag{C–5}$$

2. Lossless Medium Moving with Constant Velocity

Frequently the medium carrying the sound wave is itself in motion. Suppose the motion is steady, say at a constant velocity U_0 in the x direction. In this case let $\mathbf{U}$ stand for total particle velocity, i.e.,

$$\mathbf{U} = \mathbf{i}U_0 + \mathbf{u},$$

where $\mathbf{u}$ is the particle velocity associated with any acoustical disturbance that may be present. In place of Eqs. C–2, the zero-order equations are now

$$\rho = \rho_0, \qquad P = p_0, \qquad \mathbf{U} = \mathbf{i}U_0.$$

When an acoustical disturbance is also present, the zero-order quantities must be supplemented by the acoustical quantities $\delta\rho$, p, and $\mathbf{u}$. Substitution of p/c_0^2 for $\delta\rho$ turns the linearized continuity and momentum equations into

$$p_t + U_0 p_x + \rho_0 c_0^2 \nabla \cdot \mathbf{u} = 0,$$

$$\rho_0 \left[\mathbf{u}_t + U_0 \frac{\partial \mathbf{u}}{\partial x} \right] + \nabla p = 0.$$

Elimination of $\mathbf{u}$ yields

$$p_{tt} + 2U_0 p_{xt} + U_0^2 p_{xx} - c_0^2 \nabla^2 p = 0. \tag{C–6}$$

It is reassuring to note that the traditional form of the wave equation, Eq. C–5, may be restored by transforming to a coordinate system moving with the steady flow, i.e.,

$$x' = x - U_0 t, \tag{C–7a}$$

$$y' = y, \qquad z' = z, \qquad t' = t. \tag{C–7b}$$

Sample derivatives in the new coordinate system are $\partial p/\partial x = \partial p/\partial x'$ and $\partial p/\partial t = \partial p/\partial t' - U_0\, \partial p/\partial x'$. When the transformation is completed, Eq. C–6 assumes the standard form

$$p_{t't'} - c_0^2\, \nabla^2 p = 0, \tag{C–8}$$

where in this case ∇^2 means

$$\nabla^2 = \frac{\partial^2}{\partial x'^2} + \frac{\partial^2}{\partial y^2} + \frac{\partial^2}{\partial z^2}.$$

Equation C–8 is useful for treating sound propagation in a ventilation duct or in an atmosphere having a constant wind. Consider, for example, the effect of a steady wind on sound propagation. If the sound is traveling in the direction d, making an angle θ with the direction the wind is blowing (see Fig. 2.9), we may write

$$p = f\left(t - \frac{d}{c_0}\right) = f\left(t - \frac{x'\cos\theta + y\sin\theta}{c_0}\right),$$

or, since x' is given by Eq. C–7a,

$$p = f\left[t\left(1 + \frac{U_0}{c_0}\cos\theta\right) - \frac{x\cos\theta + y\sin\theta}{c_0}\right]$$

$$= f\left[\left(1 + \frac{U_0}{c_0}\cos\theta\right)\left(t - \frac{x\cos\theta + y\sin\theta}{c_0 + U_0\cos\theta}\right)\right].$$

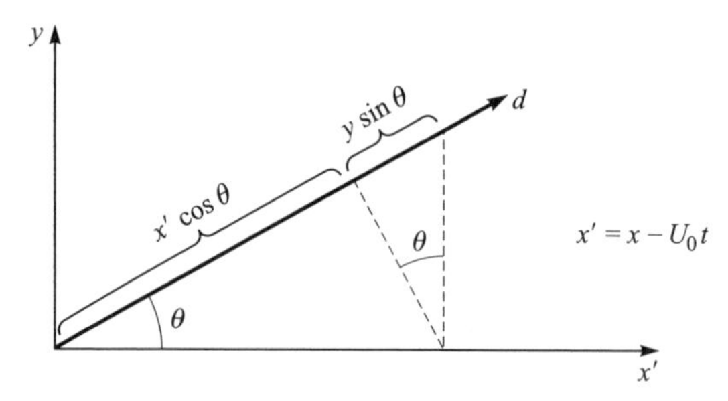

Figure 2.9 Coordinate system for a plane wave propagating in an atmosphere with wind.

The apparent propagation speed is therefore $c_0 + U_0 \cos\theta$. Sound traveling downwind ($\theta = 0$) has speed $c_0 + U_0$; if it travels crosswind ($\theta = 90°$), the speed is c_0.[7]

3. Lossless Medium in a Gravitational Field

Because this topic is covered in detail in Chap. 8, only a few aspects are noted here. First consider vertical propagation in the atmosphere or the ocean. The medium is assumed to be at rest. Given $\mathbf{u} = \mathbf{k}w$ and no dependence on x or y, the continuity and momentum equations become

$$\frac{D\rho}{Dt} + \rho w_z = 0, \tag{C–9a}$$

$$\rho \frac{Dw}{Dt} + P_z = -\rho g, \tag{C–9b}$$

where in this case $D/Dt = \partial/\partial t + w\,\partial/\partial z$. Notice that the zero-order versions of these equations are $\rho = \rho_0$ and the equation of hydrostatic equilibrium

$$(p_0)_z = -\rho_0 g. \tag{C–9c}$$

Because gravity causes stratification of the medium, p_0 and ρ_0 are functions of z. Moreover, so is the static entropy s_0. Therefore, even though the acoustic propagation is assumed to be lossless, the simple isentropic equation of state $P = P(\rho)$ cannot be used. On the other hand, because lack of dissipation does imply that the entropy of a fluid particle stays the same (the entropy may vary from particle to particle), i.e., $Ds/Dt = 0$, one may write

$$\frac{DP}{Dt} = \frac{\partial P}{\partial \rho}\frac{D\rho}{Dt} + \frac{\partial P}{\partial s}\frac{Ds}{Dt} = c^2 \frac{D\rho}{Dt} \tag{C–9d}$$

as the equation of state.

The linearized versions of Eqs. C–9a, b, and d are, respectively,

$$\delta\rho_t + (\rho_0)_z w + \rho_0 w_z = 0,$$

[7]The discerning reader may notice that our solution implies a Doppler shift. That is, if the initial expression for f were $A \sin\omega(t - d/c_0)$, the final equation would indicate an apparent angular frequency $\omega(1 + U_0/c_0 \cos\theta)$. At first this seems to be a disturbing result because wind cannot by itself induce a Doppler shift. The explanation is that our initial expression $p = f(t - d/c_0)$ implies a source that is moving with the wind (the source is located at $d = 0$). The Doppler shift is therefore due to motion of the source.

$$\rho_0 w_t + p_z = -g\delta\rho,$$

$$p_t + (p_0)_z w = c_0^2[\delta\rho_t + (\rho_0)_z w].$$

Combination of these equations and Eq. C–9c yields

$$w_{zz} + (\ln \rho_0 c_0^2)_z w_z - \frac{1}{c_0^2} w_{tt} = 0. \tag{C–10}$$

To solve this equation, one must first have more information about the static properties of the medium. The particular case of an isothermal atmosphere is considered in Chap. 8. There it is shown that the amplitude of the acoustic wave is affected by the dependence of the ambient properties on height.

Although the variation of amplitude with height is interesting, it is important only over very large changes in height. In the ocean the amplitude variation due to depth is practically insignificant. It turns out, therefore, that the ordinary wave equation, Eq. C–5, is sufficient for describing three-dimensional propagation of sound in the ocean (and also in the quiet atmosphere, provided the change in altitude is not great). This is not to say, however, that stratification has no effect. Indeed, because the sound speed varies with z, the sound rays are bent, or refracted, as they travel through the medium. A major portion of Chap. 8 is devoted to ray theory for a stratified fluid.

4. Viscous Fluid

The first dissipative fluid we consider is one that is viscous but thermally nonconducting. The linearized continuity expression continues to be Eq. C–4a. The linearized, isentropic equation of state, Eq. C–4c, also continues to hold despite the fact that viscosity implies dissipation. The explanation is that the viscous dissipation function $\Phi^{(\mathrm{visc})}$, which appears in the entropy equation, Eq. A–43, is a set of nonlinear terms, which disappear when the equation is linearized. The entropy equation therefore reduces to $DS/Dt = 0$, or, because the fluid is assumed homogeneous, $s = \mathrm{const}$.

The momentum equation requires special treatment. When the scalar potential ϕ and vector potential $\mathbf{\Psi}$ are introduced in Eq. A–24 and the equation is linearized, the various terms can be arranged as follows:

$$\nabla[\rho_0\phi_t + p - \tilde{V}\mu\nabla^2\phi] + \nabla\times[\rho_0\mathbf{\Psi}_t + \mu\nabla\times\nabla\times\mathbf{\Psi}] = 0. \tag{C–11}$$

The first group of terms constitutes an irrotational vector, the second group a rotational vector. Since neither group can contribute to the other, each vector must vanish separately. In this way the momentum equation is "split" into irrotational and rotational parts. Integrating each equation separately, we get

$$\rho_0 \phi_t + p - \tilde{V} \mu \nabla^2 \phi = 0$$
$$\rho_0 \boldsymbol{\Psi}_t + \mu \nabla \times \nabla \times \boldsymbol{\Psi} = 0,$$

where for each equation the constant of integration has been absorbed in the potential, ϕ or $\boldsymbol{\Psi}$, as the case may be. Because the $\boldsymbol{\Psi}$ equation does not involve the pressure or the density, it may be dealt with separately. Generally, in the absence of surfaces, or far away from surfaces, we may ignore the $\boldsymbol{\Psi}$ equation.

The remaining three equations in p, $\delta\rho$, and ϕ are Eqs. C–4a, A–27, and C–4c. Elimination of ϕ and $\delta\rho$ yields the "viscous wave equation"

$$\frac{\lambda + 2\mu}{\rho_0 c_0^2} \nabla^2 p_t + \nabla^2 p - \frac{1}{c_0^2} p_{tt} = 0. \tag{C–12}$$

The plane-wave form of this equation is solved in Sec. 9B.

The $\boldsymbol{\Psi}$ equation is used whenever an important viscous shearing motion is associated with the acoustic field. For example, when sound propagates through a tube, a viscous shear is set up in the immediate vicinity of the tube wall surface. The $\boldsymbol{\Psi}$ equation is then needed to describe properly the flow in the neighborhood of the wall.

5. Viscous, Thermally Conducting Fluid

The treatment for a thermoviscous medium is similar to that for a viscous medium. The momentum equation splits into irrotational and rotational parts as before. The energy equation is, however, somewhat complicated. We consider the case of a perfect gas, the energy equation for which has been given previously as

$$\frac{DP}{Dt} + \gamma P \nabla \cdot \mathbf{u} = (\gamma - 1)\Phi^{(\text{visc})} + \frac{\kappa}{C_v} \nabla^2 \left(\frac{P}{\rho} \right).$$

On the right-hand side $\Phi^{(\text{visc})}$ is nonlinear, and the linear part of $\nabla^2(P/\rho)$ is

$$\frac{1}{\gamma \rho_0} \nabla^2 (\gamma p - c_0^2 \delta\rho) = \frac{C_v}{C_p \rho_0} \nabla^2 (\gamma p - c_0^2 \delta\rho).$$

The linearized energy equation may therefore be written

$$p_t + \gamma p_0 \nabla \cdot \mathbf{u} = \frac{\kappa}{C_p \mu} \frac{\mu}{\rho_0} \nabla^2 (\gamma p - c_0^2 \delta\rho) = \frac{\nu}{\text{Pr}} \nabla^2 (\gamma p - c_0^2 \delta\rho), \tag{C–13}$$

where $\nu = \mu/\rho_0$ is the kinematic viscosity coefficient and $\text{Pr} = C_p \mu / \kappa$ is the Prandtl number. The Prandtl number gives a measure of the effects of viscosity relative to the effects of heat conduction. For gases the Prandtl number is of order 1. For many liquids the Prandtl number is of order 10 (for water Pr $\doteq$ 8).

Equations C–4a, A–27, and C–13 may be combined to yield a single "wave equation" in ϕ:

$$\frac{\gamma\tilde{V}\nu^2}{\mathrm{Pr}}\nabla^4\phi_t + c_0^2\frac{\nu}{\mathrm{Pr}}\nabla^4\phi - \nu\left(\frac{\gamma}{\mathrm{Pr}} + \tilde{V}\right)\nabla^2\phi_{tt} - c_0^2\nabla^2\phi_t + \phi_{ttt} = 0. \qquad \text{(C–14)}$$

Notice that for a thermally nonconducting medium (Pr $\to \infty$), this equation reduces to Eq. C–12.

Although Eq. C–14 has been derived for a perfect gas, it turns out to apply to other fluids—liquids and other gases—as well.[8] An approximate solution of Eq. C–14 for plane waves is given in Sec. 9B.

6. Relaxing Fluid

If the fluid is inviscid and thermally nonconducting but subject to relaxation, the equation of state takes the following form:

$$\tau(p_t - c_\infty^2\,\delta\rho_t) + (p - c_0^2\,\delta\rho) = 0 \qquad \text{(C–15)}$$

(see Chap. 9). Here c_∞ and c_0 are the speeds of sound for frequencies approaching infinity and zero, respectively, and τ is the relaxation time. It is left as an exercise for the reader to show that the wave equation for this case is

$$\tau(\phi_{tt} - c_\infty^2\nabla^2\phi)_t + (\phi_{tt} - c_0^2\nabla^2\phi) = 0. \qquad \text{(C–16)}$$

REFERENCES

1. R. Aris, *Vectors, Tensors, and the Basic Equations of Fluid Mechanics* (Prentice-Hall, Englewood Cliffs, NJ, 1961).
2. D. B. Cruikshank, "Growth of distortion in a finite-amplitude sound wave in air," *J. Acoust. Soc. Am.* **40**, 731–733(L) (1966).
3. M. F. Hamilton, "Sound Beams," Chap. 8 in M. F. Hamilton and D. T. Blackstock, Eds., *Nonlinear Acoustics* (Academic Press, New York, 1998), pp. 246–250.
4. F. V. Hunt, "Notes on the exact equations governing the propagation of sound in fluids," *J. Acoust. Soc. Am.* **27**, 1019–1038 (1955); also available as "Propagation of sound in fluids," Chap. 3c in D. E. Gray, Ed., *American Institute of Physics*

[8]For an exceedingly thorough treatment of sound propagation in thermoviscous fluids, see Ref. 9. This article contains a handy table of approximate values of γ, Pr, V, and (with some calculation) ν.

Handbook, 3rd ed. (McGraw-Hill, New York, 1972), pp. 3–37 to 3–68.

5. R. L. Panton, *Incompressible Flow*, 2nd ed. (Wiley–Interscience, New York, 1996).
6. A. D. Pierce, *Acoustics* (McGraw-Hill, New York, 1989), Chap. 10. Reprinted by Acoustical Society of America, New York, 1989.
7. G. G. Stokes, "An examination of the possible effect of radiation of heat on the propagation of sound," *Philos. Mag. Series 4*, **7**, 305–317 (1851).
8. P. A. Thompson, *Compressible-Fluid Dynamics* (McGraw-Hill, New York, 1972; reprinted by P. A. Thomson, Rensselaer Polytechnic Inst., Troy, NY, 1984).
9. C. Truesdell, "Precise theory of the absorption and dispersion of forced plane infinitesimal waves according to the Navier-Stokes equations," *J. Math. Mech.* **2**, 643–741 (1953).
10. W. G. Vincenti and S. G. Traugott, "The coupling of radiative transfer and gas motion," *Ann. Rev. Fluid Mech.* **3**, 89–116 (1971).
11. P. J. Westervelt, "Parametric acoustic array," *J. Acoust. Soc. Am.* **35**, 535–537 (1963). M. F. Hamilton, "Sound Beams." Chap. 8 in M. F. Hamilton and D. T. Blackstock, Eds., *Nonlinear Acoustics* (Academic Press, New York, 1998), pp. 246–250.

PROBLEMS

2–1. Derive the equation of continuity for a medium in which mass sources are present. If q_v represents the volume flow from the sources per unit volume, then $\rho q_v \, dV$ is the mass generated within the volume element dV.

(a) First obtain the integral form of the continuity equation for this case.

(b) Then, using the divergence theorem, convert to differential form.

ANSWER, part (b): $\rho_t + \nabla \cdot (\rho \mathbf{u}) = \rho q_v$.

2–2. This problem is a generalization of Example 2.1. Here the pressure step is an actual shock wave, still plane and steady but of unrestricted amplitude, i.e., no small-signal assumption is made. The shock travels horizontally with constant velocity v_s (not approximated by c_0) into a quiet fluid; see sketch (*a*). The properties of the flow behind the shock are $u = u_1$, $P = P_1$, $\rho = \rho_1$; the corresponding properties in front of the shock are $u = 0$, $P = p_0$, $\rho = \rho_0$. Use the integral form of the continuity equation, Eq. A–6, to obtain the relation between ρ_1, u_1, and v_s. The problem may be worked most simply by transforming to a coordinate system moving with the shock, as in Example 2.1; see sketch (*b*). To transform to the new system (in which the flow is steady), subtract v_s from every velocity in the problem. The new particle velocities v_1 and v_0 at the left and right ends of the control volume V are thus $v_1 = u_1 - v_s$ and $v_0 = -v_s$, respectively. After evaluating the integrals in the continuity equation, express the answer in terms of particle velocities in the original coordinate system.

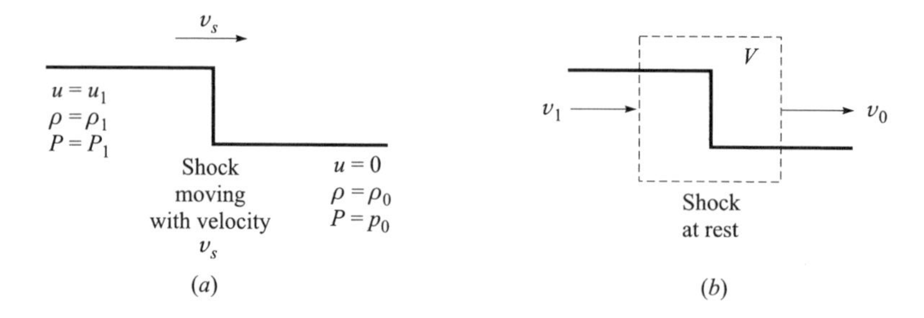

$$\text{ANSWER: } \rho_1(v_s - u_1) = \rho_0 v_s.$$

2–3. Generalization of Example 2.2: Use the integral form of the momentum equation, Eq. A–13, to obtain a conservation-of-momentum relation for the shock wave described in Problem 2–2. Then eliminate the density ρ_1 by using conservation of mass (Problem 2–2).

$$\text{ANSWER: } P_1 - p_0 = \rho_0 u_1 v_s.$$

Show that this result reduces to the small-signal characteristic impedance relation $\Delta P = \rho_0 c_0 u_1$ when the pressure jump $\Delta P\ (= P_1 - p_0)$ becomes very small. (See the following problem for an expression for v_s.)

2–4. Generalization of Example 2.3: Use the integral form of the energy equation, Eq. A–29, to obtain a conservation-of-energy relation for the shock wave described in Problem 2–2. Assume the fluid is a perfect gas. Using the perfect gas law and the results of Problems 2–2 and 2–3 to eliminate pressure, density, and temperature, obtain the following relation:

$$v_s^2 - \beta u_1 v_s - c_0^2 = 0,$$

where $\beta = \frac{1}{2}(\gamma + 1)$. Solve this equation for v_s. Obtain the so-called weak shock limit by expressing your formula for v_s as a series expansion in the quantity u_1/c_0 and then terminating the series after the linear term. The weak shock expression for v_s is useful when $u_1/c_0 \ll 1$.

2–5. Shock speed v_s is frequently expressed in terms of the pressure jump $\Delta P = P_1 - p_0$ across the shock rather than the particle velocity jump u_1. Using results from Problems 2–3 and 2–4, eliminate u_1 and obtain an exact expression for v_s as a function of ΔP. Use this expression to fill in the table below; take the medium to be air. To relate SPL to ΔP, let p_{rms} be found by centering the shock in the time interval $0 \leq t \leq t_{\text{av}}$ over which the time average (see Eq. 1E–17) is calculated.

SPL	v_s (m/s})	% Deviation of v_s from c_0
134 dB		
154 dB		
174 dB		

2–6. Another way to obtain the conservation-of-mass relation for a steady shock wave (see Problem 2–2) is to integrate the one-dimensional continuity equation

$$\rho_t + u\rho_x + \rho u_x = 0$$

directly. Because the shock is moving with uniform velocity v_s, the x, t dependence of all the field variables (ρ, u, P, etc.) is of the form $f(x, t) = f(x - v_s t)$, where f is any field variable. Let $y = x - v_s t$. The x and t derivatives may be transformed as follows:

$$f_t = f_y y_t = -v_s f_y,$$
$$f_x = f_y y_x = f_y.$$

Use of these relations reduces the continuity equation to an ordinary differential equation, which may be integrated immediately. By evaluating the constant of integration first behind the shock (i.e., as $y \to -\infty$), where $u = u_1$ and $\rho = \rho_1$, and then in front of the shock (i.e., as $y \to +\infty$), where $u = 0$ and $\rho = \rho_0$, you should be able to obtain the answer given in Problem 2–2.

2–7. Use the procedure described in the previous problem to carry out one integration of the one-dimensional momentum equation for a viscous fluid

$$\rho(u_t + uu_x) + P_x = (\lambda + 2\mu)u_{xx}$$

for the case of a steady shock. For this problem, information about the slope of the wave, as well as its value, as $y \to \pm\infty$ must be used. Obtain the answer given in Problem 2–3.

2–8. The one-dimensional energy equation for a perfect gas may be written

$$\frac{\rho}{\gamma - 1} \frac{D}{Dt}\left(\frac{P}{\rho}\right) + Pu_x = (\lambda + 2\mu)u_x^2 + \kappa T_{xx}$$

(see Eq. A–40). Use the procedure described in Problems 2–6 and 2–7 to integrate this equation once for the case of a steady shock wave. You will need to use results from Problems 2–6 and 2–7 in order to perform the integration. Show that evaluation of the integrated equation as $y \to \pm\infty$ leads to the following intermediate result:

$$\frac{\rho_0 v_s}{\gamma - 1}\left(\frac{p_0}{\rho_0} - \frac{P_1}{\rho_1}\right) + p_0 u_1 + \frac{1}{2}\rho_0 v_s u_1^2 = 0.$$

Next, by using the results from Problems 2–6 and 2–7 to eliminate P_1 and ρ_1, obtain the following quadratic equation in the shock velocity:

$$v_s^2 - \beta u_1 v_s - c_0^2 = 0,$$

where $\beta = \frac{1}{2}(\gamma + 1)$. Solve this equation for v_s. Expand your answer as a power series in the quantity u_1/c_0. Terminate the series after the linear term. The expression you obtain is the so-called weak shock limit formula for shock velocity. This formula is useful when $u_1/c_0 \ll 1$.

2–9. A duct is filled with a porous material, such as fiberglass, as shown in the sketch. The dots represent the matrix of solid particles or frame of the porous material; the fluid fills the open space. The porosity $\mathcal{P}$ is defined to be the ratio of the volume of fluid (open space) to the total volume. It may be shown from the definition that

$$\mathcal{P} = \frac{S_{\text{open}}}{S_{\text{open}} + S_{\text{closed}}} = \frac{S_{\text{open}}}{S_{\text{total}}},$$

where S_{open} is the open portion (fluid filled) of the cross-sectional area S and S_{closed} is the closed portion (solid filled). If u is the particle velocity of the fluid in the absence of solid particles, then $u/\mathcal{P}$ is the particle velocity with the solid particles present (to maintain the same volume velocity, the fluid must flow faster in a constricted space). Derive the differential form of the equation of continuity for one-dimensional flow in the porous material. Carry out the derivation directly, that is, do not derive the integral form first and then transform to the differential form. Assume that the porous material is homogeneous, that is, that $\mathcal{P}$ is a constant.

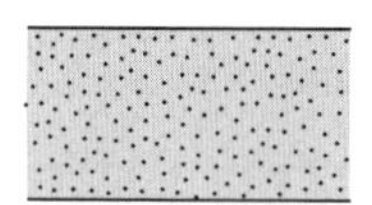

ANSWER: $\rho_t + (\rho u/\mathcal{P})_x = 0.$

2–10. Derive the differential form of the momentum conservation equation for the porous material described in the previous problem. As in the previous problem, carry out the derivation directly, that is, do not derive the integral form first and then transform to the differential form. The force exerted by the frame (solid particle matrix) on the fluid may be taken to be one of frictional drag. Assume for simplicity that the frame itself does not move. To characterize the drag force, consider the following measurement of steady flow in a duct. Fluid is forced to flow at a steady velocity u through a sample of porous material (see sketch). The frictional resistance of the porous material causes the pressure to drop an amount $P_1 - P_2$ across the sample. The pressure drop is found to be proportional to the length L of the sample and to the flow velocity in the sample. In other words, the porous material exerts a drag force per unit volume F given by

$$F \equiv -\frac{P_1 - P_2}{L} = -r\frac{u}{\mathcal{P}}$$

on the fluid. The minus sign is necessary to show that the drag force opposes the flow. The proportionality coefficient r is called the flow resistivity; at low velocities it is a constant. To include the frictional force in your basic expression for momentum balance, assume that the drag force has the same functional dependence on u when the flow is unsteady.

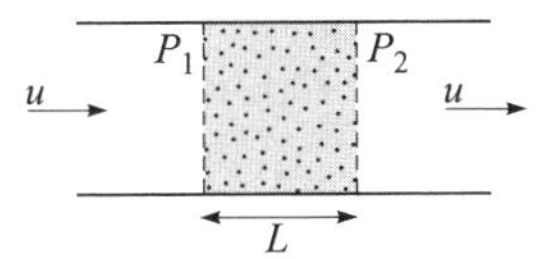

$$\text{ANSWER: } \rho\left(u_t + \frac{uu_x}{\mathcal{P}}\right) + \mathcal{P}P_x + ru = 0.$$

2–11. Consider irrotational flow ($\mathbf{u} = \nabla\phi$) in an inviscid gas that obeys the adiabatic gas law $P/p_0 = (\rho/\rho_0)^\gamma$. For this case the momentum equation Eq. A–26 may be integrated once (exactly). Obtain the relation

$$P = p_0\left[1 - \frac{\gamma - 1}{c_0^2}\left(\phi_t + \frac{1}{2}\nabla\phi\cdot\nabla\phi\right)\right]^{\gamma/(\gamma-1)}.$$

Then, by linearizing this expression, obtain the small-signal relation between p and ϕ (see Eq. 1D–6).

2–12. This problem is a variation on the method used in Sec. B.2 to obtain a first-order nonlinear equation for plane waves. Although lossless, the fluid is not restricted to being one that satisfies the adiabatic gas law.

(a) First use the definition of sound speed for a lossless fluid ($c^2 \equiv dP/d\rho$) to transform the continuity equation into

$$p_t + up_x + \rho c^2 u_x = 0.$$

The momentum equation may be put in the form

$$u_t + uu_x + \rho^{-1} p_x = 0.$$

(b) For a progressive wave, p is expected to be a function of u alone, that is, $p = p(u)$. Find the progressive wave relation, in terms of the derivative $p'(u) = dp/du$, that renders the two equations in part (a) the same.

(c) Show that use of the relation found in part (b) transforms both equations in part (a) into the following equation:

$$u_t + (u \pm c)u_x = 0.$$

What is the physical significance of the $\pm$ sign? Note that the equation you have derived is more general than Eq. B–13 because the specific form of the equation of state $p = p(\rho)$ has not been specified.

2–13. Show by direct substitution that $u = f[t - x/(c_0 + \beta u)]$ satisfies the nonlinear wave equation $u_t + (c_0 + \beta u)u_x = 0$.

2–14. Solve the piston-in-the-tube problem to second order for the case in which the piston is excited at two frequencies (see sketch). In particular, find the amplitudes of the two second-harmonic signals, the sum frequency signal, and the difference frequency signal. Sketch the amplitude of each (all on one graph) as a function of distance for the case in which the two primary angular frequencies ω_1 and ω_2 are

(a) Close together

(b) Far apart

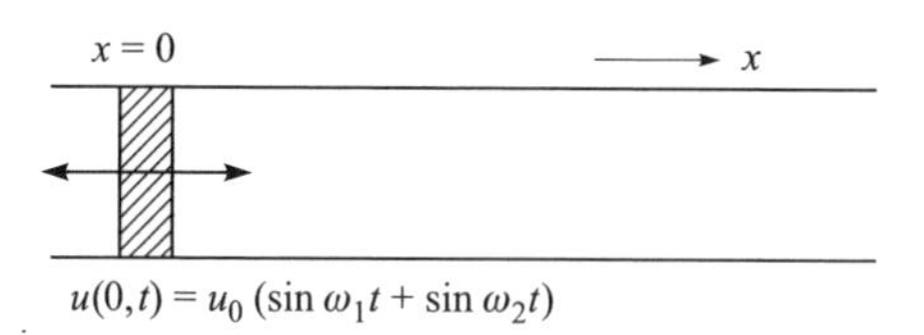

2–15. A lossless, one-dimensional flow exists in a constant-cross-section duct. Under steady conditions the flow velocity is U_0, a constant. When a disturbance is also present, the velocity of the fluid may be expressed as $U_0 + u$, where u is the particle velocity associated with the disturbance. Derive (directly) the differential form of the one-dimensional equations of

(a) continuity and

(b) conservation of momentum

for this flow.

ANSWER, part (a): $\rho_t + [\rho(u + U_0)]_x = 0$.

2–16. Starting with the results of the previous problem and an appropriate pressure-density relation, derive the small-signal wave equation for plane waves in a duct in which the fluid is moving with a constant velocity U_0. Compare your result with Eq. C–6.

2–17. Solve the viscous wave equation, C–12, for a spherically spreading radial wave. Do this by assuming a solution

$$p = \frac{A}{r} e^{j(\omega t - kr)},$$

where r is the radial distance and A is a constant. If the assumed solution is substituted in Eq. C–12, the result is a dispersion relation $k = k(\omega)$. Solve for k using the notation $\delta = \mu\omega/\rho_0 c_0^2$. Next, assuming that $\delta \ll 1$, approximate your result by using the binomial expansion. Retain only the first two terms in the expansion, i.e., put the expression for k in the form

$$k \doteq \frac{\omega}{c_0}(a_0 + a_1\delta)$$

(your task is to find the values of a_0 and a_1). From this result deduce expressions for the attenuation coefficient α and the phase velocity c^{ph} of the wave.

2–18. Consider the propagation of sound in a medium in which heat transfer is due to radiation (not conduction). One model for heat radiation is Newton's law of cooling,[9]

$$\nabla \cdot \mathbf{q} = \rho C_v h_r (T - T_0),$$

where T_0 is the ambient temperature and h_r is the heat radiation coefficient. Substitute this expression for the heat transfer term in the energy

[9]Although this is not a very accurate model for the effect of heat radiation on sound, it is of historical interest. Stokes (Ref. 7) used it to help lay to rest remaining nineteenth-century doubts that sound propagates adiabatically. For modern views on how heat radiation affects sound, see Ref. 10.

equation for an inviscid gas, Eq. A–31.

(a) Assume the medium is a perfect gas. Without further approximation, convert the energy equation into the form

$$\frac{DP}{Dt} - c^2 \frac{D\rho}{Dt} + h_r \Big(p - \frac{c_0^2}{\gamma}\delta\rho\Big) = 0,$$

where c and c_0 are given by Eq. B–3 and Eq. 1C–45, respectively.

(b) Linearize this expression and obtain

$$(p_t - c_0^2 \delta\rho_t) + h_r \Big(p - \frac{c_0^2}{\gamma}\delta\rho\Big) = 0.$$

(c) Note that this equation has the same form as the pressure-density relation for a relaxing fluid (Sec. C.6). By simple direct analysis of this equation, show that at low frequency sound propagates at speed $b_0 = c_0/\sqrt{\gamma}$ (the isothermal sound speed), while at high frequency sound propagates at speed c_0 (the adiabatic sound speed). *Suggestion*: Convert the derivative with respect to t to one with respect to a dimensionless time, say $\tau = \omega t$, where ω is angular frequency. Then consider what happens to the equation in part (b) when ω is very small or very large compared to an appropriate reference frequency. What is the reference frequency?

2–19. Derive the small-signal wave equation for sound in the porous material described in Problems 2–9 and 2–10. For the pressure-density relation, assume that the heat transfer between the fluid and the solid matrix frame is so nearly perfect that sound transmission is an isothermal process, that is, that the pressure and density are related by

$$P/p_0 = \rho/\rho_0.$$

ANSWER: $u_{tt} - b_0^2 u_{xx} + (r/\rho_0)u_t = 0$, where $b_0^2 = p_0/\rho_0$.

2–20. Find the attenuation α and the phase speed c^{ph} for small-signal sound propagating in the porous material described in the previous problem.

(a) Develop exact expressions for α and c^{ph}. The following approach is suggested: Obtain the dispersion relation $k = k(\omega)$ by substituting a time-harmonic wave solution in the wave equation. By separating k into its real and imaginary parts ($k = \omega/c^{\text{ph}} - j\alpha$), you will be able to find expressions for α and c^{ph}. However, the algebra will be simplified if you first put the dispersion relation in the dimensionless form

$$K^2 = 1 - j\theta,$$

where $K = b_0 k/\omega$ and θ is a function of ω and r. By writing $K \doteq R - jI$ and squaring as indicated by the formula, you should be able to solve for R and I and subsequently for α and c^{ph}.

(b) Starting with the results from part (a), obtain asymptotic expressions for α and c^{ph}, first for low frequency and then for high frequency. The leading term is sufficient in each case.

3

REFLECTION AND TRANSMISSION OF NORMALLY INCIDENT PLANE WAVES OF ARBITRARY WAVEFORM

Chapters 3–5 are about reflection and transmission of plane waves at a plane interface between lossless media. Normal incidence is treated in this chapter and the next, oblique incidence in Chap. 5. In the present chapter, which serves as an introduction to the subject, the waveform of the incident pressure disturbance is taken to be arbitrary. This approach is particularly well suited to problems involving pulses. Time-harmonic incident waves are considered, and many applications discussed, in Chap. 4.

A. REFLECTION AND TRANSMISSION COEFFICIENTS FOR AN INTERFACE BETWEEN TWO IDEAL FLUIDS

Consider the reflection and transmission problem sketched in Fig. 3.1. A forward traveling wave (p^+) of arbitrary waveform is normally incident on the interface between two semi-infinite media, such as air and water, which have real characteristic impedances $Z_1 = \rho_1 c_1$ and $Z_2 = \rho_2 c_2$.[1] The wave is partly reflected and partly transmitted. We should like to know the waveform and strength of the reflected (p^-) and transmitted (p^{tr}) signals. Let the pressure of each wave be denoted as follows:

$$\text{Incident wave:} \quad p^+ = p^+(t - x/c_1), \tag{A–1a}$$

$$\text{Reflected wave:} \quad p^- = p^-(t + x/c_1), \tag{A–1b}$$

$$\text{Transmitted wave:} \quad p^{tr} = p^{tr}(t - x/c_2). \tag{A–1c}$$

[1]For simplicity the subscript 0 is now omitted in the symbol for characteristic impedance, that is, Z_1 is used instead of Z_{01} and Z_2 instead of Z_{02}.

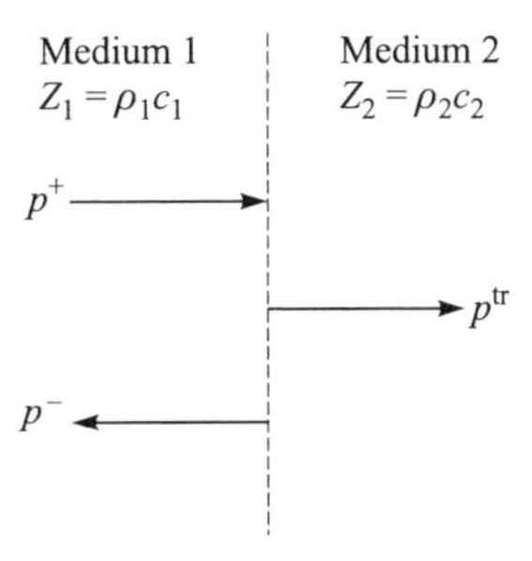

Figure 3.1 Reflection and transmission of sound at normal incidence.

If the coordinate system is chosen so that $x = 0$ coincides with the interface, the three pressure signals at the interface are simply $p^+(t)$, $p^-(t)$, and $p^{tr}(t)$, respectively. Hereafter, the argument t is usually omitted. In this section the cross-sectional areas S_1 and S_2 of the two media are assumed equal. For example, the media might be of infinite lateral extent or they might be enclosed in a pipe of constant cross section. The case of unequal areas, $S_2 \neq S_1$, is taken up in Sec. C.

1. Pressure Signals

The reflection coefficient R and transmission coefficient T are defined by

$$R = \frac{p^-}{p^+} \tag{A–2}$$

and

$$T = \frac{p^{tr}}{p^+}, \tag{A–3}$$

2 B.C.

respectively. Expressions for R and T are found by applying two conditions that must be satisfied at the interface. First, the pressure must be the same on either side of the interface (otherwise the interface, being massless, would experience an infinite acceleration),

$$p^+ + p^- = p^{tr}. \tag{A–4}$$

Divide through by p^+ and apply Eqs. A–2 and A–3:

$$1 + R = T. \tag{A–5}$$

Second, the normal component of particle velocity must be continuous across the interface (in this chapter the normal component is the only component). The total particle velocity on the left-hand side of the interface is $u^+ + u^-$, where the superscripts + and − continue to mean incident and reflected, respectively. On the right-hand side the particle velocity is just that due to the transmitted wave u^{tr}. Thus we have

$$u^{+} + u^{-} = u^{tr}. \tag{A–6}$$

Since we should also like to express this boundary condition in terms of R and T, it must first be converted to a pressure relation. Use the characteristic impedance relation but remember to include a minus sign for the backward traveling wave:

$$\frac{p^{+}}{Z_1} - \frac{p^{-}}{Z_1} = \frac{p^{tr}}{Z_2}.$$

Rearrangement and application of Eqs. A–2 and A–3 give

$$1 - R = \frac{Z_1}{Z_2} T. \tag{A–7}$$

Equations A–5 and A–7 may be solved for R and T:

$$R = \frac{Z_2 - Z_1}{Z_2 + Z_1} \tag{A–8}$$

$$T = \frac{2Z_2}{Z_2 + Z_1}. \tag{A–9}$$

Notice that these expressions are quite general in that they have been derived without assuming any special waveform for the incident wave. In many textbooks it is assumed that the waves are sinusoidal. Because the impedances here are real, such an assumption is unnecessary. The quantities R and T are sometimes called *pressure* reflection and transmission coefficients in order to distinguish them from corresponding *sound power* coefficients, which are discussed below.

Note carefully the meaning of the results obtained thus far. When an incident wave p^{+} encounters an interface, the time signal of the reflected wave as it leaves the interface is

$$p^{-}(t) = Rp^{+}(t).$$

The reflected wave therefore has the same time waveform as the incident wave, but unless $|R| = 1$, its amplitude is different. If $R > 0$, the reflected wave has the same phase as the incident wave; $R < 0$ implies a 180° phase shift of the reflected wave. Similarly, because the time signal of the transmitted wave is

$$p^{\text{tr}}(t) = Tp^{+}(t),$$

it too has the same basic waveform as the incident wave. If the propagation speeds c_2 and c_1 are different, however, the transmitted wave is either stretched ($c_2 > c_1$) or compressed ($c_2 < c_1$) spatially relative to the incident wave.

2. Sound Power

Coefficients to describe the sound power reflected and transmitted are also in common use. The sound power reflection coefficient r is the ratio of the reflected power to the incident power,

$$r \equiv \frac{W^{-}}{W^{+}} = \frac{\mathbf{I}^{-}\cdot\mathbf{S}^{-}}{\mathbf{I}^{+}\cdot\mathbf{S}^{+}} = \left|\frac{\mathbf{I}^{-}}{\mathbf{I}^{+}}\right| = \frac{(p_{\text{rms}}^{-})^2/Z_1}{(p_{\text{rms}}^{+})^2/Z_1} = \frac{(p_{\text{rms}}^{-})^2}{(p_{\text{rms}}^{+})^2}, \tag{A–10}$$

where Eq. 1E–16 has been used, and it should be noted that $\mathbf{S}^{-} = -\mathbf{S}^{+}$. Because p_{rms}^{-} is related to p_{rms}^{+} by

$$(p_{\text{rms}}^{-})^2 = \frac{1}{t_{\text{av}}}\int_0^{t_{\text{av}}} (p^{-})^2\,dt = \frac{1}{t_{\text{av}}}\int_0^{t_{\text{av}}} (Rp^{+})^2\,dt = R^2(p_{\text{rms}}^{+})^2,$$

Eq. A–10 reduces to the very simple formula

$$r = R^2. \tag{A–11}$$

The power transmission coefficient τ, defined as the ratio of the transmitted power to the incident power, is not, however, equal to T^2 except in special cases. We have

$$\tau \equiv \frac{W^{\text{tr}}}{W^{+}} = \frac{\mathbf{I}^{\text{tr}}\cdot\mathbf{S}^{\text{tr}}}{\mathbf{I}^{+}\cdot\mathbf{S}^{+}} = \frac{(p_{\text{rms}}^{\text{tr}})^2/Z_2}{(p_{\text{rms}}^{+})^2/Z_1}, \tag{A–12}$$

or, because (prove this result) $(p_{\text{rms}}^{\text{tr}})^2 = T^2(p_{\text{rms}}^{+})^2$,

$$\tau = T^2\frac{Z_1}{Z_2}. \tag{A–13}$$

When use is made of the expressions for R and T, Eqs. A–11 and A–13 reduce, respectively, to

$$r = \frac{(Z_2 - Z_1)^2}{(Z_2 + Z_1)^2} \tag{A–14}$$

and

$$\tau = \frac{4Z_1Z_2}{(Z_2+Z_1)^2}. \tag{A–15}$$

Again note that these results do not depend on any assumption about the wave shape. Is conservation of energy,

$$r + \tau = 1, \tag{A–16}$$

fulfilled?

3. Transmission Loss

A commonly used quantity in transmission problems is the transmission loss TL, which is expressed in decibels,

$$\mathrm{TL} \equiv -10 \log_{10} \tau. \tag{A–17}$$

Use of Eq. A–13 yields

$$\mathrm{TL} = -10 \log_{10} T^2 - 10 \log_{10}(Z_1/Z_2). \tag{A–18}$$

How can a transmission *loss* occur when the fluids are dissipationless? *Hint*: Where has the "lost" energy gone?

The reader is reminded that the formulas for reflection and transmission developed here are for real impedances. If Z_1 and/or Z_2 are complex, Eqs. A–8 and A–9 continue to hold, but R and T in general become complex. Some of the expressions developed here then require modification; see Chap. 4.

B. SPECIAL CASES

1. Rigid Wall

First consider sound incident on a rigid wall, an interface at which the particle velocity must vanish. For acoustical purposes "rigid" means that the impedance of the second medium is much, much larger than that of the first medium (the second medium is often described as being very hard compared to the first). In this case Eq. A–8 reduces to $R = 1$, a result that implies the reflected signal is an exact replica of the incident signal.

Next solve for the pressure field in front of the wall. In the example shown in Fig. 3.2 the rigid wall is at the position $x = x_r$, and the incident pressure wave is taken to be a boxcar pulse of amplitude A. The problem is easily solved by using the image method: the rigid wall is removed and the medium is imagined to continue in the region $x > x_r$. To retain the boundary condition that the particle velocity must vanish at $x = x_r$, we construct an image wave, a back-

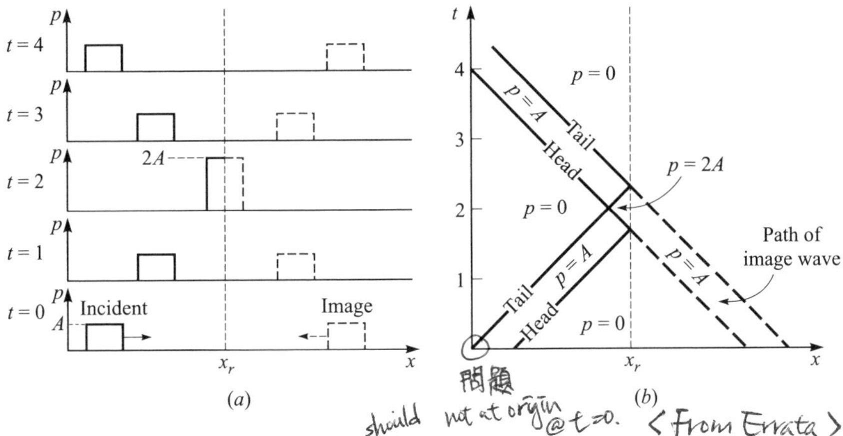

Figure 3.2 Reflection of a pulse from a rigid wall. (*a*) Spatial waveforms at a sequence of times. (*b*) Characteristics diagram.

ward propagating wave of just the right amplitude, shape, and timing, to cancel the particle velocity of the incident wave when the latter reaches the point $x = x_r$. (Why has the image been chosen to be a positive-pressure pulse rather than a negative-pressure pulse?) The x, t diagram (Fig. 3.2*b*) is used to show the paths of the various signals in the problem and to indicate the amplitudes at various places and times. Notice that the pressure at the wall ($x = x_r$) is twice its free field value,

$$p_{\text{wall}} = p^+ + p^- = (1 + R)p^+ = 2p^+.$$

This result is called pressure doubling. It occurs, for example, when a wave in air strikes a water surface or the surface of an impervious solid, such as wood or metal.

Finally, although Eq. A–9 yields $T = 2$ for this case, no power is really transferred into the wall. As Eq. A–15 shows, the sound power transmission coefficient approaches zero as Z_2 approaches infinity.

2. Pressure Release Surface

The opposite of an infinitely hard medium is one that is infinitely soft. If $Z_2 \ll Z_1$, then $T = 0$ and $R = -1$. Again no wave is transmitted, but this time the image wave has polarity opposite to that of the incident wave (see Fig. 3.3). What happens to the particle velocity at the interface? Such an interface is described as "pressure release" because the acoustic pressure falls to zero there ($p^+ + p^- = 0 = p^{\text{tr}}$). To a wave in water, for example, an air surface is a pressure release boundary. Another example (to a first approximation) is the open end of a tube. When a wave traveling down the tube reaches the open

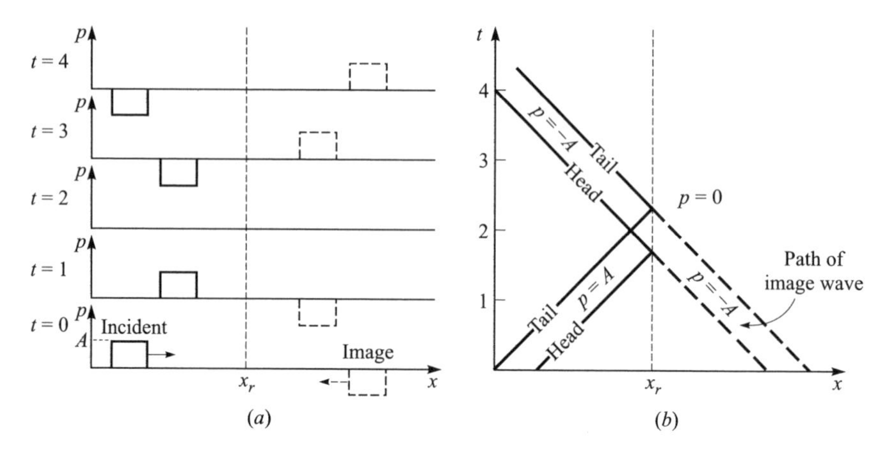

Figure 3.3 Reflection of a pulse from a pressure release surface. (*a*) Spatial waveforms at a sequence of times. (*b*) Characteristics diagram.

end, the sudden lack of confinement at the opening has the same effect as a very soft boundary. The pressure tends to fall to zero.[2]

3. Matched Impedance Interface

If $Z_1 = Z_2$, then $R = 0$ and $T = 1$. The incident wave is transmitted perfectly across the interface. Notice that it is not necessary for medium 1 and medium 2 to be identical to achieve perfect transmission; all that is required is that the products $\rho_1 c_1$ and $\rho_2 c_2$ have the same value. Inspect the tables in Appendix A to identify pairs of materials having about the same value of characteristic impedance. For example, because its impedance is nearly the same as the impedance of seawater, so-called rho-c rubber is used to cover the face of underwater sound transducers. The covering protects the transducer from the salt water but reflects little of the sound that strikes it.

C. CHANGE IN CROSS-SECTIONAL AREA

A sudden change in cross-sectional area of a pipe or duct can generate a reflection of sound even when the medium is the same in both sections of the pipe. A typical example is shown in Fig. 3.4. Here S_1 and S_2 are the initial and final cross-sectional areas, respectively. To calculate the reflection and transmission coefficients, we must apply the boundary conditions governing

[2]The impedance of an open end is actually a function of frequency. Very small for low frequencies, the impedance approaches Z_0 at high frequencies. See Secs. 4C.2 and 13D. Treating an open end as a pressure release surface is therefore a low frequency approximation.

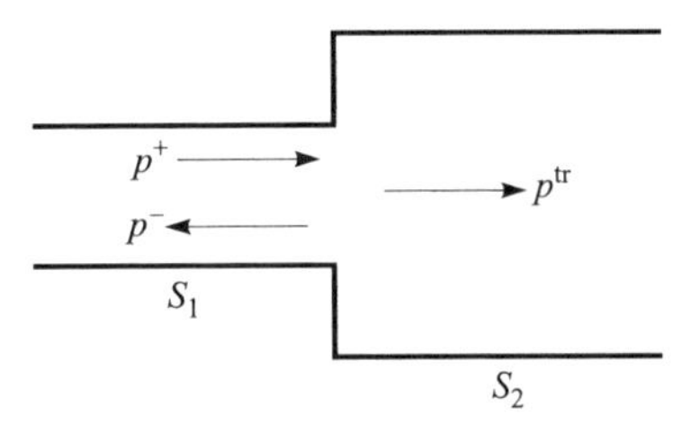

Figure 3.4 Reflection and transmission at an area change.

the flow at the junction of the two ducts. First, continuity of pressure at the junction leads to Eqs. A–4 and A–5, just as for a constant-cross-section duct. The particle velocity boundary condition used previously, Eq. A–6, is not appropriate here, however. A more general statement of the velocity boundary condition is that the *volume* velocity q of fluid just to the left of the junction must be equal to the volume velocity just to the right,

$$q^+ + q^- = q^{tr}, \tag{C–1}$$

or, if the particle velocity is assumed constant over a given cross section,

$$S_1 u^+ + S_1 u^- = S_2 u^{tr}. \tag{C–2}$$

Use of the characteristic impedance relation leads to

$$1 - R = \frac{S_2}{S_1} T. \tag{C–3}$$

Combine this equation with Eq. A–5 to obtain

$$R = \frac{S_1 - S_2}{S_1 + S_2} \tag{C–4}$$

and

$$T = \frac{2S_1}{S_1 + S_2}. \tag{C–5}$$

In the interpretation of Eqs. C–4 and C–5, the first thing to note is that S_1 and S_2 play the same roles that Z_2 and Z_1, respectively, play in the constant-cross-section case. If $S_2 < S_1$, then $R > 0$: the second duct acts as a high impedance medium. In the limit as $S_2 \to 0$, of course, the junction becomes a closed end and pressure doubling ($R = 1$) occurs. Conversely, if $S_2 > S_1$, the reflection coefficient is negative, and the junction has the properties of a soft termination. As $S_2 \to \infty$, the condition of the open pipe is approached. This analysis confirms our earlier assertion that an open end behaves as a pressure release boundary.

It is worthwhile generalizing this problem one step further. It may easily be shown that if the media on either side of the junction are different, the following result replaces Eq. C–3:

$$1 - R = \frac{S_2 Z_1}{S_1 Z_2} T = \frac{Z_{ac1}}{Z_{ac2}} T, \tag{C–6}$$

where Z_{ac1} and Z_{ac2} are the two *acoustic impedances*. The most general expressions for the R and T coefficients for normally incident waves are therefore

$$R = \frac{Z_{ac2} - Z_{ac1}}{Z_{ac2} + Z_{ac1}} \tag{C–7}$$

and

$$T = \frac{2Z_{ac2}}{Z_{ac2} + Z_{ac1}}. \tag{C–8}$$

With proper interpretation, these formulas turn out to be applicable even in the case of oblique incidence at an interface between two media. See Chap. 5.

A caution must now be added. We have assumed that the wave motion on both sides of the junction is planar. This means that the wave must, without the benefit of transverse flow, fill the corners of the junction (Fig. 3.4) instantaneously, something it cannot do. At low frequencies, however, the time required for the corner to communicate its presence to the fluid in the center of the tube is small compared with the period of the wave, that is, $kr \ll 1$, where r is the radius[3] of the larger duct. In this case violation of the plane-wave assumption is negligible. As the frequency becomes high enough that the communication time is comparable to the period, the plane-wave assumption becomes difficult to justify. Although we shall, for simplicity, continue to use the formulas derived here for all waves, the inherent limitation to lower frequencies should be kept in mind.

D. EXAMPLES

Up to this point the media on either side of the interface ($x = 0$) have been characterized as semi-infinite. In many cases at least one medium is of limited extent. Suppose, for example, that medium 2, instead of being semi-infinite, is

[3]If the duct is not circular, approximate r by half the hydraulic diameter HD, where $HD = 4S/L_{perim}$ and L_{perim} is the perimeter of the duct cross section.

terminated by a rigid plate at $x = L$, as shown in Fig. 3.5. If the incident wave in medium 1 is a short enough pulse, calculation of the reflected and transmitted pulses at the interface proceeds as though medium 2 were semi-infinite because the transmitted wave, when it is generated at the interface, does not yet know that medium 2 is finite. In other words, the impedance seen by the transmitted wave is still $Z_2 = \rho_2 c_2$. After it is launched from the interface, the transmitted wave plays the role of incident wave as it approaches the plate. Calculation of reflection at the plate then takes place as though medium 2 were semi-infinite in the $-x$ direction, again because the reflected wave, when it is launched from the plate, does not yet know about the interface at $x = 0$. The reflected wave now plays the role of incident wave as it approaches the interface. Reflection and transmission at the interface are again calculated as though the media on both sides were semi-infinite. And so on. Each time a pulse is incident on an interface, therefore, the appropriate impedances to use in calculating the reflected and transmitted pulses are the characteristic impedances on either side of the interface. In the particular example described here, the field in medium 2 is an endless sequence of pulses traveling back and forth. The reflection and transmission at each end are calculated separately for each pulse.

This approach has to be modified in the next chapter, where the waves considered are steady state and time harmonic. A steady-state wave completely fills a finite medium, and the field at one end of the medium is influenced by the field at the other end. In the present chapter, however, because the waves considered are transients, they interact only locally with an interface.

1. Rectangular Pulse in an Air-Filled Tube of Finite Length

At one end of an air-filled tube of length L, a rigid piston accelerates to a velocity u_0 instantaneously, continues at velocity u_0 for a moment, and then decelerates instantaneously to rest. The piston motion generates a rectangular pulse in the air, as indicated in Fig. 3.6. Two different terminations at $x = L$ are considered. First, let the termination be a rigid plate, as in Fig. 3.6*a*. The pulse bounces back and forth between the plate ($R = 1$) and the (now at rest) rigid piston ($R = 1$). The first reflection from the plate is just like that for the pulse shown in Fig. 3.2. In this case the amplitude of the incident pressure pulse is $A = \rho_0 c_0 u_0$. For subsequent reflections, from the piston as well as from the plate, the characteristics for the pulse simply crisscross back and forth in the

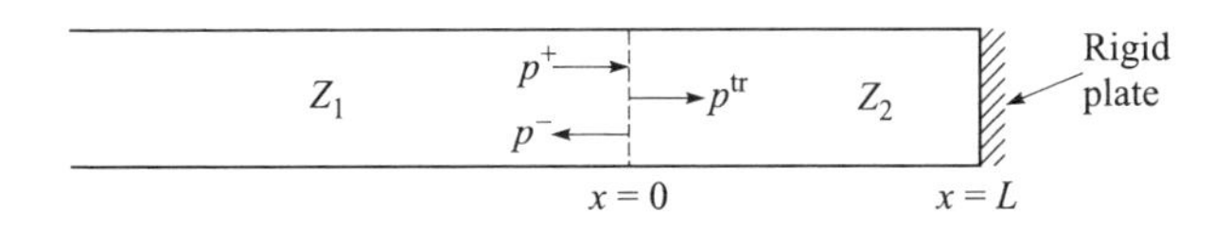

Figure 3.5 Reflection and transmission when medium 2 is finite.

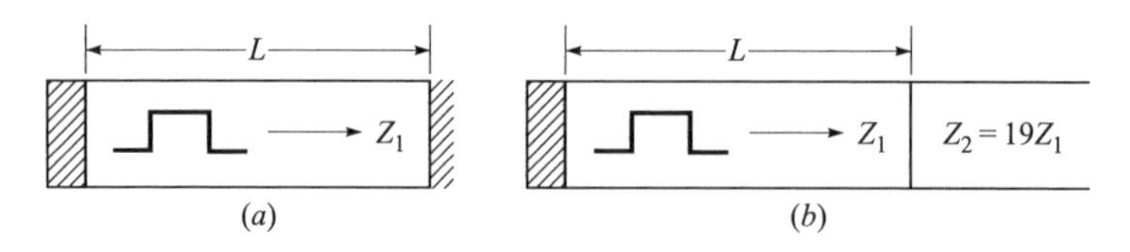

Figure 3.6 Generation of a rectangular pulse in a tube. (*a*) Rigid termination at $x = L$. (*b*) Fluid of impedance $Z_2 = 19Z_1$ at $x = L$.

x, t diagram between positions $x = 0$ and $x = L$. Because $R = 1$ at each reflection, the amplitude of the pulse never changes.

Next let the termination at $x = L$ be a fluid of impedance 19 times that of air, as shown in Fig. 3.6*b*. For the interface

$$R = \frac{19 - 1}{19 + 1} = 0.9$$

and

$$T = \frac{2(19)}{19 + 1} = 1.9.$$

At the piston the reflection coefficient is still unity. The characteristics diagram for this problem is shown in Fig. 3.7. Why do the transmitted wave characteristics have a slope different from that of the incident wave characteristics?

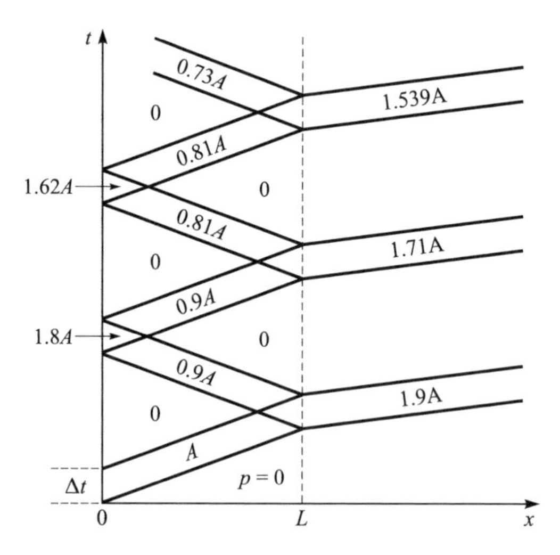

Figure 3.7 Characteristics diagram for problem shown in Fig. 3.6(*b*).

2. Shock Tube

In a shock tube (Fig. 3.8) the gas between the rigid end ($x = 0$) and the diaphragm ($x = L$) is pumped up to a total pressure $P = p_0 + A$. The gas beyond the diaphragm is at a lower total pressure, which in the example here is $P = p_0$. At time $t = 0$ the diaphragm is burst, and the problem is to find the pressure distribution at all subsequent times. The initial and boundary conditions are as follows:

$$\text{IC}: \quad p(x, 0) = A[H(x) - H(x - L)], \tag{D–1a}$$

$$u(x, 0) = 0, \tag{D–1b}$$

$$\text{BC}: \quad u(0, t) = 0. \tag{D–1c}$$

We solve this problem first analytically, then graphically. It is assumed that $A \ll \rho_0 c_0^2$ in order that the sound produced qualify as a small signal.

a. Analytical Approach

Start with the general solution of the wave equation for the pressure p,

$$p = f(x - c_0 t) + g(x + c_0 t).$$

The corresponding expression for the particle velocity is

$$u = \frac{f(x - c_0 t) - g(x + c_0 t)}{\rho_0 c_0}$$

(the characteristic impedance Z_0 is the same throughout the tube; in the small-signal approximation the fact that the static density is a little higher in the compressed region is insignificant). The requirement that u vanish at $t = 0$ (Eq. D–1b) is met by taking

$$g(x) = f(x).$$

Consequently, the expressions for p and u become

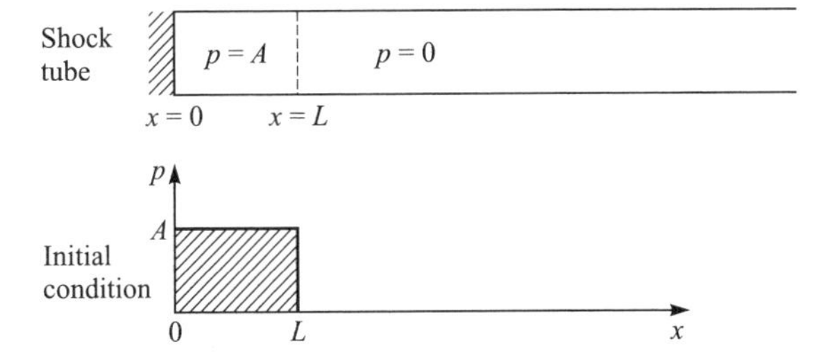

Figure 3.8 Initial pressure distribution in a shock tube.

$$p = f(x - c_0 t) + f(x + c_0 t) \tag{D–2a}$$

and

$$u = \frac{f(x - c_0 t) - f(x + c_0 t)}{\rho_0 c_0}. \tag{D–2b}$$

Next, an expression for f is found by applying Eq. D–1a to Eq. D–2a,

$$f(x) = \frac{A}{2}[H(x) - H(x - L)], \quad x \geq 0. \tag{D–3}$$

But this is not a sufficient definition of f to satisfy the last condition, Eq. D–1c. As Eq. D–2b shows, that condition can be satisfied only if

$$f(-c_0 t) = f(c_0 t).$$

In other words f must be an even function of its argument. Equation D–3, extended to make f an even function, becomes

$$f(x) = \frac{A}{2}\,\text{rect}\left(\frac{x}{2L}\right), \tag{D–4}$$

an expression valid for all values of x, negative as well as positive. In effect we have obtained the function f for the equivalent problem of an infinite tube in which the particle velocity must always be zero at $x = 0$. The final solution is found by combining Eqs. D–2a and D–4,

$$p = \frac{A}{2}\left[\text{rect}\left(\frac{x - c_0 t}{2L}\right) + \text{rect}\left(\frac{x + c_0 t}{2L}\right)\right]. \tag{D–5}$$

Waveforms and the characteristics diagram are shown in Fig. 3.9. The observable waveforms are the ones that are shaded.

b. Graphical Approach

Here it is informative to start with a slightly more general problem. The region of initial overpressure A is displaced a little from the wall, as shown in Fig. 3.10. When the constraints are removed at time $t = 0$, the early behavior is like that of a free pulse (Sec. 1B.2). The initial pressure disturbance splits into two waves of amplitude $A/2$, one traveling to the right and one to the left. However, the left-traveling pulse soon bumps into the wall and reflects. To account for the reflection, we may introduce an image wave, for example, by starting with an image distribution of amplitude A in the $-x$ region, as shown in the bottom sketch of Fig. 3.11. After being released, the image distribution splits into two half-amplitude waves, just as the real distribution in the $+x$ region does. The right-traveling image wave then combines with the left-traveling

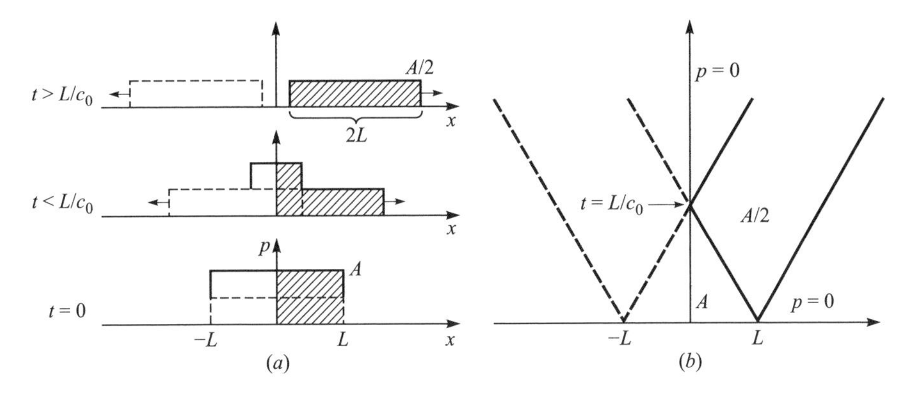

Figure 3.9 Solution of the shock tube problem. (a) Waveforms. (b) Characteristics diagram.

real wave to satisfy the boundary condition $u = 0$ at $x = 0$. The sequence of events is shown in Fig. 3.11. Of course the physical part of the problem is only that shown by the shaded waves. After the reflection at $x = 0$ has been completed, the wave motion consists of two half-amplitude, right-traveling pulses which are separated by a distance $2x_0$. To reduce this problem to the original shock tube problem, take $x_0 = 0$. The two pulses then join head to tail, and the spatial length of the resulting single pulse is $2L$. The shock tube problem may therefore be thought of as the right-hand half of an ordinary free-wave problem for an infinite tube in which the initial pressure distribution A is of extent $2L$. We thus confirm the distribution shown in the bottom sketch of Fig. 3.9*a*.

Because the graphical method is so simple, one may wonder why we bother with the analytical method at all. One reason is that the analytical approach may easily be extended to solve a more difficult problem, that of the bursting balloon (Ref. 2), which is considered next.

3. Bursting Balloon

Here we consider a transient problem involving spherical waves. The pressure inside a perfectly spherical balloon of radius r_0 is $p = A$ (Fig. 3.12*a*). The balloon bursts at time $t = 0$. The initial pressure distribution $p(r)$ is shown in

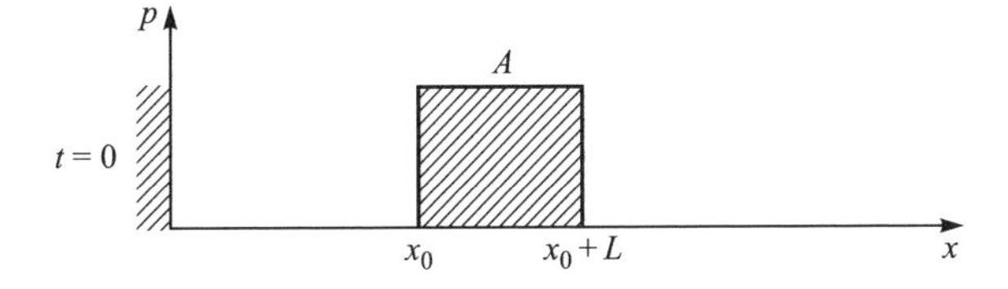

Figure 3.10 Modified shock tube problem: initial overpressure displaced from the wall.

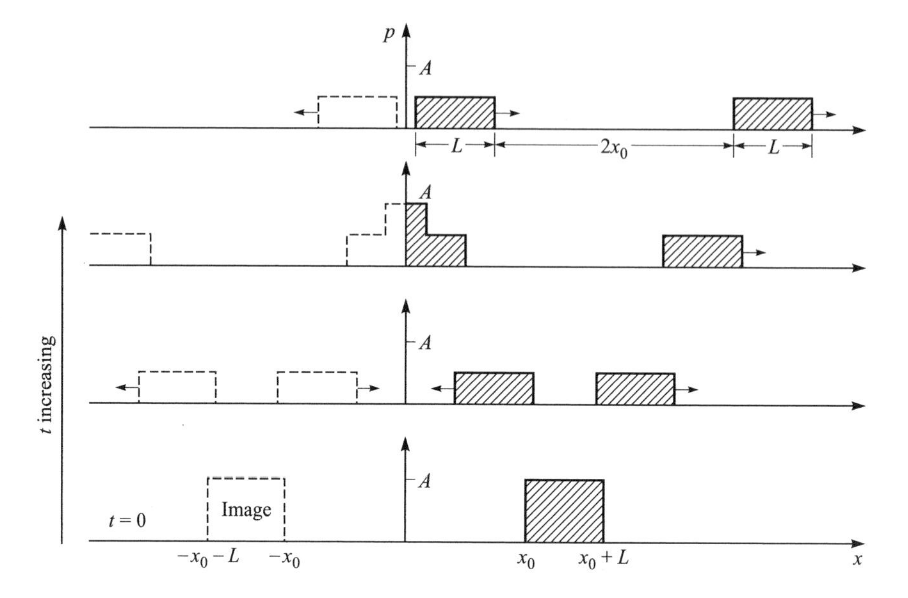

Figure 3.11 Solution of the modified shock tube problem by the image method.

Fig. 3.12*b* and the product $rp(r)$ in Fig. 3.12*c*. The gas in the balloon is at rest when the burst occurs. The initial conditions are therefore

$$p(r, 0) = A[H(r) - H(r - r_0)], \tag{D–6a}$$

$$u(r, 0) = 0. \tag{D–6b}$$

Although these conditions are similar to those of the shock tube, the balloon has no solid wall at $r = 0$. Even so, a condition at the center of the balloon must be satisfied: the *volume* velocity $q = Su$ must vanish there,[4]

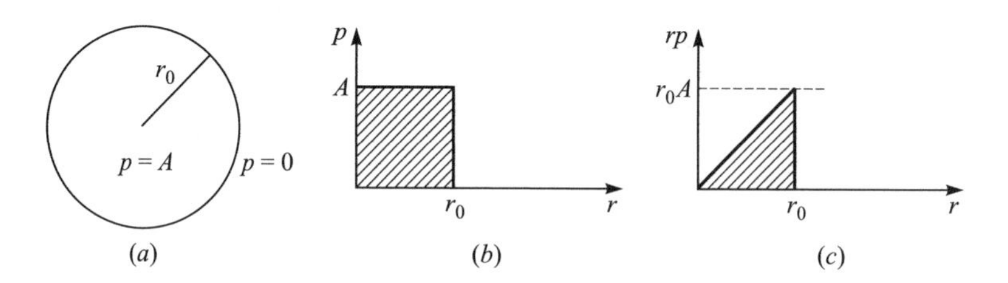

Figure 3.12 Initial conditions in the bursting balloon problem.

[4]Because of the $1/r$ geometrical factor for a spherical wave, we cannot a priori set a condition on u itself at the origin.

$$\lim_{r\to 0} 4\pi r^2 u = 0. \tag{D–6c}$$

The analysis is much the same as that for the shock tube problem. The general solution in terms of the velocity potential ϕ is Eq. 1D–13. Our task is to find the functions f and g that satisfy Eqs. D–6. The expressions for the pressure and particle velocity are

$$p = -\rho_0\phi_t = \rho_0 c_0 \frac{f'(r - c_0 t) - g'(r + c_0 t)}{r} \tag{D–7a}$$

and

$$u = \phi_r = \frac{f'(r - c_0 t) + g'(r + c_0 t)}{r} - \frac{f(r - c_0 t) + g(r + c_0 t)}{r^2}, \tag{D–7b}$$

respectively. First, Eq. D–6b is satisfied if $g(r)$ is taken to be equal to $-f(r)$.

$$\therefore\ p = \rho_0 c_0 \frac{f'(r - c_0 t) + f'(r + c_0 t)}{r}. \tag{D–8}$$

By evaluating this expression at $t = 0$ and applying Eq. D–6a, we obtain a preliminary expression for f',

$$f'(r) = \frac{rA[H(r) - H(r - r_0)]}{2\rho_0 c_0}. \tag{D–9}$$

This is not, however, a sufficient definition of f' to satisfy Eq. D–6c. Use Eq. D–7b (with $g = -f$) to compute the volume velocity $q = 4\pi r^2 u$,

$$q = 4\pi r[f'(r - c_0 t) - f'(r + c_0 t)] - 4\pi[f(r - c_0 t) - f(r + c_0 t)].$$

Satisfaction of Eq. D–6c requires that

$$f(-c_0 t) = f(c_0 t),$$

and consequently that

$$f'(-c_0 t) = -f'(c_0 t).$$

Since f' must be an *odd* function of its argument, we modify Eq. D–9 so that f' is described in the (nonphysical) $-r$ domain as well as in the $+r$ domain,

$$f'(r) = \frac{rA}{2\rho_0 c_0}\,\mathrm{rect}\left(\frac{r}{2r_0}\right). \tag{D–10}$$

This function is sketched in Fig. 3.13. Substitution of Eq. D–10 in Eq. D–8 yields the final solution,

$$p = \frac{A}{2r}\left[(r - c_0t)\mathrm{rect}\left(\frac{r - c_0t}{2r_0}\right) + (r + c_0t)\mathrm{rect}\left(\frac{r + c_0t}{2r_0}\right)\right]. \quad \text{(D–11)}$$

Now analyze the solution. Just as in the shock tube problem, the initial pressure distribution splits into an outgoing wave (the first term in Eq. D–11) and an incoming wave (the second term in Eq. D–11). But in this case the reflection of the incoming wave at the origin is more complicated. If we restrict our observations to points outside the balloon, i.e., $r > r_0$, only the outgoing wave term counts. The head of the wave is at $r = r_0 + c_0t$. As Eq. D–11 shows, the amplitude is $p_{\max} = Ar_0/2r$. Sample time and spatial waveforms are shown in Fig. 3.14. (Why is the tail amplitude greater than the head amplitude in the spatial waveform?) These waveforms are markedly different from those for the shock tube. The balloon burst produces an N wave (see Example 1.3) rather than a rectangular pulse. Particularly striking is the fact that the tail section of the wave is negative. The tail may be thought of as being produced when the incoming wave, which originates on the far side of the balloon, passes through the origin. Since the incoming wave is a positive-pressure signal, we see that passage through the origin produces a 180° phase shift. This is a general result: a converging wave experiences phase inversion when it passes through a three-dimensional focus.

The ideal balloon burst, which produces a perfect N wave, is very difficult to achieve in practice. Most real balloons (1) are not spherical and (2) tear when they explode, rather than disintegrate uniformly. The actual waveform of the pressure signal produced is very complicated, only vaguely N shaped.[5] Even so, the ideal balloon burst is useful as a model for other processes, such as explosions, electrical discharges, and exploding wires, which do produce N waves or nearly N-shaped waves.

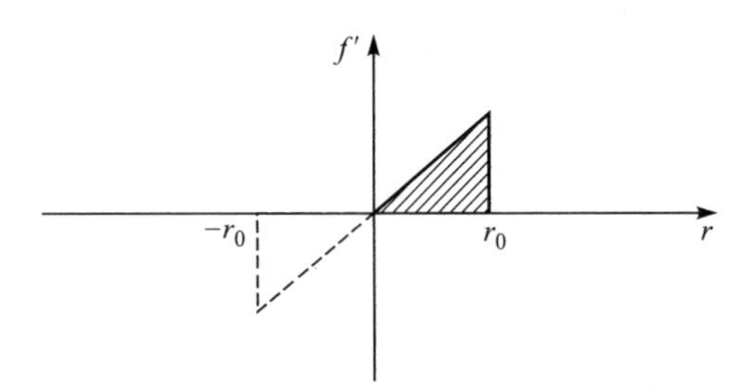

Figure 3.13 Function f' for the balloon burst problem.

[5]For an experiment on bursting balloons see Ref. 1.

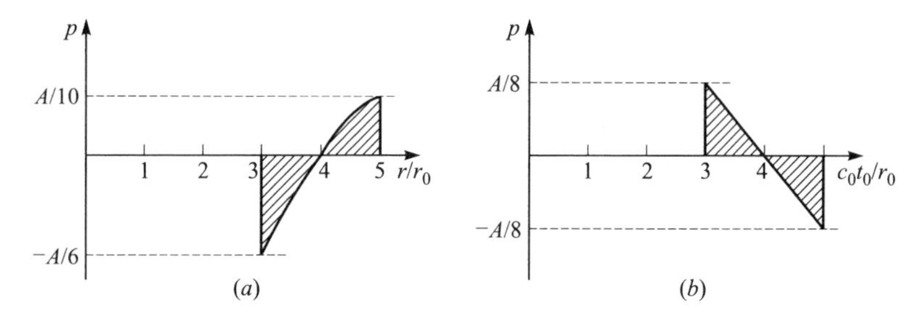

Figure 3.14 N wave produced by a balloon burst. (*a*) Spatial waveform at $t = 4r_0/c_0$. (*b*) Time waveform at $r = 4r_0$.

REFERENCES

1. D. T. Deihl and F. R. Carlson, Jr., "N waves from bursting balloons," *Am. J. Phys.* **36**, 441–444 (1968).
2. H. Lamb, *The Dynamical Theory of Sound*, 2nd ed. (Dover Publications, New York, 1960), Art. 71, pp. 208–213.

PROBLEMS

3–1. One of the concerns sometimes voiced about over-water flights by supersonic passenger aircraft is whether the sonic booms produced hurt the fish. Attempt to analyze the question by the following simple approach. Assume for simplicity that the sonic boom is normally incident on a perfectly flat ocean surface. A typical sonic boom amplitude is $A = 2$ lb/ft^2 (NOTE: lb/ft^2, not lb/in^2) when the boom (incident wave) reaches sea level. A typical half-duration T of the boom is $T = 100$ ms.

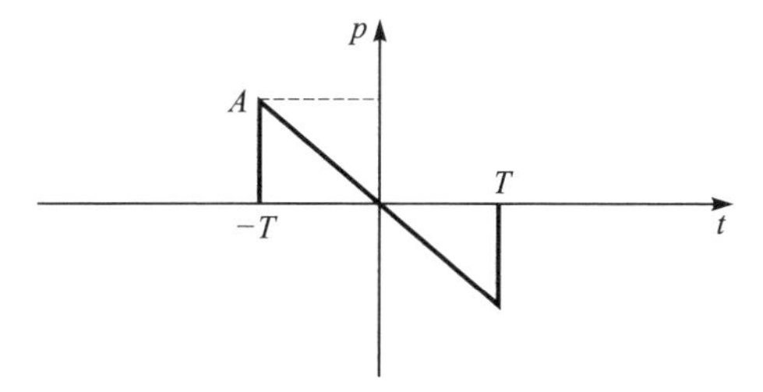

(a) Find the values of p_{rms}, I, SPL, and IL for the incident wave. (One answer for guidance: $p_{\mathrm{rms}} \doteq 55$ Pa.)

(b) Find the pressure transmission coefficient T. Then find A^{tr}, $p_{\mathrm{rms}}^{\mathrm{tr}}$, and $\mathrm{SPL}_{\mathrm{water}}$, i.e., the amplitude, rms value, and SPL of the pressure signal in the water.

(c) Find the sound power transmission coefficient τ, the intensity I^{tr} of the transmitted wave, and the transmission loss TL. Where has the "lost" energy gone?

(d) Do you think a sonic boom of this magnitude would do damage to fish? Why or why not?

3–2. The first meter of a semi-infinite pipe contains air at 100°C at atmospheric pressure. Thereafter the pipe is filled with steam (also at 100°C and at atmospheric pressure). A weightless compliant membrane separates the two gases so that they do not diffuse into each other. A piston at $x = 0$ generates a rectangular pulse of duration 1 ms and amplitude $A = 10$ Pa. After generating the pulse, the piston stays rigidly fixed. The pulse bounces back and forth between the gaseous interface and the piston.

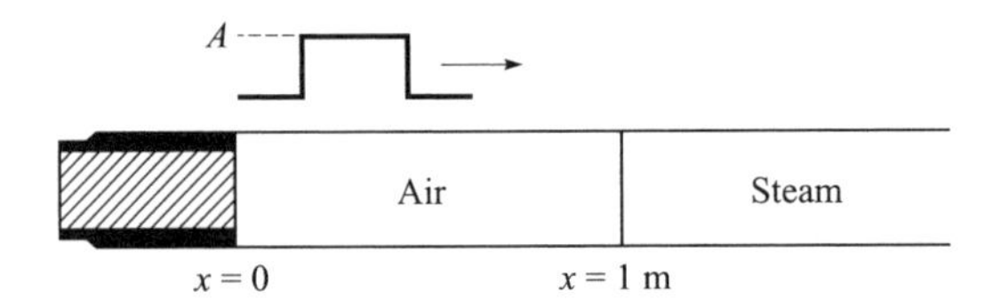

(a) Draw an x, t diagram showing the path of the pulse during the first 17 ms after it leaves the piston (let $t = 0$ be the time the leading edge of the pulse leaves the piston). Be sure to show characteristics for both leading and trailing edges of the pulse. Indicate the acoustic pressure p in all regions of the x, t diagram, including the region $x > 1$ m.

(b) Sketch both the pressure and the particle velocity time waveforms (out to $t = 17$ ms) at a point 0.75 m from the piston, that is, three-fourths of the way to the gaseous interface. Indicate all amplitudes.

3–3. A plane wave propagates vertically upward in a swimming pool. The pressure waveform is a postive rectangular pressure pulse of amplitude 3 Pa and duration t_0. The depth of the pool is $4c_0t_0$. After bouncing off the air-water interface, the pulse travels back through the water and bounces off the concrete bottom. The pulse continues to bounce back and forth between top and bottom of the pool. Sketch the x, t diagram for this problem, showing the path of the pulse and indicating the pressure in each section of the path. When (in terms of number of bounces) does the absolute value of the pressure pulse first drop below 1 Pa, not counting regions where incident and reflected pulses overlap? When the drop below 1 Pa occurs, is the pulse positive or negative? *Hint*: What is the value of the reflection coefficient at the water-concrete interface?

3–4. A rigid tube is closed at both ends. The fluid in the left half of the tube has sound speed c_1 and density ρ_1. In the right half the values are $c_2 = c_1$ and $\rho_2 = 3\rho_1$. A rectangular pulse of amplitude 12 Pa is started down the tube from the left end. Pulses then travel back and forth through the tube as shown in the characteristics diagram. Complete the diagram by showing the values of the various pressures indicated, that is, p_{11}, p_{12}, p_{m1}, and so on. Be sure to show your calculation of the various reflection and transmission coefficients.

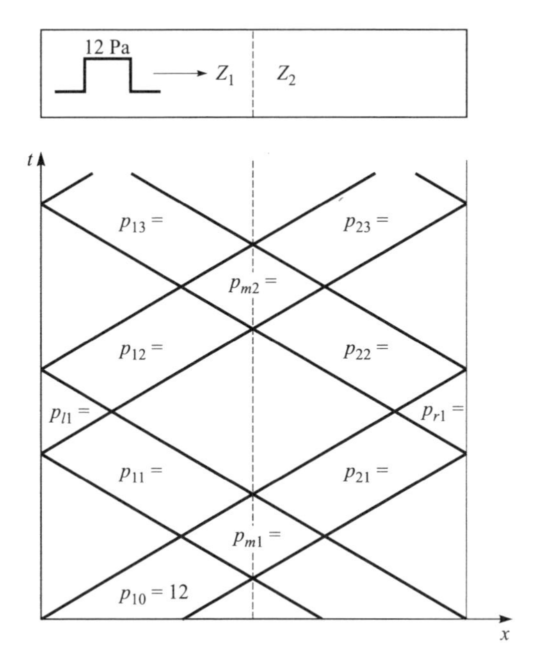

3–5. A sonic boom of amplitude A and half-duration $T = 29.15$ ms (see Problem 3–1 for the waveform) is normally incident on the ground. First show that the spatial extent of the waveform is 20 m. A microphone mounted at height h above the ground is to be used to measure the time waveform of the sonic boom. With the help of an x, t diagram sketch the time waveform recorded by the microphone for each of the following cases:

(a) $h = 0$ m (microphone mounted flush with ground)

(b) $h = 5$ m

(c) $h = 10$ m

From your sketches, what do you conclude is the best height at which to mount a microphone when you wish to record a signal coming from above? Defend your choice.

3–6. A tube is closed at one end ($x = 0$), open at the other ($x = L$). Membranes at $x = L/2 \pm 1$ separate the center section of the tube, where the acoustic pressure is $p = A$, from the other two sections, where the pressure is $p = 0$. At time $t = 0$ the membranes are ruptured.

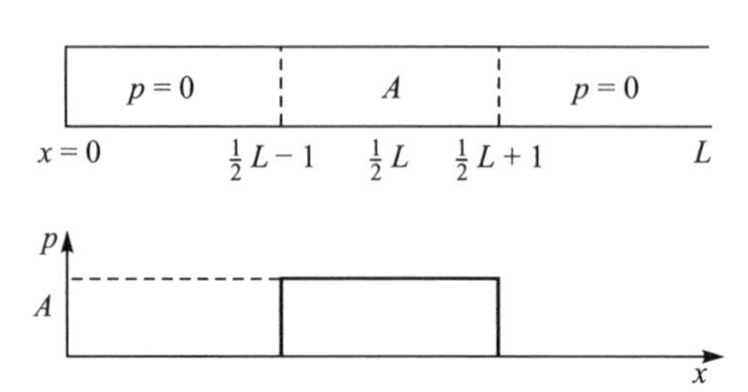

(a) Sketch the characteristics diagram for this problem, showing several bounces off the ends. Indicate the pressure in all parts of the x, t diagram.

(b) At what time (measured in terms of L and c_0), if any, does the initial pressure distribution reappear?

3–7. At a temperature of 82°C the characteristic impedance of air is 375 rayls and that of hydrogen is 100 rayls; use these values in the following problem. A plane wave of rectangular waveform, amplitude $A = 1$ Pa and time duration $t_0 = 0.5$ ms, in air propagates past a microphone M toward a layer of hydrogen 1.03 m away. The hydrogen layer is 1.45 m thick and is terminated by a rigid wall.

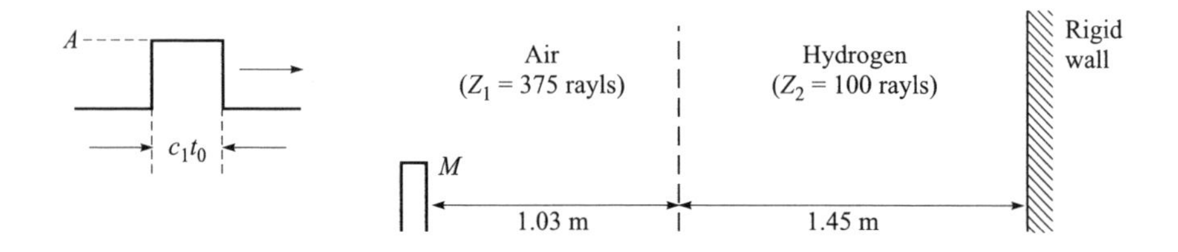

(a) Find the wave shape, amplitude, and duration of the incident and first two subsequent signals received by the microphone.

(b) Show your results in the form of a characteristics diagram and a sketch of the time waveform recorded by the microphone.

3–8. Design a tube (circular cross section) to transmit plane waves from air into chlorine gas at STP without any transmission loss ($R = 0$). Such a design might be required to radiate sound into chlorine. The air would serve as a buffer to protect the acoustic source from the chlorine, which is very corrosive. (A plastic membrane, light enough to have perfect acoustic transmission, could be used to separate the air from the chlorine.)

3–9. A pipe of diameter D_W contains water at 20°C. It is connected to a pipe of diameter D_I that contains ice. Given $D_W = 50$ mm, find the value of D_I that permits perfect transmission, i.e., no reflection, of sound,

(a) From water to ice.

(b) From ice to water.

3–10. A 25-mm-diameter pipe is connected to a second pipe of unknown diameter. An N wave is sent down the first pipe toward the juncture. A microphone in the first pipe records the time record of the incident and reflected signals, as shown in the sketch. Find the diameter of the second pipe.

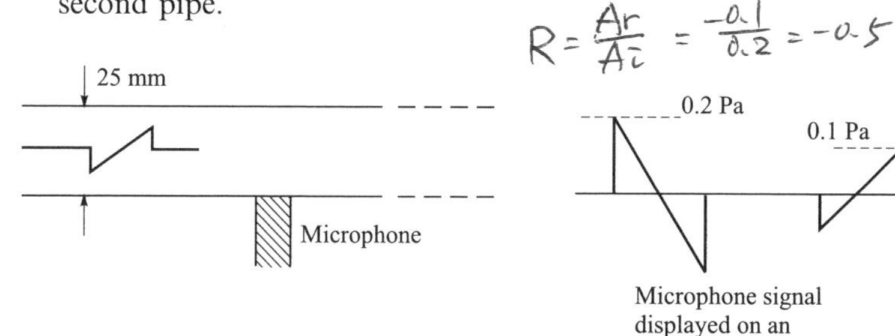

3–11. Derive power reflection and transmission coefficients r and τ for a case in which the cross-sectional areas as well as the impedances are not the same, i.e., $S_2 \neq S_1$ and $Z_2 \neq Z_1$. Check your result by determining whether $r + \tau = 1$.

3–12. A bursting balloon is to be used to simulate a full-scale sonic boom. What balloon diameter $2r_0$ and initial overpressure A are required to produce an N wave which at 200 m (from the center of the balloon) has the properties of the sonic boom described in Problem 3–1?

3–13. A spherical N wave produced by an electric spark has a duration $2T = 20$ μs and peak pressure 0.01 atm at a distance $r = 0.1$ m from the spark. If the same N wave were to be generated by an ideal balloon burst, what would the balloon radius r_0 and overpressure a have to be? Would the small-signal approximation be justified to describe the sound produced by the balloon burst?

4

NORMAL INCIDENCE CONTINUED: STEADY-STATE ANALYSIS

The time domain methods used in the previous chapter are particularly useful for problems involving transients or pulses. In many cases, however, a frequency domain approach is preferred. For example, the incident sound wave may be a pure sinusoid. Or one of the acoustical impedances in the problem may be complex, e.g., an acoustical medium may be lossy or a terminating impedance may have a reactive component, in which case the coefficients R and T are generally functions of frequency. The analysis is then simplest for sinusoidal signals. More complicated signals may be handled by breaking them down into their Fourier components (by means of either Fourier series or Fourier transforms), treating each component separately, and then, if desired, recombining all components to obtain the time waveform of the reflected and transmitted waves.

Section A of this chapter provides a general description of time-harmonic sound fields in front of a reflecting surface of arbitrary impedance. Problems involving a single reflecting surface, or terminating impedance, are treated in Sec. B. Phasor diagrams, standing wave patterns, and standing wave ratios are obtained for a variety of terminating impedances. Section C is devoted to the lumped-element approximation and its use, for example, to analyze the Helmholtz resonator. Additional examples are described in Sec. D. Three-medium problems, for instance, transmission loss through a wall, are the subject of Sec. E. Another treatment of wall transmission loss, by the lumped-element method, is given in Sec. F.

A. INTRODUCTION

In this chapter the field variables are assumed to vary time harmonically, as follows:

$$p(x, t) = P(x)e^{j\omega t}, \tag{A–1}$$

$$u(x, t) = U(x)e^{j\omega t}. \tag{A–2}$$

The amplitudes P and U, often called phasors, are generally complex. It should be kept in mind, however, that the actual expressions for the pressure and particle velocity are the real parts (or the imaginary parts) of the right-hand sides of Eqs. A–1 and A–2.[1] (What are the rms values of the two signals?)

One of the benefits of assuming time-harmonic waves is that we may now consider dissipative media, for which the characteristic impedance is generally complex (see Chap. 9). Thus when the reflection and transmission problem sketched in Fig. 3.1 is reanalyzed, Z_1 and/or Z_2 are allowed to be complex. Given Eq. A–1, the defining expressions for R and T (Eqs. 3A–2 and 3A–3) reduce to

$$R = \frac{P^-}{P^+}, \tag{A–3}$$

$$T = \frac{P^{tr}}{P^+}. \tag{A–4}$$

As in Chap. 3, superscripts +, –, and tr denote values of incident, reflected, and transmitted wave quantities, respectively, at the interface. Continuity of pressure and particle velocity at the interface again leads to

$$R = \frac{Z_2 - Z_1}{Z_2 + Z_1}, \tag{A–5}$$

$$T = \frac{2Z_2}{Z_2 + Z_1}. \tag{A–6}$$

Although these results are formally the same as the ones found in Chap. 3, Eqs. 3A–8 and 3A–9, R and T are now generally complex.

To obtain analytical expressions for P and U, we begin by substituting Eq. A–1 in the wave equation for the pressure, Eq. 1C–59. The result is the Helmholtz equation

[1] Suppose the source excitation were given as $u(0, t) = u_0 \sin \omega t$. Since this is the imaginary part of $u_0 e^{j\omega t}$, at the end of the problem one would select Im $u(x, t)$ as the component of the solution corresponding to the original source condition. However, throughout the working of the problem, complex exponentials would normally be used rather than trigonometric functions.

$$P_{xx} + k^2 P = 0, \tag{A–7}$$

($k = \omega / c_0$ is the wave number), which has the solution

$$P = Ae^{-jkx} + Be^{jkx}. \tag{A–8}$$

The constants A and B are to be determined from the boundary conditions. Note that the first term represents a wave traveling in the $+x$ direction (how can one tell?) and the second term a wave in the $-x$ direction. The expression for U is

$$U = \frac{A}{Z_0} e^{-jkx} - \frac{B}{Z_0} e^{jkx}.$$

(Why the minus sign in front of the second term?)

A sample of the type of problems we shall be considering is given in Fig. 4.1, which shows a tube with a source at one end $x = 0$ and terminated at $x = \ell$ by an impedance $Z = Z_n$. This configuration is a special case of the general one shown in Fig. 3.1. In this case Z_1 is the characteristic impedance of the fluid in the tube, here denoted Z_0, while Z_2 is the termination impedance Z_n. The termination could be considered a second medium, but for the moment we are concerned only with the field between the source and the termination.

Although the source might seem the obvious starting point, it is generally simpler to determine the behavior of P and U first at the termination and then work backward toward the source. To this end, put the expressions for P and U in terms of the distance $d = \ell - x$ from the termination. Equation A-8 becomes

$$P = \underbrace{Ae^{-jk\ell}}_{P^+} e^{jkd} + \underbrace{Be^{jk\ell}}_{P^-} e^{-jkd},$$

where the incident and reflected wave amplitudes, P^+ and P^-, at the termination are identified in the equation. Since $P^- = RP^+$, the expression for P may be written

$$P = P^+(e^{jkd} + Re^{-jkd}). \tag{A–9}$$

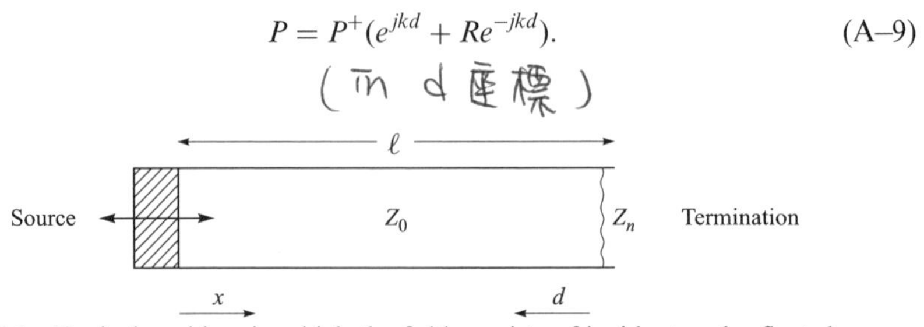

Figure 4.1 Typical problem in which the field consists of incident and reflected waves.

The two terms enclosed by the parentheses are called *phasors* (complex numbers usually in polar form). Note that the incident wave phasor is now associated with the positive exponent ($+jkd$) and the reflected wave phasor with the negative exponent ($-jkd$). The expression for U is

$$\begin{aligned} U &= U^{+}e^{jkd} + U^{-}e^{-jkd} \\ &= (P^{+}/Z_0)e^{jkd} - (P^{-}/Z_0)e^{-jkd} \\ &= (P^{+}/Z_0)\left(e^{jkd} - Re^{-jkd}\right). \end{aligned} \tag{A–10}$$

The impedance at any point in the tube is

$$Z(d) = \frac{P(d)}{U(d)} = Z_0 \frac{e^{jkd} + Re^{-jkd}}{e^{jkd} - Re^{-jkd}}. \tag{A–11}$$

In particular, when $d = 0$, the impedance must equal the terminating impedance,

$$Z_n = Z_0 \frac{1+R}{1-R}. \tag{A–12}$$

Solving for R, we find

$$R = \frac{Z_n - Z_0}{Z_n + Z_0}, \tag{A–13}$$

which is simply another form of Eq. A–5. This should not be a surprising result. The reflection of the incident wave is determined solely by the impedance seen at $d = 0$, regardless of whether it is a lumped impedance representing the termination or the characteristic impedance of a new medium.

Except when $R = 0$ ($Z_n = Z_0$), a standing wave field exists in the tube. One measure of the pattern of the field is the *pressure standing wave ratio*,

$$\mathrm{SWR} \equiv \frac{|P_{\max}|}{|P_{\min}|}, \tag{A–14}$$

where $|P_{\max}|$ is the maximum value of $|P|$ in the tube and $|P_{\min}|$ is the minimum value. For example, if a complete null occurs somewhere in the tube, i.e., $|P_{\min}| = 0$, the standing wave ratio is infinite. It may be seen from Eq. A–9 that the maximum value of $|P|$ occurs when the incident and reflected wave phasors, e^{jkd} and Re^{-jkd}, are perfectly in phase. Conversely, the minimum occurs when the two phasors are 180° out of phase. The formula for SWR may therefore be expressed as

$$\text{SWR} = \frac{1 + |R|}{1 - |R|}. \tag{A–15}$$

Since the standing wave ratio is easily measured, it is frequently used to determine the absolute value of the reflection coefficient. Solution of Eq. A–15 for $|R|$ gives

$$|R| = \frac{\text{SWR} - 1}{\text{SWR} + 1}. \tag{A–16}$$

Determination of the phase of the reflection coefficient requires additional information about the standing wave field; see Sec. B.4 below.

B. SINGLE IMPEDANCE TERMINATION

Although Eqs. A–9 and A–10 constitute the full mathematical representation of the standing wave field, a visual representation is needed for easy interpretation and analysis. The method of counterrotating phasors, long used in the study of electrical transmission lines (Ref. 6), is well suited for the purpose. To illustrate the method, we start with the simplest cases, $Z_n = 0$ and $Z_n = \infty$. Finite real impedances are then considered. The general case, Z_n complex, is taken up last.

1. Pressure Release Termination ($Z_n = 0$)

As noted in Chap. 3, at low frequencies the open end of a duct approximates a termination at which the pressure vanishes. Since $R = -1$ for this case, Eq. A–9 becomes

$$P/P^+ = e^{jkd} - e^{-jkd}. \tag{B–1}$$

Figure 4.2 contains a set of phasor diagrams that show how the two phasors on the right-hand side of Eq. B–1 may be summed graphically. The term representing the incident wave is a counterclockwise rotating phasor (it rotates counterclockwise as d increases); the reflected wave phasor rotates clockwise. Adding the two terms graphically in each case gives the resultant relative pressure P/P^+, the value of which is shown below each unit circle. Next consider the particle velocity. Equation A–10 gives

$$U/U^+ = e^{jkd} + e^{-jkd}, \tag{B–2}$$

which is represented by the sketches in Fig. 4.3. Comparison of Figs. 4.2 and 4.3 shows that when the pressure is a maximum, the particle velocity is zero, and vice versa. Moreover, the phase angle θ between the pressure and the

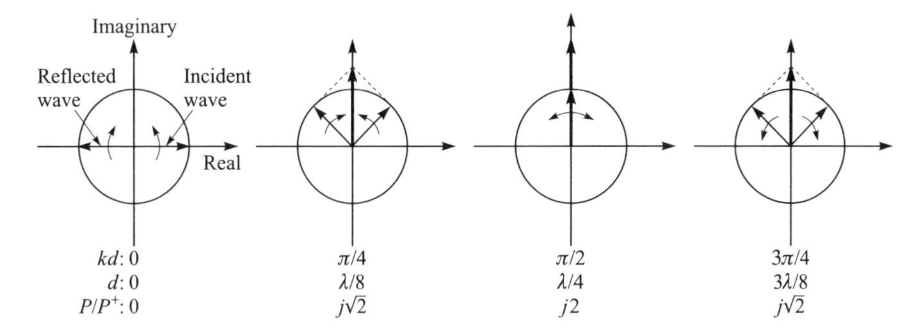

Figure 4.2 Counterrotating phasors representing the pressures of the incident and reflected waves for a pressure release termination.

particle velocity is 90° (factor of j). It turns out that θ is always 90° when Z_n is zero, infinity, or any pure imaginary number (see Problem 4B–13 for the proof when Z_n is imaginary). In these cases the intensity is zero (see Eq. 1E–18c). As much sound energy travels back toward the source as travels away from it. The net steady-state source output is therefore zero.

The information in Figs. 4.2 and 4.3 may be used to construct a standing wave pattern, which is a plot of $|P/P^+|$ or $|U/U^+|$ versus distance (or kd). The pressure and particle velocity patterns are shown in Fig. 4.4. The standing wave ratio is seen to be ∞ in this case.

Now make the connection between P^+, the incident wave pressure amplitude at the termination, and the source boundary condition. First, let the source condition be

$$p(0, t) = P_0 e^{j\omega t}. \tag{B–3}$$

By Eqs. A–1 and B–1 we have

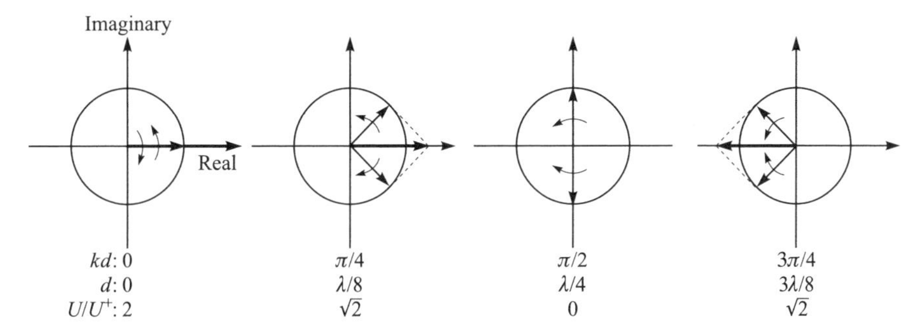

Figure 4.3 Particle velocity phasors for a pressure release termination.

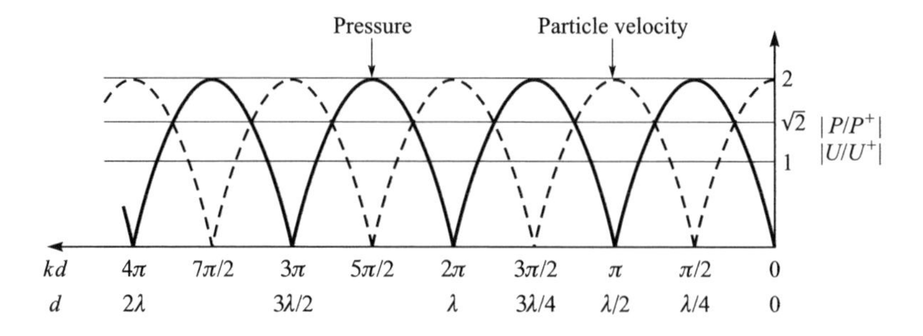

Figure 4.4 Pressure and particle velocity standing wave patterns for a pressure release termination.

$$p = P^{+}(e^{jkd} - e^{-jkd})e^{j\omega t} = 2jP^{+} \sin k(\ell - x)e^{j\omega t}.$$

Satisfaction of the source condition, Eq. B–3, requires that $P^{+} = P_0/2j \sin k\ell$. The final analytical solution is therefore

$$p = P_0 e^{j\omega t} \frac{\sin k(\ell - x)}{\sin k\ell}. \tag{B–4}$$

It is clear that the tube exhibits resonance[2] when $k\ell = n\pi$, that is, when the length is an integral number of half wavelengths. (What are the resonance frequencies?) Using the momentum equation, Eq. 1C–55, the reader should be able to show that the corresponding expression for the particle velocity is

$$u = \frac{P_0}{jZ_0} e^{j\omega t} \frac{\cos k(\ell - x)}{\sin k\ell}. \tag{B–5}$$

Next suppose the source is a vibrating piston. In this case the particle velocity, not the pressure, is prescribed:

$$u(0, t) = u_0 e^{j\omega t}. \tag{B–6}$$

What is the solution in this case? Are the resonance frequencies the same?

Finally, find the impedance at any point in the tube. Since $R = -1$, Eq. A–11 reduces to

$$Z(d) = jZ_0 \tan kd, \tag{B–7}$$

a sketch of which is given in Fig. 4.5. The impedance is purely reactive everywhere in the tube. This is an important result. It shows that despite the feverish

[2]Resonance here means that the pressure amplitude becomes unbounded. If small losses are present, the amplitude at resonance is very high but bounded.

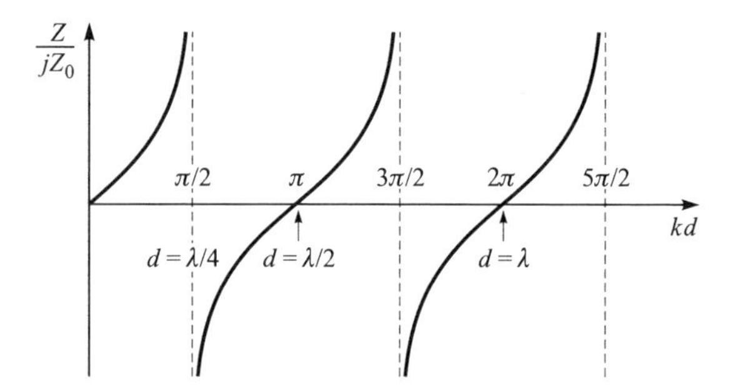

Figure 4.5 Impedance as a function of distance in front of a pressure release termination.

acoustical activity in the tube, the intensity is zero (see Eq. 1E–18a). In physical terms, the intensity associated with the backward traveling wave is equal but opposite that associated with the forward traveling wave. The net energy flow is therefore zero.

The impedance seen by the source is found by putting $d = \ell$. If the tube length is an integral number of half wavelengths, the impedance seen by the source is zero. But if $\ell = \lambda/4, 3\lambda/4, 5\lambda/4, \ldots$, the impedance seen by the source is infinite. The high impedance of a quarter-wavelength open tube is sometimes used for applications. Suppose, for example, one wants to study sound propagation in a duct in which the air is moving. An inlet port and an exhaust port may be used to maintain the air flow, as shown in Fig. 4.6. But how are the ports to admit a steady flow of air without also letting sound out and thus affecting the sound field in the duct? One way is to let the length of each port be a quarter wavelength of the sound being used in the test. When the sound wave from the source reaches the air inlet port, it has the choice of entering the port or proceeding down the duct. Since the impedance at the port entrance (i.e., looking into the port) is infinite, the sound wave prefers to take the much easier path down the main tube. For the same reason the outlet port is also an unattractive path to the sound wave. Consequently, very little sound escapes out the ports. This trick works, however, only for the frequency for which the port length is a quarter wavelength (or $3\lambda/4, 5\lambda/4, \ldots$).

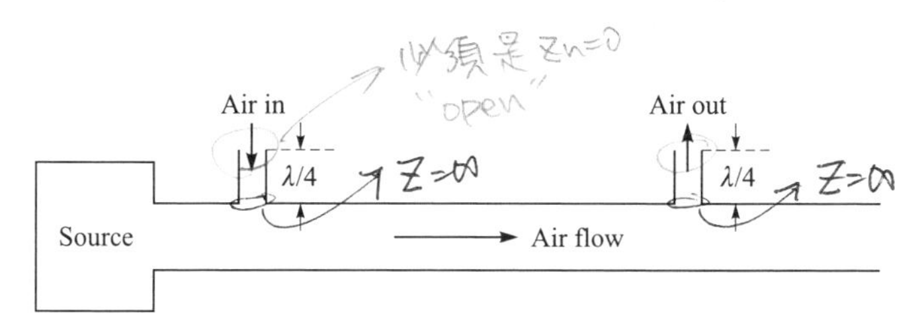

Figure 4.6 Use of a quarter-wavelength open tube to provide an infinite impedance.

2. Rigid Termination ($Z_n = \infty$)

If the end of the tube is closed, $R = 1$ and the expressions for P and U are

$$P = P^{+}(e^{jkd} + e^{-jkd}), \tag{B–8a}$$

$$U = U^{+}(e^{jkd} - e^{-jkd}). \tag{B–8b}$$

These are just the reverse of the relations that hold for the open pipe, Eqs. B–1 and B–2. The phasor diagrams for P are therefore the same as those for U for the open pipe, and vice versa. The quantity P/P^{+} is equal to 2 at $kd = 0$ while U/U^{+} is zero there. The standing wave pattern sketched in Fig. 4.4 is applicable, but now the pressure is represented by the dashed curve and the particle velocity by the solid curve. Again the standing wave ratio is ∞.

If the source is a vibrating piston, the boundary condition is given by Eq. B–6. For this case the expressions for u and p are

$$u = u_0 e^{j\omega t} \frac{\sin k(\ell - x)}{\sin k\ell}, \tag{B–9a}$$

$$p = -jZ_0 u_0 e^{j\omega t} \frac{\cos k(\ell - x)}{\sin k\ell}. \tag{B–9b}$$

Comparison of these solutions with Eqs. B–4 and B–5 shows that the closed pipe with a piston source has the same resonance frequencies as the open pipe with a pressure source.

The impedance in the closed tube (Eq. B–8a divided by Eq. B–8b) is

$$Z = -jZ_0 \cot kd. \tag{B–10}$$

See Fig. 4.7. It is seen that closed pipes of length 0, $\lambda/2$, $3\lambda/2$, …, present an infinite impedance to the source.

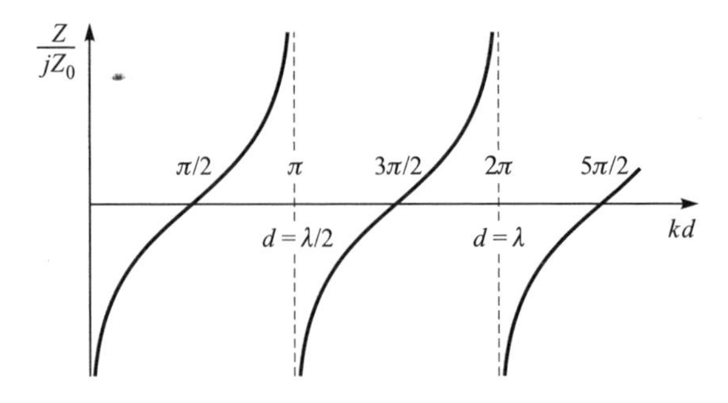

Figure 4.7 Impedance for the field in front of a rigid reflector.

3. General Resistive Termination

Now allow Z_n to be finite but real. The behavior is well explained by an example. Let $Z_n = 3Z_0$. The reflection coefficient is $R = (3-1)/(3+1) = \frac{1}{2}$, and the expressions for P and U are

$$P = P^+(e^{jkd} + \tfrac{1}{2}e^{-jkd}),$$
$$U = U^+(e^{jkd} - \tfrac{1}{2}e^{-jkd}).$$

The phasor diagrams are shown in Fig. 4.8 and the standing wave patterns in Fig. 4.9. Since the limits of the pressure variation are $\frac{3}{2}$ and $\frac{1}{2}$, the SWR is 3. Note that the standing wave patterns are not sine curves; the peaks are broader than the troughs. Moreover, the troughs are not cusps. Cusps occur only at complete nulls, which develop only when Z_n is zero, infinity, or pure imaginary.

What value of Z_n is required to make the pressure and particle velocity switch roles? It may be seen that if $R = -\frac{1}{2}$, the pressure starts out with relative amplitude $\frac{1}{2}$ and the particle velocity with relative amplitude $\frac{3}{2}$. As Eq. A–13 shows, Z_n must be $Z_0/3$ if R is to have the value $-\frac{1}{2}$. *Generalization*: If the termination impedance is changed from Z_n to $Z_n' = Z_0^2/Z_n$, the reflection coefficient reverses its sign, and thus the pressure and particle velocity standing wave patterns exchange places. In particular, if $Z_n < Z_0$, i.e., if the termination material is "softer" than the propagation medium, the particle velocity has a maximum at the termination while the pressure has a minimum. Conversely, if

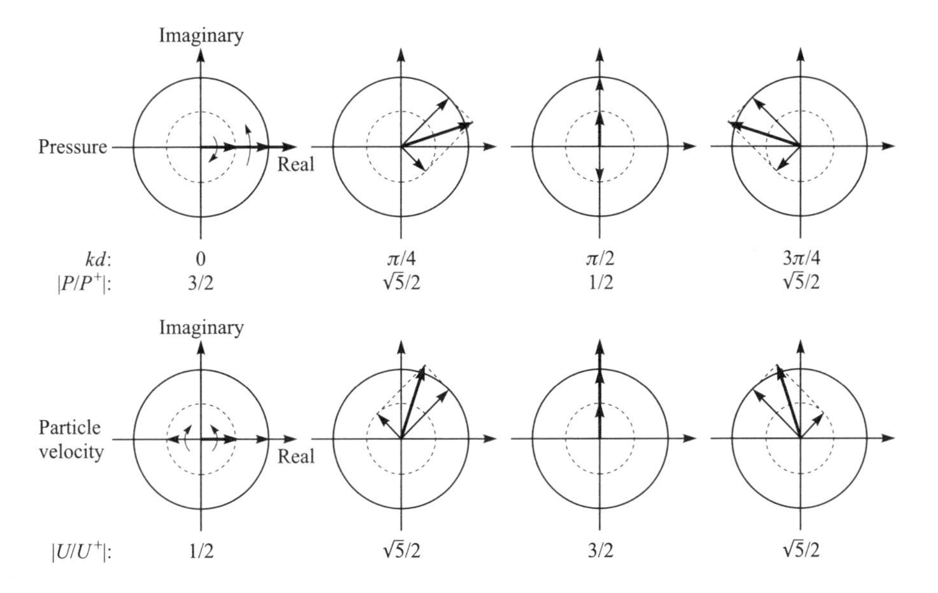

Figure 4.8 Phasor diagrams for $Z_n = 3Z_0$.

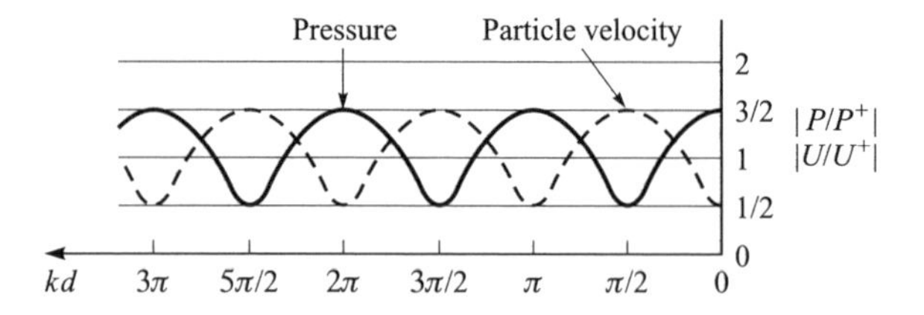

Figure 4.9 Standing wave patterns for $Z_n = 3Z_0$.

$Z_n > Z_0$, i.e., if the termination is "harder" than the medium, the pressure has a maximum at the termination.

Next consider the impedance. Since R is real, Eq. A–11 may be separated into real and imaginary parts as follows:

$$\frac{Z}{Z_0} = \frac{1 - R^2}{1 + R^2 - 2R\cos 2kd} - j\frac{2R\sin 2kd}{1 + R^2 - 2R\cos 2kd}. \tag{B–11}$$

This expression is useful for design purposes in matching a source to a given load (Z_n) so as to achieve maximum power transfer. The length of the propagation medium is chosen so that $\mathrm{Re}[Z(\ell)] = Z_0(1 - R^2)/(1 + R^2 - 2R\cos 2k\ell)$ is equal to the real part of the internal impedance of the source.

4. General Impedance Termination

If Z_n is complex, then by Eq. A–13, so is the reflection coefficient. For sketching phasor diagrams it is useful to express R in polar form,

$$R = \rho e^{j\psi}, \tag{B–12}$$

where $\rho = |R|$ is the amplitude and ψ is the phase angle.[3] Substitution in Eq. A–9 yields

$$P = P^+\left[e^{jkd} + \rho e^{j(\psi - kd)}\right], \tag{B–13}$$

and a similar formula holds for U. In the initial phasor diagram ($kd = 0$) for the pressure, the reflected wave phasor no longer lies along the real axis but rather starts at the angle ψ. When each of the two phasors has rotated half the starting angle, they point in the same direction and the first pressure maximum occurs. The distance to the maximum is thus given by

[3]The symbol ρ used here for magnitude of a complex number should not be confused with its use elsewhere to mean density.

$$(kd)_{\max} = \tfrac{1}{2}\psi. \tag{B–14}$$

(Subsequent maxima occur at increments $n\pi$ greater than this value.) Conversely the first pressure minimum, produced when the two phasors first point in opposite directions, occurs when kd has one of the two values

$$(kd)_{\min} = \tfrac{1}{2}\psi \pm \tfrac{1}{2}\pi. \tag{B–15}$$

Which comes first, the maximum or the minimum? The answer depends on the size of ψ. If ψ is in the first or second quadrant ($\psi < \pi$), the plus sign in Eq. B–15 is required in order that kd be positive. It is clear that in this case the maximum occurs first, the minimum 90° later. If, on the other hand, ψ is in the third or fourth quadrant ($\psi > \pi$), the minus sign in Eq. B–15 is possible, and the minimum occurs first. The picture is reversed when the particle velocity maxima and minima are considered. As usual, pressure maxima are paired with particle velocity minima, and vice versa.

As an example, consider a termination for which $Z_n = (1 + j)Z_0$. The reflection coefficient for this case is

$$R = \frac{1 + j - 1}{1 + j + 1} = \frac{j}{2 + j} = \frac{e^{j90^\circ}}{\sqrt{5}} e^{-j26.6^\circ} = 0.45e^{j63.4^\circ}.$$

Since $\psi = 63.4^\circ$ is an angle in the first quadrant, the first pressure extremum is a maximum, given by $(kd)_{\max} = 63.4^\circ/2 = 31.7^\circ$. Phasor diagrams for the pressure are shown in Fig. 4.10, and the pressure standing wave pattern is sketched in Fig. 4.11. The standing wave pattern for U may be obtained by translating the pressure pattern along the kd axis by an amount $\pi/2$. The standing wave ratio is

$$\text{SWR} = \frac{1 + \rho}{1 - \rho} = \frac{1 + 1/\sqrt{5}}{1 - 1/\sqrt{5}} = \frac{1.45}{0.55} = 2.62.$$

An interesting special case occurs when the terminating impedance is a pure reactance, $Z_n = jX_n$, where X_n is real. It may be shown (see Problem 4B–13) that in this case

$$R = -e^{-j2\theta}, \tag{B–16}$$

where $\theta = \tan^{-1} X_n/Z_0$, and that

$$Z = jZ_0 \tan(kd + \theta). \tag{B–17}$$

The sound field is therefore much like that in a tube having either a rigid end or a pressure release end. The reflected and incident waves at the termination have

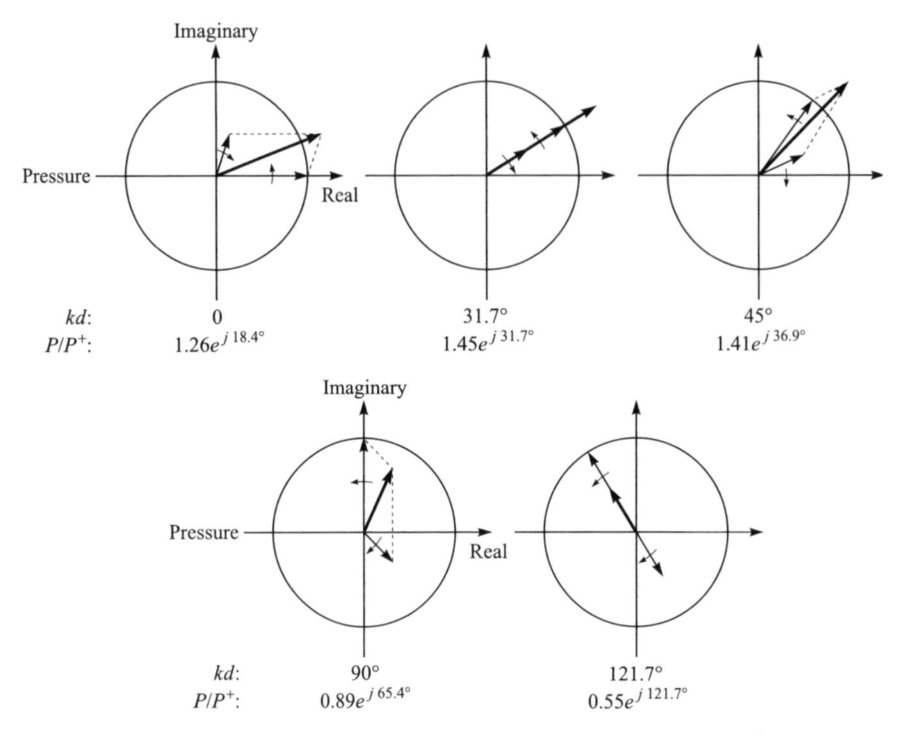

Figure 4.10 Pressure phasor diagrams for the complex terminating impedance $Z_n = (1 + j)Z_0$.

the same magnitude (but different phase), the pressure and particle velocity are everywhere 90° out of phase (and so no power is transferred), and the impedance plot is the same as that in Fig. 4.5, only shifted to the left along the *kd* axis by an amount θ.

An impedance tube is an acoustical instrument used to measure the specific acoustic impedance of materials. See Fig. 4.12. The readings of the probe microphone, as it is moved along the axis of the tube, map out the pressure standing wave pattern. From the position of the maximum or minimum nearest

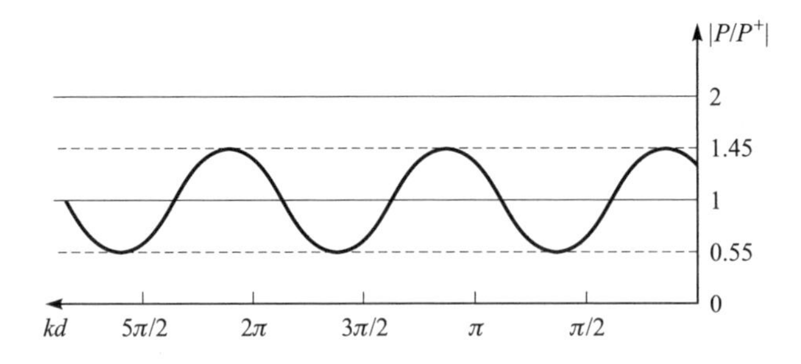

Figure 4.11 Pressure standing wave pattern for $Z_n = (1 + j)Z_0$.

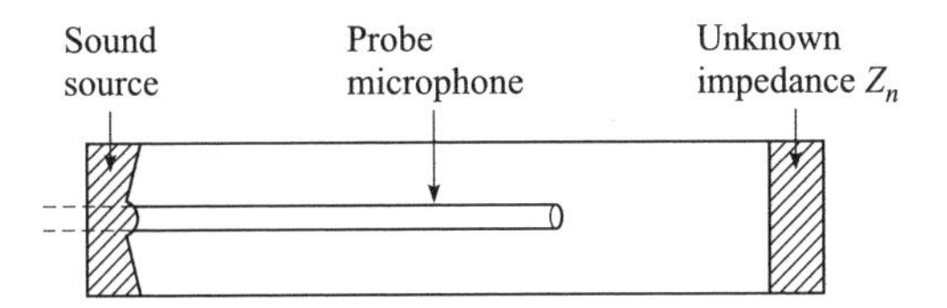

Figure 4.12 Impedance tube.

the termination and from the value of SWR, one may determine Z_n. The procedure is as follows. If the extremum closest to the termination is a maximum, it is convenient to utilize Eq. B–14, which yields $\psi = 2(kd)_{\text{max}}$. Since the magnitude ρ of the reflection coefficient is given by Eq. A–16, we have

$$R = \frac{\text{SWR} - 1}{\text{SWR} + 1} e^{j2(kd)_{\text{max}}}, \tag{B–18}$$

and Z_n may be found by applying Eq. A–12. If, on the other hand, the first extremum is a minimum, conditions are such that the initial position ($kd = 0$) of the reflected wave phasor is in the third or fourth quadrant. In this case one uses Eq. B–15 with the minus sign, and

$$R = -\frac{\text{SWR} - 1}{\text{SWR} + 1} e^{j2(kd)_{\text{min}}}. \tag{B–19}$$

Example 4.1. Impedance tube calculation. An air-filled standing wave tube is terminated with an unknown impedance Z_n. The SWR is measured and found to be 1.2. The first extremum is a pressure maximum, which is located a distance $\lambda/8$ from the termination. Find Z_n. The sequence of calculations is as follows:

1. Find $\rho : \rho \quad = \dfrac{\text{SWR} - 1}{\text{SWR} + 1} = \dfrac{0.2}{2.2} = \dfrac{1}{11}.$
2. Find $\psi : \psi \quad = 2(kd)_{\text{max}} = 2\left(\dfrac{2\pi}{\lambda}\right)\left(\dfrac{\lambda}{8}\right) = \dfrac{\pi}{2}.$
3. Find $R : R \quad = \rho e^{j\psi} = \dfrac{1}{11} e^{j\pi/2} = \dfrac{j}{11}.$
4. Find $Z_n : Z_n \quad = Z_0 \dfrac{1 + R}{1 - R} = Z_0 \dfrac{1 + j/11}{1 - j/11} = Z_0 e^{j10.39^\circ}$

$$= Z_0(0.984 + j0.180)$$
$$= 408 + j74.8 \text{ MKS rayls.}$$

The two-microphone method provides an alternative way to measure acoustic impedance. Given the pressure signals measured by two fixed microphones and their distances from the unknown impedance surface, one can compute the reflection coefficient and subsequently the impedance (see Problem 4B–14). In one version of this method, use of the transfer function between the two microphones readily yields a measurement of the magnitude and phase of the reflection coefficient (and related quantities) as a function of frequency (Ref. 3).

5. Change in Cross-Sectional Area

The change in impedance produced by a change in cross-sectional area of a tube has the same general effect on a time-harmonic wave as on any other wave; see Sec. 3C. To account for cross-sectional area changes, simply replace specific acoustic impedance with acoustic impedance in the reflection and transmission formulas. Problems 4B–9 to 4B–11 provide examples.

C. LUMPED-ELEMENT APPROXIMATION

In this section the acoustical behavior of very short tubes is considered. When the tube is short enough or the frequency low enough that $k\ell \ll 1$, the fluid in the tube may be treated as a lumped mechanical element rather than as a distributed system. Because electrical circuit analogs are often useful in solving problems involving lumped acoustical elements, we begin with a discussion of electrical analogs of acoustical quantities.

1. Electrical Analogs

The notion that acoustic pressure is analogous to voltage and that particle velocity is analogous to current was introduced in Chap. 1. Here we expand on the analog and modify it slightly. When cross-sectional area changes are involved, the volume velocity q is a more convenient flow variable to use than particle velocity u. We therefore take q as the acoustical analog of current. It follows that the acoustic impedance p/q, rather than the specific acoustic impedance p/u, is the convenient analog of electrical impedance.

We first develop the acoustical analogs of inductance and capacitance. Since the force f_{mass} associated with a mechanical mass M is $M\dot{u}$, the acoustic impedance of the mass is

$$\frac{p}{q} = \frac{f_{\text{mass}}/S}{Su} = \frac{M\dot{u}}{S^2 u} = \frac{j\omega M}{S^2} = j\omega M_{\text{ac}}, \tag{C–1}$$

where $M_{\text{ac}} = M/S^2$ is called the *acoustic inertance*. Comparison of Eq. C–1 with the formula $e/i = j\omega L$ for the electrical impedance of an inductance L indicates that L and M_{ac} may be taken to be analogs. Note that the impedance of each is a positive reactance. In the case of a spring of mechanical stiffness

K ($= 1/C_m$, where C_m is the mechanical compliance), an applied force f_{spring} results in a displacement $\xi = f_{\text{spring}}/K$. Consequently, the acoustic impedance is

$$\frac{p}{q} = \frac{f_{\text{spring}}/S}{Su} = \frac{K\xi}{S^2\dot{\xi}} = \frac{K}{j\omega S^2} = \frac{K_{\text{ac}}}{j\omega} = \frac{1}{j\omega C_{\text{ac}}}, \tag{C–2}$$

where $K_{\text{ac}} = K/S^2$ is the acoustic stiffness and $C_{\text{ac}} = 1/K_{\text{ac}}$ is the *acoustic compliance*. Since the electrical impedance of a capacitance C is $e/i = 1/j\omega C$, it is clear that C and C_{ac} are analogs. Finally, the acoustic resistance R_{ac}, which may be due to one of several different damping mechanisms not yet discussed or to radiation (see below), is analogous to electrical resistance R.

2. Short Closed Cavity

Consider a short closed tube of length ℓ, cross-sectional area S, and volume $V = \ell S$ (see Fig. 4.13), in which a sound field is set up by a source such as a vibrating piston. The input (specific acoustic) impedance of the tube Z_{in}, that is, the impedance seen by the piston, is, from Eq. B–10,

$$Z_{\text{in}} = \frac{Z_0}{j\tan k\ell}. \tag{C–3}$$

If $k\ell \ll 1$, the first term in the Maclaurin series expansion for the tangent function

$$\tan k\ell = k\ell + \tfrac{1}{3}k^3\ell^3 + \cdots, \tag{C–4}$$

is sufficient to be substituted in Eq. C–3. The result is

$$Z_{\text{in}} = \frac{\rho_0 c_0}{jk\ell} = \frac{\rho_0 c_0^2}{j\omega\ell}. \tag{C–5}$$

The acoustic impedance of the cavity is therefore

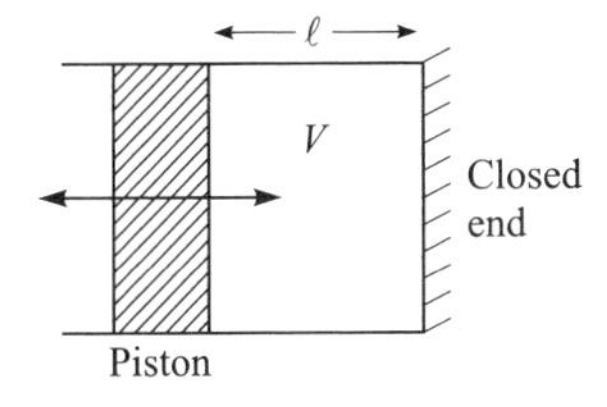

Figure 4.13 Short closed tube or cavity.

$$Z_{ac} = \frac{\rho_0 c_0^2}{j\omega \ell S} = \frac{\rho_0 c_0^2}{j\omega V}. \tag{C–6}$$

Comparison with Eq. C–2 shows that the cavity behaves as a spring of acoustic compliance

$$C_{ac} = V/\rho_0 c_0^2. \tag{C–7}$$

If the fluid in the cavity is a gas, then $\rho_0 c_0^2 = \gamma p_0$, in which case

$$C_{ac} = V/\gamma p_0. \tag{C–8}$$

This result confirms our general experience: pushing a piston into a long cylinder is easier than pushing it into a short one. The gas is more compliant if its container is deep.

Since the confined fluid acts like a spring, vibration of a piston (of mass M) against it opens up the possibility of resonance behavior. First consider free oscillations of the piston. If the piston is disturbed and then released, it oscillates against the cavity "spring" at a resonance frequency $f_0 = \omega_0/2\pi$ given by

$$\omega_0 = 2\pi f_0 = \sqrt{\frac{K_{ac}}{M_{ac}}} = \sqrt{\frac{S^2 \rho_0 c_0^2}{MV}} = \sqrt{\frac{S^2 \gamma p_0}{MV}}, \tag{C–9}$$

where the last form is limited to gases.[4] The dependence of the resonance frequency on the ratio of specific heats γ is the basis of a method used for measuring γ.[5] A ball of known mass M is allowed to oscillate freely in a closed glass tube containing the gas in question, as shown in Fig. 4.14. (A lubricant between the ball and the glass may be used to cut down friction and prevent leakage of gas past the ball.) The oscillation frequency is measured and Eq. C–9 is used to calculate γ.

Forced oscillations of a mass in a closed cavity are also of interest. Suppose, for example, that the piston in Fig. 4.13 is made to oscillate sinusoidally by an external force $f = F_{in}e^{j\omega t}$ and we wish to know the resultant pressure in the cavity $p_{cav} = P_{cav}e^{j\omega t}$. The calculation is easily made with the help of the equivalent electrical circuit shown in Fig. 4.15*a*. The interpretation is as follows: As usual, capital letters for dependent variables stand for (complex) amplitudes of time-harmonic signals, e.g., the expression for i the electrical current is $i = Ie^{j\omega t}$. Table 4.1 gives the electrical-acoustical analogs used. The output voltage is

[4]Notice that the resonance considered here has nothing to do with the closed pipe resonances that occur when the pipe is an integral number of half wavelengths (Eq. B–9b).

[5]Sometimes called Ruchardt's method (Refs. 11, 4).

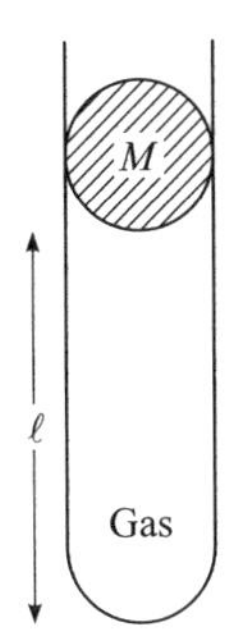

Figure 4.14 Method of measuring the ratio of specific heats.

$$E_{\text{out}} = Z_C I = \frac{Z_C}{Z_L + Z_C} E_{\text{in}}.$$

The response function $E_{\text{out}}/E_{\text{in}}$, that is, the ratio of the cavity pressure to the external force per unit area, is thus given by

$$\frac{P_{\text{cav}}}{F_{\text{in}}/S} = \frac{E_{\text{out}}}{E_{\text{in}}} = \frac{1/j\omega C}{j\omega L + 1/j\omega C} = \frac{1}{1 - \omega^2 LC} = \frac{1}{1 - (\omega/\omega_0)^2}, \tag{C–10}$$

where $\omega_0^2 = 1/LC = \rho_0 c_0^2 S^2/MV$. The response $|E_{\text{out}}/E_{\text{in}}|$ is graphed in Fig. 4.15*b*. The flat response at low frequencies gives way to a rapid rise as the resonance frequency is approached. At high frequencies, well beyond resonance, the proportionality of the response to ω^{-2} implies a rolloff of 12 dB/octave, i.e., a drop of 12 dB for each doubling of frequency, or 40 dB/decade. If desired, the response near the resonance frequency may be flattened by adding some damping, for example, by using a narrow capillary tube to connect the cavity to the outside.

A good example of a mass-cavity system is the condenser microphone, which is shown in Fig. 4.16*a*. Figure 4.16*b* shows the electrical components of the microphone (the diagram is the actual electrical circuit, not an analog of the microphone's mechanical parts). The thin metallic diaphragm and rigid metal backplate shown in Fig. 4.16*a* are the two plates of the condenser giving

Table 4.1 Analogs for Calculating Cavity Response

Acoustical	Electrical
M/S^2	L
$C_{\text{ac}} = V/\rho_0 c_0^2$	C
F_{in}/S	E_{in}
P_{cav}	E_{out}
Q	I

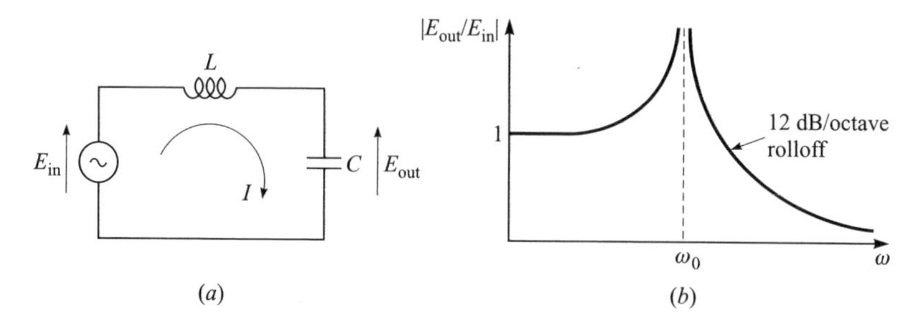

Figure 4.15 Forced oscillation of the pressure in a closed cavity. (*a*) Equivalent electrical circuit. (*b*) Cavity pressure response.

the microphone its name; their capacitance C (Fig. 4.16*b*) depends on their separation. The mass in this case is the diaphragm, and the spring is the shallow air volume separating the diaphragm from the backplate. The pressure of an acoustic wave on the outside surface of the diaphragm displaces the diaphragm, changing C and in turn producing a voltage signal E_{out} across the resistance R in the electrical network. The microphone response function is much like the one shown in Fig. 4.15*b*. In this case, however, E_{out} is a true electrical voltage while E_{in} represents the pressure of the external sound field. Damping, introduced by venting the cavity to the outside with a narrow capillary tube, counteracts the rise in the response near the resonance frequency. Indeed, with proper damping the flat region of the response curve (the pass band of the microphone) can be made to extend from zero frequency to near $f_0\,(=\omega_0/2\pi)$. The high frequency limit of the pass band is thus set by the mechanical resonance frequency f_0 of the diaphragm-cavity system. Above f_0 the response decreases very rapidly, 12 dB/octave. As Eq. C–9 shows, a very high resonance frequency, and therefore a very wide microphone bandwidth, requires that the mass per unit area $m = M/S$ of the diaphragm and/or the cavity depth $\ell = V/S$ be very small.

A quasi-static analysis may also be used to compute the compliance of a short closed cavity and at the same time generalize the result. Assume that a

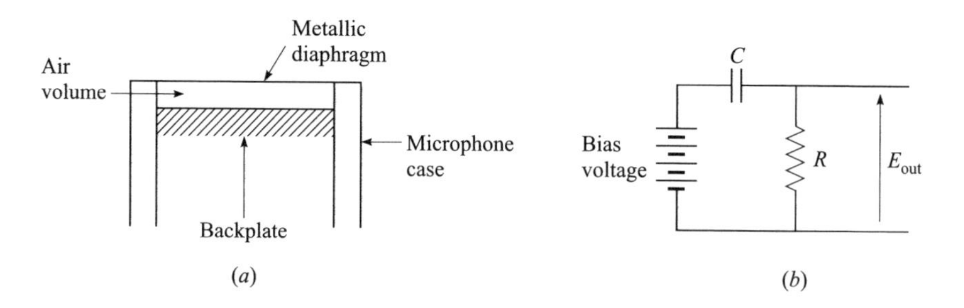

Figure 4.16 Condenser microphone.

piston moves slowly into an irregular cavity of initial volume V, as shown in Fig. 4.17. The intrusion of the piston decreases the volume to a new value $V' = V - \Delta V$ and increases the total pressure from p_0 to $p_0 + p$. The change in total pressure is assumed to be uniform throughout the volume (because the process is quasi-static) and also to be adiabatic. If the fluid in the cavity is a gas, the adiabatic gas law Eq. 1C–39 may be used to relate p to ΔV:

$$\frac{p + p_0}{p_0} = \left(\frac{\rho}{\rho_0}\right)^{\gamma} = \left(\frac{V}{V'}\right)^{\gamma} = \left(1 - \frac{\Delta V}{V}\right)^{-\gamma} = 1 + \frac{\gamma \Delta V}{V} + \cdots.$$

Dropping higher order terms in the expansion yields $p = (\gamma p_0/V)\,\Delta V$, or, since $\gamma p_0 = \rho_0 c_0^2$,

$$p = (\rho_0 c_0^2/V)\Delta V. \tag{C–11}$$

(The extension of this result to cover liquids is easily obtained by using the general isentropic equation of state, Eq. 1C–40, in place of the adiabatic gas law.) The ratio of the pressure change to the volume change is the acoustic stiffness (just as the force per unit displacement is the mechanical stiffness), and the reciprocal is the acoustic compliance,

$$C_{\text{ac}} = \frac{\Delta V}{p} = \frac{V}{\rho_0 c_0^2}.$$

Equation C–7 is thus obtained anew. The derivation here is more general in the sense that no particular shape has been assumed for the cavity. The requirement that the process be quasi-static is closely related to the requirement in the earlier derivation that $k\ell$ be small.

An improved lumped element description of the closed cavity may be found by retaining the cubic term in the tangent expansion, Eq. C–4. The improved expression for the input (specific acoustic) impedance is

$$Z_{\text{in}} = \frac{Z_0}{jk\ell(1 + k^2\ell^2/3)} \doteq \frac{Z_0}{jk\ell}\left(1 - \frac{k^2\ell^2}{3}\right) = \frac{Z_0}{jk\ell} + j\frac{k\ell Z_0}{3},$$

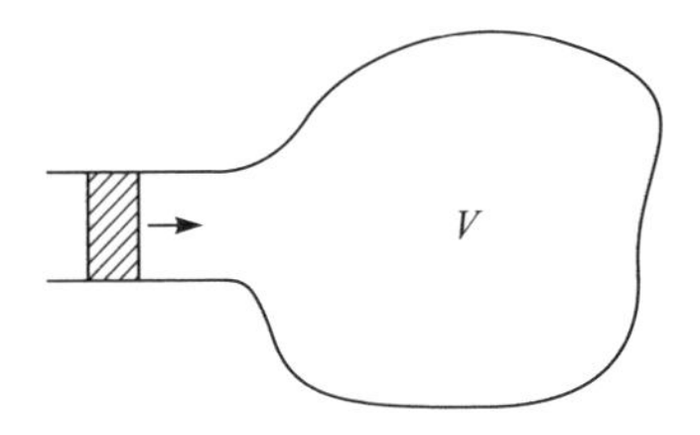

Figure 4.17 Irregular cavity.

or, in terms of the acoustic impedance,

$$Z_{\text{ac}} = \frac{\rho_0 c_0^2}{j\omega V} + j\omega \frac{\rho_0 V}{3S^2}. \tag{C–12}$$

The first term is the one shown in Eq. C–6. The correction term is a positive reactance associated with one-third the mass of the fluid in the cavity. An improved representation of the cavity is therefore a compliance in series with a mass element, as shown in Fig. 4.18, where L_f represents the inertance of the cavity fluid itself. This result is not unexpected. Advanced treatments of the motion of a bob on a spring show that the effective vibrating mass of the system is $M + M_s/3$, where M is the mass of the bob and M_s is the mass of the spring.

The variation of p and u in the short closed cavity (simple representation as a stiffness only) is also of interest. If the piston in Fig. 4.13 moves with velocity given by Eq. B–6, the expressions for u and p are those given by Eqs. B–9a and B–9b, respectively. For $k\ell \ll 1$, these formulas reduce to

$$u = u_0(1 - x/\ell)e^{j\omega t}$$

and

$$p = -j\frac{u_0 Z_0}{k\ell} e^{j\omega t} = \frac{u_0 \rho_0 c_0^2}{j\omega \ell} e^{j\omega t},$$

respectively. The velocity amplitude U is seen to decrease linearly from the velocity u_0 at the piston to zero at the closed end. The pressure amplitude P, on the other hand, is independent of x, that is, it is the same throughout the cavity (the assumption to this effect made in the quasi-static analysis is therefore justified). These results are consistent with the shapes of the standing wave curves near $d = 0$; see Fig. 4.4, but remember that for the case we are considering the dashed curve represents pressure and the solid curve particle velocity.

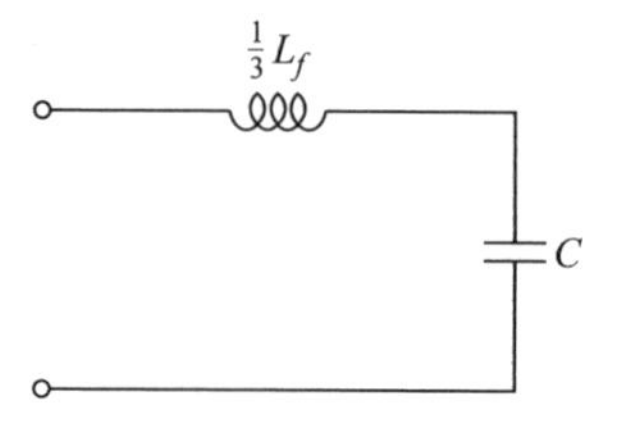

Figure 4.18 Improved lumped-element representation of a closed cavity.

3. End Correction for an Open Tube

Before the field in a short open tube is analyzed, account must be taken of the sound radiated from the end into the fluid outside. Heretofore, we have assumed that the acoustic pressure drops to zero at an open end. In fact, however, although the pressure is greatly reduced when the tube wall is no longer present to confine the wave, the pressure does not drop quite to zero. In other words, not all the sound is reflected back down the tube; some is radiated into the fluid outside. The effect of radiation from the end may be approximated by considering the final layer of fluid at the open end to vibrate back and forth as a piston. To make the problem more tractable, assume that the end of the tube is flanged by a large rigid plate, which limits radiation to the forward hemisphere of the surrounding fluid. Vibrating piston radiation theory (Chap. 13) may now be used to determine the impedance seen by the final layer. From results given in Chap. 13 (see Eq. 13D–5), the acoustic impedance seen by a piston vibrating as a disk in an otherwise rigid plane is

$$Z_{\mathrm{ac}} = \rho_0 c_0 S^{-1}[R_1(2ka) + jX_1(2ka)], \tag{C–13}$$

where a is the radius of the piston (in this case the tube opening), $S = \pi a^2$ is the area of the piston, and R_1 and X_1 are functions similar to the real and imaginary parts, respectively, of the equivalent expression for a pulsating sphere (see Eq. 1D–21 and Problem 1D–2). Graphs of R_1 and X_1 appear in Fig. 13.10, and series expansions in ka are given by Eqs. 13D–6 and 13D–7. In accordance with the lumped-element approximation (dimensions small compared to the wavelength), it is appropriate to assume $ka \ll 1$. Then only the leading term in each expansion is important:

$$R_1 = \frac{k^2 a^2}{2}, \tag{C–14}$$

$$X_1 = \frac{8ka}{3\pi}. \tag{C–15}$$

Notice that because of the assumption $ka \ll 1$, the real part of Z_{ac} is much smaller than the imaginary part. Equation C–13 may now be written

$$Z_{\mathrm{ac}} = \rho_0 c_0 \frac{k^2}{2\pi} + \frac{j\omega\rho_0 S(8a/3\pi)}{S^2}. \tag{C–16}$$

The right-hand side is dominated by the positive reactance, which represents a mass equal to $\rho_0 S(8a/3\pi)$, i.e., a fluid cylinder having the same radius as the tube and length $8a/3\pi$. In other words, when the wave comes to the open end, it sees, not zero load, but rather a load consisting mainly of a short continuation of the tube,

$$\Delta\ell = 8a/3\pi = 0.85a. \tag{C–17}$$

This is the famous end correction for an open tube. Besides the mass reactance due to the extra length $\Delta\ell$, the wave also sees a small resistance (the first term), which is called the *radiation resistance*. Its magnitude is smaller than the magnitude of the reactance by a factor of about ka. The effective length of an open tube is therefore, for acoustical purposes

$$\ell' = \ell + \Delta\ell \tag{C–18}$$

rather than just ℓ. Of course, if $a \ll \ell$, the correction to the length is very small. For very short open tubes, however, $\Delta\ell$ can be the largest part of ℓ'. For example, a hole in a thin plate of thickness ℓ may be modeled as a very short tube having two end corrections, one for each end of the tube, i.e., $\ell' = \ell + 2\,\Delta\ell \simeq 2\,\Delta\ell$.

Results from calculations based on other models show that Eq. C–17 is only an upper limit in the estimate for $\Delta\ell$ for flanged tubes (the lower limit is $\Delta\ell = \pi a/4 = 0.7854a$).[6] For our purposes we shall rely on Eq. C–17. However, Eq. C–17 cannot be used if the tube has no flange. For an unflanged tube the end correction is (Ref. 5)

$$\Delta\ell = 0.6133a. \tag{C–19}$$

Tubes having a flange of finite radius have also been treated (Ref. 2).

Although we have derived the concept of the end correction by modeling the end of the tube as a piston radiating sound into a semi-infinite medium, the concept is more general than the derivation. For example, even if the flow out of the tube is incompressible (no sound radiated), the fluid at the end still behaves as a mass reactance. An end correction is therefore appropriate even when the space beyond the end of the tube is not semi-infinite, for example, when it consists of a cavity. The radiation resistance, on the other hand, *is* associated with radiation into a semi-infinite medium and should therefore be used only when the tube radiates in that fashion. In the absence of radiation resistance, viscous friction may serve as the damping mechanism. See, for example, Ref. 7.

[6]See Ref. 10. For a modern review and analysis of the end correction problem, see Ref. 9.

4. Short Open Cavity

A flanged open cavity with an end correction indicated is shown in Fig. 4.19 (if the flange were absent, $\Delta\ell$ would be $0.6133a$). First neglect the radiation resistance. In this case the input impedance seen by the source is, from Eq. B–7,

$$Z_{\text{in}} = jZ_0 \tan k\ell',$$

which, for $k\ell' \ll 1$, reduces to

$$Z_{\text{in}} = jZ_0 k\ell' = j\omega\rho_0\ell'.$$

This formula, when put in terms of the acoustic impedance

$$Z_{\text{ac}} = \frac{j\omega(\rho_0\ell' S)}{S^2}, \tag{C–20}$$

suggests that a short open cavity behaves as a mass equal to $\rho_0\ell' S$: its inertance is that of the fluid in the extended cavity. The model may be improved, if the cavity faces an open fluid, by adding the radiation resistance in series,

$$Z_{\text{ac}} = \frac{\rho_0 c_0 k^2}{2\pi} + \frac{j\omega\rho_0\ell'}{S}. \tag{C–21}$$

The equivalent circuit is shown in Fig. 4.20.

5. Helmholtz Resonator

Figure 4.21 shows a Helmholtz resonator, which is a combination of the two lumped elements just discussed. The chamber is a short closed cavity, the neck a short open cavity. Because the fluid in the chamber acts like a spring and that in the neck acts like a mass, the system behaves as a simple oscillator. Since the neck and chamber are in series, their total acoustic impedance is

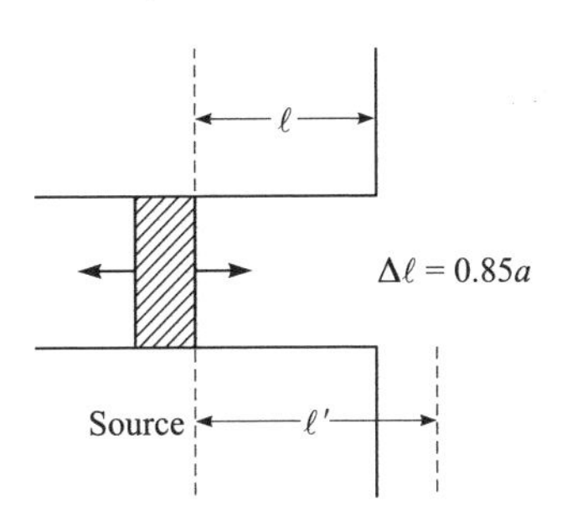

Figure 4.19 Short open cavity.

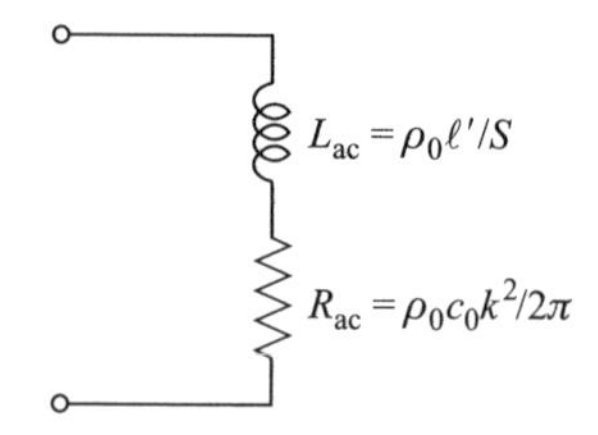

Figure 4.20 Equivalent electrical circuit for an open cavity.

$$
\begin{aligned}
Z_{ac} &= Z_{ac}^{(\text{neck})} + Z_{ac}^{(\text{ch})} \\
&= j\omega \frac{\rho_0 \ell'}{S} + \frac{\rho_0 c_0 k^2}{2\pi} + \frac{\rho_0 c_0^2}{j\omega V}.
\end{aligned}
\tag{C–22}
$$

Note that S is the cross-sectional area of the neck, not the chamber. The natural, or resonance, frequency of the resonator is the frequency at which the reactive part of the impedance vanishes:

$$\omega_0 = c_0 \sqrt{\frac{S}{\ell' V}}, \tag{C–23}$$

or

$$f_0 = \frac{c_0}{2\pi} \sqrt{\frac{S}{\ell' V}}. \tag{C–24}$$

Equation C–22 therefore takes the form

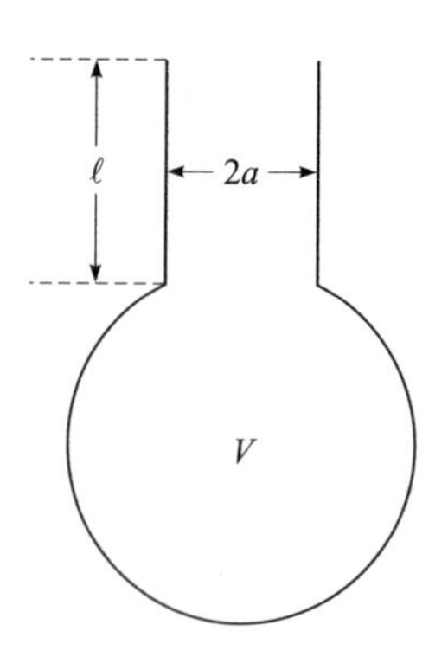

Figure 4.21 Helmholtz resonator.

$$Z_{\text{ac}} = \frac{\rho_0 c_0 k^2}{2\pi} + j\omega\rho_0 \frac{\ell'}{S}\left[1 - \frac{\omega_0^2}{\omega^2}\right], \tag{C–25}$$

which shows that so long as $\lambda \gg S/\ell'$, the real part of the impedance is important only for frequencies near resonance. The real part is often neglected in simple calculations. When it is taken into account, the usual assumption is that the system has only one resistance $\rho_0 c_0 k^2/2\pi$, not two, because only one end radiates into an open medium. The inner end does not radiate and therefore has no radiation damping.

In estimating ℓ', one must usually exercise judgment. For example, for the Helmholtz resonator shown in Fig. 4.21, we have

$$\ell' = \ell + \Delta\ell_{\text{in}} + \Delta\ell_{\text{out}},$$

where the subscripts "in" and "out" denote the inner and outer ends, respectively, of the neck. One may argue that since the inner end most closely resembles a flanged opening, Eq. C–17 should be used for it, whereas Eq. C–19 is more appropriate for $\Delta\ell_{\text{out}}$. However, see the experimental measurements, discussion, and references given by Ref. 8.

Finally, let us find the pressure amplification A produced by the Helmholtz resonator. If the resonator is driven by an external, or input, pressure $P_{\text{in}}e^{j\omega t}$ and the chamber pressure is denoted by $P_{\text{ch}}e^{j\omega t}$, the pressure amplification is $A = |P_{\text{ch}}/P_{\text{in}}|$. This factor may be obtained quite easily from the analog electrical circuit shown in Fig. 4.22. The input "voltage" P_{in}, which is impressed across the acoustic impedance represented by the neck in series with the chamber, causes a volume velocity "current" $Q = P_{\text{in}}/Z_{\text{ac}}$ to flow. Because the "voltage" P_{ch} is equal to $Q/j\omega C_{\text{ac}}$ and $Q = P_{\text{in}}/[j\omega L + R + 1/j\omega C]$, the response function is

$$\frac{P_{\text{ch}}}{P_{\text{in}}} = \frac{1/j\omega C}{R + j\omega L + 1/j\omega C} = \frac{1}{(1 - \omega^2/\omega_0^2) + j\omega RC}. \tag{C–26}$$

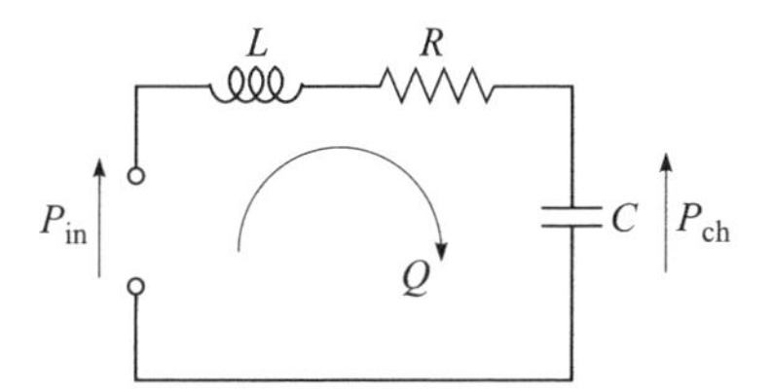

Figure 4.22 Equivalent circuit for a Helmholtz resonator driven by an external sound field.

Note that the frequency dependence of R makes the analysis of this circuit a little different from that of the ordinary RLC circuit. Nevertheless, as long as the resistance term is small, the maximum amplification is attained at the natural resonance frequency ω_0 (see Problem 4C–8),

$$A_{\text{max}} = \frac{1}{\omega_0 R_0 C} = \frac{\omega_0 L}{R_0} = 2\pi\sqrt{\frac{(\ell')^3 V}{S^3}}, \tag{C–27}$$

where R_0 is the value of R at resonance. The frequency response of the cavity (A vs. frequency) is sketched in Fig. 4.23, where it has been assumed that the frequency dependence of R has little effect on the high frequency rolloff.

6. Orifice

An orifice is just a hole in a plate (Fig. 4.24). Acoustically, it is a very short cavity open at both ends. Its impedance is therefore

$$Z_{\text{ac}} = 2\frac{\rho_0 c_0 k^2}{2\pi} + j\omega\frac{\rho_0 \ell'}{S}, \tag{C–28}$$

where $\ell' = \ell + 2\,\Delta\ell$ and ℓ is the plate thickness. In this case the "tube" is flanged at both ends. Because both sides of the orifice radiate, the radiation resistance is twice that for an ordinary open flanged tube. (The effect of viscous damping may also be considered; see Ref. 7.)

D. EXAMPLES

1. Side Branches, Filters

A "Y" tube junction is shown in Fig. 4.25. If the incident sound wave is offered a choice between transmission paths, how much will be transmitted down each

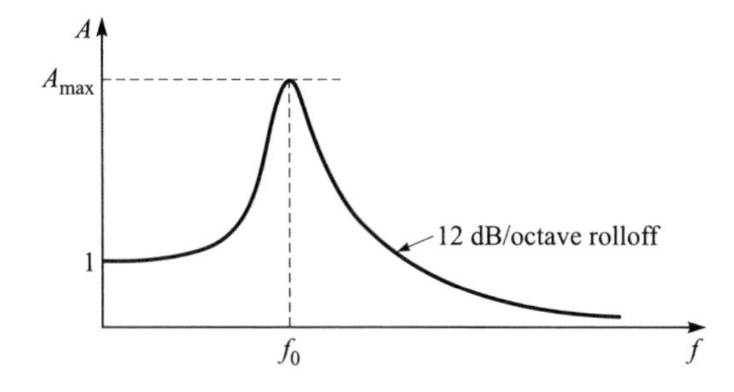

Figure 4.23 Frequency response curve for a Helmholtz resonator.

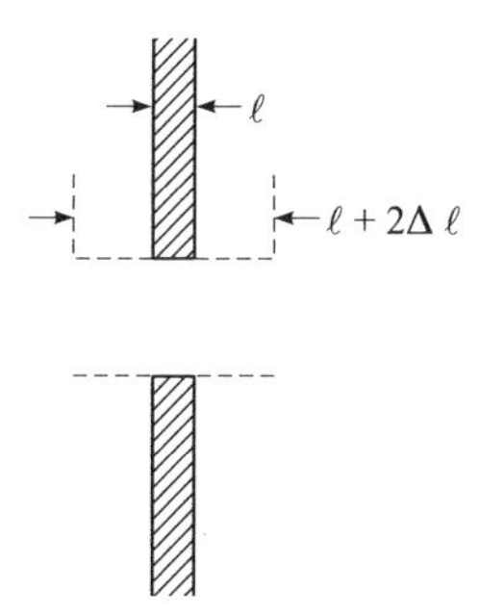

Figure 4.24 Orifice and its end correction.

path and how much will be reflected? The pressure boundary condition is that the pressure must be the same on both sides of the junction[7]:

$$p^+ + p^- = p^{\mathrm{tr}2} = p^{\mathrm{tr}3},$$

or

$$1 + R = T_2 = T_3. \tag{D–1}$$

The particle velocity boundary condition is that the volume velocity q into the junction must equal that out:

$$q^+ + q^- = q^{\mathrm{tr}2} + q^{\mathrm{tr}3},$$

or, if the acoustic impedance relations are used,

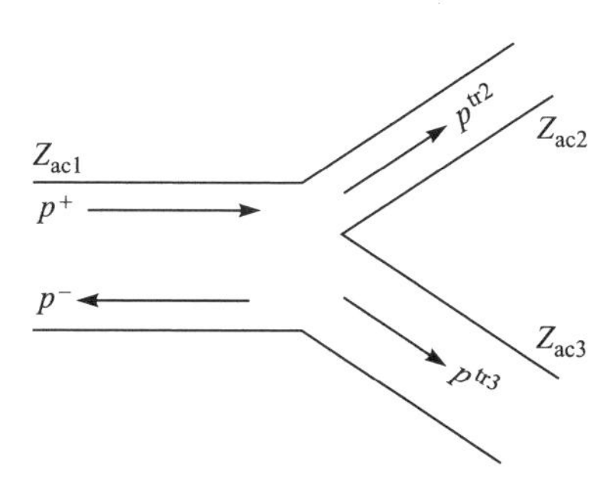

Figure 4.25 Reflection and transmission for a tube branch.

[7]Here lowercase letters are used for pressure and volume velocity because the boundary conditions at the junction are independent of time waveform. If any impedance is complex, however, time-harmonic signals are called for, in which case a switch should be made to capital letters.

$$1 - R = Z_{\text{ac1}}\left[Z_{\text{ac2}}^{-1} + Z_{\text{ac3}}^{-1}\right]T_2 \tag{D–2}$$

(recall from Eq. D–1 that $T_2 = T_3$). Combining Eqs. D–1 and D–2 yields

$$R = \frac{Z_{\text{ac1}}^{-1} - Z_{\text{ac2}}^{-1} - Z_{\text{ac3}}^{-1}}{Z_{\text{ac1}}^{-1} + Z_{\text{ac2}}^{-1} + Z_{\text{ac3}}^{-1}} = \frac{Y_{\text{ac1}} - Y_{\text{ac2}} - Y_{\text{ac3}}}{Y_{\text{ac1}} + Y_{\text{ac2}} + Y_{\text{ac3}}} \tag{D–3}$$

and

$$T_2 = T_3 = \frac{2Z_{\text{ac1}}^{-1}}{Z_{\text{ac1}}^{-1} + Z_{\text{ac2}}^{-1} + Z_{\text{ac3}}^{-1}} = \frac{2Y_{\text{ac1}}}{Y_{\text{ac1}} + Y_{\text{ac2}} + Y_{\text{ac3}}}, \tag{D–4}$$

where Y_{ac} is the acoustic admittance ($Y_{\text{ac}} = 1/Z_{\text{ac}}$).

What if the wave motion in branch 2 or branch 3 (or both) is not progressive? The formula for the reflection coefficient, Eq. D–3, continues to hold. To the wave field in branch 1, the junction simply represents a terminating impedance equal to the parallel combination of Z_{ac2} and Z_{ac3},

$$Z_{\text{acn}} = \frac{Z_{\text{ac2}}Z_{\text{ac3}}}{Z_{\text{ac2}} + Z_{\text{ac3}}},$$

regardless of the conditions that determine the particular values of Z_{ac2} and Z_{ac3} (i.e., whether the fields in branches 2 and 3 are progressive or standing). As for the transmission formula, Eq. D–4, it is meant for a branch in which the field is progressive. Suppose, for example, that branch 3 contains a progressive field but branch 2 a standing field. Then Eq. D–4 (for T_3) continues to hold, but Z_{ac2} now represents the actual acoustic impedance at the entrance to branch 2, not the characteristic acoustic impedance of that branch.

Two applications follow.

Example 4.2. Paralleling horn drivers to the throat of a horn. See Fig. 4.26. How should the cross-sectional areas S_1, S_2, and S_3 be chosen so that all sound from the drivers is transmitted into the horn? The arrangement is the reverse of the one indicated in Fig. 4.25, but reciprocity should be expected to hold. Therefore, if R given by Eq. D–3 is zero, no reflected waves should be present in pipe sections 2 and 3. One therefore puts $Y_{\text{ac1}} = Y_{\text{ac2}} + Y_{\text{ac3}}$, or since the fluid is the same in all three sections,

$$S_1 = S_2 + S_3.$$

Furthermore, if the sections have circular cross sections, and if sections 2 and 3 have the same diameter D, then $D_1^2 = 2D_2^2$, or

$$D_1 = \sqrt{2}D_2.$$

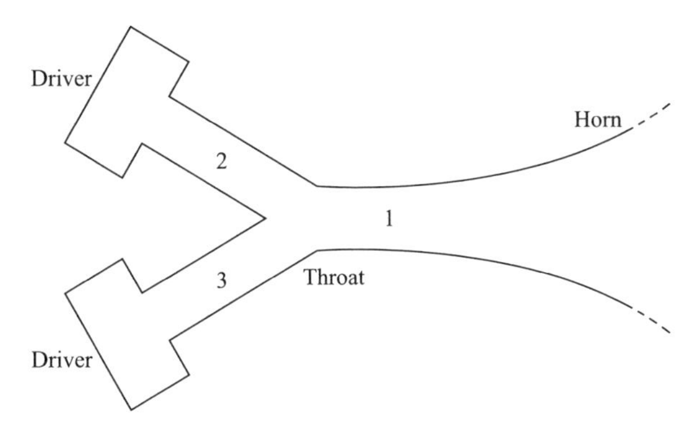

Figure 4.26 Use of two drivers to feed sound into a horn.

For example, two horn drivers having 1-in. throats (a common size) should be coupled to a common throat 1.4 in. in diameter, not to a 2-in. throat.

Example 4.3. Helmholtz resonator used as a filter. Let a Helmholtz resonator be used as a sidebranch to a tube, as shown in Fig. 4.27. The incident sound is traveling in branch 1 toward the junction. Let the acoustic impedance of the resonator be $Z_{ac2} = R_2 + jX_2$, where X_2 is given by Eq. C–25 and R_2 is whatever damping is appropriate for a resonator used in this configuration.[8] Because $Z_{ac1} = Z_{ac3}$ in this case, Eq. D–3 reduces to

$$R = -\frac{Y_{ac2}}{Y_{ac2} + 2Y_{ac1}} = -\frac{Z_{ac1}}{Z_{ac1} + 2Z_{ac2}} = -\frac{Z_{ac1}}{Z_{ac1} + 2(R_2 + jX_2)}. \tag{D–5}$$

When the frequency of the incident wave is equal to the resonance frequency of the Helmholtz resonator, the reactance X_2 vanishes, and, since in general $R_2 \ll Z_{ac1}$, the reflection coefficient becomes

$$R = -1.$$

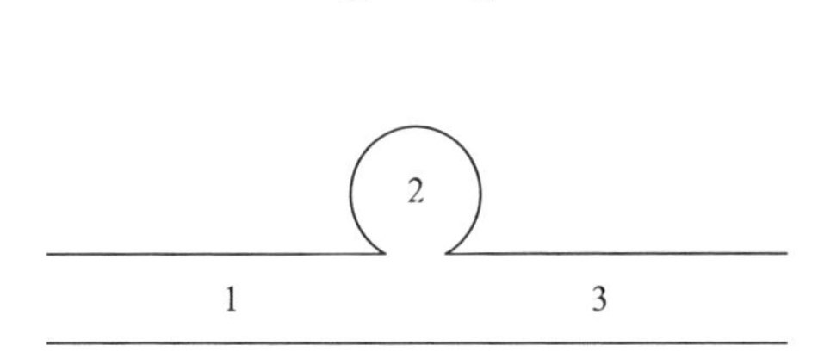

Figure 4.27 Helmholtz resonator used as a filter.

[8]The main damping mechanism in this case is probably not due to radiation resistance (the real term in Eq. C–16) because the neck of the resonator does not radiate into an open medium. The dominant mechanism is expected to be viscosity. See, for example, Ref. 7.

In other words all the sound is reflected back into section 1. The Helmholtz resonator is therefore an effective "band stop" filter at its resonance frequency.

It is also interesting to compute the transmission factor T_3. The computation is left as an exercise. It is suggested that after the expression for T_3 is obtained, it be analyzed for the cases in which $\omega \ll \omega_0$, $\omega \gg \omega_0$, and $\omega \simeq \omega_0$.

2. Probe Tube Microphone

It is frequently not feasible to insert a normal microphone directly into the sound field one wants to measure. The microphone may be too bulky to fit in a small space, the microphone may be so large compared with a wavelength that it disturbs the sound field, or the sound field may be in a corrosive, high temperature, or otherwise severe environment. In such cases a probe tube may be attached to the microphone (Fig. 4.28) to allow the microphone itself to be removed from the immediate measurement region. The effect of the probe tube on the frequency response of the microphone must, however, be considered.

We should like to find an expression for P_{cav}/P_{in}, that is, the ratio of the pressure in the cavity (which is the pressure to which the microphone itself responds) to the pressure in the sound field. Let the cavity volume be V_c. Because the cavity is very short, it may be considered a lumped stiffness of acoustic impedance given by Eq. C–6, where in this case $V = V_c$. If the microphone diaphragm is slightly resilient, its compliance may be included with that of the cavity by adding a small equivalent volume (one way of specifying the diaphragm compliance) to V_c. For purposes of computing P_{cav}, therefore, the acoustical system is the simple one shown in Fig. 4.29.[9] Notice that we have

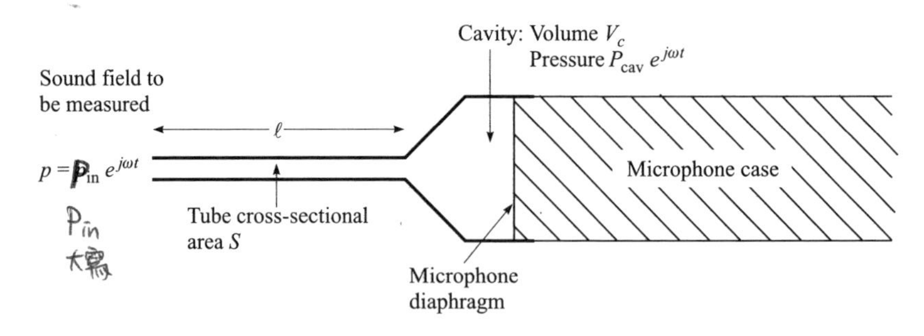

Figure 4.28 Probe tube attached to a microphone.

[9]Notice that in specifying the pressure at the entrance to the probe tube to be the true sound field pressure P_{in}, we are assuming that the probe tube itself does not disturb the sound field. This is normally a good assumption. A typical probe tube may have an outside diameter of 1 or 2 mm and an inside diameter of a fraction of a millimeter.

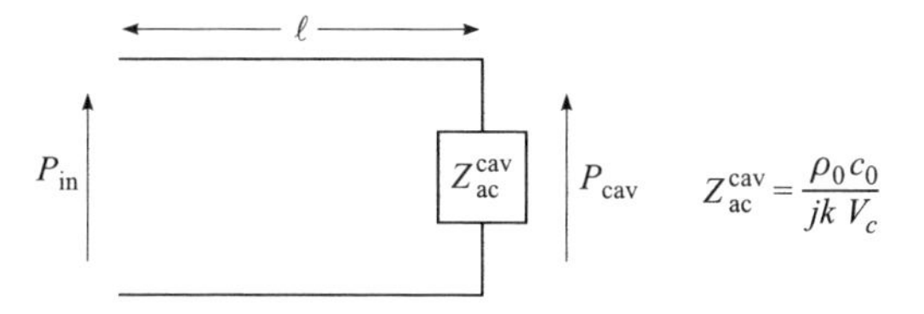

Figure 4.29 Simple model of a probe microphone.

here an example of a system with a pure imaginary termination impedance. Using Eq. A–9, we see that

$$P_{in} = P^+\left(e^{jk\ell} + Re^{-jk\ell}\right) \tag{D–6}$$

and

$$P_{cav} = P^+(1 + R). \tag{D–7}$$

Because of the change in cross-sectional area where the tube enters the cavity, it is best to use acoustic impedances when computing R. Use of Eq. 3C–7 leads to

$$R = \frac{\rho_0 c_0/jkV_c - \rho_0 c_0/S}{\rho_0 c_0/jkV_c + \rho_0 c_0/S} = \frac{1 - jk\ell\alpha}{1 + jk\ell\alpha} = e^{-j2\theta}, \tag{D–8}$$

where $\alpha = V_c/V_t$, $V_t = \ell S$ is the tube volume, and $\theta = \arctan k\ell\alpha$. Combining Eqs. D–6, D–7, and D–8 yields

$$\begin{aligned}\frac{P_{cav}}{P_{in}} &= \frac{1 + R}{e^{jk\ell} + Re^{-jk\ell}} = \frac{e^{j\theta} + e^{-j\theta}}{e^{j(k\ell+\theta)} + e^{-j(k\ell+\theta)}} = \frac{\cos\theta}{\cos(\theta + k\ell)} \\ &= \frac{1}{\cos k\ell - \alpha k\ell \sin k\ell}.\end{aligned} \tag{D–9}$$

Now analyze the result. At very low frequencies ($k\ell \ll 1$), the probe tube has the ideal response $P_{cav}/P_{in} = 1$. As frequency increases, however, resonances are encountered. Resonance occurs when the denominator in Eq. D–9 vanishes, i.e., when

$$\cot k\ell = \alpha k\ell. \tag{D–10}$$

Roots of this equation are tabulated. (See, for example, Ref. 1, p. 225.) As the graphical solution in Fig. 4.30 shows, the roots are found from the intersections of the straight lines $\alpha k\ell$ with the various branches of the cotangent curve. When $\alpha = 1$ ($V_c = V_t$), the first root occurs at $k\ell = 0.86$. Consider, for example, a 10-cm-long probe tube in air. If $\alpha = 1$, the probe

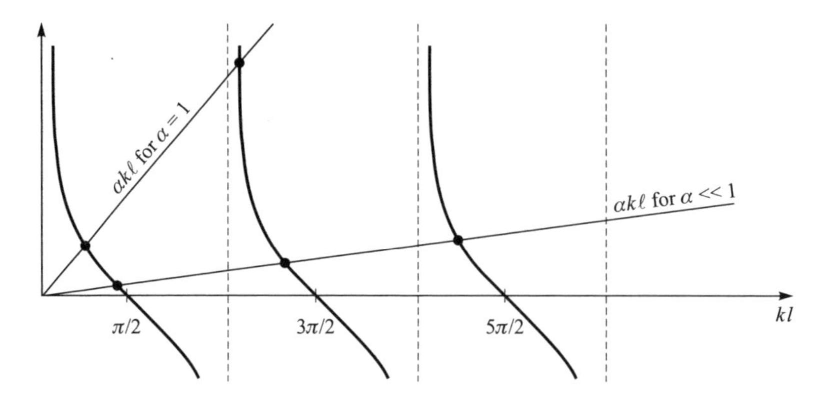

Figure 4.30 Graph showing roots of Eq. D–10.

tube response has a resonance peak at about 470 Hz; succeeding resonances appear at 1870 Hz, 3514 Hz, 5202 Hz, etc. The first resonance can be pushed up in frequency by making $\alpha \ll 1$, as shown in Fig. 4.30, that is, by making $V_c \ll V_t$. In the limit as $\alpha \to 0$, the first root occurs at $k\ell = \pi/2$, that is, at $\ell = \lambda/4$. For the 10-cm probe tube this means a resonance frequency of 860 Hz—better than 470 Hz, but still nothing to crow about. To extend the flat region of the response to higher frequency, one would have to shorten the probe tube. In any case, our analysis shows that the uniform flat response at low frequency is broadest when the cavity volume is made as small as possible.

In practice, damping material, such as fine steel wool, is inserted in the probe tube to smooth out the resonances. Typical response curves for a probe tube used with a Brüel and Kjær $\frac{1}{2}$-in. microphone are shown in Fig. 4.31 for (i) no damping, (ii) just the right amount of damping, and (iii) too much damping.

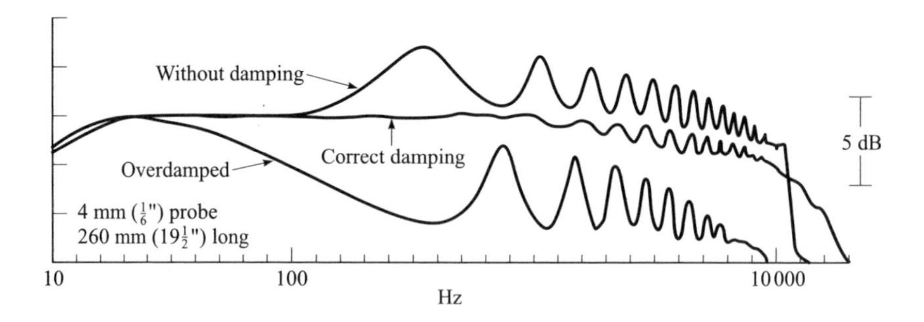

Figure 4.31 Frequency responses for a probe microphone with different degrees of damping. (From instruction manual for Brüel and Kjær $\frac{1}{2}$-in. microphones.)

E. THREE-MEDIUM PROBLEMS

Up to now it has generally been assumed that when a wave is incident on an interface, the second medium is of infinite thickness. If the second medium is of finite thickness ℓ, at which point a second interface marks the beginning of a third medium, the wave transmitted into the second medium from the first is reflected back and forth between the two interfaces. In other words a standing wave field is set up in the second medium. Interesting phenomena occur when the thickness of the second medium is a certain fraction of a wavelength and when the impedances of the three media have a special relation to each other.

Of course, any change in *acoustic impedance* Z_{ac}—whether it be due to a change in medium, a change in cross-sectional area, or both—is sufficient to cause a reflection. Whatever applies to three slabs of different material therefore also applies to tube problems in which three tubes of different diameter are connected in series. For simplicity, the analysis is done first for three different media all having the same cross section. The results are then generalized by substituting acoustic impedances for specific acoustic impedances.

1. Three Different Media, Constant Cross Section

Refer to Fig. 4.32. A time-harmonic incident wave enters medium I from the left. Reverting to the x coordinate system, we take the origin at the first interface. The pressure fields in media I and II are as follows:

$$\text{I}: \quad P_{\text{I}} = A_1 e^{-jk_1 x} + B_1 e^{jk_1 x}, \tag{E–1}$$

$$\text{II}: \quad P_{\text{II}} = A_2 e^{-jk_2 x} + B_2 e^{jk_2 x}. \tag{E–2}$$

For medium III it is convenient to take $x = \ell$ as the origin for the transmitted wave:

$$\text{III}: \quad P_{\text{III}} = A_3' e^{-jk_3 x} = A_3 e^{-jk_3 (x-\ell)} \tag{E–3}$$

(A_3', the amplitude in medium III referred to the origin $x = 0$, is used in Sec. E.2 below). Continuity of pressure and particle velocity at the first interface leads to

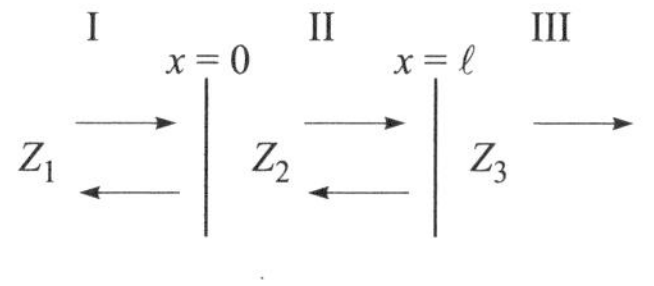

I II III
$x = 0$ $x = \ell$
Z_1 Z_2 Z_3

Figure 4.32 Incident, reflected, and transmitted wave fields for two interface problems.

$$A_1 + B_1 = A_2 + B_2, \tag{E–4}$$

and

$$A_1 - B_1 = \frac{Z_1}{Z_2}(A_2 - B_2), \tag{E–5}$$

respectively. The same requirements at the second interface ($x = \ell$) lead to

$$A_2 e^{-jk_2\ell} + B_2 e^{jk_2\ell} = A_3 \tag{E–6}$$

and

$$A_2 e^{-jk_2\ell} - B_2 e^{jk_2\ell} = \frac{Z_2}{Z_3} A_3, \tag{E–7}$$

respectively. Adding Eqs. E–4 and E–5 eliminates B_1:

$$2A_1 = (1 + Z_1/Z_2)A_2 + (1 - Z_1/Z_2)B_2. \tag{E–8}$$

Equations E–6 and E–7 are simplified by first adding them and then subtracting them:

$$\begin{aligned} \text{Add:} \quad 2A_2 e^{-jk_2\ell} &= (1 + Z_2/Z_3)A_3, \\ \text{Subtract:} \quad 2B_2 e^{jk_2\ell} &= (1 - Z_2/Z_3)A_3. \end{aligned}$$

These two expressions are used to eliminate A_2 and B_2 from Eq. E–8. One may then solve for A_3/A_1, which is the overall transmission coefficient T:

$$\begin{aligned} T &= \frac{4}{(1 + Z_2/Z_3 + Z_1/Z_2 + Z_1/Z_3)e^{jk_2\ell} + (1 - Z_2/Z_3 - Z_1/Z_2 + Z_1/Z_3)e^{-jk_2\ell}} \\ &= \frac{2}{(1 + Z_1/Z_3)\cos k_2\ell + j(Z_2/Z_3 + Z_1/Z_2)\sin k_2\ell}. \end{aligned} \tag{E–9}$$

Let the denominator of Eq. E–9 be called Δ. To find the primary reflection coefficient $R = B_1/A_1$, we first establish a relation between R and T. This may be done by using Eqs. E–6 and E–7 to eliminate A_2 and B_2 from Eq. E–4; the result is

$$1 + R = [\cos k_2\ell + j(Z_2/Z_3)\sin k_2\ell]T.$$

Notice that this is a much more complicated relation than the one that holds for the two-medium problem, Eq. 3A–5. Elimination of T between the last two equations yields

$$R = \frac{(1 - Z_1/Z_3)\cos k_2\ell + j(Z_2/Z_3 - Z_1/Z_2)\sin k_2\ell}{\Delta}. \tag{E–10}$$

Various special cases are now discussed.

(*a*) $k_2\ell = n\pi$ (ℓ an integral number of half wavelengths). In this case the sin $k_2\ell$ terms vanish. The results are

$$T = (-1)^n \frac{2}{1 + Z_1/Z_3} = (-1)^n \frac{2Z_3}{Z_3 + Z_1}, \tag{E–11}$$

$$R = \frac{1 - Z_1/Z_3}{1 + Z_1/Z_3} = \frac{Z_3 - Z_1}{Z_3 + Z_1}, \tag{E–12}$$

which, except for the factor $(-1)^n$, which is associated with the time it takes the wave to pass through medium II, are the same as the expressions for a single interface between two media. In other words, when medium II is exactly n half wavelengths long, its presence is not felt. If in addition $Z_3 = Z_1$, e.g., if media I and III are the same, $T = (-1)^n$ and $R = 0$. Perfect transmission through a barrier may be achieved in this way, but notice that it is only at frequencies for which the barrier thickness ℓ is $n\lambda/2$.

(*b*) $k_2\ell = (n - 1/2)\pi$ (ℓ an odd number of quarter wavelengths). Now the cosine terms drop out, and Eqs. E–9 and E–10 reduce to

$$T = j(-1)^n \frac{2}{Z_2/Z_3 + Z_1/Z_2}, \tag{E–13}$$

$$R = \frac{Z_2/Z_3 - Z_1/Z_2}{Z_2/Z_3 + Z_1/Z_2} = \frac{(Z_2)^2 - Z_1 Z_3}{(Z_2)^2 + Z_1 Z_3}. \tag{E–14}$$

An important special case occurs when $Z_2 = \sqrt{Z_1 Z_3}$, that is, when the impedance of medium II is the geometric mean of the impedances of media I and III. Equation E–14 reduces to

$$R = 0,$$

which implies that all the incident energy passes through to medium III. Although the transmission coefficient T is not unity [in particular, $T = j(-1)^n\sqrt{Z_3/Z_1}$], the sound power transmission coefficient τ is equal to 1; see Eq. 3A–12, applied to this problem. This result is of considerable practical significance. Perfect transmission from material A to material B can be achieved by inserting a quarter-wavelength thickness of a still different material between A and B, provided that the impedance of the inserted material is equal to the geometric mean of the impedances of A and B.

(c) $k_2\ell \ll 1$ (thickness of medium II very small). For small arguments $\cos k_2\ell = 1$ and $\sin k_2\ell = k_2\ell$. Equations E–9 and E–10 then reduce to

$$T = \frac{2}{(1 + Z_1/Z_3) + j(Z_2/Z_3 + Z_1/Z_2)k_2\ell}, \tag{E–15}$$

$$R = \frac{(1 - Z_1/Z_3) + j(Z_2/Z_3 - Z_1/Z_2)k_2\ell}{(1 + Z_1/Z_3) + j(Z_2/Z_3 + Z_1/Z_2)k_2\ell}. \tag{E–16}$$

Further reductions now depend on the size of the impedance ratios relative to $k_2\ell$. If $k_2\ell$ is so small that the imaginary terms are negligible regardless of the size of the impedance ratios, we find

$$T = \frac{2Z_3}{Z_3 + Z_1}, \tag{E–17a}$$

$$R = \frac{Z_3 - Z_1}{Z_3 + Z_1}, \tag{E–17b}$$

which are the coefficients for an ordinary two-medium problem. Two gases or two liquids can therefore be separated by a very thin membrane of solid material, such as Mylar, without affecting the (low frequency) acoustical reflection and transmission properties of the interface (see Problem 4E–1).

Of more practical importance is the case in which $Z_1 = Z_3$ but $Z_2 \gg Z_1$. An example is a flexible panel or wall that separates two rooms; one wishes to know the transmission loss through the wall. For this case Eq. E–15 reduces to

$$T = \frac{2Z_1}{2Z_1 + jZ_2k_2\ell} = \frac{1}{1 + j\omega m/2Z_1} \tag{E–18}$$

(since $Z_2k_2\ell = \omega\rho_2\ell = \omega m$), where m is the mass per unit area, or surface density, of the panel. Notice that the sound speed of the second medium no longer appears in the formula. It is important now only in the restriction $k_2\ell \ll 1$. The form of Eq. E–18 suggests an inductance-resistance circuit like that shown in Fig. 4.33. The wall is seen to behave as a low pass filter. At very low frequency, or for very light walls, $T \to 1$, i.e., the wall permits sound to pass freely from one room to another. At high frequency, or for heavy walls, on the other hand, $\omega m/2Z_1 \gg 1$, and Eq. E–18 reduces to

$$T = \frac{2Z_1}{j\omega m}.$$

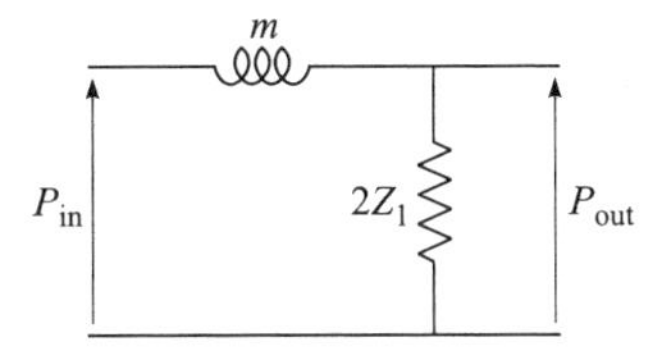

Figure 4.33 Equivalent electrical circuit for transmission through a wall or panel at normal incidence.

In this case only a small fraction of the sound passes through the wall.

The effectiveness of a wall as a sound barrier is conveniently discussed in terms of the transmission loss, which is defined in Sec. 3A.3. Since media I and III are the same in this case, we have (replacing T^2 by $|T|^2 = TT^*$ because T is here complex)

$$\mathrm{TL} = -10 \log_{10} |T|^2. \tag{E–19}$$

Use of Eq. E–18 leads to the formula

$$\mathrm{TL} = 10 \log_{10}[1 + (\omega m/2Z_1)^2] \tag{E–20}$$
$$= 10 \log_{10}[1 + (\pi f m/Z_1)^2]. \tag{E–21}$$

Since the transmission loss is 3 dB when the frequency is

$$f_{3\mathrm{dB}} = Z_1/\pi m,$$

Eq. E–21 may be expressed as

$$\mathrm{TL} = 10 \log_{10}[1 + (f/f_{3\mathrm{dB}})^2]. \tag{E–22}$$

See Fig. 4.34. At high frequency the transmission loss reduces to

$$\mathrm{TL} = 20 \log_{10}(\omega m/2Z_1)$$
$$= 20 \log_{10}(\pi f m/Z_1). \tag{E–23}$$

This is one statement of the so-called *mass law* of sound barriers. As Fig. 4.34 shows, each doubling of either the frequency or the mass of the wall produces 6 dB more transmission loss. In practice the 6-dB/octave slope at high frequencies is not often achieved, for reasons given in Chap. 5.

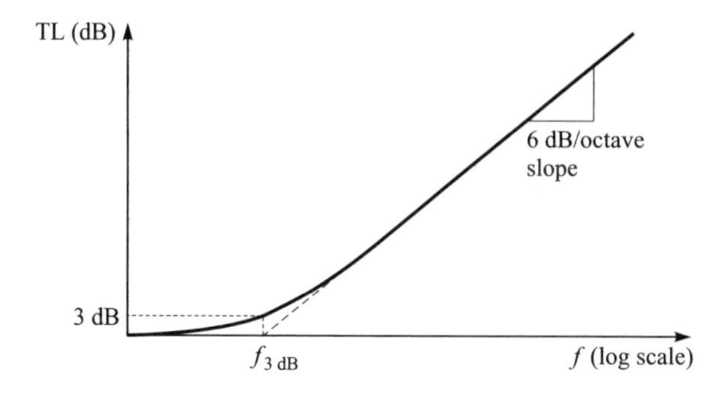

Figure 4.34 Normal-incidence transmission loss through a wall or panel.

2. Cross Sections Different for the Three Media

If the cross-sectional area changes as well as the specific acoustic impedance (see Fig. 4.35), the only modification of the previous analysis is to require that the volume velocity $Q = SU$ be continuous across each interface. Equations E–4 and E–6, which are deduced from the pressure boundary condition, remain the same, but Eqs. E–5 and E–7 are replaced by

$$A_1 - B_1 = \frac{Z_1 S_2}{Z_2 S_1}(A_2 - B_2) = \frac{Z_{\text{ac1}}}{Z_{\text{ac2}}}(A_2 - B_2), \tag{E–24}$$

$$A_2 e^{-jk_2\ell} - B_2 e^{jk_2\ell} = \frac{Z_2 S_3}{Z_3 S_2} A_3 = \frac{Z_{\text{ac2}}}{Z_{\text{ac3}}} A_3, \tag{E–25}$$

respectively. Because the form of these equations is exactly the same as that of Eqs. E–5 and E–7, the previous expressions for T and R (Eqs. E–9 and E–10, respectively) may be generalized to cover the present case simply by replacing Z_i with $Z_{\text{ac}i}$.

When the medium is the same in all three sections, the formula for the transmission coefficient may be simplified by redefining it in terms of A_3' (see Eq. E–3),

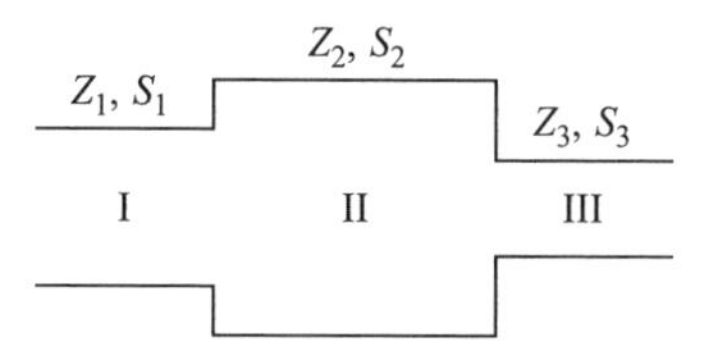

Figure 4.35 Three-medium problem in which the cross sections change.

$$T' \equiv \frac{A_3'}{A_1} = \frac{A_3 e^{jk_3\ell}}{A_1} = Te^{jk\ell}$$

(note that $k_3 = k_2 = k_1 = k$ in this case). The expression for T' is then

$$T' = \frac{2e^{jk\ell}}{(1 + S_3/S_1)\cos k\ell + j(S_3/S_2 + S_2/S_1)\sin k\ell}. \tag{E–26}$$

The formula for the reflection coefficient is

$$R = \frac{(1 - S_3/S_1)\cos k\ell + j(S_3/S_2 - S_2/S_1)\sin k\ell}{\Delta}, \tag{E–27}$$

where, as before, we have let Δ stand for the denominator in the expression for the transmission coefficient. The convenience of using T' instead of T may now be seen. In the special cases $k\ell = n\pi$ and $k\ell = (n - 1/2)\pi$, the factors $(-1)^n$ and $j(-1)^n$, respectively, do not appear; that is, in place of Eqs. E–11 and E–13 one has

$$T' = \frac{2S_1}{S_1 + S_3} \qquad (\ell = \text{integral number of half wavelengths}),$$

$$T' = \frac{2}{S_3/S_2 + S_2/S_1} \qquad (\ell = \text{odd number of quarter wavelengths}),$$

respectively.

Two special cases are of particular interest. First, when $k\ell = (n - \frac{1}{2})\pi$ (ℓ an odd number of quarter wavelengths), Eq. E–27 reduces to

$$R = \frac{S_1S_3 - S_2^2}{S_1S_3 + S_2^2}, \tag{E–28}$$

which vanishes when $S_2 = \sqrt{S_1S_3}$. To obtain perfect transmission from pipe A to pipe B, all that is needed is to insert an intermediate section whose length is $\lambda/4$ and whose cross-sectional area is the geometric mean of the cross-sectional areas of A and B.

Second, let us treat an orifice in a duct (Fig. 4.36) as a three-duct problem in which $S_3 = S_1$, the orifice is the middle section, and ℓ', which includes the end correction, replaces ℓ. Given the assumptions that $k\ell' \ll 1$ and $S_2 \ll S_1$, Eq. E–26 becomes

$$T' = \frac{2}{2 + j(S_1/S_2)k\ell'} = \frac{1}{1 + (j\omega\rho_0\ell'/S_2)/(2\rho_0c_0/S_1)} = \frac{1}{1 + Z_{ac2}/2Z_{ac1}}. \tag{E–29}$$

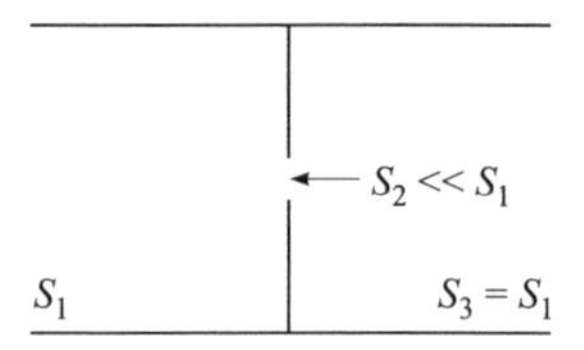

Figure 4.36 Orifice in a duct as a three-medium problem.

From this expression, which is analogous to Eq. E–18, we see that the effective acoustic impedance of the orifice is

$$Z_{ac}^{(\text{orifice})} = j\omega\rho_0\ell'/S_{\text{orifice}}.$$

This is the same as Eq. C–28, except that, because the orifice does not radiate sound into an open fluid, the radiation resistance term is omitted. If a damping term is required, see Ref. 7.

3. Sound Power Reflection and Transmission Coefficients

The power transmission coefficient τ is found as follows:

$$\begin{aligned}\tau &\equiv \frac{\text{power transmitted into medium III}}{\text{power incident}} = \frac{S_3 I_{\text{III}}}{S_1 I_{\text{I}}^{+}} \\ &= \frac{S_3 (p_{\text{III}_{\text{rms}}})^2/Z_3}{S_1 (p^{+}_{\text{I}_{\text{rms}}})^2/Z_1} \\ &= \frac{S_3 Z_1}{S_1 Z_3}\,|T|^2. \end{aligned} \tag{E–30}$$

Notice that it does not matter here whether T or T' is used because $|T'|^2 = |T|^2$. Using Eq. E–9, with Z_{aci} substituted for Z_i, we obtain

$$\tau = \frac{Z_{ac1}}{Z_{ac3}} \frac{4}{(1 + Z_{ac1}/Z_{ac3})^2 \cos^2 k_2\ell + (Z_{ac2}/Z_{ac3} + Z_{ac1}/Z_{ac2})^2 \sin^2 k_2\ell}. \tag{E–31}$$

The power reflection coefficient $r = |R|^2$ is

$$\begin{aligned} r &= \frac{\text{Power reflected back into medium I}}{\text{Power incident}} \\ &= \frac{(1 - Z_{ac1}/Z_{ac3})^2 \cos^2 k_2\ell + (Z_{ac2}/Z_{ac3} - Z_{ac1}/Z_{ac2})^2 \sin^2 k_2\ell}{(1 + Z_{ac1}/Z_{ac3})^2 \cos^2 k_2\ell + (Z_{ac2}/Z_{ac3} + Z_{ac1}/Z_{ac2})^2 \sin^2 k_2\ell}. \end{aligned} \tag{E–32}$$

F. WALL TRANSMISSION LOSS: LUMPED-ELEMENT APPROACH

In Sec. E the transmission through a wall is calculated in the limit of small values of $k\ell$. In this limit the wall behaves as a lumped mass element. Here we calculate the transmission by assuming at the outset that the wall is a lumped mass. To generalize the treatment, we take the wall to have inherent stiffness and damping as well. For example, a wall fixed at its edges acts like a spring when its midsection is displaced.

First find the appropriate pressure and particle velocity boundary conditions. Let the (mechanical) mass, stiffness, and damping resistance of the wall be M_w, K_w, and R_w, respectively. The incident (p^+), reflected (p^-), and transmitted (p^{tr}) waves are shown in Fig. 4.37. Newton's law for the motion of the wall is

$$M_w \dot{u}_w = -R_w u_w - K_w \xi_w + S(p^+ + p^- - p^{\mathrm{tr}}),$$

where ξ_w is the wall displacement and u_w is the wall velocity. Now express the displacement as $\xi_w = \int u_w\,dt$ and introduce the "per-unit-area" quantities $m = M_w/S$, $r = R_w/S$, and $k = K_w/S$.[10] Rearrangement yields

$$m\dot{u}_w + ru_w + k\int u_w\,dt = p^+ + p^- - p^{\mathrm{tr}}. \tag{F–1}$$

Notice that when the wall has no mass, resistance, or stiffness, Eq. F–1 reduces to the ordinary pressure boundary condition for an interface. Next apply the particle velocity boundary condition. The layer of fluid just to the left of the wall must have the same velocity as the wall, as must the layer of fluid just to the right of the wall, i.e.,

$$u^+ + u^- = u_w = u^{\mathrm{tr}}. \tag{F–2}$$

Up to this point no assumption has been made about the waveform of the incident sound wave. We now assume that the incident wave is time harmonic,

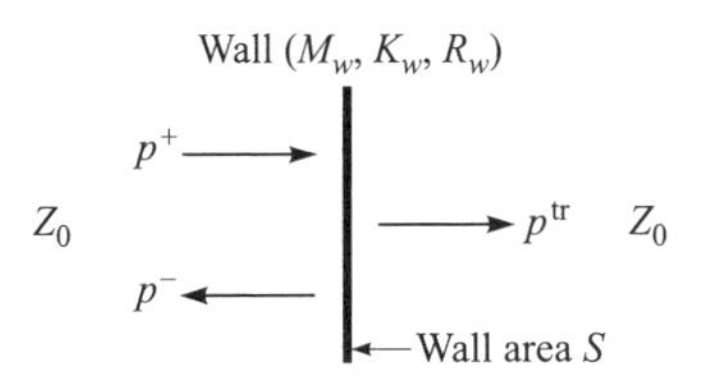

Figure 4.37 Lumped-element model of wall transmission.

[10]Notice that the symbol k used here is not to be confused with wave number ω/c_0.

i.e., $p^+ = P^+ e^{j\omega t}$, and the steady-state solution is sought. Because all the other dependent variables must now also be time harmonic, e.g., $u^+ = U^+ e^{j\omega t}$, $u_w = U_w e^{j\omega t}$, and so on, Eq. F–1 becomes

$$(j\omega m + r + k/j\omega)U_w = P^+ + P^- - P^{\text{tr}}.$$

Moreover, since $U_w = U^{\text{tr}} = P^{\text{tr}}/Z_0$, the equation reduces to

$$(Z_w/Z_0)P^{\text{tr}} = P^+ + P^- - P^{\text{tr}},$$

where $Z_w = j\omega m + r + k/j\omega$ is the specific acoustic impedance of the wall. Division by P^+ yields

$$1 + R = (1 + Z_w/Z_0)T. \tag{F–3}$$

Since the characteristic impedance Z_0 is the same on both sides of the wall, the velocity boundary condition, Eq. F–2, reduces to

$$1 - R = T. \tag{F–4}$$

The solution of Eqs. F–3 and F–4 for the transmission and reflection coefficients is

$$T = \frac{2Z_0}{(2Z_0 + r) + j(\omega m - k/\omega)} = \frac{1}{(1 + r/2Z_0) + (j\omega m/2Z_0)[1 - (\omega_0/\omega)^2]}, \tag{F–5}$$

where $\omega_0 = \sqrt{k/m}$ is the resonance (angular) frequency of the wall based on its own mass and stiffness, and

$$R = \frac{r/2Z_0 + (j\omega m/2Z_0)[1 - (\omega_0/\omega)^2]}{(1 + r/2Z_0) + (j\omega m/2Z_0)[1 - (\omega_0/\omega)^2]}. \tag{F–6}$$

If r and k vanish, Eqs. F–5 and F–6 reduce to expressions found in Sec. E. For example, Eq. F–5 reduces to Eq. E–18. If $r \ll 2Z_0$, then at resonance ($\omega = \omega_0$), $T = 1$ and $R = 0$. In other words at the resonance frequency, if the damping is small, sound travels unimpeded through the wall. A qualitative plot of the transmission loss $\text{TL} = 10 \log_{10} |T|^{-2}$ is given in Fig. 4.38. The mass law of sound transmission loss is represented by the right-hand branch of the TL curve.

Example 4.4. Transmission loss for a door. Find T and TL at 100 Hz for a hollow-core door that is 1 in. thick, 78 × 36 in. in area, and weighs 39 lb. The value of m for this door is

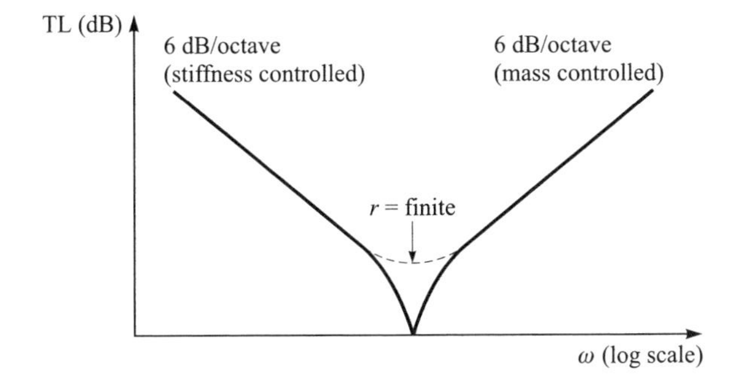

Figure 4.38 Transmission loss for a wall having mass, resistance, and stiffness.

$$m = \frac{39}{(3)(6.5)} = 2\ \text{lb/ft}^2.$$

If one assumes that $f_0 \ll 100$ Hz, the terms involving k and r in Eq. F–5 may be ignored. When m is given in English units, it is convenient to express Z_0 in English units. For air $Z_0 = 85.0\ \text{lb/ft}^2\text{s}$.[11] We obtain

$$T = \left(1 + \frac{j\omega m}{2Z_0}\right)^{-1} = \left(1 + \frac{j2\pi(100)(2)}{2(85)}\right)^{-1} = (1 + j7.39)^{-1}.$$

The transmission loss is therefore

$$\text{TL} = 20 \log_{10} |T| = 20 \log_{10} 7.46 = 17.5\ \text{dB}.$$

In this case the quantity 1 could have been ignored relative to $j\omega m/2Z_0$. The result would have been $T \doteq (j\omega m/2Z_0)^{-1} = 1/j7.39$, or

$$\text{TL} = 20 \log_{10} 7.39 = 17.4\ \text{dB}.$$

Thus at 100 Hz the mass law applies very well to this particular door.

REFERENCES

1. M. Abramowitz and I. A. Stegun (Eds.), *Handbook of Mathematical Functions*, National Bureau of Standards, Applied Mathematics Series 55 (U.S. Government Printing Office, Washington, DC, 1964).
2. Y. Ando, "On the sound radiation from semi-infinite circular pipe of certain wall thickness," *Acustica*, **22**, 219–225 (1969/1970).

[11] The value of Z_0 in English units for fresh water is $3.03 \times 10^5\ \text{lb/ft}^2\text{s}$.

3. J. Y. Chung and D. A. Blaser, "Transfer function method of measuring in-duct acoustical properties I. Theory, II. Experiment," *J. Acoust. Soc. Am.* **68**, 907–921 (1980).
4. J. L. Hunt, "Accurate experiment for measuring the ratio of specific heats of gases using an accelerometer," *Am. J. Phys.* **53**, 696–697 (1985).
5. H. Levine and J. Schwinger, "On the radiation of sound from an unflanged circular pipe," *Phys. Rev.* **73**, 383–406 (1948).
6. R. K. Moore, *Traveling-Wave Engineering* (McGraw-Hill, New York, 1960), Chap. 5.
7. P. M. Morse and K. U. Ingard, *Theoretical Acoustics* (McGraw-Hill Book Co., New York, 1968), pp. 482–483. Reprinted by Princeton University Press, Princeton, NJ, 1986.
8. R. L. Panton and J. M. Miller, "Resonant frequencies of cylindrical Helmholtz resonators," *J. Acoust. Soc. Am.* **57**, 1533–1535(L) (1975).
9. A. D. Pierce, *Acoustics: An Introduction to Its Physical Principles and Applications* (McGraw-HIll, New York, 1981), Sec. 7–6. Reprinted by the Acoustical Society of America, New York, 1989.
10. Lord Rayleigh, *Theory of Sound* (Dover Publications, New York, 1945, first American ed.), Vol. 2, Secs. 307 and 312 and Appendix A.
11. E. Ruchardt, *Phys. Z.* **30**, 58–59 (1929).

PROBLEMS

4B–1. A tube is terminated with an impedance $Z_n = jZ_0$.

(a) Sketch phasor diagrams for the relative pressure P/P^+ and relative particle velocity U/U^+ to indicate how these quantities vary with kd, where d is the distance from the termination. For each diagram show the resultant and give its magnitude and phase.

(b) Working backward from the termination, sketch the standing wave patterns for both $|P/P^+|$ and $|U/U^+|$ as a function of kd.

(c) Sketch the normalized impedance $Z/j\rho_0 c_0$ as a function of kd.

4B–2. A pipe two wavelengths long is terminated by an impedance $Z_n = -j2Z_0$, where Z_0 is the specific acoustic impedance of the fluid in the pipe.

(a) Using phasor diagrams, show how the relative pressure $|P/P^+|$ and relative particle velocity $|U/U^+|$ vary as functions of kd, where d is distance from the termination. For each diagram show the resultant and give its magnitude and phase.

(b) Sketch the standing wave patterns for both $|P/P^+|$ and $|U/U^+|$.

(c) Where could the pipe be cut off and left open without disturbing the standing wave pattern in the region between the source and the opening? (*Hint*: Recall that $p = 0$ at an open end.)

(d) What can you say in general about the standing wave pattern whenever the termination Z_n is pure imaginary?

4B–3. Consider the semi-infinite pipe shown in the figure. The reflection coefficient of the gaseous interface is $R = \frac{1}{3}$. A piston at $x = 0$ generates a 686-Hz sinusoidal acoustic signal, and we are concerned with the standing wave field in the first meter (air-filled portion) of the pipe. Use 343 m/s as the value for c_0.

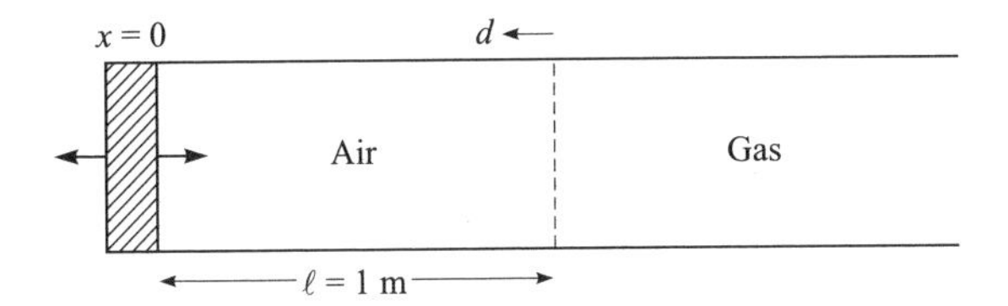

(a) Use phasor diagrams to determine magnitude and phase of the relative pressure P/P^+ and relative particle velocity U/U^+ at various points along the air-filled portion of the pipe.

(b) Sketch the standing wave pattern for both $|P/P^+|$ and $|U/U^+|$. Where do the pressure maxima and minima occur (in terms of wavelengths from the point $d = 0$)? If R were $-\frac{1}{3}$ instead of $+\frac{1}{3}$, where would the maxima and minima occur?

(c) What is the pressure SWR? What would it be if R were $-\frac{1}{3}$ instead of $+\frac{1}{3}$?

4B–4. A pipe is terminated by an impedance Z_n. The reflection coefficient, measured by a pulse method, is $R = -\frac{1}{3}$.

(a) Find Z_n.

(b) Sketch phasor diagrams for P/P^+ and U/U^+. For each diagram show the resultant and give its magnitude and phase.

(c) Sketch the standing wave patterns for the pressure and particle velocity.

(d) Find the standing wave ratio.

4B–5. An underwater ocean research vessel sends a collimated beam of plane waves upward toward the sea surface. The reflected plane wave beam superposes on the incident beam to form a standing wave pattern. The pressure SWR is measured continuously as the vessel travels (horizontally) through the water. For a long time the SWR has a very large value (approaching ∞). Then suddenly the SWR drops to a value of 1.9 and stays that way as the vessel continues to travel through the water. What is the most likely explanation for the initial value of SWR and the change in SWR? Support your answer with quantitative calculations. Assume all pertinent impedances are real.

4B–6. A 750-Hz plane sound wave in seawater (13°C) is normally incident on a stone seawall. The sound field is found to have a maximum at the wall; the SPL there is 190 dB re 1 μPa. The SPL 0.5 m from the wall is 177 dB.

(a) Use the information given about the sound field to find the impedance of the wall.

(b) Find the SPL of the incident wave.

4B–7. A plane-wave tube is to be used to study progressive waves in air. An absorptive termination is needed to prevent reflections from the end. The maximum standing wave ratio, expressed in decibels, that can be tolerated is 2 dB, i.e.,

$$20\log_{10} \mathrm{SWR} \leq 2 \text{ dB}.$$

A candidate for the termination is a tough fibrous material whose impedance over a wide frequency range is

$$Z = 540 - j166 \text{ MKS rayls}.$$

Determine whether this material satisfies the SWR criterion.

4B–8. An air-filled standing wave tube is terminated with an unknown specific acoustic impedance Z_n. At a frequency of 1143 Hz the SWR is 1.4. As the distance d from the termination is increased, the pressure amplitude increases monotonically until it reaches a maximum at $d = 0.05$ m. Find the following:

(a) Wavelength

(b) Reflection coefficient R (in polar form, $R = \rho e^{j\psi}$)

(c) Impedance Z_n (real and imaginary parts)

4B–9. A round tube is connected to a tube of square cross section. The diameter of the round tube is exactly equal to the length ℓ of one side of the square tube. What are the values of the reflection and transmission coefficients for sound waves traveling from the square tube into the round tube? For sound waves traveling the opposite direction? What is the standing wave ratio for the two cases?

4B–10. Sketch the pressure standing wave pattern for the sound field

(a) In the small tube.

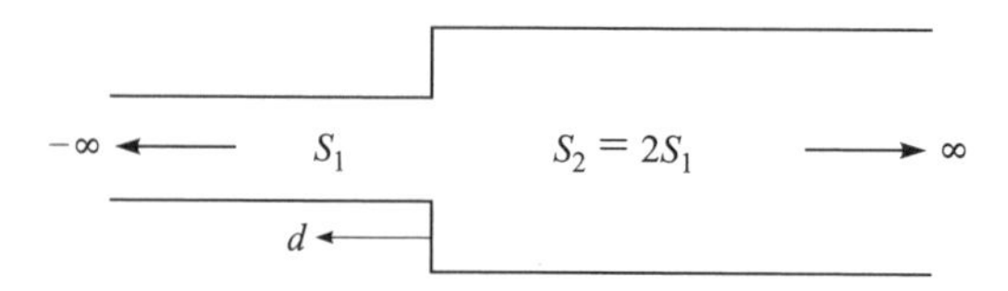

(b) In the large tube.

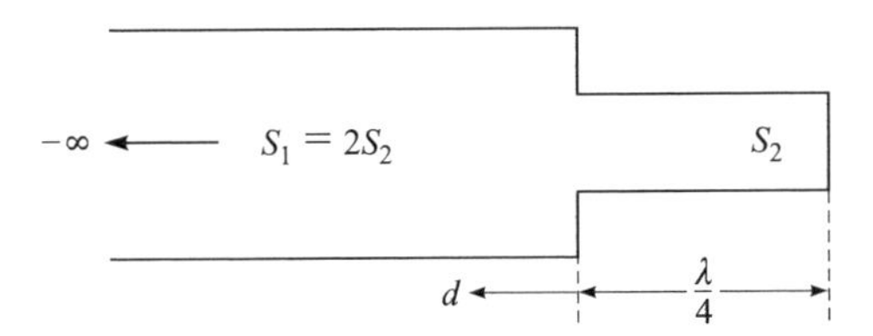

4B–11. Two tubes containing air and an unknown gas, both at 0°C, are connected as shown (a massless membrane at the junction prevents diffusion of one gas into the other). When a periodic sound field is established in the air-filled tube, the standing wave ratio is found to be very nearly equal to unity. Identify the unknown gas.

$S_1 = 10\ \text{cm}^2$ | $S_2 = 12\ \text{cm}^2$

Air (0°C) | Unknown gas (0°C)

4B–12. A rigid conical horn is connected to a tube as shown. The problem is to determine how well sound is transmitted across the connection. The quantity r_0 is the distance from the connection to the virtual apex of the cone. The terminating impedance Z_n for the tube in this case is the impedance presented by the conical horn. Since the waves propagating in the horn are simply segments of spherical waves which seem to emanate from the virtual apex, Z_n is given to good approximation by Eq. 1D–21.

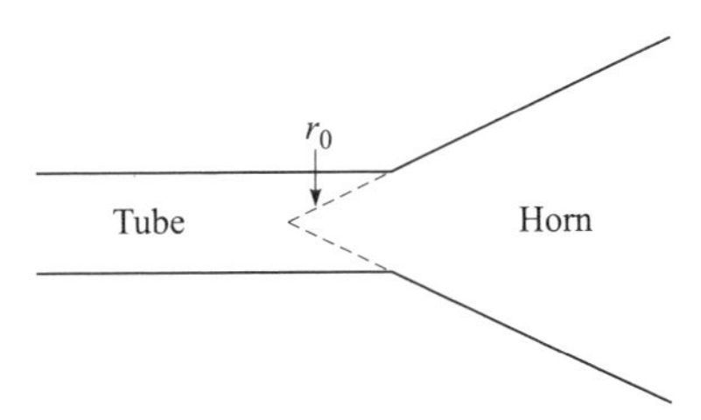

(a) Derive the following expression for the reflection coefficient for waves traveling down the tube toward the connection:

$$R = -\frac{1}{1 + j2kr_0}.$$

(b) Find the SWR in the tube when the wavelength of the incident sound is $4\pi r_0$.

✓ **4B–13.** Let the terminating impedance Z_n at the end of a tube be pure imaginary, $Z_n = jX_n$, where X_n is real. Show that in this case the reflection coefficient has magnitude 1. Moreover show that the impedance at any point d in front of the termination is pure imaginary. In particular, prove Eqs. B–16 and B–17.

4B–14. Two microphones are mounted in the sidewall of an air-filled tube that is terminated with a material of unknown impedance Z_n. The frequency of a test signal is 1000 Hz. The distance of microphone 1 from the material is 54.59 mm ($kd_1 = 1$ rad). Microphone 2 is mounted next to the material ($kd_2 = 0$). The ratio of the pressures measured by the two microphones is

$$\frac{P_2}{P_1} = 1.885\, e^{-j1.157},$$

where the phase is in radians. Find the value of Z_n.

4C–1. A steel ball of density 7700 kg/m^3 and diameter 2 cm fits closely in a gas-filled glass tube, as in Fig. 4.14. The column of gas is 1.0132 m long and the pressure inside is standard atmospheric. When the ball is disturbed, it is found to oscillate with frequency 5.672 Hz.

(a) What is the value of γ for the gas?

(b) If the gas is pure, i.e., not a mixture, what gas is it?

4C–2. A ball made of an unknown metal (pure metal, not an alloy) fits closely in a glass tube containing air at STP, as shown in Fig. 4.14. The diameter of the ball is $2a = 11.68$ mm. When the length of the tube is adjusted to $\ell = 2.475$ m, the resonance frequency of the ball-cavity system is found to be 10 Hz.

(a) Find the ball density ρ_{ball}.

(b) From the value of ρ_{ball}, identify the metal.

4C–3. A soft drink can is to be used as a Helmholtz resonator. Although the metal used in most soft drink cans is so thin that the wall and ends do not act as rigid surfaces, nevertheless, for purposes of this problem, assume that the can surfaces are rigid.

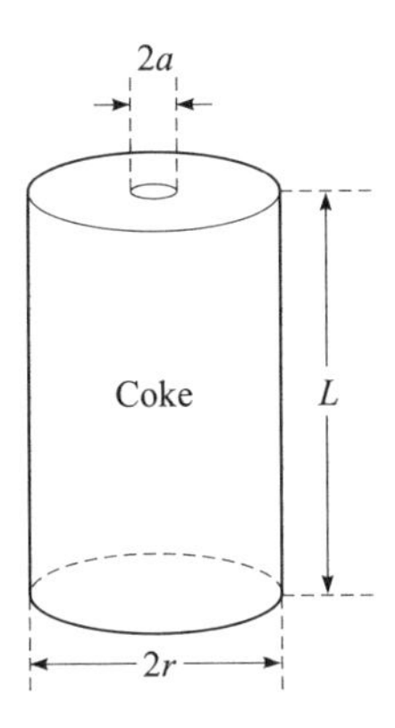

(a) Measure the dimensions indicated in the sketch from an actual can. Idealize the opening in the top as a round hole of radius $a = \sqrt{S_1/\pi}$, where S_1 is the measured area of the actual opening. (*Suggestion*: To measure S_1, trace the tab on graph paper and then count squares.) Assume the hole is centered in the top of the can.

(b) Find the resonance frequency f_0 by using the usual formula for a Helmholtz resonator.

(c) Is the answer found in part (b) accurate? (*Hint*: What is the fundamental restriction on which the lumped-element model is based?)

(d) Find and carry out an improved calculation of the resonance frequency f_0' of the can. By what percentage does your new answer differ from that in part (b)?

4C–4. Two holes, 1 and 2, of radii a_1 and a_2 ($a_2 \neq a_1$), respectively, are drilled in a hollow metal sphere. The volume of the sphere is V, and the thickness ℓ of the metal wall is very small. When only hole 1 is open, i.e., hole 2 is covered up, the resonance frequency is found to be 100 Hz (call this frequency f_1). When only hole 2 is open, the resonance frequency is found to be 200 Hz (call this frequency f_2). Neglect the effect of radiation resistance.

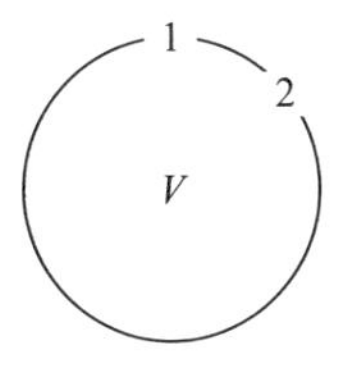

(a) What is the ratio of the radii, a_2/a_1?

(b) What is the resonance frequency when both holes are left open?

4C–5. The mass-cavity systems shown in sketches (*a*) and (*b*) contain the same gas and have the same mass M_m, volume V, and cavity length ℓ. Assume that $k\ell \ll 1$, where k is the wave number of sound generated at resonance.

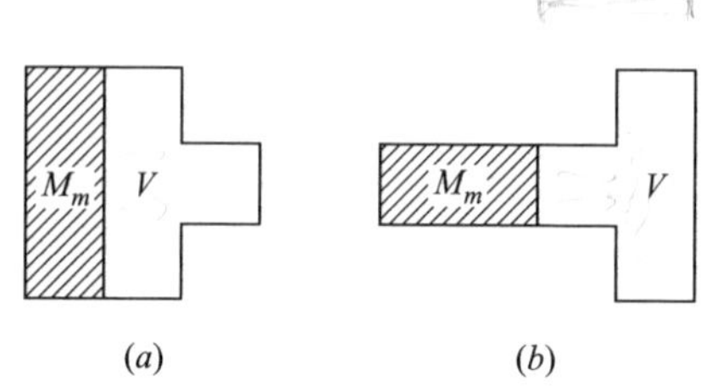

(a) Are the two resonance frequencies f_a and f_b the same?

(b) If so, explain why. If not, find the ratio f_b/f_a.

4C–6. An air-filled semi-infinite tube is terminated by a rigid plate that has a small hole (radius a) in it. Find the standing wave ratio for the sound field in the tube.

(a) First neglect the effect of the radiation resistance.

(b) Then discuss qualitatively how including the radiation resistance would affect your result.

4C–7. An air-filled tube is terminated by a Helmholtz resonator. The source S establishes a sound field in the tube at the resonance frequency of the Helmholtz resonator. Using kd as the scale along the horizontal axis, sketch the pressure standing wave pattern in the tube.

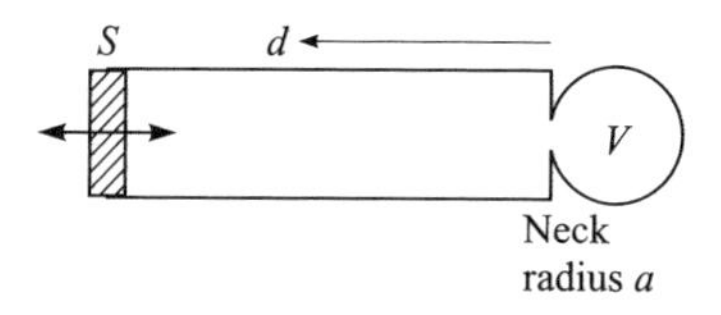

(a) First neglect the effect of damping.

(b) Then indicate qualitatively how a small amount of damping (added to the resonator impedance) would change the pattern found in part (a).

4C–8. The pressure response function for a Helmholtz resonator is given by Eq. C–26.

(a) Show that the response function may be expressed as

$$\frac{P_{\text{ch}}}{P_{\text{in}}} = \frac{1}{1 - \chi^2 + j\delta\chi^3},$$

where $\chi = k/k_0 = \omega/\omega_0$ and $\delta = k_0^3 V/2\pi$. The lumped-element approximation implies that $\delta \ll 1$.

(b) Let $A = |P_{\text{ch}}/P_{\text{in}}|$ stand for the pressure amplification factor. By maximizing A with respect to χ, show that, given $\delta \ll 1$, $A_{\max}$ occurs at the natural resonance (angular) frequency ω_0.

(c) Find $A_{\max}$.

4D–1. An 8-driver feed for a horn is desired (see sketch). If the diameter of each of the smallest connecting tubes (1–1, 1–2, etc.) is 25 mm, what should be the diameters of the remaining branches and the horn throat in order that none of the sound be reflected back to any of the drivers? Assume that the length of each branch is the same.

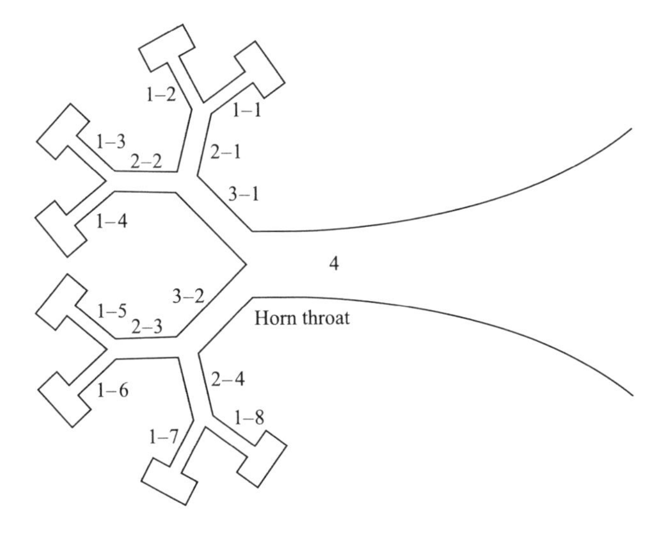

4D–2. Design a Helmholtz resonator to be used as an acoustic filter against low frequency noise in a rectangular air ventilating duct for which the cross section is 12 × 6 in. The center of the stop band of the filter is to be 30 Hz, and the transmission loss (TL $= -10 \log_{10} \tau$) at 60 Hz must be 3 dB, i.e., $|T| = 1/\sqrt{2}$. Use a design like that shown in the sketch. The neck of the resonator is a small round hole of radius a; take the effective neck length to be $\ell' = 2\,\Delta\ell$. To design the resonator, you must specify the neck radius a and the chamber volume V.

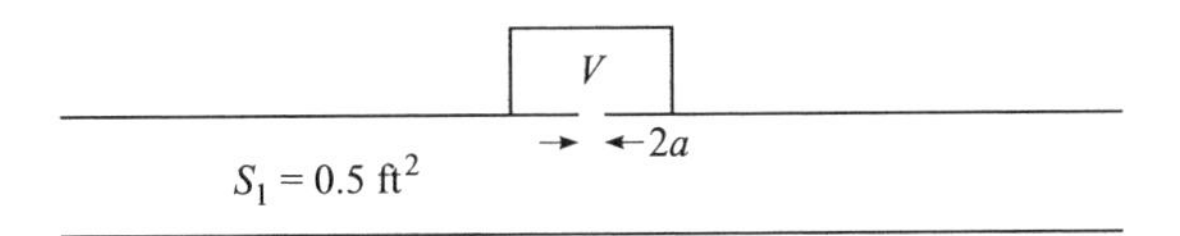

4D–3. The end section of a duct is divided into two equal parts by a rigid septum. The length of the divided section is ℓ. At the end, one of the "half ducts" is closed, the other open (neglect the end correction in this case). Show that the reflection coefficient for waves incident on the place where the septum begins is $R = -\exp(-j4k\ell)$.

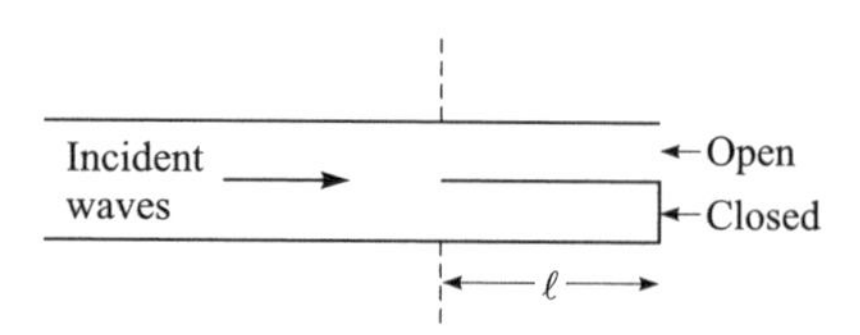

4D–4. A pipe of inside radius r_0 and wall thickness ℓ_0 has a small hole of radius a drilled in the wall. Assume that the frequency of the incident sound is low enough that $ka \ll 1$. For simplicity neglect the resistive component of the hole impedance.

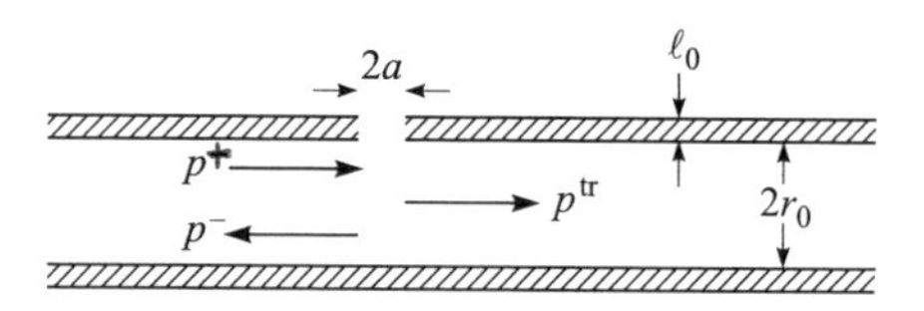

(a) Derive the expression for the transmission coefficient.

(b) Sketch the curve of TL vs. frequency (log scale). From the sketch determine whether the pipe-hole system behaves as a low pass, band pass, or high pass filter.

(c) Find the 3-dB-down frequency f_{3dB} (frequency at which TL = 3 dB). Then put your expression for TL in terms of f and f_{3dB}.

(d) As an example, find f_{3dB} for a $\frac{1}{8}$-in.-diameter hole drilled in the $\frac{1}{16}$-in.-thick wall of a pipe of inside diameter 2 in.

4E–1. An experiment on sound transmission from helium ($c_0 = 1000$ m/s, $Z_0 = 166$ MKS rayls) to air is to be performed. The two gases are separated by a sheet of Mylar ($\rho_0 = 1395$ kg/m^3) of thickness 0.125 mil (1 mil = 0.001 in.). Is it proper to ignore the effect of the Mylar on the transmission? To answer the question, compare the values of $|T|$ found (1) by ignoring the presence of the Mylar entirely and (2) by taking the presence of the Mylar into account. In the latter case it is sufficient to model the Mylar as a thin sheet ($k_2\ell \ll 1$) of high impedance material ($Z_2 \gg Z_1, Z_3$). Calculate the frequency range over which the two values of $|T|$ differ by no more than 1%.

4E–2. Seawater at 0°C ($c_0 = 1449$ m/s, $Z_0 = 1.49 \times 10^6$ MKS rayls) is overlain by ice. Plane monochromatic sound waves in the water are normally incident on the ice interface.

(a) Assume that the ice is infinitely thick.

(i) Sketch the pressure standing wave pattern of the field in the seawater.

(ii) Find the SWR.

(b) The ice layer is really finite in thickness (call the thickness L). Above the ice is air at 0°C. Compute the SWR for this case.

4E–3. A rigid plate with an orifice in it is placed in a 4-in.-(0.1016-m-) diameter air duct to attenuate sound in the duct. Find the diameter of the orifice necessary to provide 20 dB of transmission loss at 1000 Hz.

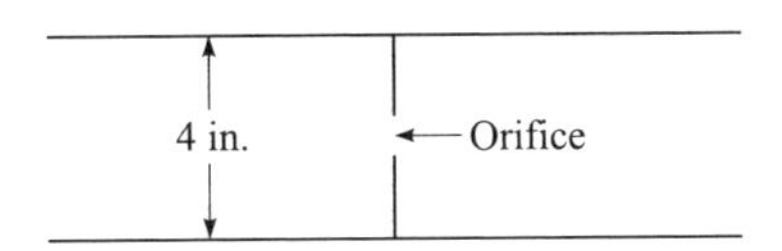

4E–4. An electrical transformer in a large factory room emits objectionable noise, largely at frequencies 60 Hz, 120 Hz, and 240 Hz. A large box is to be built to cover the transformer so as to reduce the noise that gets into the room. The attenuation requirement is that 120-Hz sound waves normally incident on any side of the box (from the inside) are to suffer a transmission loss of 20 dB.

(a) If the box is made of lead sheet, what thickness is required? What is the TL at 60 Hz? At 240 Hz?

(b) Reanswer part (a) for a box made of pine wood.

(c) If the interior dimensions are the same, which box is heavier, lead or pine?

4E–5. By using cork as an intermediate material, find a way to obtain perfect transmission of sound from air to some metal.

(a) What metal? (Use bulk values for Z_0.)

(b) What thickness of cork allows perfect transmission at 1 kHz?

(c) How good is your design? For the field established in the air find $\Delta\mathrm{SPL} = \mathrm{SPL}^+ - \mathrm{SPL}^-$.

4E–6. Show that irrespective of frequency, transmission is perfect (no reflection) in going from medium I through medium II to medium III (see sketch) if the acoustic impedance Z_i/S_i is the same in all three sections. Several means of obtaining this result are possible, including the long, hard way entailed by a complete derivation. Seek a shorter way, such as considering two sections at a time. Another short way would be to take the results for R, T, τ, and R^2 given in the text for three media and to generalize these results by plausible reasoning.

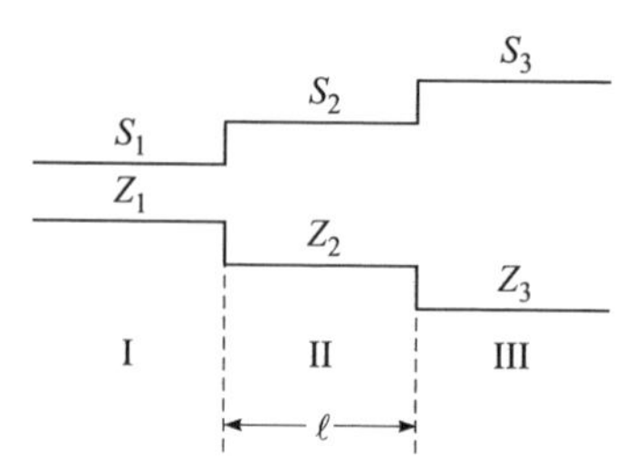

4E–7. It is necessary to propagate sound pulses through a pipe containing air at 20°C in one section, hydrogen at 20°C (same ambient pressure) in a second section, and air at 20°C again in the third section. The pipe diameter of the first and third sections is 5 cm. Design the second section to achieve perfect sound transmission through the entire system.

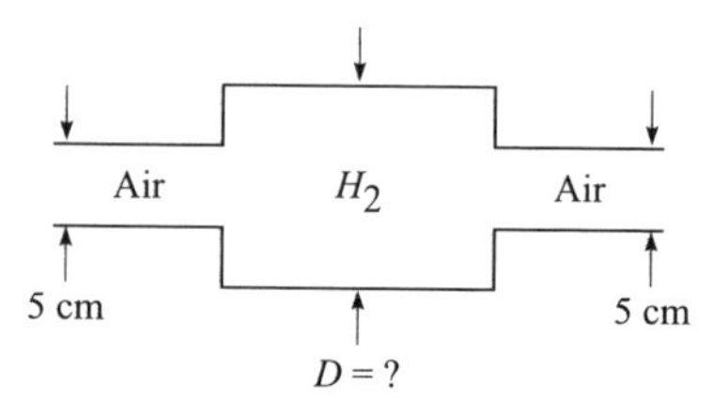

4E–8. Design a $\lambda/4$ matching section to minimize reflection of 8-kHz plane waves transmitted from fresh water into steel. Limit your choice of materials to those listed in Appendix A (for impedance, use bulk values, not bar values). In particular,

(a) Choose the material, i.e., the one that gives the best matching.

(b) Select the thickness of the material.

(c) Check your design by calculating SWR and $20 \log_{10}$ SWR for the sound field in the water.

4E–9. It is desired to transmit sound perfectly from aluminum into inconel nickel ($Z_0 = 64.5 \times 10^6$ MKS rayls). Design a quarter-wavelength matching section to go between the aluminum and the nickel that will allow perfect transmission at a frequency of 100 kHz. Limit your choice of materials to those listed in Appendix A (use bulk values for impedance).

(a) Select the material and its thickness.

√ **(b)** Check your design by computing τ for the material you selected.

4E–10. A fire siren of frequency 4032 Hz is located in a room containing a large freshwater aquarium. The glass wall of the aquarium is 1 in. thick. When the siren goes off, the SPL in the air next to the aquarium wall is measured and found to be 100 dB (re 20 μPa). Assume for simplicity that the incident sound in the air is plane and normally incident on the glass wall of the aquarium and that the 100 dB is the SPL of the combined incident and reflected fields at the wall.

(a) Find SPL^{+}, i.e., the SPL of the incident wave.

(b) Find SPL^{tr} (re 1 μPa), i.e., the SPL of the wave transmitted into the water.

5

TRANSMISSION PHENOMENA: OBLIQUE INCIDENCE

It is now time in our study of reflection and transmission to drop the restriction to normal incidence. When the angle of incidence is oblique, a new wave phenomenon is possible: refraction. In most treatments of refraction, time-harmonic waves are assumed. Many important results may, however, be obtained without making any assumption about the waveform. In Sec. A a simple physical derivation is given of the two most important laws, Snell's law and the law of specular reflection. Section B contains a more mathematical treatment of the two laws, a derivation of the various reflection and transmission coefficients, and a discussion of special cases, such as perfect transmission and total internal reflection. The remainder of the chapter is devoted to transmission loss through panels and walls. The mass law (and its relatives) and the coincidence effect are treated in Sec. C. Section D is about composite walls.

A. SIMPLE DERIVATION OF SNELL'S LAW AND SPECULAR REFLECTION

A plane wave obliquely incident on a plane interface between two fluids is partly reflected and partly transmitted. Transmission is taken up first. Figure 5.1 shows the incident and transmitted fields (for clarity the reflected field is omitted; it appears in Fig. 5.2). The y axis marks the interface that separates medium 1 (properties ρ_1 and c_1) from medium 2 (properties ρ_2 and c_2). The x axis is normal to the interface. The angle of incidence θ_i is the angle the incoming ray makes with the normal to the interface. An arbitrary time waveform $p(t)$ of the incident wave is shown in the inset of Fig. 5.1. Focus attention

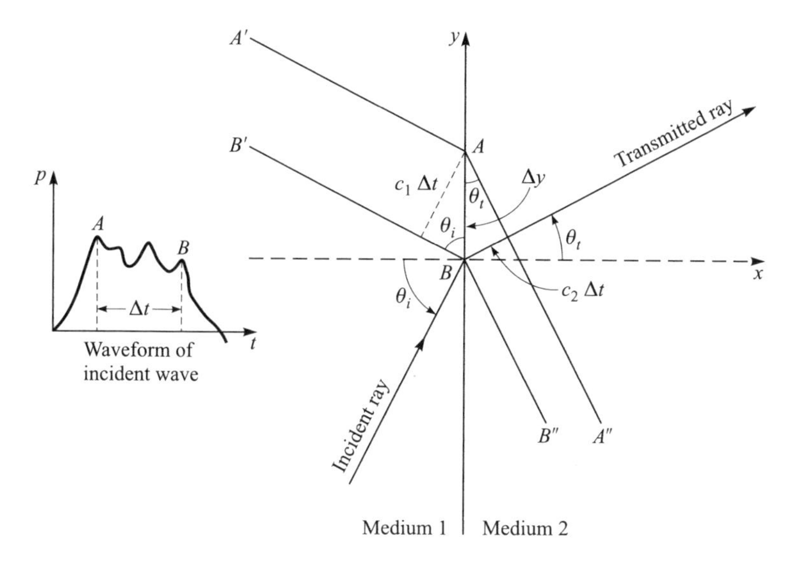

Figure 5.1 Transmission through an interface at oblique incidence. Only the incident and transmitted rays and wavefronts are shown. The reflected ray and wavefronts are shown in Fig. 5.2.

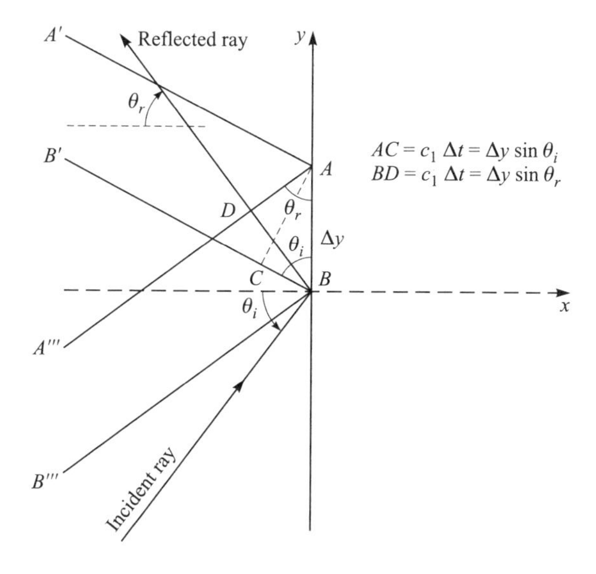

Figure 5.2 Reflected rays and wavefronts for oblique incidence. The transmitted ray and wavefronts are shown in Fig. 5.1.

on points A and B of the waveform, separated in time by Δt (which need not be small). Point B follows the straight line marked "Incident ray." Ahead of the B wavefront (line BB') is the A wavefront, line AA' (the incident ray associated with point A is not shown). The two wavefronts are separated in space by distance

$$c_1 \, \Delta t = \Delta y \sin \theta_i, \tag{A–1}$$

where Δy is the section of the interface intercepted by the A and B wavefronts. In medium 2 the transmitted ray makes an angle θ_t with the normal to the interface, the A wavefront is the line AA'', and the B wavefront is BB''. The spatial separation of the two wavefronts is

$$c_2 \, \Delta t = \Delta y \sin \theta_t. \tag{A–2}$$

Dividing Eq. A–2 by Eq. A–1, we obtain

$$\frac{\sin \theta_t}{\sin \theta_i} = \frac{c_2}{c_1}. \tag{A–3}$$

This is the law of refraction, or Snell's law.

Next consider the reflected field, which is shown along with the incident field in Fig. 5.2. Observe that the reflected wavefronts AA''' and BB''' are separated by the distance BD, given by

$$c_1 \, \Delta t = \Delta y \sin \theta_r. \tag{A–4}$$

Comparison of this equation with Eq. A–1 shows that the angle of reflection must equal the angle of incidence,

$$\theta_r = \theta_i. \tag{A–5}$$

This statement is known as the law of specular reflection.

Snell's law and specular reflection result from the way the traces of the incident, reflected, and transmitted waves track together along the interface. In fact the trace velocity along the interface, i.e., the apparent speed with which a point on the waveform (such as B) travels along the interface, is

$$c_{\text{trace}} = \frac{\Delta y}{\Delta t} = \frac{c_1}{\sin \theta_i} = \frac{c_1}{\sin \theta_r} = \frac{c_2}{\sin \theta_t}. \tag{A–6}$$

In other words, the two laws are simply statements that the trace velocities of all three waves must be the same. The trace velocity is sometimes called the phase speed along the interface because it is the travel speed associated with a given phase point on the waveform, for example, point A or point B on the waveform in Fig. 5.1.

Because A and B are any two points on an arbitrary waveform, it follows that Snell's law and the law of specular reflection hold for any plane wave incident on a plane interface. In the next section a slightly more mathematical derivation of the two laws is given. In addition, application of the pressure and particle velocity boundary conditions at the interface leads to expressions for the reflection and transmission coefficients R and T, respectively.

Note that the angles have been defined by the intersection of the rays with the normal to the interface. Sometimes the *grazing* angles, which are the complements of the angles used here, are employed. In this case Snell's law is given in terms of cosines rather than sines, i.e.,

$$\frac{\cos\theta_t'}{c_2} = \frac{\cos\theta_i'}{c_1},$$

where $\theta' = \pi/2 - \theta$. In the present chapter all angles are with respect to the normal. In underwater sound, however, it is conventional to use grazing angles, and we follow that convention in Chap. 8 in the section on ray theory.

B. PLANE INTERFACE SEPARATING TWO FLUIDS

In this section a more detailed analysis of reflection and transmission at oblique incidence is given. Because the wave field is now a function of two spatial coordinates, x and y, we must first obtain the representation of a plane wave traveling at an arbitrary angle θ in an x, y coordinate system. If d is the direction of propagation (see Fig. 5.3), the wave may be represented as follows:

$$\begin{aligned} p &= f\left(t - \frac{d}{c_1}\right) \\ &= f\left(t - \frac{x\cos\theta + y\sin\theta}{c_1}\right), \end{aligned}$$

where c_1 is the speed of sound in the medium.

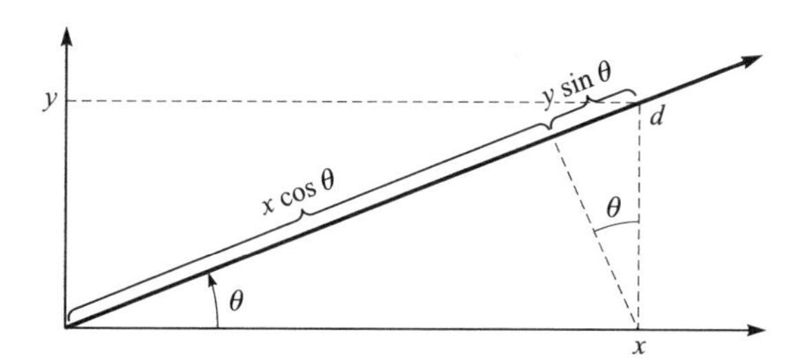

Figure 5.3 Plane wave traveling at an oblique angle θ.

1. Alternative Derivation of Snell's Law; *R*, *T*, and τ Coefficients

Figure 5.4 shows a simplified diagram of the incident, reflected, and transmitted rays, which make angles θ_i, θ_r, and θ_t, respectively, with the normal (the x axis) to the interface (the y axis). The three waves may be represented as follows:

$$\text{Incident wave:} hskip1emp^{+} = p^{+}\left(t - \frac{x\cos\theta_i + y\sin\theta_i}{c_1}\right), \qquad \text{(B–1)}$$

$$\text{Reflected wave:} \quad p^{-} = p^{-}\left(t + \frac{x\cos\theta_r - y\sin\theta_r}{c_1}\right), \qquad \text{(B–2)}$$

$$\text{Transmitted wave:} \quad p^{\text{tr}} = p^{\text{tr}}\left(t - \frac{x\cos\theta_t + y\sin\theta_t}{c_2}\right). \qquad \text{(B–3)}$$

The sign in each argument is determined by the direction the wave is traveling relative to the x and y axes. First apply the pressure boundary condition, which is, as usual, that the pressure must be continuous across the interface ($x = 0$):

$$p^{+}\left(t - \frac{y\sin\theta_i}{c_1}\right) + p^{-}\left(t - \frac{y\sin\theta_r}{c_1}\right) = p^{\text{tr}}\left(t - \frac{y\sin\theta_t}{c_2}\right). \qquad \text{(B–4)}$$

This relation must hold everywhere along the interface, that is, for arbitrary values of y.[1] To suppress the y dependence in the p^+ term, define a new time variable $t' = t - (y\sin\theta_i)/c_1$. Equation B–4 then becomes

$$p^{+}(t') + p^{-}\left[t' - \frac{y}{c_1}(\sin\theta_r - \sin\theta_i)\right] = p^{\text{tr}}\left[t' - y\left(\frac{\sin\theta_t}{c_2} - \frac{\sin\theta_i}{c_1}\right)\right]. \qquad \text{(B–5)}$$

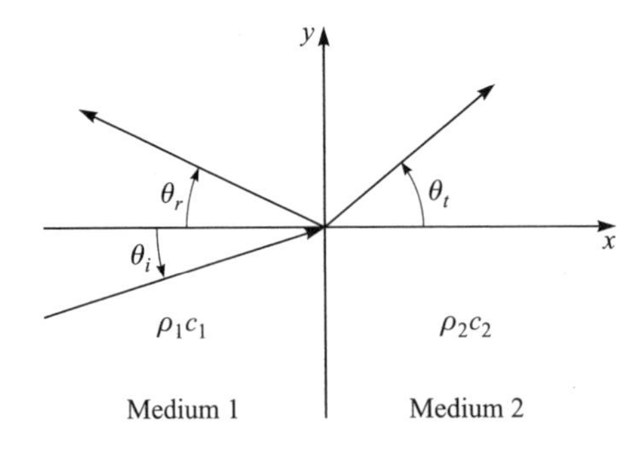

Figure 5.4 Incident, reflected, and transmitted rays.

[1] This is another way of saying that the trace velocity of each wave along the interface must be the same.

The only way this expression can hold for arbitrary values of y is for the y dependence to disappear, which happens when $\sin\theta_r = \sin\theta_i$, or

$$\theta_r = \theta_i$$

and

$$\frac{\sin\theta_t}{c_2} = \frac{\sin\theta_i}{c_1}.$$

These are, of course, the same results derived previously, Eqs. A–5 and A–3, respectively. Equation B–5 reduces to

$$p^+(t') + p^-(t') = p^{\mathrm{tr}}(t'),$$

or, in terms of the reflection and transmission coefficients,

$$1 + R = T. \tag{B–6}$$

Although Eq. B–6 repeats the result found for normal incidence, Eq. 3A–5, something new develops when the particle velocity boundary condition is applied. The condition is that the *normal component* of particle velocity must be continuous across the interface. The tangential component need not be continuous because in our model (inviscid fluids) a velocity shear generates no force. Since the normal components of **u** are found by multiplying by the cosine of the appropriate angle, we have

$$u^+\cos\theta_i + u^-\cos\theta_r = u_{\mathrm{interface}} = u^{\mathrm{tr}}\cos\theta_t, \tag{B–7}$$

where $u_{\mathrm{interface}}$ is the normal velocity of the interface. Recall that $\theta_r = \theta_i$ and apply the usual characteristic impedance relations; the result is

$$1 - R = \frac{Z_1\cos\theta_t}{Z_2\cos\theta_i}\,T, \tag{B–8}$$

which for normal incidence reduces to Eq. 3A–8. Combining Eqs. B–6 and B–8 yields

$$R = \frac{Z_2\cos\theta_i - Z_1\cos\theta_t}{Z_2\cos\theta_i + Z_1\cos\theta_t}. \tag{B–9a}$$

$$T = \frac{2Z_2\cos\theta_i}{Z_2\cos\theta_i + Z_1\cos\theta_t}. \tag{B–9b}$$

To obtain the corresponding sound power coefficients r and τ, we must take account of the fact that the refraction at the interface changes the intensity because of a change in cross-sectional area. Consider a bundle of incident rays that intercepts an area $\ell_z \, \Delta y$ of the interface (ℓ_z is the depth of the bundle in the z direction), as shown in Fig. 5.5. The cross-sectional area of the incident bundle S_i and that of the corresponding transmitted bundle S_t are given by

$$S_i = \ell_z \, \Delta y \cos \theta_i, \tag{B–10a}$$

$$S_t = \ell_z \, \Delta y \cos \theta_i, \tag{B–10b}$$

Of course S_r (the cross-sectional area of the reflected bundle) is equal to S_i. Using these expressions, we obtain

$$r = \frac{W_r}{W_i} = \frac{S_r I_r}{S_i I_i} = \frac{(p_{\mathrm{rms}}^-)^2/Z_1}{(p_{\mathrm{rms}}^+)^2/Z_1} = R^2, \tag{B–11a}$$

$$\tau = \frac{S_t I_T}{S_i I_i} = \frac{S_t (p_{\mathrm{rms}}^{\mathrm{tr}})^2/Z_2}{S_i (p_{\mathrm{rms}}^+)^2/Z_1} = \frac{Z_1 \cos \theta_t}{Z_2 \cos \theta_i}. \tag{B–11b}$$

Conservation of energy requires that $r + \tau = 1$; can you show that Eqs. B–11a and B–11b satisfy this relation?

The form of Eqs. B–10 prompts us to reexamine several of the formulas derived previously for normally incident waves. First notice that Eq. B–7 may be expressed as

$$u^+ S_i + u^- S_r = u^{\mathrm{tr}} S_t,$$

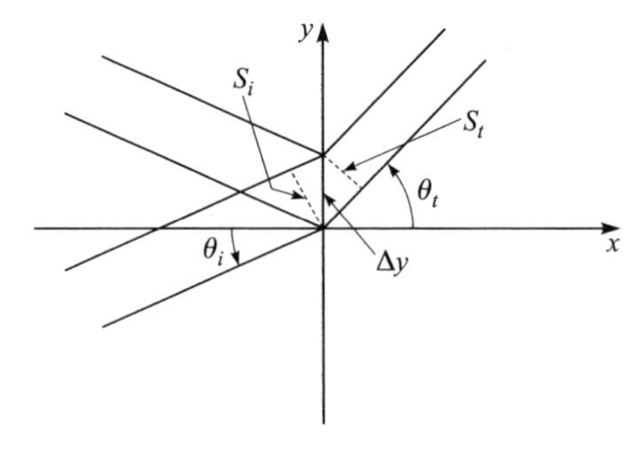

Figure 5.5 Cross-sectional areas of comparable segments of plane incident, reflected, and transmitted waves.

which is simply a statement that the volume velocity on the left-hand side of the interface must equal the volume velocity on the right-hand side. Moreover, the factor multiplying T in Eq. B–8 is really the ratio of the acoustic impedances,

$$\frac{Z_1 \cos \theta_t}{Z_2 \cos \theta_i} = \frac{Z_1/S_i}{Z_2/S_t} = \frac{Z_{ac1}}{Z_{ac2}}.$$

Therefore, alternative forms of Eqs. B–9a and B–9b are

$$R = \frac{Z_{ac2} - Z_{ac1}}{Z_{ac2} + Z_{ac1}}, \tag{B–12a}$$

$$T = \frac{2Z_{ac2}}{Z_{ac2} + Z_{ac1}}. \tag{B–12b}$$

These formulas appear in Chap. 3 (see Eqs. 3C–7 and 3C–8) for normally incident waves. Here we see that the formulas hold regardless of the angle of incidence. It is simply necessary to calculate the relative cross-sectional area for each wave.

2. Special Cases

Three special topics in refraction are now discussed: perfect transmission, total internal reflection, and grazing incidence.

a. Perfect Transmission: Angle of Intromission

Perfect transmission occurs when no reflected wave is produced. As Eq. B–9a shows, R vanishes when

$$Z_2 \cos \theta_i = Z_1 \cos \theta_t.$$

This result, combined with Snell's law, leads to the following relation for the angle of incidence at which all the sound is transmitted into the second medium, sometimes called the *angle of intromission* θ_0:

$$\sin^2 \theta_0 = \frac{(Z_2/Z_1)^2 - 1}{(Z_2/Z_1)^2 - (c_2/c_1)^2} \tag{B–13}$$

$$= \frac{(Z_1/Z_2)^2 - 1}{(\rho_1/\rho_2)^2 - 1}. \tag{B–14}$$

Not every pair of media has such an angle. In order for θ_0 to be real, both numerator and denominator must have the same sign.

Example 5.1. Water-castor oil interface. Let a wave in castor oil be incident on a water interface. Find θ_0, if it exists. For castor oil the pertinent values are $\rho_1 = 950$ kg/m^3 and $c_1 = 1540$ m/s; the values for water are $\rho_2 = 998$ kg/m^3 and $c_2 = 1540$ m/s. Evaluate Eq. B–14:

$$\sin^2\theta_0 = \frac{[(950)(1540)/(998)(1481)]^2 - 1}{[950/998]^2 - 1} = 0.2157.$$

$$\therefore \quad \theta_0 = 27.7^\circ, \qquad \text{and} \qquad \theta_t = 26.5^\circ \quad \text{(from Snell's law)}.$$

Question: Is perfect transmission possible in the opposite direction, i.e., from water to castor oil?[2]

Although perfect transmission might seem a curious mathematical accident, it has a quite simple physical explanation. We have already seen that the degree of transmission depends on the *acoustic impedances* of the two media. First consider two media for which $Z_2 > Z_1$ and $c_2 < c_1$. In this case medium 2 is acoustically harder than medium 1 *at normal incidence*. But refraction when $c_2 < c_1$ makes the effective cross-sectional area larger in medium 2. The larger area tends to make medium 2 acoustically softer than medium 1. Since the effect due to an increase in Z opposes that due to a decrease in c, might the two effects possibly cancel? Yes, at the angle of intromission. Second, let the two media have properties $Z_2 < Z_1$ and $c_2 > c_1$. Although the respective effects on acoustic impedance are now reversed, cancellation is still possible. Finally, if the properties of the two media are such that the two effects reinforce each other—$Z_2 > Z_1$ and $c_2 > c_1$, or $Z_2 < Z_1$ and $c_2 < c_1$—the acoustic impedances can never be equal, and no possibility for perfect transmission exists.

b. Total Internal Reflection

Snell's law gives

$$\theta_t = \sin^{-1}\left(\frac{c_2}{c_1}\sin\theta_i\right). \tag{B–15}$$

But θ_t is a real angle only if $(c_2/c_1)\sin\theta_i \leq 1$. The value of θ_i for which $(c_2/c_1)\sin\theta_i = 1$ is called the critical angle θ_{cr},

$$\theta_{\text{cr}} = \sin^{-1}(c_1/c_2). \tag{B–16}$$

Below critical incidence ($\theta_i < \theta_{\text{cr}}$) the transmitted wave emerges at the angle given by Eq. B–15. At critical incidence ($\theta_i = \theta_{\text{cr}}$) the transmitted wave travels parallel to the interface, i.e., $\theta_t = 90^\circ$. Beyond critical incidence ($\theta_i > \theta_{\text{cr}}$) the

[2] *Warning*: The value of θ_0 given by Eqs. B–13 and B–14 is very sensitive to changes in values of Z, ρ, and c. For instance, if all the data for density and sound speed in Example 5.1 were rounded off to two significant digits, the result would be $\theta_0 = 90^\circ$.

transmitted wave cannot emerge at any real angle. The result is *total internal reflection*, that is, no real wave can get into medium 2.

What happens to the transmitted wave when $\theta_i > \theta_{cr}$? Because the answer depends on frequency, the analysis proceeds most simply by specializing to time-harmonic waves. Equations B–1, B–2, and B–3 become

$$p^+ = A_1 e^{j(\omega t - k_1 x \cos\theta_i - k_1 y \sin\theta_i)}, \tag{B–17a}$$

$$p^- = B_1 e^{j(\omega t + k_1 x cos \theta_i - k_1 y \sin\theta_i)}, \tag{B–17b}$$

$$p^{tr} = A_2 e^{j(\omega t - k_2 x \cos\theta_t - k_2 y \sin\theta_t)}, \tag{B–17c}$$

respectively. Now focus attention on Eq. B–17c. Snell's law may be used to change the last term in the exponent to $k_1 y \sin\theta_i$. The middle term in the exponent turns out to be particularly interesting. The cosine function may be written

$$\begin{aligned}\cos\theta_t &= \pm\sqrt{1-\sin^2\theta_t} \\ &= \pm\sqrt{1-(c_2/c_1)^2\sin^2\theta_i} \quad \text{(by Snell's law)} \\ &= \pm j\sqrt{(c_2/c_1)^2\sin^2\theta_i - 1}.\end{aligned}$$

In other words, beyond critical incidence $\cos\theta_t$ is imaginary. Choose the minus sign, for physical reasons explained below, and let $b = \sqrt{(c_2/c_1)^2\sin^2\theta_i - 1}$:

$$\cos\theta_t = -jb. \tag{B–18}$$

Substitution in Eq. B–17c yields

$$p^{tr} = A_2 e^{-k_2 b x} e^{j(\omega t - k_1 y \sin\theta_i)}. \tag{B–19}$$

This kind of signal is called an *evanescent wave*. It is not a true wave at all. Although the signal is time varying and moves along the interface with the trace velocity $c_1/\sin\theta_i$, it does not *propagate* in medium 2. Rather, the amplitude of the signal simply decreases exponentially, as $e^{-k_2 b x} = e^{-\alpha_e x}$, where

$$\alpha_e = \sqrt{(c_2/c_1)^2\sin^2\theta_i - 1}. \tag{B–20}$$

The reason for choosing the minus sign for Eq. B–18 is now apparent. The plus sign would have led to an exponentially growing signal, an unacceptable result.

Of particular interest is the dependence of the attenuation on frequency. The higher the frequency, the more poorly the evanescent wave penetrates into medium 2. An interesting application is to transmission of sonic booms into the ocean when the angle of incidence exceeds θ_{cr} (see Problem 5–4). The evanescent sonic boom generated at the water surface has the same waveform as the incident sonic boom. As water depth increases, however, the evanescent boom loses more and more of its high frequency content and thus becomes progressively more rounded; see, for example, Ref. 6.

How does the reflected wave behave when total internal reflection occurs? Since the acoustic impedance of medium 2 is now pure imaginary, i.e., $Z_{ac2} \propto Z_2/\cos\theta_t = jZ_2/b$, one expects results similar to those found in Chap. 4 for a pure reactive termination (see Problems 4B–1, 4B–2, and 4B–13). Equation B–9a becomes

$$R = \frac{Z_2 \cos\theta_i + jZ_1 b}{Z_2 \cos\theta_i - jZ_1 b} = e^{j2\psi}, \tag{B–21}$$

where $\tan\psi = Z_1 b / Z_2 \cos\theta_i$. Moreover, from Eq. B–6 we find that

$$T = 1 + e^{j2\psi} = 2\cos\psi e^{j\psi}. \tag{B–22}$$

Thus although $|R| = 1$ over the entire region $\theta_i > \theta_{cr}$, $|T| < 2$ except at $\theta_i = \theta_{cr}$ itself.

These results, in particular $|R| = 1$, hold only if medium 2 is of infinite thickness. If medium 2 is finite in extent, for example, if it is a wall or a panel, beyond which a second section of medium 1 exists (Fig. 5.6), the evanescent wave is reflected from the second interface. The reflection is in turn reflected from the first interface, and so on. The "bouncing back and forth" of evanescent waves, albeit with exponentially decreasing amplitude, between the two interfaces sets up a standing evanescent field inside the panel. As Eq. B–19 shows, the time-varying signal at the second interface travels along the interface with the same trace velocity ($c_1/\sin\theta_i$) as that at the first interface. Although of diminished amplitude at the second interface, the traveling excitation there radiates a true wave into the second section of medium 1. By this

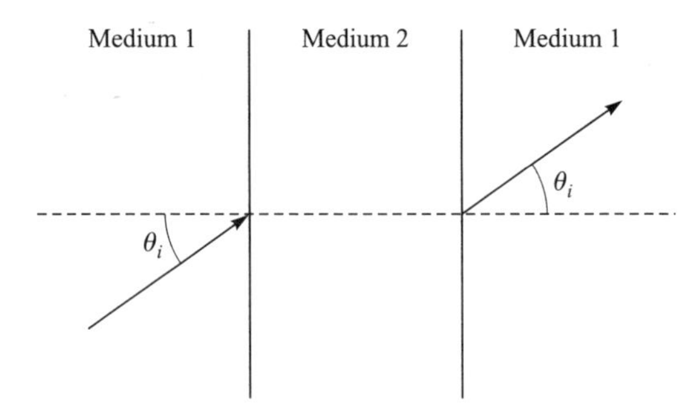

Figure 5.6 Transmission through a wall or panel in spite of total internal reflection.

means is sound transmitted through a wall or panel even when θ_i exceeds θ_{cr} for the wall or panel material.[3] In fact, since most walls and panels are solid ($c_2 \gg c_1$), most of the sound in a room falls on a wall or panel at angles beyond critical incidence.

c. *Grazing Incidence*

At grazing incidence ($\theta_i = 90°$) Eqs. B–9a and B–9b reduce to

$$R = \frac{-Z_1 \cos \theta_t}{Z_1 \cos \theta_t} = -1, \quad T = 0. \tag{B–23}$$

This is a rather surprising result: Even when Z_2/Z_1 is large, at grazing incidence medium 2 behaves as though it were a pressure release material. (The effect of total internal reflection is considered below.) Again the acoustical impedance is the key property. For a fixed cross-sectional area of the transmitted wave, the corresponding area of the incident wave approaches zero as $\theta_i \to 90°$, i.e., Z_{ac2} becomes small compared to Z_{ac1}. See, for example, Fig. 5.7, which shows a source in air above a concrete surface and the receiver far away. As long as θ_i is not close to 90°, the fact that $Z_2 \gg Z_1$ keeps R close to the value +1. As the receiver approaches the concrete surface ($H \to 0$), pressure doubling occurs. But when the distance D becomes large enough that $\theta_i \to 90°$, R becomes −1. The received pressure then drops to zero as the receiver height approaches zero.

If medium 2 is a solid, as in Fig. 5.7, the wave fields are generally more complicated than we have indicated because shear waves as well as longitudinal waves can be excited at the interface. However, if the shear waves may be ignored (i.e., the solid treated as a fluid), reflection and refraction are relatively simple. Because of the fact that $c_2 \gg c_1$, total internal reflection is the rule except for very small angles. We therefore use Eq. B–21 but rearranged as follows:

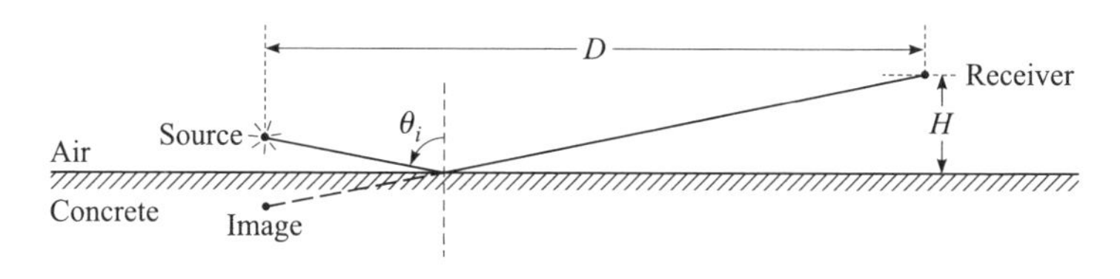

Figure 5.7 Reflection near grazing incidence above a concrete surface.

[3] It is interesting to note that transmission of sound by means of evanescent waves is a mathematical analog of quantum mechanical tunneling through a potential barrier.

$$R = \frac{1 + j(\rho_1/\rho_2)\tan\theta_i\sqrt{1 - (\sin\theta_{cr}/\sin\theta_i)^2}}{1 - j(\rho_1/\rho_2)\tan\theta_i\sqrt{1 - (\sin\theta_{cr}/\sin\theta_i)^2}}.$$

A useful simplification for angles large enough that $(\sin\theta_{cr}/\sin\theta_i)^2 \ll 1$ is

$$R = \frac{1 + j(\rho_1/\rho_2)\tan\theta_i}{1 - j(\rho_1/\rho_2)\tan\theta_i}. \tag{B–24}$$

This expression allows us to estimate how close θ_i must approach 90° in order that the interface act as a pressure release surface.

Example 5.2. Air-concrete interface. See Figure 5.7. For air over concrete the critical angle is (concrete's properties vary; the values used here are $\rho_2 = 2600\ \text{kg/m}^3$, $c_2 = 3100\ \text{m/s}$)

$$\theta_{cr} = \sin^{-1}\frac{343}{3100} = 6.35°.$$

Use of Eq. B–24 shows that θ_i must be well over 85° before R deviates noticeably from $1e^{j0°}$. At $\theta_i = 89.973°$, R becomes $(1+j)/(1-j) = j$. Only when θ_i is within a few millidegrees of 90° is the limiting value $R = -1$ approached.[4]

The reader may notice that although our entire treatment of reflection and transmission has been for plane incident waves, the discussion of Fig. 5.7 implies that the plane-wave results apply equally well to spherical waves. In most cases the extension to spherical waves is only a useful approximation. The plane-wave reflection formula, Eq. B–9a, is valid for spherical waves only if the second medium is infinitely hard ($Z_2/Z_1 \to \infty$) or infinitely soft ($Z_2/Z_1 \to 0$). For all other cases the image source is "fuzzy," as explained by Embleton (Ref. 3). The portion of the reflected field not described by the plane-wave reflection formula is called the *ground wave*. The ground wave can be particularly important in outdoor sound propagation; see, for example, Refs. 5 and 3.

C. TRANSMISSION THROUGH PANELS AT OBLIQUE INCIDENCE

Next consider the problem of sound transmission through a panel when the incoming wave is incident at an oblique angle. Transmission occurs because the pressure due to the incident sound wave forces the panel to move, and in turn

[4]In practice, atmospheric effects, such as turbulence and wind, make the theoretical limit difficult to observe.

the panel motion radiates sound into the fluid on the other side. In the case of normal incidence, analyzed in Sec. 4F, the whole panel moves as a unit because the driving pressure at any instant is the same over the entire surface of the panel. When the angle of incidence is oblique, however, the excitation pressure varies spatially over the panel surface. As a result, the panel is set in flexural, not uniform, motion.

The problem of transmission through a panel is similar to refraction at a fluid-fluid interface (Sec. A). However, a material layer, not a massless interface, now separates the two fluids. In the most common situation, the one considered here, the fluids are the same on both sides of the panel.[5] Because no refraction occurs in this case, we drop the subscripts on θ. Transmission is determined solely by the response of the panel, which depends strongly on frequency. The analysis is therefore carried out with time-harmonic waves. Incident signals that are not time harmonic may be broken up into their Fourier components and dealt with separately.

The panel motion is shown in Fig. 5.8. The y axis lies in the plane of the panel. In our model the panel thickness ℓ is small compared with the wavelength of sound in the panel material, i.e., $k\ell \ll 1$. The wavefronts of the peaks and troughs are labeled P and T, respectively. For simplicity neither the

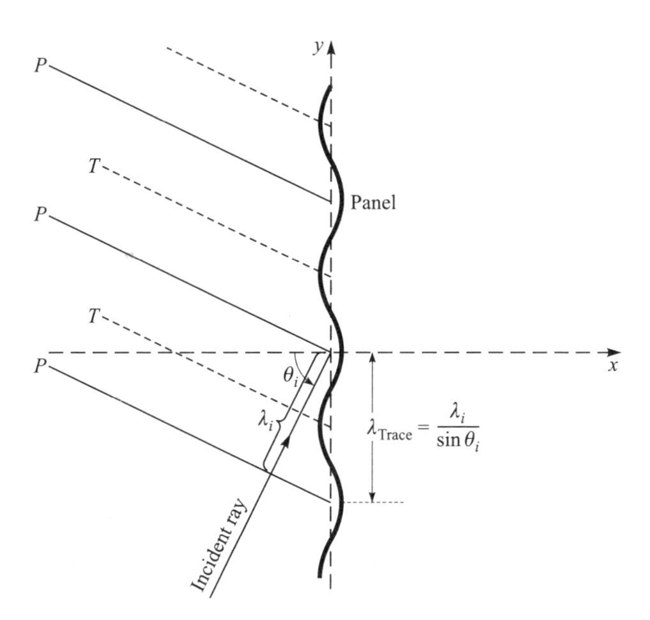

Figure 5.8 Bending deformation of a panel caused by an obliquely incident sound field. For simplicity the reflected and transmitted sound fields are not shown.

[5]Transmission through a panel when the two fluids are different is taken up in Problems 5–10 and 5–12.

reflected nor the transmitted rays and wavefronts are shown.[6] It can be seen, however, that the panel is deformed alternately outward and inward by the pressure peaks and troughs. As the peaks and troughs move over the panel, they push the deformation along. In this way the panel is driven into flexural wave motion. Figure 5.8 shows that the flexural wave travels with a speed equal to the trace velocity of the incident wave along the surface of the panel,

$$c_{\text{trace}} = c_1 / \sin\theta. \tag{C–1}$$

Equation C–1 is interesting in its own right. Suppose that regardless of what caused it, a transverse wave propagates over the surface of a panel with speed V_{panel}. Does the traveling surface wave generate a sound wave in the adjacent fluid? If so, the angle θ at which the sound is radiated into the fluid is given by Eq. C–1, rewritten for this case as

$$\sin\theta = c_1 / V_{\text{panel}}. \tag{C–2}$$

It is clear that θ is a physically realizable angle only if $V_{\text{panel}} \geq c_1$, that is, only if the surface wave represents a supersonically moving disturbance to the fluid. If the surface wave is subsonic, no real angle of radiation exists. Therefore, no sound is radiated into the fluid.[7] This result is of great importance in determining the extent to which a vibrating structure generates sound in the fluid surrounding it.

1. Transmission Dominated by Panel Mass: The Mass Law

Figure 5.9 shows the incident (p^+), reflected (p^-), and transmitted (p^{tr}) rays for a panel being irradiated by a plane sound wave. If the panel is assumed flexible but not stiff, the panel's paramount property is its mass M. More important is the mass per unit area $m = M/S = \rho_{\text{panel}}\,\ell$, where S, ρ_{panel}, and ℓ are the panel's surface area, density, and thickness, respectively.

As usual, two equations in R and T are sought. First, in place of the usual pressure boundary condition, such as Eq. B–4, a force balance that includes the panel motion must be used. The mass of a small panel element of surface area dS is $m\,dS$, and the net force imposed on it by the three sound fields is $(p^+ + p^- - p^{\text{tr}})\,dS$. Newton's law for the element is therefore

$$m\dot{u}_{\text{panel}} = p^+ + p^- - p^{\text{tr}}, \tag{C–3}$$

where the dot signifies $\partial/\partial t$. Because the panel acceleration may be expressed as

[6] Many texts contain figures that show all three wave fields (incident, reflected, and transmitted). See, for example, Refs. 1 and 2.

[7] However, an evanescent wave is produced in the fluid. The situation is just like that for the transmitted wave in total internal reflection. See Sec. B.2.b above.

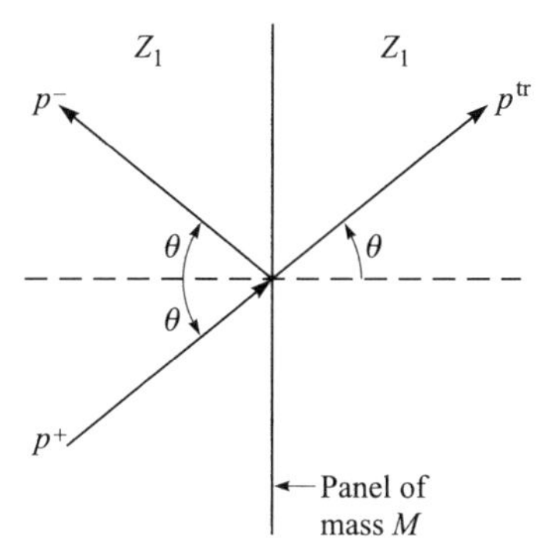

Figure 5.9 Sound transmitted through a massive panel when the media are the same on both sides of the panel.

$$\dot{u}_{\text{panel}} = \dot{u}^{\text{tr}} \cos\theta = \frac{\cos\theta}{Z_1}\dot{p}^{\text{tr}},$$

Eq. C–3 reduces to

$$p^+ + p^- = p^{\text{tr}} + \frac{m\cos\theta}{Z_1}\dot{p}^{\text{tr}}.$$

To simplify this expression further, we assume time-harmonic waves, for which case $\dot{p}^{\text{tr}} = j\omega p^{\text{tr}}$. The result is

$$1 + R = \left(1 + \frac{j\omega m\cos\theta}{Z_1}\right)T. \tag{C–4}$$

At normal incidence ($\theta = 0$) this expression reduces to the one derived in Sec. 4F. The second R, T relation comes from the particle velocity boundary condition,

$$(u^+ + u^-)\cos\theta = u_{\text{panel}} = u^{\text{tr}}\cos\theta.$$

Because the impedance is the same on both sides of the panel, this expression reduces to $p^+ - p^- = p^{\text{tr}}$, or

$$1 - R = T \tag{C–5}$$

(compare with Eq. B–8).

Solution of Eqs. C–4 and C–5 for R and T gives

$$T = \frac{1}{1 + j\omega m\cos\theta/2Z_1}, \tag{C–6a}$$

$$R = \frac{j\omega m \cos\theta/2Z_1}{1 + j\omega m \cos\theta/2Z_1}. \tag{C–6b}$$

Equation C–6a is the same as Eq. 4E–18, except that the ratio of specific acoustic impedances $j\omega m/Z_1$ is replaced by the ratio of acoustic impedances $j\omega m/(Z_1/\cos\theta)$. Putting the result in terms of acoustic impedances helps explain why the transmission improves with angle of incidence. As θ increases, the fluid seems to get "harder," that is, less soft with respect to the panel.

Because the acoustic impedance is the same on both sides of the panel, the power transmission coefficient is $\tau = |T|^2$, or, given Eq. C–6a,

$$\tau = 1/[1 + (\omega m \cos\theta/2Z_1)^2]. \tag{C–7}$$

The transmission loss (Eq. 3A–17) is therefore

$$\mathrm{TL} = 10\log_{10}[1 + (\omega m \cos\theta/2Z_1)^2]. \tag{C–8}$$

The normal-incidence transmission loss is

$$\mathrm{TL}_0 = 10\log_{10}[1 + (\omega m/2Z_1)^2]. \tag{C–9}$$

For "high frequencies" ($\omega m \cos\theta/2Z_1 \gg 1$) Eq. C–8 reduces to

$$\mathrm{TL} \doteq 10\log_{10}(\omega m \cos\theta/2Z_1)^2 = 20\log_{10}(\omega m \cos\theta/2Z_1). \tag{C–10}$$

This result is the *mass law* of sound barriers for oblique incidence.[8] For normal incidence Eq. C–10 reduces to Eq. 4E–23.

Although its simplicity makes it very appealing, the mass law often needs to be modified in practice. In many applications, for example, room acoustics and noise control, the sound field is random, or diffuse. Sound is incident at all angles on the wall or panel. For such a field an average power transmission coefficient may be derived by integrating over all angles of incidence from 0° to 90°. The following expression for what is called the random-incidence transmission loss $\mathrm{TL}_{\mathrm{ran}}$ is obtained:

$$\mathrm{TL}_{\mathrm{ran}} = \mathrm{TL}_0 + 6.4 - 10\log_{10}\mathrm{TL}_0, \tag{C–11}$$

[8] Transmission through the panel may also be approached as a three-medium problem in which the middle medium is very thin, as in Sec. 4E.1.c. To generalize Eq. 4E–15 for oblique incidence, first replace ratios of specific acoustic impedances by their respective ratios of acoustic impedances, e.g., replace Z_1/Z_2 by $(Z_1/\cos\theta_1)/(Z_2/\cos\theta_2)$. Second, replace k_2 by $k_2\cos\theta_2$. Approximations similar to those in Sec. 4E.1.c then lead to Eq. C–6a, *even when the field in medium 2 is evanescent*, i.e., even when $\theta_i > \theta_{\mathrm{cr}}$.

where the frequency is assumed high enough that the factor 1 in Eq. C–9 is ignored. This equation, although in theory preferable to the ordinary mass law, is still not a sufficiently accurate predictor. Further tinkering is required. Field measurements of transmission loss TL_{field} are generally well fit by the following empirical law:

$$TL_{field} \doteq TL_0 - 5. \tag{C–12}$$

As Fig. 5.10 shows, the curve for TL_{field} lies between those for TL_0 and TL_{ran}. Why is the transmission loss given by Eq. C–11 too little? An ad hoc explanation is that integration over all angles $0° \leq \theta \leq 90°$, as required by the random-incidence model, gives too much weight to grazing rays (θ near 90°), which don't have a high degree of probability. When the upper limit of the integration is taken to be 78°, the final result is found to agree relatively well with Eq. C–12.[9]

As a predictor of transmission loss, the mass law, even in its modified forms, may be further limited by the phenomenon of coincidence, which is taken up next.

2. Panel Stiffness: The Coincidence Effect

Up to now mass has been considered the only means by which the panel resists the undulating deformation initiated by the incident sound wave (Fig. 5.8). Here we include the effect of panel bending stiffness (not to be confused with the wall stiffness considered in Sec. 4F)—and a spectacular effect it has. For a thin panel (thickness ℓ) modeled as an Euler-Bernoulli plate (see, for example, Ref. 4), the bending stiffness is

$$B = \frac{E\ell^3}{12(1-\sigma^2)}, \tag{C–13}$$

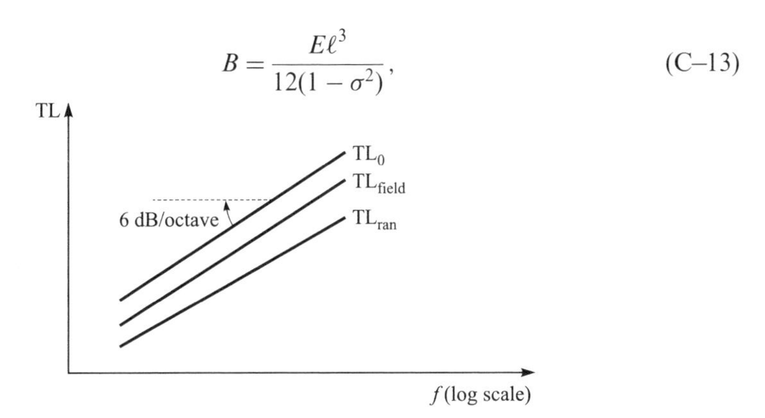

Figure 5.10 Comparison of transmission loss for normal, field, and random incidence.

[9]For a derivation of Eq. C–11 and a discussion of random- and field-incidence transmission loss, see Ref. 2, pp. 289–291.

where E is Young's modulus and σ is Poisson's ratio. A useful material property of panels is the speed of propagation of longitudinal waves in the panel, in any direction in the plane of the panel:

$$c_L = \sqrt{\frac{E}{\rho_{\text{panel}}(1-\sigma^2)}}, \tag{C–14}$$

where $\rho_{\text{panel}} = m/\ell$ is the density of the panel material.[10] In terms of c_L, B may be expressed as

$$B = \frac{m\ell^2 c_L^2}{12}. \tag{C–15}$$

Before assessing the effect of bending stiffness on transmission loss, we examine the vibration of a so-called unloaded panel, that is, one free of any forces imposed by any surrounding fluid (and therefore not subject to a sound field). The equation of flexural motion of such a panel is

$$m\xi_{tt} + B\nabla^4\xi = 0, \tag{C–16}$$

where ξ is the transverse displacement of the panel (the displacement in the x direction; see Fig. 5.8). If ξ is assumed to vary only in the y direction, $\nabla^4\xi$ reduces to ξ_{yyyy} and Eq. C–16 becomes

$$m\xi_{tt} + B\xi_{yyyy} = 0. \tag{C–17}$$

This expression resembles the equation for a mass on a spring, except that the spring restoring force is proportional to ξ_{yyyy} rather than to ξ.

To determine whether bending wave motion may be described by Eq. C–17, try a solution that represents a time-harmonic traveling wave,

$$\xi = \xi_0 e^{j\omega(t-y/c_b)}, \tag{C–18}$$

where c_b is the speed of the presumed bending wave. Substitution in Eq. C–17 yields

$$m(j\omega)^2 + B(-j\omega/c_b)^4 = 0.$$

An expression of this sort, which shows how the wave speed depends on frequency, is called a *dispersion relation*. Dispersion relations are common in acoustics, as shown in ensuing chapters. Solving for c_b, we find

[10]Longitudinal wave speed in solids is given in Appendix A for bars c_{bar} and bulk material c_{bulk}, neither of which is c_L. However, c_L and c_{bar} are related by $c_L = c_{\text{bar}}/\sqrt{1-\sigma^2}$.

$$c_b = (B\omega^2/m)^{1/4} = \sqrt{\ell c_L f \pi/\sqrt{3}} \tag{C–19}$$

(the factor $\pi/\sqrt{3}$ is often approximated by the numeric 1.8). The meaning of this result is that bending waves of the type described by Eq. C–18 are proper solutions of Eq. C–17 provided the travel speed c_b is given by Eq. C–19. Notice that the speed varies as the square root of the frequency of the wave. When wave speed depends on frequency, the wave motion is said to be *dispersive*.

Now let the panel be surrounded by fluid and let a sound wave impinge on the panel from one side. An interesting possibility may be anticipated. When the sound wave forces the panel into flexural motion, as shown in Fig. 5.8, the speed at which the flexural wave is driven is the trace velocity c_{trace}, given by Eq. C–1. In general c_{trace} is different from the natural speed c_b at which the flexural wave would like to travel. It might be possible, however, given just the right angle of incidence, for the two speeds to coincide, i.e., $c_{\text{trace}} = c_b$. At this angle, if it exists, coincidence is said to occur. What happens to the transmission loss at coincidence?

To carry out the analysis, first modify Eq. C–17 to account for the loading of the fluid. When the fluid and an incident sound field are present, forces due to the pressures of the incident, reflected, and transmitted sound waves must be included:

$$m\xi_{tt} + B\xi_{yyyy} = p^+ + p^- - p^{\text{tr}} \tag{C–20}$$

(compare with Eq. C–3). The flexural motion of the panel is again a time-harmonic traveling wave. This time, however, since the flexural wave is driven by the sound field, the speed of propagation of the flexural wave is c_{trace}. Thus

$$\xi = \xi_0 e^{j\omega(t - y/c_{\text{trace}})}.$$

Substitute this expression in Eq. C–20, noting that $j\omega\xi$ is the normal velocity of the panel u_{panel}, and arrange the result as follows:

$$u_{\text{panel}}\left[j\omega m + \frac{B}{j\omega(c_{\text{trace}}/\omega)^4}\right] = p^+ + p^- - p^{\text{tr}}.$$

The first term in brackets represents the mass reactance of the panel, the second term the stiffness reactance (but notice that the apparent stiffness, $B(\omega/c_{\text{trace}})^4$, is frequency dependent). Further manipulation of this equation yields

$$j\omega m u_{\text{panel}}[1 - (c_b/c_{\text{trace}})^4] = p^+ + p^- - p^{\text{tr}}. \tag{C–21}$$

Because u_{panel} is related to p^{tr} by

$$u_{\text{panel}} = u^{\text{tr}} \cos\theta = \frac{p^{\text{tr}} \cos\theta}{Z_1},$$

Eq. C–21 becomes

$$\frac{j\omega m \cos\theta}{Z_1}\left[1 - \left(\frac{c_b}{c_{\text{trace}}}\right)^4\right] p^{\text{tr}} = p^+ + p^- - p^{\text{tr}},$$

or, if the coefficients R and T are introduced,

$$\left(1 + \frac{j\omega m \cos\theta}{Z_1}\left[1 - \left(\frac{c_b}{c_{\text{trace}}}\right)^4\right]\right) T = 1 + R. \quad \text{(C–22)}$$

Since the existence of panel stiffness does not affect the particle velocity boundary condition, Eq. C–5 provides the second R, T relation. Elimination of R yields

$$T = \frac{1}{1 + (j\omega m \cos\,\theta/2Z_1)[1 - (c_b/c_{\text{trace}})^4]}, \quad \text{(C–23)}$$

We now see how coincidence affects transmission through the panel. When coincidence occurs, i.e., when the trace speed c_{trace} equals the natural bending wave speed c_b of the panel, the transmission coefficient becomes unity (the transmission loss drops to zero). A sort of resonance occurs in that the panel impedance drops to zero when the mass and stiffness reactances cancel. In physical terms the panel becomes so easy to drive that the sound wave hardly "sees" it. As Fig. 5.11 shows, coincidence causes an undesirable notch in the transmission loss curve. For frequencies low enough that $c_b \ll c_{\text{trace}}$ (recall that $c_b \propto \sqrt{f}$), Eq. C–23 reduces to the mass law expression, Eq. C–6a. As frequency increases, the coincidence condition is approached, and TL begins to fall away from the mass law curve. At exact coincidence TL falls to zero. At still higher frequency the TL curve rebounds, and then exceeds the mass law curve, as the panel motion becomes stiffness controlled.

Coincidence is usually characterized by the frequency f_{coinc} at which it occurs. By setting $c_b = c_{\text{trace}}$, we find

$$f_{\text{coinc}} = \frac{1}{2\pi}\left(\frac{c_1}{\sin\theta}\right)^2 \sqrt{\frac{m}{B}} = \frac{c_1^2}{(\pi/\sqrt{3})\ell c_L \sin^2\theta}. \quad \text{(C–24)}$$

Note the dependence on the angle of incidence. The *lowest* frequency f_0 at which coincidence is possible occurs at $\theta = \pi/2$. This is the "critical frequency" (not to be confused with "critical incidence," which is associated with refraction),

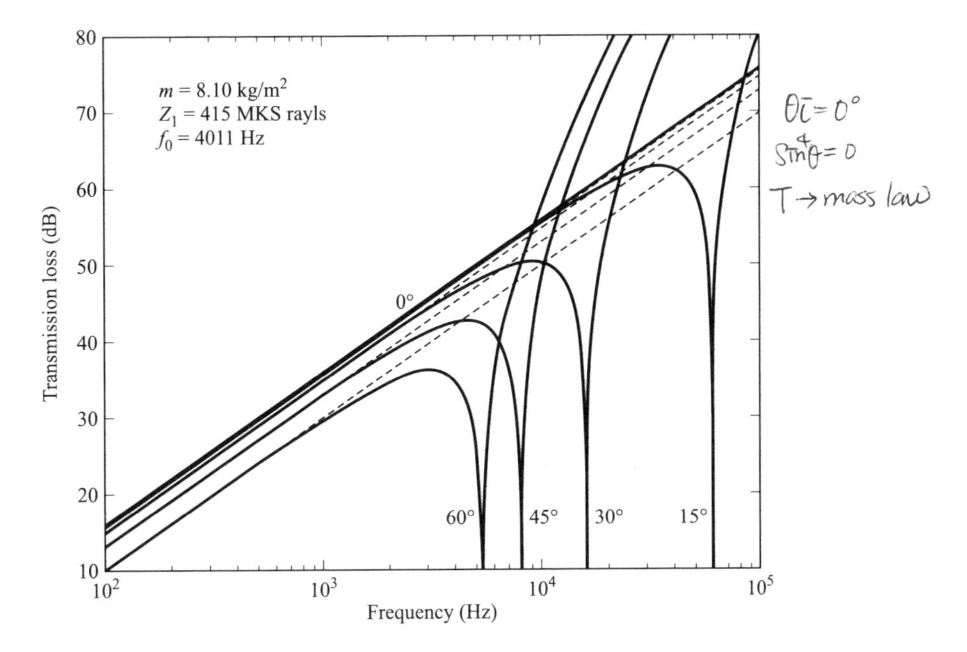

Figure 5.11 Coincidence effect, at various angles of incidence, for an aluminum panel in air.

$$f_0 = \frac{c_1^2}{2\pi}\sqrt{\frac{m}{B}} = \frac{c_1^2}{(\pi/\sqrt{3})\ell c_L}. \tag{C–25}$$

Equation C-23 may thus be recast in terms of frequency as follows:

$$T = \frac{1}{1 + (j\omega m \cos\theta/2Z_1)\left[1 - (f/f_0)^2 \sin^4\theta\right]}. \tag{C–26}$$

The roles played by frequency and angle of incidence are illustrated by the curves in Fig. 5.11, which shows the transmission loss for a 3-mm-thick aluminum panel in air. The parameter is angle of incidence θ. Note how the coincidence notch drops as θ increases. Note too the low frequency asymptote set by the mass law.[11]

Example 5.3. Transmission loss through glass. Sound of frequency 1200 Hz is incident at an angle of 65° on a $\frac{1}{8}$-in.-thick pane of window glass. How much

[11]Only the simplest model of coincidence has been discussed here. The depth of the coincidence notch is limited by internal frictional loss inside the plate material. At high frequency, where the thin-plate model is no longer valid, wave motion inside the plate, both shear and compressional, becomes important. For details and references, see Ref. 2, pp. 284–289.

transmission loss is afforded by the glass? How can an additional 12 dB of loss be obtained? First use the simple mass law model. For $\frac{1}{8}$-in. glass plate that has a density of 2500 kg/m^3, the surface density is $m = 7.94$ kg/m^2. At $\theta =$ 65° Eq. C–10 yields TL = 29.7 dB. If an additional loss of 12 dB is desired, simply quadruple the glass thickness to $\frac{1}{2}$ in. Before accepting these numbers at face value, however, we should determine whether coincidence is important. If the value of c_L for this particular glass plate is 5200 m/s, the critical frequency for the $\frac{1}{8}$-in. pane is, by Eq. C–25,

$$f_0 = \frac{(343)^2}{(\pi/\sqrt{3})(5200)(0.0254/8)} = 3930\,\text{Hz}.$$

However, this is just the lowest possible coincidence frequency ($\theta = 90°$) for the glass. For the angle of incidence given, $\theta = 65°$, we obtain

$$f_{\text{coinc}} = f_0/\sin^2 65° = 4785\,\text{Hz}.$$

Since this is about two octaves above the frequency of the incident sound (1200 Hz), coincidence would seem to be of little importance in this problem. Indeed, evaluation of Eq. C–26 shows that the effect of coincidence is to reduce the mass law value (TL = 29.7 dB) by about 0.6 dB. However, what about the $\frac{1}{2}$-in. pane? Will using it result in the extra 12 dB hoped for? No, because increasing the thickness to $\frac{1}{2}$ in. moves the coincidence notch from 4785 to 1196 Hz, right where it does the most damage. The actual calculation based on Eq. C–26 gives TL = 2.1 dB, a *reduction* of 27.6 dB rather than the desired 12-dB increase. This example shows the drastic effect coincidence can have on noise control design.

D. COMPOSITE WALLS

Up to now our calculation of wall transmission loss has been based on known solutions of the wave equation. This approach quickly becomes very complicated if the wall is composed of several elements, for example, gypsum board on studs, windows, and a door. For one thing diffraction or scattering at the edges of the elements, on both sides of the wall, greatly complicates the wave fields. A useful simpler approach is to assume that the sound fields on both sides are diffuse and uncorrelated. The total power transmitted is then simply the sum of the individual power transmissions. The composite transmission loss TL_{comp} for the wall is found as follows. First find the average power transmission coefficient $\bar{\tau}$:

$$\bar{\tau} = \frac{S_1\tau_1 + S_2\tau_2 + \cdots + S_n\tau_n}{\sum S_i}, \tag{D–1}$$

where τ_i is the power transmission coefficient associated with the area S_i. It is customary to use the normal-incidence, high frequency value for τ_i (ignore the factor 1 and take $\cos\theta = 1$ in Eq. C–7). The composite transmission loss TL_{comp} is then

$$TL_{comp} = -10 \log_{10} \bar{\tau}.$$

Example 5.4. Brick wall with glass windows. Figure 5.12 shows a brick wall 4 in. thick that has two 5×3-ft windows in it. The window glass is $\frac{1}{8}$ in. thick. Find the transmission loss at 500 Hz for each material and for the composite wall.[12]

1. Glass $\frac{1}{8}$ in. ($m = 1.5$ lb/ft^2):

$$\frac{m\omega}{2Z_1} = \frac{(1.5)(2\pi \times 500)}{2(85)} = 27.72,$$

$$\tau = (27.72)^{-2} = 1.3 \times 10^{-3}.$$

$$\therefore \quad TL_{glass} = 28.9 \text{ dB}.$$

2. Brick 4 in. thick ($m = 38$ lb/ft^2):

$$\frac{m\omega}{2Z_1} = 703,$$

$$\tau = 2.02 \times 10^{-6}.$$

$$\therefore \quad TL_{brick} = 57 \text{ dB}.$$

3. Composite:

$$\bar{\tau} = \frac{(2.02 \times 10^{-6})130 + (1.3 \times 10^{-3})30}{160} = 2.45 \times 10^{-4}$$

$$\therefore \quad TL_{comp} = 36 \text{ dB}.$$

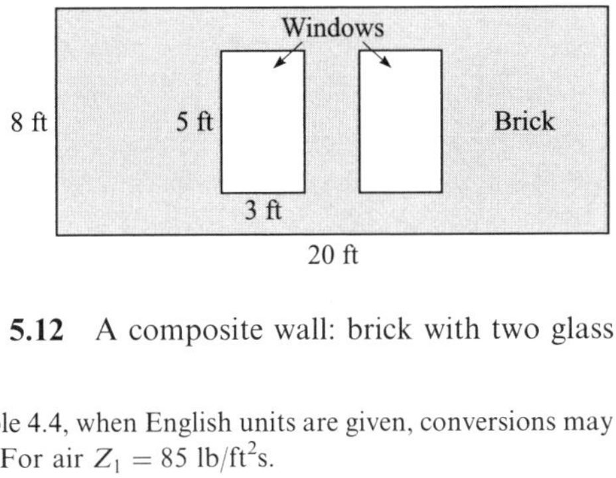

Windows
8 ft
5 ft
Brick
3 ft
20 ft

Figure 5.12 A composite wall: brick with two glass windows.

[12]As noted in Example 4.4, when English units are given, conversions may be avoided by expressing Z_1 in English units. For air $Z_1 = 85$ lb/ft^{2}s.

Therefore, even though the glass occupies only 19% of the wall area, it dominates the transmission characteristics of the wall.

The composite wall model is also useful for assessing the effect of openings in a wall, for example, the space at the threshold at the bottom of a door or the open spaces in a louvered door. As an example, Fig. 5.13 shows a glass window (frame length L_{total}, glass length L_{glass}) that is partly open at the bottom (length of open section L_o). Remember that diffraction effects are being ignored. We obtain

$$\bar{\tau} = \frac{S_o \tau_o + S_{\text{glass}} \tau_{\text{glass}}}{S_{\text{total}}} = \frac{L_o}{L_{\text{total}}} + \frac{L_{\text{glass}}}{L_{\text{total}}} \tau_{\text{glass}}, \tag{D–2}$$

since $\tau_o = 1$ and the width w is the same for both glass and opening. It can be seen that no matter how small τ_{glass} is, the composite power transmission coefficient cannot be made smaller than the fraction L_o/L_{total}. Unless the window is completely shut and virtually sealed, therefore, the opening dominates the sound transmission characteristics.

Example 5.5. Partially open glass window. Given a glass window $\frac{1}{8}$ in. thick, $f = 2000$ Hz, $L_{\text{total}} = 40$ in., and $L_o = 1$ in. The value of τ_{glass} is 8.13×10^{-5}.

$$\bar{\tau} = \frac{1}{40} + \frac{39}{40}(8.13 \times 10^{-5}) \doteq \frac{1}{40},$$
$$\text{TL}_{\text{comp}} = 16\,\text{dB}.$$

If the window were completely shut, the transmission loss would be $\text{TL}_{\text{glass}} = 41$ dB. Thus 25 dB of transmission loss is lost by leaving the window 2.5% open. What if the opening were just a 0.1-in. crack?

$$\bar{\tau} = \frac{0.1}{40} + \frac{39.9}{40}(8.13 \times 10^{-5}) = 0.00258,$$
$$\text{TL}_{\text{comp}} = 26\,\text{dB}.$$

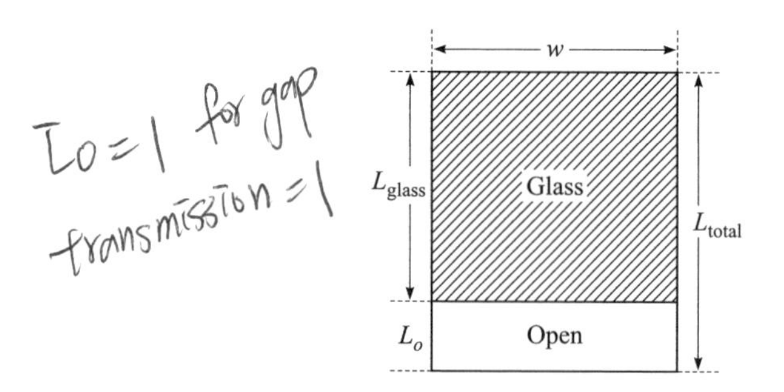

Figure 5.13 Window partly open at the bottom.

How small must the crack be for TL_{comp} to be within 3 dB of TL_{glass}? This means $\bar{\tau} = 2\tau_{glass}$, that is, the two terms in Eq. D–2 must be equal. Since $L_{glass}/L_{total} \doteq 1$ in this case, $L_o/L_{total} = 8.13 \times 10^{-5}$, or

$$L_o = 0.00325 \text{ in.} = 0.0826 \text{ mm},$$

which is a very small crack indeed. This example shows that cracks around windows and doors are the enemy when high transmission loss is desired.

REFERENCES

1. L. L. Beranek (Ed.), *Noise and Vibration Control*, rev. ed. (Institute of Noise Control Engineering, Washington, DC, 1988), Fig. 11.4, p. 280.
2. L. L. Beranek and I. L. Vér (Eds.), *Noise and Vibration Control Engineering* (John Wiley & Sons, New York, 1992).
3. T. F. W. Embleton, "Tutorial on sound propagation outdoors," *J. Acoust. Soc. Am.* **100**, 31–48 (1996).
4. A. D. Pierce, *Acoustics: An Introduction to Its Physical Principles and Applications* (McGraw-Hill, New York, 1981), pp. 144–146. Reprinted by Acoustical Society of America, 1989, pp. 144–146.
5. J. E. Piercy, T. F. W. Embleton, and L. C. Sutherland, "Review of noise propagation in the atmosphere," *J. Acoust. Soc. Am.* **61**, 1403–1418 (1977).
6. V. W. Sparrow, "The effect of supersonic aircraft speed on the penetration of sonic boom noise into the ocean," *J. Acoust. Soc. Am.* **97**, 159–162 (1995).

PROBLEMS

5–1. A ship sonar is being used to locate objects buried in the sediment at the bottom of the ocean. Assume the sonar signal to be a collimated plane-wave pulse. At point O the pulse strikes the sediment interface at angle θ_i. Because of the roughness of the interface, the reflected sound is scattered in all directions. In particular some is "backscattered," that is, it is reflected back toward the source. The backscattered wave retraces the path of the incident wave back to the ship and is detected by the sonar receiver. Similarly, if the wave transmitted into the sediment at angle θ_t strikes an object buried at depth d, scattering also occurs. The backscattered portion of the sound returns to the interface at point O, passes back into the ocean by refraction, and eventually finds its way back to the ship. The sonar receiver thus records two echoes, one from the sediment and one from the buried object. Let Δt be the difference between the arrival times of the two echoes.

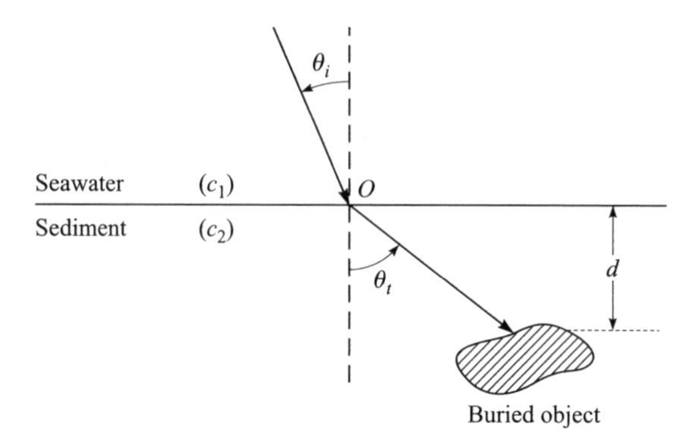

(a) Derive a formula for time difference Δt in terms of c_1, c_2, θ_i, and d.

(b) Let the depth be $d = 1$ m in a sediment for which the sound speed is 1700 m/s. What is the time delay Δt for a 30° angle of incidence?

(c) What is the time delay if the angle of incidence is changed to 65°?

5–2. A point source of sound S in a freshwater lake sends out sound in all directions. A listener in air observes that sound is being transmitted into the air at an angle of 5° at his location (see sketch). Although the listener does not know the depth of the source or his horizontal range R from the source, he finds that by moving 1 m further from the source, he increases the transmission angle to 6°. Find the range R and the depth d.

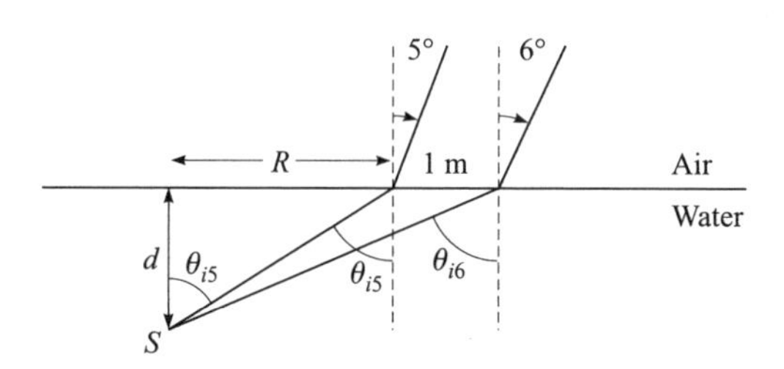

5–3. An unknown gas is separated from air by a membrane of negligible mass. At a temperature of 0°C, sound from a source in air falls on the interface. Measurement shows that $\theta_t = 41.67°$ when $\theta_i = 10°$.

(a) What is the unknown gas?

(b) At what angle θ_i does the transmission coefficient have the value $T = 2$?

5–4. If sound is to be transmitted from air (20°C) into seawater (13°C), the angle of incidence θ_i must be less than the critical angle θ_{cr}. See sketch (*a*).

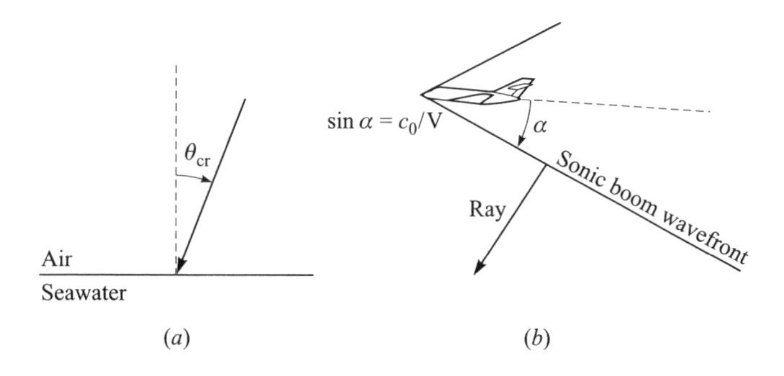

(a) What is the value of θ_{cr}?

(b) Now refer to sketch (*b*). If the sonic boom from a supersonic airplane (velocity V, in horizontal flight) is to penetrate the ocean surface, how fast must the airplane fly? Assume a flat ocean surface.

(c) How, qualitatively, would your answer to part (b) change if you allow for the fact that the ocean surface is actually rough, not flat, because of crests and troughs of ocean waves?

5–5. A plane wave in air at 20°C is incident on an air-helium interface. The properties of helium are as follows: $\rho_0 = 0.166\ \mathrm{kg/m^3}$, $c_0 = 1000\ \mathrm{m/s}$, $Z_0 = 166$ MKS rayls.

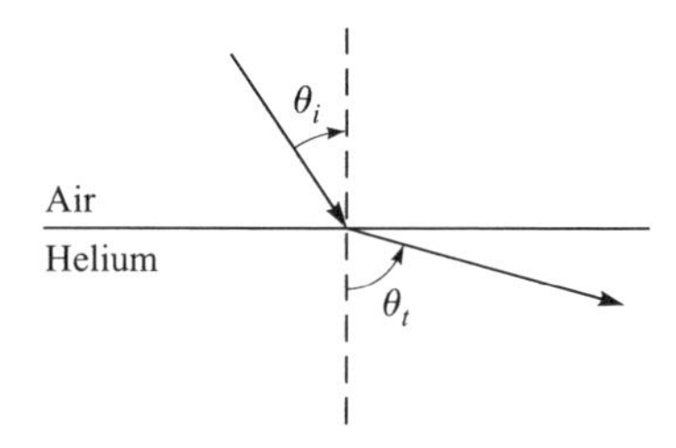

(a) Find the reflection coefficient R for normal incidence ($\theta_i = 0$).

(b) Find the angle of intromission θ_0 if one exists.

(c) Find the critical angle θ_{cr}.

(d) At what angle θ_i does R change sign (R real).

(e) Find the angle $\theta_i > \theta_{cr}$ at which $R = j$.

5–6. Consider propagation from glycerine into mercury.

(a) Find the R and T coefficients for normal incidence. Does the mercury boundary appear "hard" or "soft" to the incident acoustic wave?

(b) Find the angle of intromission θ_0 and the value of θ_t when $\theta_i = \theta_0$.

(c) Find R and T when $\theta_i = \theta_0$.

(d) For angles $\theta_i > \theta_0$, determine whether the mercury boundary appears "hard" or "soft" to the incident acoustic wave.

5–7. An airborne plane sound wave of frequency 1881 Hz is incident at an angle 45° on the calm surface of a freshwater lake. Assume the temperature is 20°C for the water and the air. The sound pressure level (SPL) of the incident sound wave is 100 dB (re 20μPa). What is the SPL of the sound in the water (re 1μPa) 0.1 m below the surface?

5–8. Wood paneling of density 36 lb/ft^3 and thickness $\frac{1}{4}$ in. is to be used for sound isolation in air at a frequency of 1000 Hz. Find the transmission loss TL

(a) At normal incidence ($\theta_i = 0°$).

(b) At 60° incidence ($\theta_i = 60°$).

The acoustical medium is changed to water.

(c) Reanswer part (a), making the computation using the mass law formula. But is the mass law formula you used in part (a) appropriate? If not, correct your calculation accordingly.

(d) Your answer to part (c) should show that the panel is much less effective as a sound barrier in water. Why physically is this so?

5–9. A hollow-core wooden door 1.75 in. thick has a surface density of 3.5 lb/ft^2. On a single sheet plot transmission loss vs. frequency for the door for three cases: (a) normal incidence, (b) 60° incidence, and (c) random incidence. The frequency range for the plots is 31.5–8000 Hz. Use semi-log paper (or scales) with TL along the linear scale (ordinate) and f along the log scale (abscissa). Show a sample calculation for each of the three cases.

5–10. Two fluids, medium 1 and medium 2, are separated by a thin solid panel of mass per unit area m. Derive the formula for the pressure transmission coefficient T for plane waves passing from medium 1 (angle of incidence θ_i) into medium 2 (angle of transmission θ_t). (Ignore coincidence.)

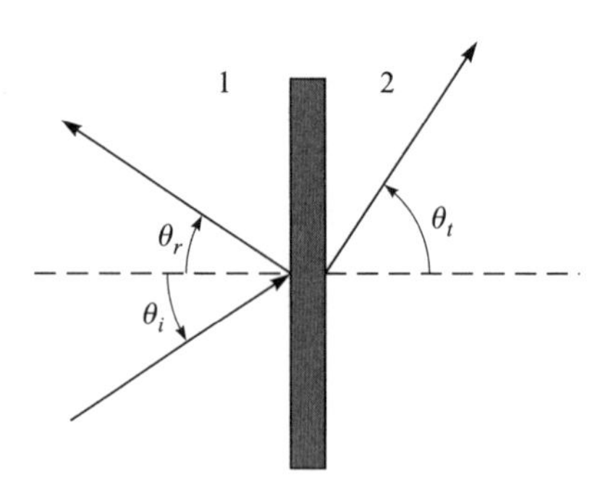

5–11. A sheet of gypsum board (Sheetrock) is $\frac{1}{2}$ in. thick. The speed c_L of longitudinal waves in this material is 6800 m/s and the density ρ_0 is 43 lb/ft^3.

(a) Find the critical frequency f_0, that is, the lowest frequency at which coincidence occurs.

(b) Find the TL at normal incidence when $f = 1500$ Hz.

(c) Find the TL for a diffuse sound field when $f = 1500$ Hz.

5–12. Plane sound waves in air (medium 1) are incident on a steel plate 10 mm thick. The angle of incidence is $\theta_i = 30°$. On the other side of the plate is fresh water (medium 2). The arrangement is like that in Problem 5–10.

(a) Find the coincidence frequency f_{coinc} for this problem.

(b) At the coincidence frequency, what is the value of $|R|$, where R is the reflection coefficient?

5–13. A $\frac{3}{4}$-in.-thick metal plate of unknown composition is immersed in fresh water (20°C).

(a) A collimated beam of plane sound waves of frequency 1.854 kHz is normally incident on the plate, and the standing wave ratio for the field in front of the plate is found to be 3. Find the metal or metals of which the plate could be made.

(b) You should find that two metals are possible. A transmission experiment is devised to decide between the two. The angle of incidence is set at 60°. As the frequency is slowly increased, the transmitted sound on the other side of the plate is measured. The transmission is found to be maximum at a frequency of 16.15 kHz. From these data determine which of the metals is the more likely candidate.

5–14. A 1-mm-thick sheet of lead is used as a sound barrier in air at a frequency of 6225 Hz.

(a) Find the TL for a normally incident sound field.

(b) Find the TL for a diffuse sound field (random incidence).

(c) Repeat parts (a) and (b) for a sheet of lead 1 cm thick.

5–15. Sound waves of frequency 5000 Hz in air are incident on a 5-mm-thick pane of glass. How much will the SPL in the air on the other side of the glass change when the angle of incidence is changed from 0° to 45°?

5–16. A noise control engineer is given the choice of steel or aluminum to make a panel to provide 40 dB of transmission loss for 7-kHz sound in air, $\theta_i = 70^\circ$. After calculating the thicknesses to be $\ell_{\text{steel}} = 0.716$ mm and $\ell_{\text{alum}} = 2.04$ mm, the engineer concludes that since both metal panels provide TL = 40 dB, both are equally good for the job.

(a) Confirm or deny the engineer's calculations.

(b) Confirm or deny the engineer's conclusion.

The required value of TL is changed to 50 dB.

(c) Using the methods of part (a), find the new thicknesses required.

(d) Does it make any differerence, solely on the basis of acoustical properties, which metal is used? Explain.

5–17. A gas turbine that emits intense sound at about 2000 Hz is to be tested in a test cell. An instrumentation room is located on the other side of one of the test cell walls. If the wall is made of steel $\frac{1}{2}$ in. thick, how much protection in the form of transmission loss (in decibels) will the wall afford against the turbine noise

(a) If the noise is normally incident on the wall?

(b) If the noise is incident at an angle of 45°?

(c) If the noise field in the test cell is diffuse (random)?

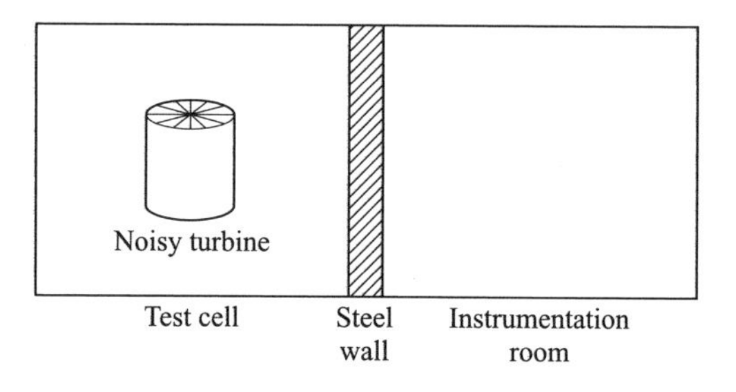

5–18. An optically transparent door is to be installed in the doorway to a room containing noisy equipment. The door must have the best possible sound transmission loss in the frequency range 1800–2700 Hz, where the noise is most intense. A manufacturer offers two transparent doors, a Lucite door 10 mm thick and a glass door 5.22 mm thick. Each weighs 50 lb. On the basis of acoustical properties, which door should be selected and why?

The following problems are about transmission loss of doors. As shown in the figure, the door frame is L_{frame} in length, w in width. The dimension of the air gap at the bottom, between the door and the floor, is L_o.

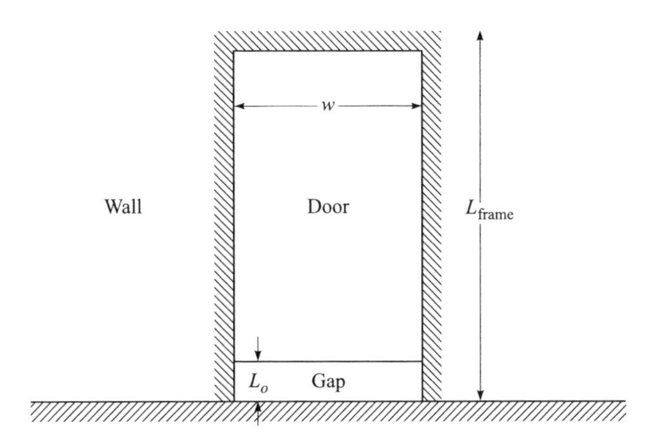

5–19. A door frame 7 ft × 3 ft is to be fit with a 1-in.-thick door made of wood of density 36 lb/ft^3. Transmission loss (normal incidence) is to be measured at a frequency of 902 Hz.

(a) Find the transmission loss TL_0 for a door that completely fills the frame, i.e., $L_o = 0$.

(b) Find the composite transmission loss $\mathrm{TL}_{\mathrm{comp}}$ when the gap is $L_o = 1.2$ in.

(c) How small must L_o be if $\mathrm{TL}_{\mathrm{comp}}$ is to be within 3 dB of TL_0?

5–20. A 7 × 3-ft door frame is to be fitted with a wooden door. The smallest possible gap at the bottom is $L_o = \frac{1}{8}$ in. The normal-incidence transmission loss when the door is closed, i.e., door plus gap, must be at least 25 dB in the frequency range 300–6000 Hz. Find the minimum mass M of the door.

5–21. The composite transmission loss for a door, that is, door plus air gap at the bottom, is specified to be $\mathrm{TL}_{\mathrm{comp}}$ in the frequency range $f_1 > f > f_2$.

(a) Derive a formula for the total mass M of the door to meet this specification.

(b) What is the maximum value of TL that can be achieved for given values of L_{frame} and L_o?

6

NORMAL MODES IN CARTESIAN COORDINATES: STRINGS, MEMBRANES, ROOMS, AND RECTANGULAR WAVEGUIDES

The general topic of this chapter is wave motion in bounded regions. This subject is treated from one viewpoint in Chaps. 3 and 4, but only one-dimensional problems are considered. In the present chapter use of a new approach, separation of variables, leads to the classical eigenfunction-eigenvalue solutions of the wave equation. We restrict ourselves, however, to rectangular geometry (spherical and cylindrical geometries are taken up in Chaps. 10 and 11, respectively). Except for the waveguide, the wave motion is assumed to be due to an initial disturbance in one or more of the field variables, such as displacement, particle velocity, or pressure.

One-dimensional problems, e.g., strings and plane-wave tubes, are the subject of Sec. A. Two-dimensional problems (vibrating membrane) are taken up in Sec. B and three-dimensional problems (sound in rectangular enclosures) in Sec. C. Section D is a brief introduction to rectangular waveguides. Waveguides are treated again, in broader scope, in Chap. 12.

A. VIBRATING STRING (AND OTHER ONE-DIMENSIONAL PROBLEMS)

Fundamental aspects of wave motion on strings—the wave equation derivation, characteristic impedance, boundary conditions, and energy—are discussed in Sec. 1C.2. The wave equation is repeated here for convenience,

$$\xi_{xx} - \frac{1}{c^2}\,\xi_{tt} = 0, \tag{A–1}$$

where ξ is the displacement of the string from its equilibrium position, $c = \sqrt{\mathcal{T}/\rho_\ell}$ is the wave speed on the string, $\mathcal{T}$ is the string tension, and ρ_ℓ is the mass/unit length, or *linear density*, of the string. Let the string be of length L. The ends $x = 0, L$ are subject to certain boundary conditions, which are to be specified presently. Initial conditions must also be specified. Here the initial string displacement and velocity are assumed known,

$$\xi(x, 0) = f(x), \tag{A–2a}$$

$$\xi_t(x, 0) = g(x). \tag{A–2b}$$

The method of solution, separation of variables, is described first for a string with fixed ends. Then other common boundary conditions are considered.

1. String with Fixed Ends

The boundary conditions for a string with fixed ends (Fig. 6.1) are

$$\xi(0, t) = 0, \tag{A–3a}$$

$$\xi(L, t) = 0. \tag{A–3b}$$

a. Method of Separation of Variables

The classical method for solving the wave equation for problems of this sort is *separation of variables*. A product solution

$$\xi = X(x)T(t) \tag{A–4}$$

is assumed and substituted in Eq. A–1, which then becomes

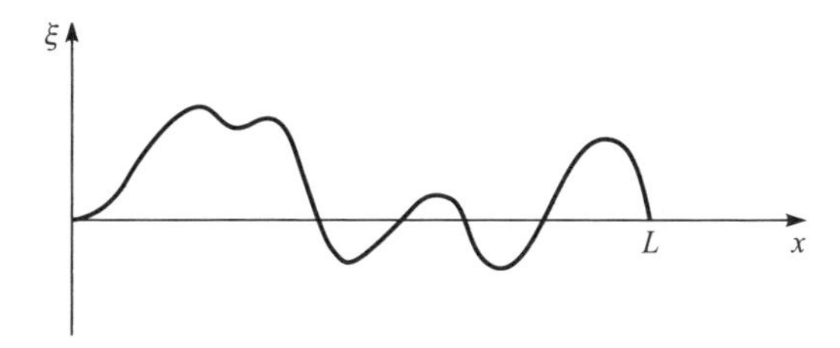

Figure 6.1 Wave motion on a string fixed at its end points $x = 0, L$.

$$X''T - \frac{1}{c^2} XT'' = 0,$$

or, after division by XT and rearrangement,

$$\frac{X''}{X} = \frac{1}{c^2} \frac{T''}{T}.$$

But how can a function of x (the left-hand side) be equal to a function of t (the right-hand side)? The only way is for both sides of the equation to be equal to a constant. Although the constant could be called anything, the symbol $-k^2$ turns out to be a convenient choice, i.e.,

$$\frac{X''}{X} = -k^2 = \frac{1}{c^2} \frac{T''}{T}; \tag{A–5}$$

k is sometimes called the separation constant (it would be more precise to speak of $-k^2$ as the separation constant).

The wave equation has now been separated into two ordinary differential equations, which have classical solutions. The x-dependent part of Eq. A–5 is recognized as the Helmholtz equation

$$X'' + k^2 X = 0. \tag{A–6}$$

The solution is

$$X = A_1 \cos kx + A_2 \sin kx, \tag{A–7}$$

where A_1 and A_2 are arbitrary constants. Similarly the time-dependent part of Eq. A–5,

$$T'' + k^2 c^2 T = 0, \tag{A–8}$$

has the solution

$$T = B_1 \cos \omega t + B_2 \sin \omega t, \tag{A–9}$$

where $\omega^2 = k^2 c^2$, or $k = \pm\omega/c$. It is now clear why $-k^2$ was chosen for the constant in Eq. A–5: k turns out to be the wave number.[1] Substitution of Eqs. A–8 and A–9 in Eq. A–4 yields

[1] Choice of $-k^2$ led us to the solution in terms of ordinary sine and cosine functions, which are ideal for satisfying typical boundary conditions. Had the choice been $+k^2$, hyperbolic sine and cosine functions would have resulted.

$$\xi = (A_1 \cos kx + A_2 \sin kx)(B_1 \cos \omega t + B_2 \sin \omega t). \tag{A–10}$$

A convenient shorthand way of writing this expression is

$$\xi = \begin{Bmatrix} \cos kx \\ \sin kx \end{Bmatrix} \begin{Bmatrix} \cos \omega t \\ \sin \omega t \end{Bmatrix}, \tag{A–11}$$

which means the product of any linear combination of the two space functions with any linear combination of the two time functions. At this point the amplitudes and the separation constant k are still unspecified.

b. Application of the Boundary Conditions

Because Eq. A–3a requires that ξ vanish at $x = 0$, we must discard the cos kx solution (the sin kx solution automatically satisfies Eq. A–3a) by setting $A_1 = 0$. The boundary condition at $x = L$, Eq. A–3b, then requires

$$\sin kL = 0,$$

which may be satisfied by choosing special values, called *eigenvalues*, for k,

$$k = \frac{\pi}{L}, \quad \frac{2\pi}{L}, \quad \frac{3\pi}{L}, \ldots, \frac{n\pi}{L}, \tag{A–12}$$

where n is any positive integer. Moreover, since $\omega = 2\pi f = kc$, the *eigenfrequencies* f_n are given by

$$f_n = \frac{k_n c}{2\pi} = \frac{nc}{2L}. \tag{A–13}$$

A string with fixed ends may vibrate freely at these frequencies and only these frequencies.

At this point our solution equation Eq. A–11 has become

$$\xi_n = \sin \frac{n\pi x}{L} [a_n \cos \omega_n t + b_n \sin \omega_n t], \tag{A–14}$$

where $a_n = A_2 B_1$ and $b_n = A_2 B_2$. The subscript n is used to indicate a different solution for each different integer. A superposition of all such solutions (the wave equation is linear) is the most general or total solution,

$$\xi = \sum_{n=1}^{\infty} \sin \frac{n\pi x}{L} [a_n \cos \omega_n t + b_n \sin \omega_n t]. \tag{A–15}$$

Although the eigenvalue $k = 0$ (i.e., $n = 0$) satisfies $\sin kL = 0$, the summation begins with the index $n = 1$, not $n = 0$, because the eigenfunction for $n = 0$ is trivial.

The functions $\sin(n\pi x/L)$ that satisfy the given boundary conditions are called *eigenfunctions*. The fact that they constitute a complete orthogonal set makes them very useful, as the example below shows. The standing wave functions ξ_n (Eq. A–14), which are eigenfunctions combined with their associated time functions, are called *normal modes* of vibration of the string. Often the spatial parts by themselves, i.e., $X_n = \sin(n\pi x/L)$, are referred to for short as the normal modes.

c. Initial Conditions

The coefficients a_n and b_n are determined from the initial conditions, Eqs. A–2a,b. Evaluation of the full solution, Eq. A–15, at $t = 0$ yields

$$f(x) = \sum a_n \sin\frac{n\pi x}{L}. \tag{A–16}$$

This is a Fourier series. To determine the coefficients a_n, use the orthogonality of the eigenfunctions. Multiply both sides of the equation by $\sin(m\pi x/L)\,dx$, where m is an integer, and integrate over the length of the string (0 to L):

$$\int_0^L f(x)\sin\frac{m\pi x}{L}\,dx = \sum a_n \int_0^L \sin\frac{n\pi x}{L}\sin\frac{m\pi x}{L}\,dx. \tag{A–17}$$

But each integral on the right-hand side vanishes for every term for which $n \neq m$. The infinite series therefore reduces to a single term, namely the one for which $n = m$. Since

$$\int_0^L \sin^2\frac{m\pi x}{L}\,dx = \frac{L}{2},$$

Eq. A–17 becomes

$$\frac{L}{2}a_m = \int_0^L f(x)\sin\frac{m\pi x}{L}\,dx,$$

or, if the index m is replaced by the index n,

$$a_n = \frac{2}{L}\int_0^L f(x)\sin\frac{n\pi x}{L}\,dx. \tag{A–18a}$$

Matching the string velocity, that is, the time derivative of Eq. A–15, with Eq. A–2b at time $t = 0$ yields another Fourier series, from which the coefficients b_n may be found,

$$b_n = \frac{2}{L\omega_n} \int_0^L g(x) \sin \frac{n\pi x}{L} dx. \tag{A–18b}$$

Together, Eqs. A–15 and A–18 constitute the final solution.

d. Normal Modes

We have referred to the functions ξ_n (Eq. A–14) as the normal modes of the string. For example, $\sin(\pi x/L)$ is the first mode, $\sin(2\pi x/L)$ the second mode, and so on. Equation A–16 shows that the initial string displacement is made up of a collection of the normal modes. Before proceeding with the detailed mathematics, let us analyze the situation in physical terms, making as much use of symmetry as possible. To be definite, suppose the initial string displacement is the rooftop function shown in Fig. 6.2. To see how this function, which is symmetric about the midpoint $x = L/2$, may be synthesized by the normal modes, consider the shapes of the modes. If just the first mode by itself is used to approximate the rooftop, the shape is roughly correct. In particular, the curve has the correct symmetry about the midpoint. See Fig. 6.3. But the $n = 1$ curve is smooth, not pointed, at the top. Can a more peaked curve be gotten by adding other modes? As Fig. 6.3 shows, adding the second mode would hurt matters more than it would help because the symmetry about the midpoint would be destroyed. Adding some of the third mode, if it is first inverted, does, however, produce a better approximation because the center is pushed up while at the same time the sides are reduced. Moreover, the third mode has the right symmetry, as do all the odd modes. We decide against the fourth mode, and in fact against all the other even modes, because they are anti-symmetric about the midpoint. Attainment of the rooftop shape must therefore be left to the odd modes. The purpose of this discussion is to show that a great deal may be determined about which modes participate in the motion without going deeply into the mathematics. Symmetry is a powerful property. If $f(x)$ has a certain symmetry, then only those modes possessing that property are excited. This principle is illustrated in the following example.

Example 6.1. Release of a string from rest. Consider the rooftop initial displacement shown in Fig. 6.2. In other words, let the string be pulled up at

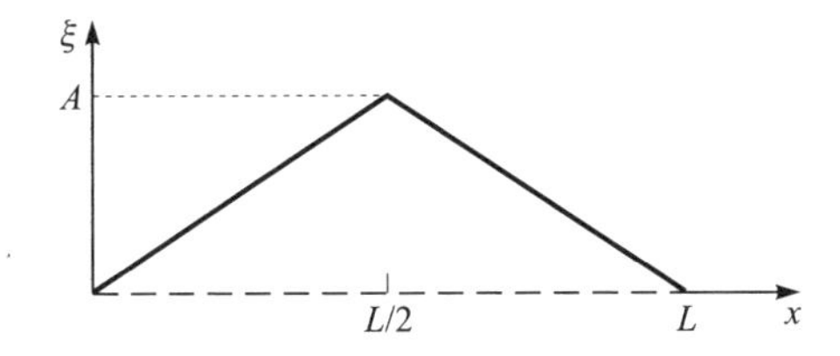

Figure 6.2 Sample initial displacement of a string.

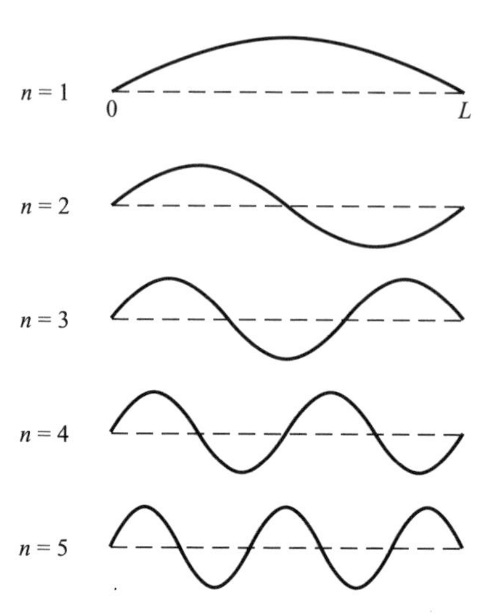

Figure 6.3 Shapes of first five normal modes for a string fixed at its ends.

its center and then released from rest at time $t = 0$. For this case the initial conditions $f(x)$ and $g(x)$ are as follows:

$$\text{Displacement:}\quad f(x) = \begin{cases} 2Ax/L, & 0 \le x \le L/2, \\ -2A(x/L - 1), & L/2 \le x \le L, \end{cases}$$

$$\text{Velocity:}\quad g(x) = 0.$$

The velocity initial condition yields $b_n = 0$. Equation A–18a becomes

$$\begin{aligned} a_n &= \frac{2}{L}\left\{\int_0^{L/2} \frac{2A}{L}\, x \sin\frac{n\pi x}{L}\, dx + \int_{L/2}^{L}\left(-\frac{2A}{L}\right)(x - L)\sin\frac{n\pi x}{L}\, dx\right\} \\ &= \frac{4A}{L^2}\left\{\int_0^{L/2} x\sin\frac{n\pi x}{L}\, dx - \int_{L/2}^{L} x\sin\frac{n\pi x}{L}\, dx + L\int_{L/2}^{L} \sin\frac{n\pi x}{L}\, dx\right\}. \end{aligned}$$

When the formula $a^2 \int x \sin ax\, dx = \sin ax - ax \cos ax$ is used and the various expressions are evaluated at the limits indicated, the resulting expression is quite simple,

$$a_n = \frac{8A}{\pi^2 n^2} \sin\frac{n\pi}{2}.$$

We observe that, as expected, $a_2 = a_4 = a_6 = \cdots = 0$, that is, the even modes do not participate in the string vibration. The solution is

$$\xi = \sum_{n=1}^{\infty} a_n \sin \frac{n\pi x}{L} \cos \omega_n t$$
$$= \frac{8A}{\pi^2}\left[\sin \frac{\pi x}{L} \cos \omega_1 t - \frac{1}{9} \sin \frac{3\pi x}{L} \cos \omega_3 t + \frac{1}{25} \sin \frac{5\pi x}{L} \cos \omega_5 t + \cdots\right].$$

Since the amplitudes of the higher order modes decrease rapidly, the series converges quickly. If the summation index is changed to $N = (n+1)/2$, the solution may be written

$$\xi = \frac{8A}{\pi^2} \sum_{N=1}^{\infty} \frac{(-1)^{N+1}}{(2N-1)^2} \sin \frac{(2N-1)\pi x}{L} \cos \omega_{2N-1} t.$$

If in the preceding example the initial displacement had been a function that is odd about the midpoint of the string, none of the odd modes ($n = 1, 3, 5, \ldots$) would have been excited. Only the even modes, which are anti-symmetric about the midpoint, would be able to participate in the motion. Symmetry considerations should be used whenever possible to gain physical insight about which modes are excited by a given initial condition and which are not.

e. Discussion

We have focused our attention on free-standing-wave solutions of the wave equation, that is, solutions of the form of Eq. A–11. If the string is infinite in length, all solutions of this type are possible because the wave number k and therefore the frequency f are unrestricted. If the string is of finite length, however, the number of free-wave solutions is severely limited. In particular, if the string has fixed ends, only sine waves having nodes at the ends of the string are allowed. The allowed solutions are called normal modes. Finally, the number of normal modes may be further restricted by the initial conditions. If a normal mode is not needed to satisfy the initial conditions, it is not excited and therefore never appears in the subsequent motion of the string.

Notice that our entire discussion has been about *free waves*. If the string is *forced* (driven) at any frequency, it must vibrate at that frequency. The amplitude of the response depends on the proximity of the driving frequency to one of the normal-mode frequencies (eigenfrequencies). In particular, if the driving frequency is equal to one of the normal-mode frequencies, the response turns out to be infinite (no losses have been considered), that is, resonance occurs. Therefore, although we have not studied forced waves per se here, a knowledge of the results for free waves is very useful in analyzing the response of the string to a driving force. Indeed, now we see the main difference between the approach used in Chap. 4 and that used here. Here the waves are free (they are not generated by a vibrating source) whereas in Chap. 4 the primary concern is with forced waves. However, see Problem 6–5 for an opportunity to study waves excited by an external force applied over the entire length of the string.

2. Other Boundary Conditions

Although the fixed end is the most common boundary condition considered for a string, other end conditions may be treated. Some have practical counterparts in problems involving plane sound waves in tubes.

a. Free end

The boundary condition at a free end of a string is that the slope ξ_x must vanish (see Sec. 1C.2). The cosine function is ideally suited to satisfy this condition because it has zero slope when its argument is 0, π, 2π, ... For example, if the string has a free end at $x = 0$, we must discard the $\sin kx$ solution in Eq. A–11 and retain the $\cos kx$ solution. If furthermore the end at $x = L$ is free, the relation determining the eigenvalues is again $\sin kL = 0$, and the eigenfunctions are $\cos n\pi x/L$. In this case the $n = 0$ eigenfunction is not trival (a string free at both ends may take on a fixed dc bias). If, on the other hand, the end at $x = L$ is fixed, the eigenvalues are determined by the relation $\cos kL = 0$, i.e., $kL = (2n + 1)\pi/2$, where $n = 0, 1, 2, \ldots$.

b. Impedance Boundary Condition

A more general boundary condition is that a linear combination of the field variable and its gradient vanishes at the boundary, for example,

$$A\xi + B\xi_x = 0. \qquad \text{(A–19)}$$

This is sometimes called an impedance boundary condition. To see why, consider the analogous problem of a plane sound wave in a tube, one end of which is terminated by an impedance Z_n. For time-harmonic waves the linear momentum equation, Eq. 1C–55, reduces to

$$j\omega\rho_0 u + p_x = 0. \qquad \text{(A–20)}$$

Evaluation of this relation at the termination, where $u = p/Z_n$, yields

$$j\omega\rho_0 p + Z_n p_x = 0. \qquad \text{(A–21)}$$

Specifying the impedance at a termination thus leads to a condition having the form of Eq. A–19.

Example 6.2. Tube having a mass at one end. The left end ($x = 0$) of the tube shown in Fig. 6.4 is rigid. Here $u = 0$, or, by application of Eq. A–20,

$$p_x = 0 \qquad (x = 0). \qquad \text{(A–22a)}$$

At the right end ($x = L$) is a piston of mass M. The specific acoustic impedance of the piston is $j\omega m$, where $m = M/S$ and S is the cross-sectional area. Equation A–21 for this case is

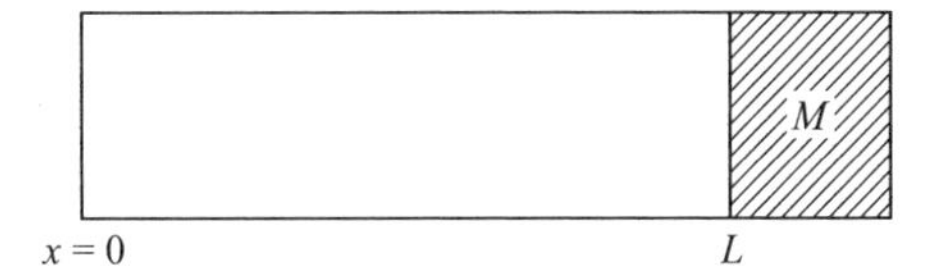

Figure 6.4 Wave motion in a tube with one end rigid and a mass at the other end.

$$\rho_0 p + m p_x = 0 \qquad (x = L). \tag{A–22b}$$

When the wave equation in p (Eq. 1C–59) is solved by the method of separation of variables, a solution having the form of Eq. A–11 is obtained. To satisfy Eq. A–22a, discard the sin kx solution. Satisfaction of Eq. A–22b requires

$$\rho_0 \cos kL - km \sin kL = 0,$$

or

$$\cot kL = \alpha kL,$$

where $\alpha = m/\rho_0 L$ ($= M/M_g$; M_g is the mass of the gas in the tube). This equation is discussed in Sec. 4D (see Eq. 4D–10 and Fig. 4.30). Its roots $kL = \beta_n$ (see Ref. 1, p. 225, for numerical values of β_n) give us the eigenvalues for this problem. Notice that because the eigenvalues are not related to each other by integers, the eigenfrequencies are not harmonically related to each other. Harmonically related eigenfrequencies generally occur only when the boundary condition at $x = L$ is that either a field variable or its gradient vanishes.

3. The Struck String

To complete our discussion of strings, we revisit the struck string problem, solved in Chap. 1 (Example 1.1) for a string of infinite length. Here the string is finite and fixed at its ends. The string is struck uniformly from below so as to give it an initial velocity u_0. See Fig. 6.5. The initial conditions are

$$\xi(x, 0) = 0, \qquad \xi_t(x, 0) = u_0.$$

Because the ends are fixed, the general solution is Eq. A–15. The coefficients a_n are zero in this case because the initial displacement is zero. Equation A–18b is used to find the coefficients b_n; the result is

$$b_n = \frac{2u_0 L}{(n\pi)^2 c}[1 - (-1)^n]$$

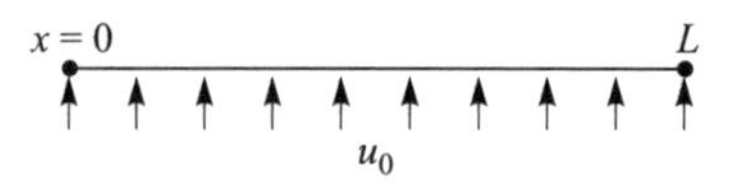

Figure 6.5 Struck string problem for a fixed-end string.

$$= \begin{cases} \dfrac{4u_0 L}{(n\pi)^2 c} & \text{if } n \text{ is odd,} \\ 0 & \text{if } n \text{ is even.} \end{cases}$$

Only the odd modes participate in the motion. This was to be anticipated because the even modes are anti-symmetric about the string midpoint whereas the initial velocity distribution is symmetric about the midpoint. The final solution is

$$\xi = \frac{4u_0 L}{\pi^2 c} \sum_{n=\text{odd}} \frac{\sin n\pi x/L}{n^2} \sin \frac{n\pi c}{L} t.$$

Next, let the initial velocity distribution be that shown in Fig. 6.6. What modes participate in the subsequent motion in this case? Because the initial velocity distribution is now anti-symmetric about the string midpoint, no odd mode is excited. This leaves just the even modes, $n = 2, 4, 6, 8, \ldots$, but are all of them excited? It turns out that every other even mode is missing. Why?

Finally, suppose the striking hammer is very narrow so that it hits the string at a single point $x = x_0$, as shown in Fig. 6.7. The initial velocity is then given by

$$\dot{\xi}(x, 0) = u_0\, \delta(x - x_0),$$

where δ is the delta or impulse function. Equation A–18b for this case is

$$\begin{aligned} b_n &= \frac{2}{L\omega_n} \int_0^L u_0\, \delta(x - x_0) \sin \frac{n\pi x}{L}\, dx \\ &= \frac{2u_0}{n\pi c} \sin \frac{n\pi x_0}{L}, \end{aligned}$$

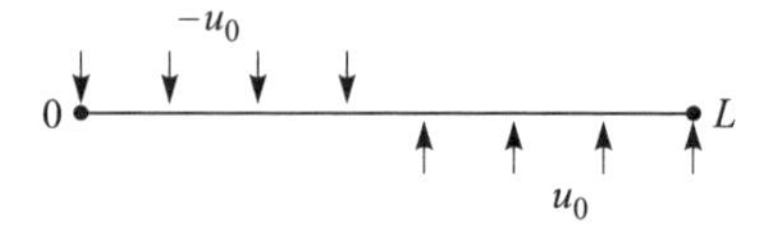

Figure 6.6 Variation on the struck string problem.

Figure 6.7 String struck at a point.

where we have used the filtering property of the delta function,

$$\int_{x_1}^{x_2} \delta(x - x_0) f(x)\, dx = f(x_0), \quad x_1 < x_0 < x_2.$$

It will be seen that b_2 vanishes if the striking point is in the middle of the string, $x_0 = L/2$. Similarly $b_3 = 0$ if $x_0 = L/3$ or $x_0 = 2L/3$. In general, $b_n = 0$ if $x_0 = mL/n$, where m is an integer ($m < n$). These striking points are precisely the places where the nth mode has a node. In other words a normal mode cannot be excited by striking the string (or plucking it) at one of the nodes of the normal mode. This is an important general result.

B. VIBRATING MEMBRANE

A rectangular membrane of length L and width W is shown in Fig. 6.8.

The wave equation for the displacement η on this membrane is $\nabla^2\eta - c^{-2}\eta_{tt} = 0$, or since the coordinate system is rectangular,

$$\eta_{xx} + \eta_{yy} - \frac{1}{c^2}\eta_{tt} = 0. \tag{B–1}$$

The wave speed c on the membrane is given by the formula $c = \sqrt{\mathcal{T}_\ell/m}$, where $\mathcal{T}_\ell$ is the tension per unit length and m is the mass per unit area (surface density).[2]

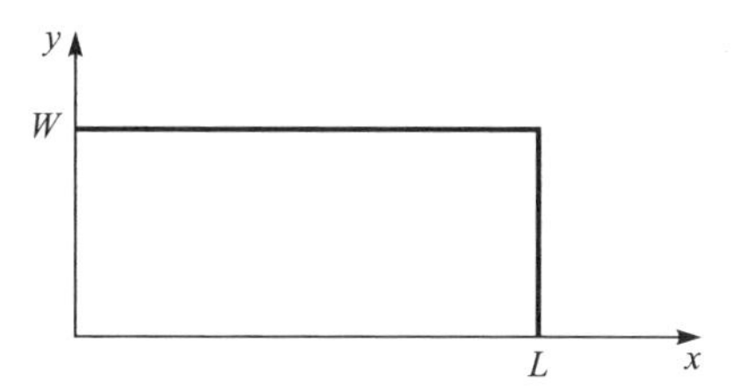

Figure 6.8 Rectangular membrane of length L and width W.

[2]The symbol m is used later in this section to mean an integer index.

The solution by means of separation of variables begins with the assumption of a product solution,

$$\eta = X(x)\,Y(y)\,T(t). \tag{B–2}$$

Substitution into Eq. B–1, followed by division by XYT, leads to

$$\frac{X''}{X} + \frac{Y''}{Y} = \frac{1}{c^2}\frac{T''}{T}.$$

Just as in the case of the vibrating string, we set both sides equal to the constant $-k^2$. The T equation is the same as the one for the string and therefore has the same solution, Eq. A–9. The equations for X and Y are obtained by separating variables again. We have

$$\frac{X''}{X} = -k^2 - \frac{Y''}{Y}.$$

The left-hand side is a function of x only, the right-hand side a function of y only. The equation can hold only if both sides are equal to a constant, $-p^2$, say.[3] The X equation becomes

$$X'' + p^2 X = 0,$$

the solution of which is

$$X = A_1 \cos px + A_2 \sin px. \tag{B–3}$$

The Y equation is

$$Y'' + (k^2 - p^2)Y = 0,$$

which, if we put $k^2 - p^2 = q^2$, has the solution

$$Y = C_1 \cos qy + C_2 \sin qy. \tag{B–4}$$

Equations B–3, B–4, and A–9 may now be multiplied together as indicated by Eq. B–2, but the shorthand way of expressing the product solution is

$$\eta = \begin{Bmatrix} \cos px \\ \sin px \end{Bmatrix} \begin{Bmatrix} \cos qy \\ \sin qy \end{Bmatrix} \begin{Bmatrix} \cos \omega t \\ \sin \omega t \end{Bmatrix}, \tag{B–5}$$

where

[3]Here p is a separation constant; it does not stand for pressure.

$$p^2 + q^2 = k^2 = \omega^2/c^2. \tag{B–6}$$

Restrictions on p and q, and therefore on k and ω, are set by the boundary conditions. If the membrane is clamped at its edges, the boundary conditions are

$$\eta(0, y, t) = 0, \tag{B–7a}$$
$$\eta(L, y, t) = 0, \tag{B–7b}$$
$$\eta(x, 0, t) = 0, \tag{B–7c}$$
$$\eta(x, W, t) = 0. \tag{B–7d}$$

Equations B–7a and B–7c require that the cos px and cos qy solutions, respectively, be discarded. Equations B–7b and B–7d may then be satisfied by choosing $pL = m\pi$, or

$$p = \frac{m\pi}{L}, \tag{B–8a}$$

and $qW = n\pi$, or

$$q = \frac{n\pi}{W}, \tag{B–8b}$$

respectively, where m and n are integers. These two results, when combined with Eq. B–6, provide us with an expression for the allowed frequencies (eigenfrequencies),

$$f_{mn} = \frac{c}{2}\sqrt{\left(\frac{m}{L}\right)^2 + \left(\frac{n}{W}\right)^2}. \tag{B–9}$$

The expression for the (mn)th normal mode of a rectangular membrane clamped at its edges is therefore

$$\eta_{mn} = \sin\frac{m\pi x}{L}\sin\frac{n\pi y}{W}(a_{mn}\cos\omega_{mn}t + b_{mn}\sin\omega_{mn}t), \tag{B–10}$$

where $\omega_{mn} = 2\pi f_{mn}$. The complete solution is the sum of all the normal-mode solutions, in this case a double sum,

$$\eta = \sum_{m,n=1}^{\infty}\sin\frac{m\pi x}{L}\sin\frac{n\pi y}{W}(a_{mn}\cos\omega_{mn}t + b_{mn}\sin\omega_{mn}t). \tag{B–11}$$

As in the case of the fixed-end vibrating string, $m = 0$ and $n = 0$ are allowed indices but their associated eigenfunctions are trivial.

To determine the coefficients a_{mn} and b_{mn}, apply the initial conditions. Suppose the initial displacement η_0 and initial velocity u_0 are given, i.e.,

$$\eta(x, y, 0) = \eta_0(x, y), \tag{B–12a}$$

$$\dot{\eta}(x, y, 0) = u_0(x, y). \tag{B–12b}$$

When combined with the solution, Eq. B–11, these equations give

$$\eta_0 = \sum_{m,n} a_{mn} \sin\frac{m\pi x}{L} \sin\frac{n\pi y}{W}, \tag{B–13a}$$

$$u_0 = \sum_{m,n} b_{mn}\omega_{mn} \sin\frac{m\pi x}{L} \sin\frac{n\pi y}{W}. \tag{B–13b}$$

Again, the orthogonality properties of the sine functions are used to obtain expressions for a_{mn} and b_{mn}. Multiply Eq. B–13b by $\sin(m'\pi x/L)\sin(n'\pi y/W)$ $dx\,dy$ and integrate over the domain of the membrane (0 to L in x, 0 to W in y). The right-hand side reduces to the single term for which $m = m'$ and $n = n'$. One obtains

$$a_{mn} = \frac{4}{LW}\int_0^W \int_0^L \eta_0(x, y) \sin\frac{m\pi x}{L} \sin\frac{n\pi y}{W}\, dx\, dy. \tag{B–14a}$$

The expression for b_{mn} is found similarly,

$$b_{mn} = \frac{4}{LW_{mn}}]\int_0^W \int_0^L u_0(x, y) \sin\frac{m\pi x}{L} \sin\frac{n\pi y}{W}\, dx\, dy. \tag{B–14b}$$

The shapes of the first three modes are shown in Fig. 6.9. (See Figs. 36 and 37 in Ref. 2 for additional mode shapes.) Why is no 1, 0 or 0, 1 mode shown? What would the lowest (nontrivial) mode be if the sides $x = 0$ and $x = L$ were free instead of clamped?

Suppose the membrane were set into driven vibration by setting a sound source, such as a loudspeaker, in front of the membrane. If the excitation frequency were slowly swept through a broad frequency range and the amplitude of the corresponding membrane vibration recorded, the frequency response of the membrane could be obtained. Although we have not analyzed the problem of forced motion, it can be appreciated that the membrane, just as any other resonant system, will have its greatest response at its natural frequencies. These are, of course, the eigenfrequencies given by Eq. B–9. A frequency response of the sort shown in Fig. 6.10 may therefore be anticipated. The magnitude of the peak at each resonance frequency depends on the damping (which we have not considered) and on the driving conditions. For example, suppose the membrane is driven at a point, say by the end of a rod that is

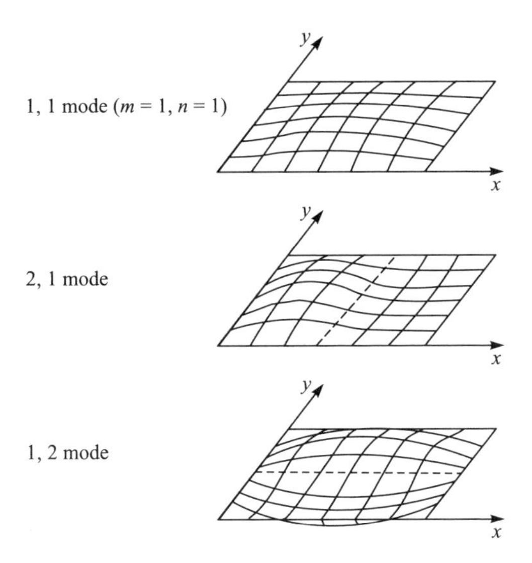

Figure 6.9 Three simplest modes of vibration of a rectangular membrane.

attached to a shaker. In general, a mode is excited whenever the rod is driven at the appropriate eigenfrequency. The membrane does not respond at all, however, if the driving point happens to be at a node. For instance, if the driving point is at the center of the membrane, only modes with odd indices (1, 1; 1, 3; 3, 1; 3, 3; etc.) can be excited.

C. SOUND IN A RECTANGULAR ENCLOSURE

In this section let us use the wave equation in the velocity potential ϕ, Eq. 1D–7, which, for rectangular coordinates, becomes

$$\phi_{xx} + \phi_{yy} + \phi_{zz} - \frac{1}{c_0^2}\phi_{tt} = 0. \tag{C–1}$$

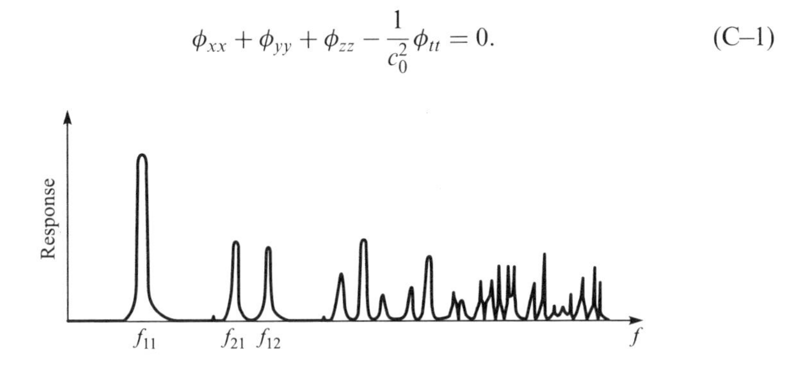

Figure 6.10 Frequency response of a driven membrane.

The particle velocity and pressure are given by Eqs. 1D–5 and 1D–6, respectively, repeated here for convenience,

$$\mathbf{u} = \nabla\phi, \tag{C–2}$$

$$p = -\rho_0\phi_t. \tag{C–3}$$

If the x, y, and z components of $\mathbf{u}$ are denoted u, v, and w, respectively, the components of Eq. C–2 are

$$u = \phi_x, \tag{C–4a}$$

$$v = \phi_y, \tag{C–4b}$$

$$w = \phi_z. \tag{C–4c}$$

For boundary conditions, we limit consideration to surfaces at which either the pressure vanishes or the normal component of particle velocity vanishes. For example, if the bottom surface of the enclosure shown in Fig. 6.11 is rigid, the proper boundary condition is $\phi_z = 0$ at $z = 0$. If the top is a pressure release surface, Eq. C–3 implies $\phi = 0$ at $z = H$ (if ϕ_t vanishes at a surface, ϕ is expected to vanish also).

Example 6.3. Normal modes of a fish tank. Determine the motion of free waves in a rectangular container for which the top is a pressure release surface while all other surfaces are rigid. A fish tank filled with water approximates these conditions provided the sides and bottom are made of sufficiently heavy, hard material. By the method of separation of variables, we obtain the solution of the wave equation in the form

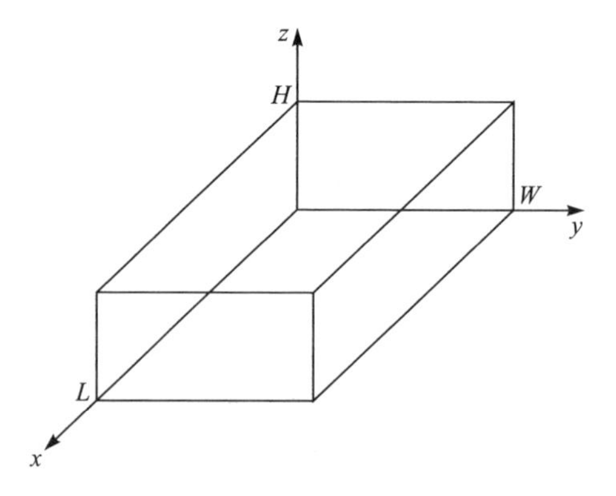

Figure 6.11 Coordinate system for a rectangular enclosure.

$$\phi = \begin{Bmatrix} \cos qx \\ \sin qx \end{Bmatrix} \begin{Bmatrix} \cos ry \\ \sin ry \end{Bmatrix} \begin{Bmatrix} \cos sz \\ \sin sz \end{Bmatrix} \begin{Bmatrix} \cos \omega t \\ \sin \omega t \end{Bmatrix}, \tag{C–5}$$

where the various separation constants are related to each other by

$$q^2 + r^2 + s^2 = k^2 = \omega^2/c_0^2. \tag{C–6}$$

This equation may be obtained by substituting Eq. C–5 directly in Eq. C–1. Now apply the six boundary conditions:

- $\phi_x(0, y, z) = 0$: Requires that the sin qx solution be discarded.
- $\phi_x(L, y, z) = 0$: Requires that $\sin qL = 0$.

$$\therefore \quad q = \frac{\ell\pi}{L}, \quad \ell = 0, 1, 2, 3, \ldots. \tag{C–7a}$$

- $\phi_y(x, 0, z) = 0$: Requires that the sin ry solution be discarded.
- $\phi_y(x, W, z) = 0$: Requires that $\sin rW = 0$.

$$\therefore \quad r = \frac{m\pi}{W}, \quad m = 0, 1, 2, 3, \ldots. \tag{C–7b}$$

- $\phi_z(x, y, 0) = 0$: Requires that the sin sz solution be discarded.
- $\phi(x, y, H) = 0$: Requires that $\cos sH = 0$.

$$\therefore \quad s = \frac{(2n+1)\pi}{2H}, \quad n = 0, 1, 2, 3, \ldots. \tag{C–7c}$$

The (ℓmn)th normal mode is therefore given by

$$\phi_{\ell mn} = \cos\frac{\ell\pi x}{L}\cos\frac{m\pi y}{W}\cos\frac{(2n+1)\pi z}{2H}\begin{Bmatrix} \cos \omega_{\ell mn} t \\ \sin \omega_{\ell mn} t \end{Bmatrix}, \tag{C–8}$$

or, in terms of the pressure,

$$p_{\ell mn} = \cos\frac{\ell\pi x}{L}\cos\frac{m\pi y}{W}\cos\frac{(2n+1)\pi z}{2H}\begin{Bmatrix} \sin \omega_{\ell mn} t \\ \cos \omega_{\ell mn} t \end{Bmatrix}, \tag{C–9}$$

where the constant $\pm\rho_0\omega_{\ell mn}$ has been absorbed by the coefficients of the two time functions. Notice that in this case any or all of the indices ℓ, m, n may be zero. For example, the expression for the 0, 0, 0 normal mode is

$$p_{000} = \cos \frac{\pi z}{2H}(a_{000} \cos \omega_{000} t + b_{000} \sin \omega_{000} t).$$

This mode represents a time-harmonic pressure field having an amplitude that is constant over any horizontal plane ($z =$ const) in the tank. The amplitude decreases monotonically from its maximum at the bottom ($z = 0$) to zero at the top ($z = H$). The eigenfrequencies are obtained from Eq. C–6 by substituting the values found for q, r, and s,

$$f_{\ell mn} = \frac{c_0}{2}\sqrt{\left(\frac{\ell}{L}\right)^2 + \left(\frac{m}{W}\right)^2 + \left(\frac{2n+1}{2H}\right)^2}. \qquad \text{(C–10)}$$

For example, the frequency of the 0, 0, 0 mode is $f_{000} = c_0/4H$; at this frequency the depth of the tank is one-quarter wavelength.

In the case of sound in a rectangular room, the boundary conditions are that the normal component of particle velocity vanish at all six surfaces. What is the expression for the (ℓmn)th normal mode in this case? Are the restrictions on the indices any different from those for the fish tank?

D. RECTANGULAR WAVEGUIDE

As the name suggests, a waveguide is a device for guiding the propagation of waves. The wave motion in the guide is usually partly progressive and partly standing. The progressive wave motion is in the direction of the desired transmission and is generally initiated by a source at one end of the guide. The standing waves are caused by the presence of the confining surfaces that do the guiding. This section may be read as an introduction to Chap. 12, which is devoted entirely to waveguides.

1. Membrane Waveguide

A very simple waveguide is a long membrane strip clamped at two rails $y = 0$ and $y = d$, as shown in Fig. 6.12. The strip stretches indefinitely far in the $+x$ direction. At the edge $x = 0$ is a source, which, by vibrating time harmonically, is give as

$$\eta(0, y, t) = G(y)e^{j\omega t}, \qquad \text{(D–1)}$$

launches progressive waves that advance down the guide. However, they must travel in such a way as to satisfy the clamp conditions ($\eta = 0$) at the two rails.

The wave equation to be solved is Eq. B–1. The method of separation of variables may be used even though the membrane is semi-infinite in the x

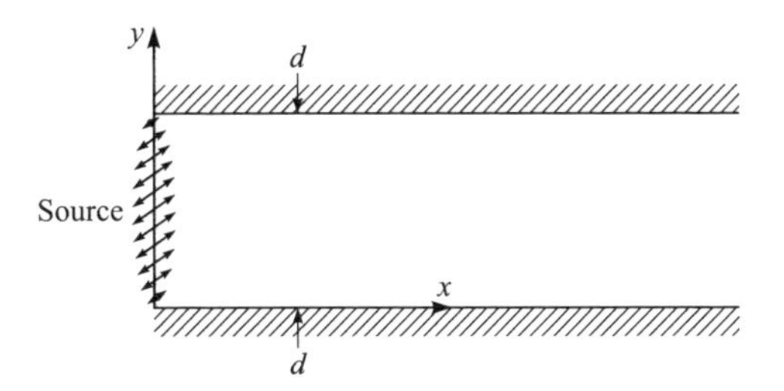

Figure 6.12 Membrane waveguide. The membrane is clamped at $y = 0$, d.

direction. Instead of Eq. B–5, however, we use the following, equally valid, product solution[4]:

$$\eta = \left\{ \begin{matrix} e^{j\beta x} \\ e^{-j\beta x} \end{matrix} \right\} \left\{ \begin{matrix} \cos \gamma y \\ \sin \gamma y \end{matrix} \right\} \left\{ \begin{matrix} e^{j\omega t} \\ e^{-j\omega t} \end{matrix} \right\}. \tag{D–2}$$

The reason for using complex exponentials rather than sines and cosines for the X and T functions is that we are anticipating progressive waves, not standing waves, in the x direction. We must now decide which exponentials to pick. Because the solution must ultimately satisfy the source condition, Eq. D–1, our choice for the time function is $T = e^{j\omega t}$. Next, since the X function must combine with the T function to represent a forward traveling wave (a backward traveling wave has no way of getting started on the waveguide), the choice for X must be $e^{-j\beta x}$. Equation D–2 therefore reduces to

$$\eta = \left\{ \begin{matrix} \cos \gamma y \\ \sin \gamma y \end{matrix} \right\} e^{j(\omega t - \beta x)}, \tag{D–3}$$

which describes a progressive wave having an amplitude that depends on y. As can be shown by substituting Eq. D–3 (or D–2) in the wave equation (Eq. B–1), the various separation constants are related to each other by

$$\beta^2 = (\omega/c)^2 - \gamma^2. \tag{D–4}$$

This expression is a dispersion relation. It tells us how the wave number β depends on frequency (the dispersion relation found in Sec. 5D for bending waves on a plate gives the frequency dependence of the wave speed).

To complete the solution, we apply the boundary conditions

$$\eta(x, 0, t) = 0, \tag{D–5a}$$

[4]In place of p and q, β and γ have been chosen as symbols for the separation constants. The switch is partly to agree with traditional waveguide nomenclature and partly to make it easy to distinguish between the waveguide and rectangular membrane solutions.

$$\eta(x, d, t) = 0. \tag{D–5b}$$

The first condition requires that the cos γy solution be discarded, the second that γ be an eigenvalue

$$\gamma = \frac{n\pi}{d}, \quad n = 1, 2, 3, \ldots \tag{D–6}$$

(why is $n = 0$ not included?). The assumed solution, often called the nth mode, therefore has the form

$$\eta_n = A_n \sin \frac{n\pi y}{d} e^{j(\omega t - \beta_n x)}. \tag{D–7}$$

The wave number β_n is found by substituting Eq. D–6 in Eq. D–4,

$$\beta_n = \pm\sqrt{(\omega/c)^2 - (n\pi/d)^2}. \tag{D–8}$$

The full solution is the sum of all the possible modes,

$$\eta = \sum_n A_n \sin \frac{n\pi y}{d} e^{j(\omega t - \beta_n x)}. \tag{D–9}$$

As explained in Sec. 4 below, the coefficients A_n are chosen to satisfy the source condition, Eq. D–1.

2. Forward Traveling Waves, Phase Velocity, and Cutoff

The plus sign in Eq. D–8 signifies a forward traveling wave, the minus sign a backward traveling wave. From now on only forward traveling waves are considered.

The phase velocity is the rate at which the phase $\varphi = \omega t - \beta_n x$ of a given mode advances along the waveguide. If the expression for φ is differentiated,

$$d\varphi = \omega\, dt - \beta_n\, dx,$$

it is seen that a point of constant phase ($d\varphi = 0$) is distinguished by the relation

$$\left(\frac{dx}{dt}\right)_{\varphi = \text{const}} = \frac{\omega}{\beta_n},$$

which is called the *phase velocity*[5] and denoted c_n^{ph} to indicate its dependence on the mode number n. Use of Eq. D–8 leads to

[5]A very elementary way of arriving at the same result is to write the exponential in Eq. D–7 as $\exp \omega[t - x/(\omega/\beta_n)]$. The form of this expression makes it clear that ω/β_n is the apparent speed with which the signal (of frequency f) travels along the x axis.

$$c_n^{\mathrm{ph}} \equiv \frac{\omega}{\beta_n} = \frac{c}{\sqrt{1-(n\pi c/\omega d)^2}} = \frac{c}{\sqrt{1-(nc/2fd)^2}}. \tag{D–10}$$

It is seen that waveguides are dispersive, that is, the phase velocity depends on frequency.[6] A graph of the phase velocity as a function of frequency is shown in Fig. 6.13. At very high frequencies, c_n^{ph} approaches the free-medium speed c. As frequency is reduced, the phase velocity increases until it becomes infinite at the frequency

$$f_n^c = \frac{nc}{2d}, \tag{D–11}$$

which is called the *cutoff frequency*. In terms of the cutoff frequency, the expression for the phase velocity is

$$c_n^{\mathrm{ph}} = \frac{c}{\sqrt{1-(f_n^c/f)^2}}. \tag{D–12}$$

At frequencies below the cutoff frequency the wave number is imaginary,

$$\beta_n = \pm j\frac{\omega}{c}\sqrt{\left(\frac{f_n^c}{f}\right)^2 - 1}. \tag{D–13}$$

If the minus sign is selected, the expression for the nth mode becomes

$$\eta_n = A_n \sin\frac{n\pi y}{d}\, e^{j\omega t} \exp\left[-kx\sqrt{\left(\frac{f_n^c}{f}\right)^2 - 1}\right], \tag{D–14}$$

which does not represent a progressive wave at all, only a time-harmonic signal that decays exponentially with distance down the waveguide. Here we have

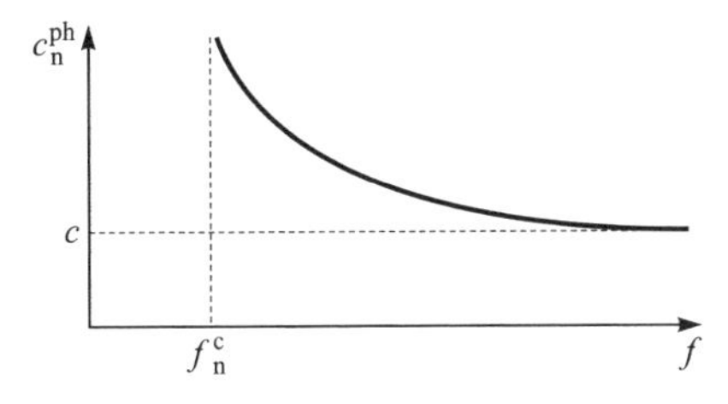

Figure 6.13 Phase velocity as a function of frequency.

[6]Dispersion is termed "normal" if phase velocity decreases with frequency, "anomalous" if phase velocity increases with frequency. Figure 6.13 shows an example of normal dispersion.

another example of an *evanescent wave*. The reason for choosing the minus sign in Eq. D–13 is now clear: choice of the plus sign would have led to a signal that grows exponentially with distance.

3. Physical Interpretation

Figure 6.14 shows the wave field in the waveguide at a particular instant. The field is really two plane-wave fields.[7] Each field is indicated by a set of parallel lines that represent the wavefronts. The solid lines are the wavefront peaks, the dashed lines the wavefront troughs. The wavefronts moving forward and downward are incident on the lower rail; their incident ray is the heavy solid line making an angle θ_i with the normal (y axis). The waves moving forward and upward are the waves reflected from the lower rail; they are at the same time the waves that are incident on the upper rail. Note that where each incident wavefront ends at a rail, a reflected wavefront of opposite phase begins; in this way the boundary condition at the rail is satisfied.

The physical meaning of phase velocity may be found by analyzing Fig. 6.14. First, focus attention on the forward, downward plane waves, which are incident at angle θ_i on the lower rail. As a wavefront peak moves one wavelength λ, it traces out a length $\lambda_{\text{trace}} = \lambda / \sin \theta_i$ along the lower rail, as

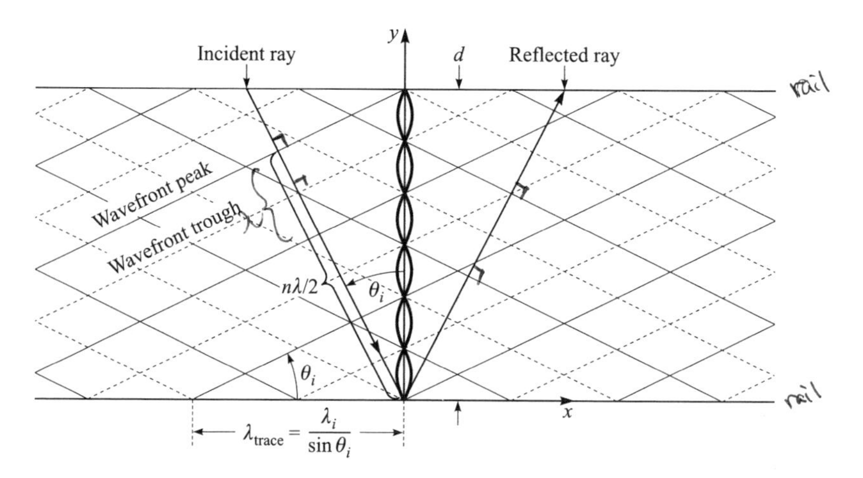

Figure 6.14 Geometry of the wave field in the waveguide. The n half wavelengths marked $n\lambda/2$ along the incident ray project n loops (on the y axis) that fit exactly in the waveguide. The fit determines the angle θ_i at which the wave bounces along the waveguide.

[7] To see this, rewrite the sine function in Eq. D–7 in terms of two complex exponentials. Equation D–7 then divides into two terms, one proportional to $\exp[j(\omega t - \beta_n x - n\pi y/d)]$, the other proportional to $\exp[j(\omega t - \beta_n x + n\pi y/d)]$. The first term represents a plane wave traveling forward and upward, the second term a plane wave traveling forward and downward.

indicated by the expression in the figure. The phase velocity is the apparent speed with which one of the wavefronts sweeps along the rail. This is simply the trace velocity, the trace wavelength divided by one period $(1/f)$:

$$c_n^{\text{ph}} = f\lambda_{\text{trace}} = \frac{f\lambda}{\sin\theta_i} = \frac{c}{\sin\theta_i} = \frac{c}{\sqrt{1-\cos^2\theta_i}}.$$

Next notice that the superposition of the two plane-wave fields creates a vertical pattern of nodes and antinodes (displacement nulls and maxima, respectively), as indicated by the loops drawn on the y axis. The total number of loops across the guide is n ($n = 6$ in Fig. 6.14), and the n loops have a projection equal to $n\lambda/2$ on the incident ray. This means that $\cos\theta_i = n\lambda/2d = nc/2fd$. The expression for the phase velocity therefore reduces to

$$c_n^{\text{ph}} = \frac{c}{\sqrt{1-(nc/2fd)^2}},$$

which is Eq. D–10, this time, however, derived by purely geometrical arguments.

We can now see why the phase velocity increases as frequency is reduced. If the wavelength is increased (frequency lowered), the angle θ_i must decrease in order that the same number of loops (six in Fig. 6.14) fit between the two rails. However, because decreasing θ_i increases the trace wavelength, c_n^{ph} goes up. Finally, at the frequency at which n half wavelengths fit into the guide only perpendicularly ($\theta_i = 0$), a single wavefront strikes all points on a rail simultaneously, i.e., the apparent propagation speed in the x direction is infinite. We have now reached the cutoff frequency. If the frequency is further reduced (wavelength further increased), the required number of half wavelengths cannot be fit in the guide at all, and that particular mode cannot exist.

The *group velocity* is another velocity associated with propagation in waveguides. It is defined by $c^{\text{gr}} \equiv d\omega/d\beta$. From Eq. D–8 and the geometrical relations just developed, it is clear that

$$c_n^{\text{gr}} = (d\beta/d\omega)^{-1} = c\sqrt{1-(nc/2fd)^2} = c\sin\theta_i. \tag{D–15}$$

The group velocity is therefore just the component of the true propagation velocity c along the x axis (the axis of the guide). Note that the group and phase velocities are related by

$$c_n^{\text{ph}} c_n^{\text{gr}} = c^2. \tag{D–16}$$

4. Source Conditions

Whether a particular mode of the waveguide is excited depends on source conditions. The source excitation is given by Eq. D–1. The function $G(y)$ represents the amplitude of the source, which may be constant across the guide ($G = \text{const}$) or may vary according to circumstance or design. Notice the way the frequency enters the problem: it is imposed by the source. The problem is therefore one of forced waves, not free waves. Evaluating the solution equation, Eq. D–9 at $x = 0$ gives

$$G(y) = \sum_n A_n \sin \frac{n\pi y}{d},$$

which is a Fourier series. The modes of the guide are therefore chosen so that their sum equals the source amplitude function $G(y)$. In particular,

$$A_n = \frac{2}{d} \int_0^d G(y) \sin \frac{n\pi y}{d}\, dy. \qquad \text{(D–17)}$$

Example 6.4. Mode excitation and cutoff for a membrane waveguide. Analyze propagation on a membrane waveguide for which $d = 10$ cm and $c = 100$ m/s. At $x = 0$ the membrane is vibrated at a frequency of 1700 Hz. The vibration amplitude η_0 is not a function of y, i.e., it is uniform.[8] First of all, since $G(y) = \eta_0$ in this case, Eq. D–17 gives

$$\begin{aligned} A_n &= \frac{2\eta_0}{d} \int_0^d \sin \frac{n\pi y}{d}\, dy \\ &= \frac{4\eta_0}{n\pi} \quad \text{if } n \text{ is odd,} \\ &= 0 \quad \text{if } n \text{ is even.} \end{aligned}$$

Only the odd modes of the guide are excited. This does not mean, however, that an observer further down the guide will detect all of the odd modes. To find out what reaches the observer, consider the cutoff frequencies. Equation D–11 gives

$$f_n^c = \frac{n(100)}{2(0.1)} = 500n,$$

[8]Obviously the membrane displacement cannot be both η_0 and zero at the rails $y = 0, d$. The impasse may be resolved by a limiting process where the amplitude remains η_0 to as close to the rails as one pleases but not at the rails themselves.

e.g., $f_1^c = 500$ Hz, $f_2^c = 1000$ Hz, $f_3^c = 1500$ Hz, $f_4^c = 2000$ Hz, $f_5^c = 2500$ Hz, and so on. We need not concern ourselves with the even modes because they are not excited. As for the odd modes, the drive frequency exceeds the cutoff frequency for the first and third modes. Propagation in these modes is therefore possible. Propagation in the higher odd modes is not possible, however, because the drive frequency is below their cutoff frequencies. For instance, the fifth-mode signal, having a cutoff frequency of 2500 Hz, is an evanescent wave. Its attenuation factor (see Eq. D–14) is

$$\exp\left[-\frac{2\pi(1700)}{100}\sqrt{\left(\frac{2500}{1700}\right)^2 - 1}\; x\right] = e^{-115x}.$$

The decay of this mode is therefore extremely rapid. Modes higher than the fifth decay even more rapidly. Accordingly, an observer only a short distance from the source perceives a membrane displacement consisting of only two modes,

$$\eta = \frac{4\eta_0}{\pi}\sin\frac{\pi y}{d}\, e^{j\omega(t - x/c_1^{\mathrm{ph}})} + \frac{4\eta_0}{3\pi}\sin\frac{3\pi y}{d}\, e^{j\omega(t - x/c_3^{\mathrm{ph}})}.$$

The phase velocities c_1^{ph} and c_3^{ph} of the two modes are found from Eq. D–12 to be 105 and 213 m/s, respectively.

The variation in mode phase velocity in the example above illustrates the dispersion property of waveguides. A broadband signal such as a pulse suffers distortion as it travels down a waveguide because the different frequency components of the pulse travel at different speeds.

REFERENCES

1. M. Abramowitz and I. A. Stegun (Eds.), *Handbook of Mathematical Functions*, National Bureau of Standards, Applied Mathematics Series 55 (U.S. Government Printing Office, Washington, DC, 1964).
2. C. A. Coulson, *Waves*, 7th ed. (Interscience, New York, 1961), Chaps. 1, 2, and 3. This text has been revised and enlarged and issued under the same title but with A. Jeffrey added as coauthor (Longman, London and New York, 1978), 2nd ed.
3. L. E. Kinsler, A. R. Frey, A. B. Coppens, and J. V. Sanders, *Fundamentals of Acoustics*, 3rd ed. (John Wiley & Sons, New York, 1982), Arts. 2.10, 4.3, 9.7, and 9.8.
4. P. M. Morse, *Vibration and Sound*, 2nd ed. (McGraw-Hill Book Co., New York, 1948), 2nd ed., pp. 314–319. Reprinted by Acoustical Society of America, New York, 1981.
5. P. M. Morse, "Waves and targets, generalizations and specifics," *Am. J. Phys.* **53**, 25–40 (1985).

6. P. M. Morse and K. U. Ingard, *Theoretical Acoustics* (McGraw-Hill Book Co., New York, 1968), pp. 332–338. Reprinted by Princeton University Press, Princeton, NJ 1986.

PROBLEMS

6–1. A string with fixed ends at $x = 0, L$ is initially displaced as shown in the sketch and indicated by the accompanying equation. At time $t = 0$, the string is released.

$$\xi(x, 0) = \begin{cases} A, & 0 < x < L/2 \\ -A, & L/2 < x < L. \end{cases}$$

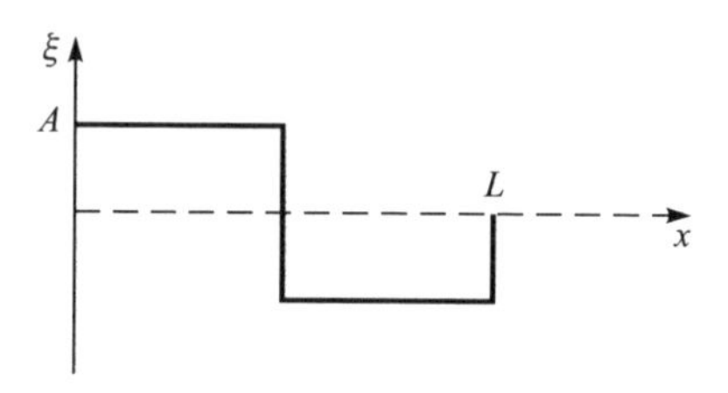

(a) Use separation of variables and the boundary conditions to find the eigenfunctions and the eigenfrequencies.

(b) Apply the initial conditions to determine the complete series solution of the problem.

(c) Which of the normal modes are missing, that is, which do not participate in the motion of the string? Using symmetry arguments, explain why these modes are missing.

6–2. Find the eigenfunctions for a string that has one end (say $x = 0$) fixed and the other end ($x = L$) free.

(a) Find the frequency f_n of the nth normal mode.

(b) Write the complete solution as an infinite series.

(c) Given the initial conditions $\xi(x, 0) = f(x)$ and $\xi_t(x, 0) = g(x)$, provide formulas for the modal coefficients in terms of f and g.

(d) Sketch the shapes of the first few modes.

6–3. Consider a pipe of length L open at both ends ($x = 0, L$). Assume that the boundary condition at the ends is that the acoustic pressure vanishes there.

(a) Find the eigenfunctions for this problem. It is suggested that you start with the wave equation in p, not u.

(b) Find the eigenfrequencies.

(c) Sketch the first several modes (pressure, not particle velocity). Identify each mode by its index number n.

(d) Initially, the gas is quiet and the acoustic pressure is constant, $p = A$, everywhere in the pipe. At time $t = 0$ the constraint on the pressure is released. Give the complete series solution of the problem (all coefficients determined). Which modes do not participate in the motion? Use symmetry arguments to explain why they are missing.

6–4. A string with fixed ends ($x = 0$ and $x = L$) is struck at time $t = 0$ as shown in Fig. 6.6. That is, the initial conditions are

$$\xi(x, 0) = 0, \quad 0 < x < L,$$

$$\xi_t(x, 0) = \begin{cases} -u_0, & 0 < x < L/2 \\ u_0, & L/2 < x < L. \end{cases}$$

(a) Obtain the expression for the string displacement $\xi(x, t)$.

(b) You should find that certain modes do not take part in the motion of the string. Which modes? Use symmetry arguments to explain why they are missing.

6–5. The wave equation for a string driven by an external force is Eq. 1C–26. Obtain the steady-state solution for the following case: The length of the string is L, the ends are fixed, and the external force per unit length is $F_\ell(x, t) = F_{\ell 0} e^{j\omega t}$. Can you identify resonance behavior? At what frequency (or frequencies) does resonance occur?

6–6. A tube filled with air is closed with a lucite piston of length 1.73 mm. When the piston is held rigidly in place by set screws, the eigenfrequencies f_n of the tube are measured and found to be 100, 200, 300 Hz, ..., i.e., $f_n = 100\,n$, $n = 1, 2, 3, \ldots$. The set screws are then removed so that the piston is free to move subject only to the constraint of its own inertia. Find the new eigenfrequencies f_n' (give the first four numerical values). Is there a one-to-one correspondence between f_n' and f_n? One way to check is to take the limit as the piston is made rigid and determine whether $f_n' \to f_n$.

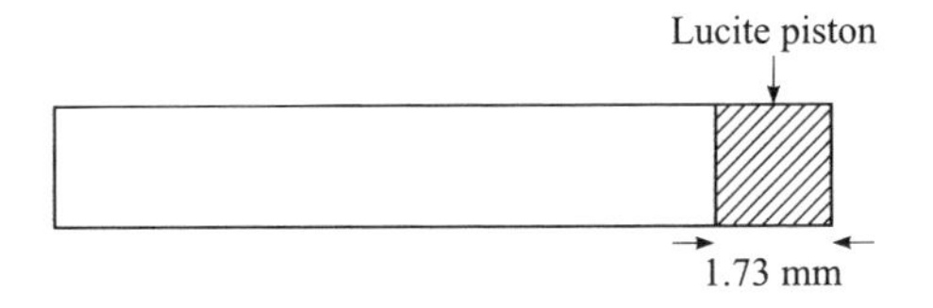

6–7. A flexible membrane is stretched between two circular hoops like a lamp shade. Find the eigenfunctions and eigenfrequencies of this membrane. *Hint*: Uncurl the membrane and lay it out flat. It is then obvious that the boundary conditions are $\eta = 0$ at $y = 0, d$ and $\eta(x, y, t) = \eta(x + 2\pi a, y, t)$, i.e., η must be periodic in x with period $2\pi a$.

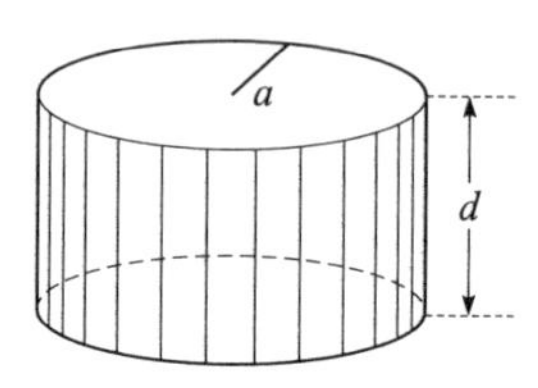

6–8. A flexible rectangular membrane is fixed at its edges $x = 0, L$, and $y = 0, W$. Because the tensions in the x and y directions are different, the propagation speed is c_1 in the x direction, c_2 in the y direction. The wave equation for this case is

$$c_1^2 \eta_{xx} + c_2^2 \eta_{yy} - \eta_{tt} = 0.$$

(a) Find the expression for the (mn)th normal mode η_{mn}. What values of the indices m and n are allowed and nontrivial?

(b) Find the expression for the frequencies f_{mn} of the normal modes.

6–9. Investigate the normal modes of sound in a rectangular room of length $x = L$, width $y = W$, and height $z = H$ (see Fig. 6.11). Start with the wave equation in the velocity potential, $\nabla^2\phi - c_0^{-2}\phi_{tt} = 0$. Assume that the walls of the room are rigid. Let ℓ, m, n be the integers that are associated with the eigenvalues.

(a) Find the expression for the frequency $f_{\ell mn}$ of the (ℓmn)th mode.

(b) The measured values for a certain classroom are $L = 27$ ft, $W = 20$ ft, and $H = 10$ ft. Find the values of the lowest six eigenfrequencies of the room, not counting $f_{000} = 0$ (the frequency of the "dc mode," i.e., the mode that has no spatial variation). List the six frequencies, with their indices, in order.

(c) If you put an acoustic pressure source at the center of the room $(L/2, W/2, H/2)$, what modes would never be excited? Where could you locate the pressure source so as to excite *all* the modes?

6–10. Find the normal-mode solution for sound in a hall with rigid walls (normal component of particle velocity vanishes at the wall surfaces) and open ends $y = 0, L$ (assume the pressure p vanishes there).

(a) Starting with the general solution,

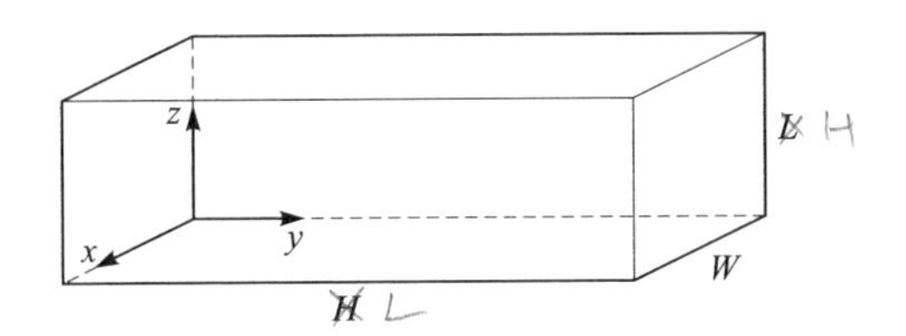

$$p = \begin{Bmatrix} \cos qx \\ \sin qx \end{Bmatrix} \begin{Bmatrix} \cos ry \\ \sin ry \end{Bmatrix} \begin{Bmatrix} \cos sz \\ \sin sz \end{Bmatrix} \begin{Bmatrix} \cos \omega t \\ \sin \omega t \end{Bmatrix},$$

apply the boundary conditions to specialize this solution to the hall.

(b) Find the formula for the frequencies $f_{\ell mn}$ of the normal modes.

(c) If $L = 20$ ft, $W = 4$ ft, and $H = 8$ ft, what are the lowest three eigenfrequencies (nontrivial modes) for the hall?

6–11. The boundary conditions for sound in a water tank with rigid ends and bottom, pressure release sides, and an open top are as follows: $p = 0$ at $x = 0, W$, $v = 0$ at $y = 0, L$, $w = 0$ at $z = 0$, and $p = 0$ at $z = H$ (the geometry is shown by the sketch in the previous problem).

(a) Find the expression for $p_{\ell mn}$, the pressure of the (ℓmn)th mode.

(b) Give the formula for the frequencies $f_{\ell mn}$ of the normal modes. What values of ℓ, m, n are allowed and nontrivial?

(c) The medium in the tank is fresh water at 20°C, and the dimensions are $L = 2$ ft, $W = 0.5$ ft, $H = 1$ ft. List the six lowest eigenfrequencies (nontrivial modes) and their indices in order.

6–12. Sound propagates down a rigid-wall duct of square cross section. The boundary conditions and the source condition are

$$u = 0 \quad \text{at} \quad x = 0, a \tag{P–1}$$

$$v = 0 \quad \text{at} \quad y = 0, a \tag{P–2}$$

$$p = f(x, y)e^{j\omega t} \quad \text{at} \quad z = 0, \tag{P–3}$$

respectively. Start with the general solution

$$p = \begin{Bmatrix} \cos qx \\ \sin qx \end{Bmatrix} \begin{Bmatrix} \cos ry \\ \sin ry \end{Bmatrix} e^{j(\omega t - \beta z)}. \tag{P–4}$$

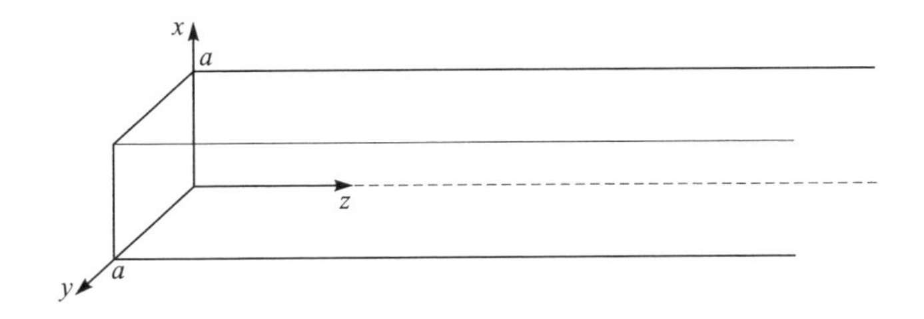

(a) What expression relates the quantities q, r, β, and k ($k = \omega/c_0$)?

(b) Specialize the solution given by Eq. P–4 so that it satisfies the boundary conditions, Eqs. P–1 and P–2. Use the symbols m and n when you solve for the eigenvalues for q and r, respectively.

(c) Write the complete solution as the sum of all the individual modal solutions p_{mn}. Find an expression for the modal coefficients A_{mn} by applying the source condition, Eq. P–3.

(d) Find the phase velocity and the cutoff frequency for the (mn)th mode.

(e) What is the lowest order mode (mode with the lowest cutoff frequency) in which sound can propagate in this duct?

(f) What is the next lowest cutoff frequency? Identify the mode(s).

(g) An air-filled duct has a side dimension $a = 6$ in., and the source frequency is 2000 Hz. Identify the modes in which wave propagation can take place.

6–13. The ocean acts as a waveguide for underwater sound. For an elementary analysis, idealize the ocean as having a pressure release surface at $z = 0$ and a rigid bottom at $z = D$. Waves of angular frequency ω can travel through the water in the x direction by bouncing back and forth obliquely between the surface and the bottom.

Solve for the wave motion. It is suggested that you start with the wave equation in p, i.e.,

$$p_{xx} + p_{zz} - \frac{1}{c_0^2} p_{tt} = 0,$$

and seek a solution of the form

$$p = Z(z) e^{j(\omega t - \beta x)}.$$

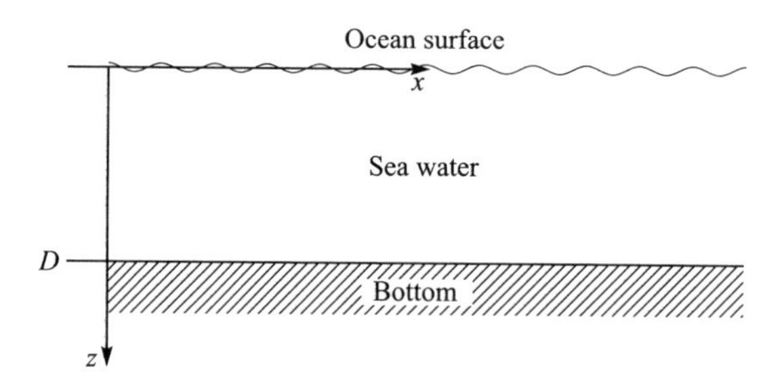

(a) Find the mode function $Z(z)$ and sketch the shapes of the first few modes.

(b) Obtain the expression for the cutoff frequency of the nth mode.

(c) If the depth of the water is $D = 15$ m and a 100-Hz source is placed in the water, in what modes can sound propagate in the waveguide?

(d) What are the angles θ_i (the angle an upward or downward ray makes with the z axis) associated with the modes found in part (c)?

(e) Give the phase and group velocities for the modes found in part (c).

7

HORNS

The author knows of no better way to approach the subject of horns than to quote the following passage from an unpublished set of notes by F. V. Hunt[1]:

> Crandall [Ref. 2] introduces the subject of horns with the challenging remark that "... theory has now reached the point where, if horns were not available, it would be our plain duty to invent them ..." The grounds for this obligation are to be found in conclusions drawn from the two preceding chapters. From Chapter V one infers that the effectiveness of a sound source is affected in an important way by the phase relation between the normal velocity of the radiating surface and the force reaction of the medium on the surface. The force reaction, or sound pressure, only becomes in phase with the velocity when the perimeter of the source becomes comparable with the wavelength ($2\pi R \geq \lambda$, or $kR \geq 1$). [For Hunt's Chap. V, "Radiation from Sound Sources," the reader may, for purposes here, substitute Secs. 1D.3 and 1E.3 of this book.] Then if sounds having a wide range of frequency are to be generated, the designer is faced with a dilemma: a surface large enough to radiate effectively at low frequency will almost certainly be too massive to be driven with useful amplitude at high frequency. What is obviously needed is a device that will permit the sound to be *generated* by the vibration of a small light diaphragm, but *radiated* from a surface or aperture large enough to keep pressure in phase with particle velocity even at the lower end of the useful frequency range.
>
> An ideal flaring horn performs just this function. It accepts sound energy at the "small end," or throat, with pressure in phase with particle velocity, just as though the throat were one end of a very long tube of uniform cross-section.

[1]The passage quoted here is from the first two pages of Hunt's Chap. VII, "Horns," dated December 1948 (Ref. 4).

Then the horn flares to a large mouth in such a way that the shape of the wavefronts is not altered, and the sound energy is radiated from the horn mouth with pressure still substantially in phase with particle velocity. If the medium in the horn is nondissipative (and better it be so), all the net sound energy accepted at the throat must be radiated at the mouth: in fact, the net sound energy flux must be the same through *any* cross-section of the horn. The horn thus performs the function of transforming a bounded sound wave of relatively high pressure and high particle velocity into an equally energetic wave, of lower pressure and lower velocity, distributed over a correspondingly larger area.

Section A of this chapter gives a derivation of the Webster horn equation, which is the classical equation governing propagation of sound in horns. The exponential horn is analyzed as an example in Sec. B. Impedance, power, and the transmission factor are treated in Sec. C. Section D is devoted to a more general way of solving the Webster horn equation, the WKB method.

A. WEBSTER HORN EQUATION

The Webster horn equation is a one-dimensional wave equation for sound in ducts of slowly varying cross section. Strictly speaking, fluid motion in a duct of nonconstant cross section cannot be one-dimensional. As fluid flows longitudinally through the duct (see Fig. 7.1), it must, in response to changing area, have a transverse component of motion, either outward if the duct is expanding or inward if the duct is contracting. If the area change is gradual, however, the transverse component may be small enough in comparison to the longitudinal component to be ignored. A one-dimensional description of the flow is thus possible as an approximation.

1. Continuity Equation

The verbal statement of conservation of mass for a variable-cross-section duct is no different from that for a duct of fixed cross section: the time rate of

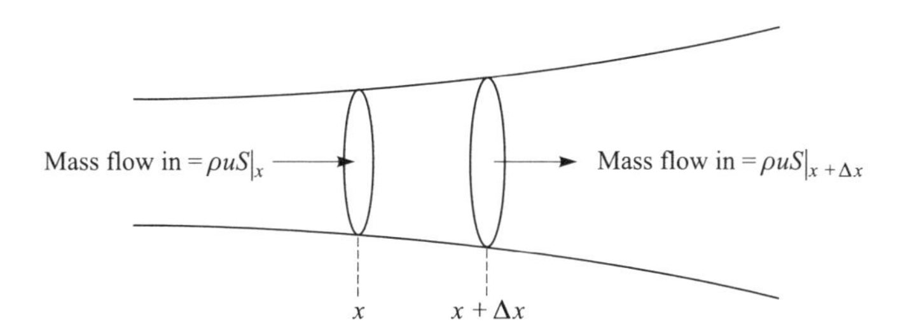

Figure 7.1 Mass flux in a duct of slowly varying cross section.

increase of mass inside the elemental volume is equal to the net mass inflow through the surfaces of the element (see Fig. 7.1), i.e.,

$$\frac{\partial}{\partial t}(\rho S\,\Delta x) = \rho u S\Big|_{x} - \rho u S\Big|_{x+\Delta x}.$$

(As usual, each property in the interior of the volume element, e.g., ρ in the term on the left-hand side, represents the mean value in the element. In the limit as $\Delta x \to 0$, the quantity becomes a true point variable.) Rearranging and taking the limit as $\Delta x \to 0$ yield

$$S\rho_t = \lim_{\Delta x \to 0} \frac{\rho u S\Big|_{x} - \rho u S\Big|_{x+\Delta x}}{\Delta x} = -(\rho u S)_x,$$

which may be put in the form

$$\rho_t + \rho u(\ln S)_x + (\rho u)_x = 0, \tag{A–1}$$

where the logarithmic derivative $(\ln S)_x$ equals S_x/S. When the duct cross section is constant, the middle term disappears and the ordinary one-dimensional continuity equation, Eq. 1C–28, is obtained.

2. Momentum Equation

It turns out that the final form of the momentum equation is no different from that for a duct of fixed cross section. The derivation is a little more complicated, however. In this section the symbol P again stands for total pressure. The longitudinal forces $PS|_x$ and $-PS|_{x+\Delta x}$ at the two ends of the control volume are supplemented by the longitudinal component of the force exerted by the duct wall. This component is equal to the pressure times the projected area $\Delta S_\perp$ of the wall; see Fig. 7.2. Setting the time rate of change of momentum inside the volume element equal to the net momentum inflow across the boundaries plus the sum of the (longitudinal) forces acting on the element, we have

$$(\rho u S \Delta x)_t = \rho u^2 S\Big|_{x} - \rho u^2 S\Big|_{x+\Delta x} + \underbrace{PS\Big|_{x} - PS\Big|_{x+\Delta x} + \Delta S_\perp P\Big|_{x+\Delta x/2}}_{F_s}. \tag{A–2}$$

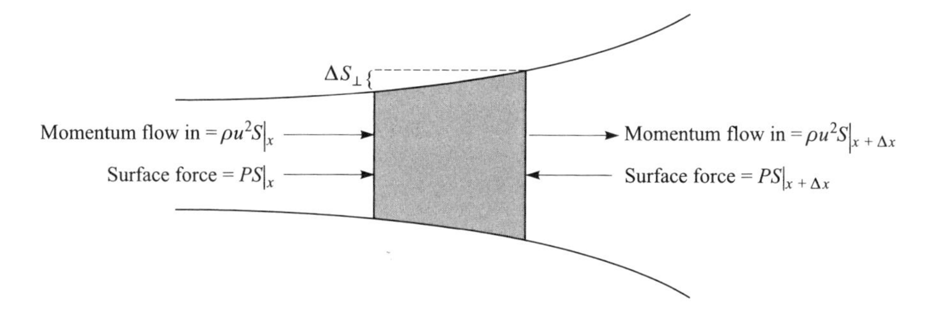

Figure 7.2 Momentum flux and longitudinal surface forces in a duct of varying cross section.

The last three terms on the right-hand side are surface forces F_s. Note that $\Delta S_\perp = S(x+\Delta x) - S(x)$ and $P(x + \frac{1}{2}\Delta x) = P(x) + \frac{1}{2}\Delta x P'(x) + \cdots$. Collecting terms allows F_s to be written as follows:

$$F_s = P(x)S(x+\Delta x) - P(x+\Delta x)S(x+\Delta x) + \tfrac{1}{2}\Delta x\,\Delta S_\perp\,P'(x)\cdots.$$

If Eq. A–2 is now divided by Δx, the result is

$$S(\rho u)_t = \frac{\rho u^2 S\Big|_x - \rho u^2 S\Big|_{x+\Delta x}}{\Delta x} + S(x+\Delta x)\frac{P(x) - P(x+\Delta x)}{\Delta x} + \frac{1}{2}\Delta S_\perp\,P'(x)\cdots,$$

or, in the limit as $\Delta x \to 0$,

$$S(\rho u)_t = -(\rho u^2 S)_x - SP_x.$$

When this equation is expanded and the continuity equation Eq. A–1 is used, several terms cancel. The result is

$$\rho(u_t + uu_x) + P_x = 0, \tag{A–3}$$

which is the same as the momentum equation for a tube of constant cross section, Eq. 1C–33.

3. Webster Horn Equation

The reduction of Eqs. A–1, A–2, and 1C–40 (the equation of state) to a wave equation proceeds as in Sec. 1C.3.d. First, linearization yields

$$\text{Continuity: } \delta\rho_t + \rho_0 u(\ln S)_x + \rho_0 u_x = 0, \tag{A–4}$$

$$\text{Momentum: } \rho_0 u_t + p_x = 0, \tag{A–5}$$

$$\text{State: } p = c_0^2 \delta\rho. \tag{A–6}$$

Equation A–6 is used to eliminate $\delta\rho$ from Eq. A–4, which is then differentiated with respect to t. Equation A–5 may be used to eliminate the particle velocity from the differentiated equation. The final result is

$$p_{xx} + (\ln S)_x p_x - \frac{1}{c_0^2} p_{tt} = 0, \tag{A–7}$$

which is known in acoustics as the Webster horn equation, despite its use much earlier by investigators in other fields (Ref. 3).

B. EXAMPLE: EXPONENTIAL HORN

The exponential horn (Fig. 7.3) is widely used in audio engineering. It serves as a good vehicle for discussing horn behavior and design.

1. Exponential Horn Equation and Solution

When the cross-sectional area varies exponentially, i.e., when

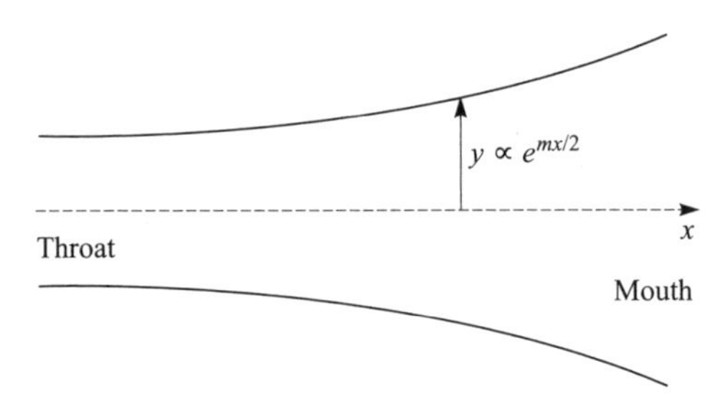

Figure 7.3 Exponential horn.

$$S = S_0 e^{mx}, \tag{B–1}$$

where m is called the flare constant, the factor $(\ln S)_x$ in Eq. A–7 is equal to m, and the horn equation reduces to

$$p_{xx} + mp_x - \frac{1}{c_0^2} p_{tt} = 0. \tag{B–2}$$

As a trial solution, use the time-harmonic wave function

$$p = Ae^{j(\omega t - kx)}. \tag{B–3}$$

Substitution in Eq. B–2 yields a dispersion relation,

$$k^2 + jmk - k_0^2 = 0, \tag{B–4}$$

where $k_0 = \omega / c_0$ is the wave number for a constant-cross-section tube. The solution for k is

$$k = \pm k_0 \sqrt{1 - \left(\frac{m}{2k_0}\right)^2} - j\frac{m}{2}. \tag{B–5}$$

In other words, the trial solution Eq. B–3 satisfies the (exponential) horn equation provided k is given by Eq. B–5. The plus sign signifies an outgoing wave, the minus sign an incoming wave.

2. Amplitude Decay and Phase Velocity

Consider now the case of an outgoing wave. Substituting Eq. B–5 in Eq. B–3, we have

$$p = A\mathrm{e}^{-mx/2} \exp j\omega \left[t - \frac{x}{c_0} \sqrt{1 - \left(\frac{m}{2k_0}\right)^2} \right] \tag{B–6}$$

or

$$p = A\mathrm{e}^{-\alpha x}\mathrm{e}^{j\omega(t - x/c^{\mathrm{ph}})},$$

where the attenuation coefficient α and the phase velocity c^{ph} are identified as

$$\alpha = \frac{1}{2}m \tag{B–7}$$

and

$$c^{\mathrm{ph}} = \frac{c_0}{\sqrt{1 - (mc_0/2\omega)^2}}, \tag{B–8}$$

respectively.

The exponential diminution of the wave amplitude is due to spreading of the wave as the horn gets larger, not to dissipation. The diminution law may also be inferred from energy considerations. If the acoustical power is to remain constant as the wave propagates through the horn, then SI must remain constant. As shown in Sec. C.1 below, the impedance of an exponential horn does not change with distance. For such cases the proportionality of the intensity to p_{rms}^2 (Eq. 1E–18b) means that Sp_{rms}^2 is constant, or

$$p_{\mathrm{rms}} \propto 1/\sqrt{S}, \tag{B–9}$$

which for our exponential horn becomes

$$p_{\mathrm{rms}} \propto \mathrm{e}^{-mx/2},$$

in agreement with Eqs. B–6 and B–7. A more general demonstration of the validity of Eq. B–9 is given in Sec. D below.

The phase velocity formula Eq. B–8 is similar to the one found for waveguides, Eq. 6D–10. At high frequencies the phase velocity approaches the open medium value c_0, as shown in Fig. 7.4. But as the frequency is decreased, the phase velocity increases, eventually becoming infinite when $mc_0/4\pi f = 1$, or

$$f^c = \frac{mc_0}{4\pi}. \tag{B–10}$$

As in the case of waveguides, f^c is called the cutoff frequency. At frequencies lower than f^c, c^{ph} is imaginary, and Eq. B–6 becomes

$$p = Ae^{-(m/2 - k_0\sqrt{1-(f^c/f)^2})x}e^{j\omega t}.$$

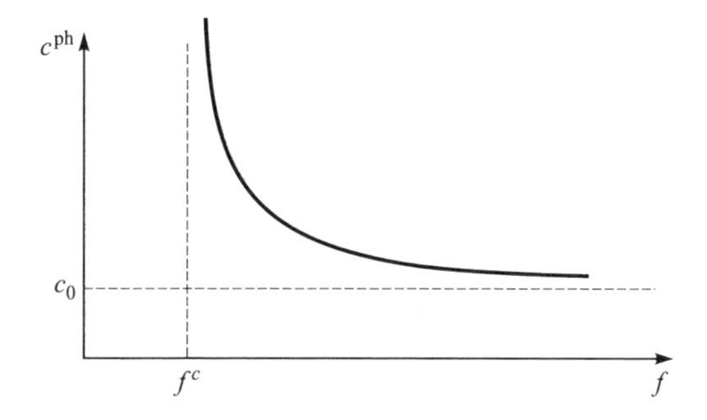

Figure 7.4 Phase velocity and cutoff for an exponential horn.

The pressure disturbance is thus an evanescent wave, that is, not a true wave at all, only a time-harmonic signal that decays with distance. The rapidity of the decay increases as the frequency decreases below cutoff. In terms of the cutoff frequency, the expression for k may be written

$$k = k_0\left[\sqrt{1 - (f^c/f)^2} - j(f^c/f)\right], \tag{B–11}$$

which is of the form $k = \beta - j\alpha$. The polar form is also useful:

$$k = k_0 e^{-j\theta}, \tag{B–12}$$

where

$$\tan\theta = \frac{\alpha}{\beta} = \frac{f^c/f}{\sqrt{1 - (f^c/f)^2}}. \tag{B–13}$$

C. IMPEDANCE, POWER TRANSMITTED, AND TRANSMISSION FACTOR

To introduce this subject, for simplicity we restrict our attention to horns for which the wave number, although complex, is not distance dependent, i.e., $k \neq k(x)$. Exponential and conical horns, for example, have this property. The more general case is taken up in Sec. D.

1. Impedance and Power

The impedance for a forward traveling time-harmonic wave is readily calculated from the momentum equation, Eq. A–5. For such a wave $u_t = j\omega u$ and, provided $k \neq k(x)$, $p_x = -jkp$, Eq. A–5 reduces to

$$\omega\rho_0 u = kp,$$

which results in

$$Z \equiv \frac{p}{u} = \frac{\rho_0\omega}{k}. \tag{C–1}$$

Recall that in general Z has a real part R and an imaginary part X; of particular importance is R. (*Note*: Here the symbol R does not stand for reflection coefficient.) The throat impedance is the load seen by the source at the throat of the horn, $x = 0$,

$$Z_{\text{th}} = R_{\text{th}} + jX_{\text{th}}.$$

The power, found by combining Eqs. 1E–18a and 1E–20,

$$W = Ru_{\text{rms}}^2 S,$$

is most easily evaluated at the throat of the horn, where $u = u_0 e^{j\omega t}$, $S = S_{\text{th}}$, and $R = R_{\text{th}}$:

$$W = \tfrac{1}{2} R_{\text{th}} u_0^2 S_{\text{th}}. \tag{C–2}$$

As an example, calculate the impedance and power for an exponential horn. Substitution of Eq. B–11 in Eq. C–1 yields

$$\begin{aligned} Z &= \frac{\rho_0 c_0}{\sqrt{1 - (f^c/f)^2} - jf^c/f} \\ &= \rho_0 c_0 \left[\sqrt{1 - (f^c/f)^2} + j(f^c/f) \right], \end{aligned} \tag{C–3}$$

Notice that $Z \neq Z(x)$, that is, the impedance does not change as the wave propagates through the horn. The exponential horn is unusual in this regard. The throat resistance is

$$R_{\text{th}} = \rho_0 c_0 \sqrt{1 - (f^c/f)^2}, \tag{C–4}$$

and consequently the power transmitted is

$$W_{\text{exp}} = \tfrac{1}{2}\rho_0 c_0 S_0 u_0^2 \sqrt{1 - (f^c/f)^2}. \tag{C–5}$$

At high frequencies ($f \gg f^c$), the power is the same as that transmitted through a tube of constant cross-sectional area S_{th} ($=S_0$). Hunt's description is confirmed: the horn provides an efficient means of transmitting power from a small area S_{th} to a large area S_{m} (mouth cross section). As frequency decreases, however, the power drops, becoming zero at the cutoff frequency. No power is transmitted below cutoff. Good horn design therefore requires a cutoff frequency well below the frequency range of interest.

2. Conical Horn

For comparison purposes consider the conical horn, which may be thought of as the ideal waveguide for sound radiated by a spherical cap located at the throat of the horn $r = r_0$ (see Fig. 7.5). Propagation behavior may be deduced either from the horn equation, which, for S proportional to r^2, reduces to the ordinary spherical wave equation, Eq. 1D–10, or from the realization that radial wave propagation in a cone is no different from free-space radial wave propagation in the same conical sector. If the spherical cap (surface area S_0 in Fig. 7.5) vibrates with velocity $u_0 e^{j\omega t}$, the pressure of the wave generated is

$$p = \frac{r_0 P_0}{r} e^{j[\omega t - k_0(r - r_0)]}, \tag{C–6}$$

where $P_0 = Z_{\text{th}} u_0$, and the throat impedance Z_{th} is given by Eq. 1D–21 (there called Z_s). Because in this case $k = k_0$, the phase velocity is c_0 for all frequencies. That is, wave propagation in a conical horn is nondispersive, and the horn has no cutoff.

To obtain the power transmitted, we first find the real part of the throat impedance from Eq. 1D–21,

$$R_{\text{th}} = \frac{\rho_0 c_0 k_0^2 r_0^2}{1 + k_0^2 r_0^2}, \tag{C–7}$$

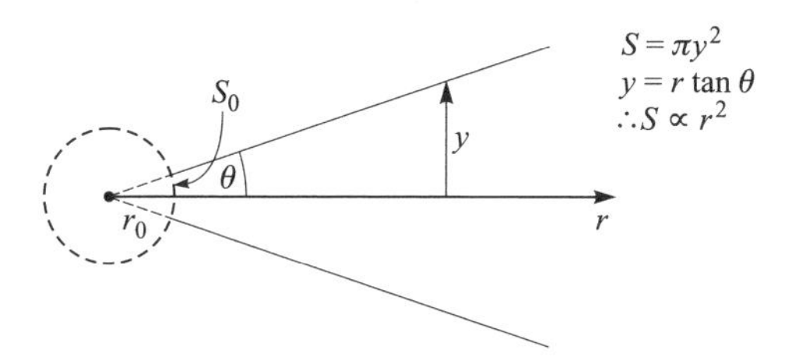

Figure 7.5 Conical horn.

and then employ Eq. C–2:

$$W_{\text{con}} = \frac{1}{2}\rho_0 c_0 S_0 u_0^2 \frac{k_0^2 r_0^2}{1 + k_0^2 r_0^2}. \tag{C–8}$$

It is seen that at low frequencies ($k_0 r_0 \ll 1$) the power falls as ω^2. Therefore, although the conical horn has no cutoff frequency, it has poor transmission at low frequency.

3. Transmission Factor

The power transmission modulus TF is introduced to facilitate comparison of transmission characteristics of horns.[2] The modulus is simply the power transmitted for a horn relative to that transmitted by a tube having a constant cross section ($S_{\text{tube}} = S_{\text{th}}$),

$$\text{TF} \equiv \frac{W_{\text{horn}}}{W_{\text{tube}}} = \frac{\frac{1}{2}R_{\text{th}}|U|^2 S_{\text{th}}}{\frac{1}{2}\rho_0 c_0 |U|^2 S_{\text{th}}} = \frac{R_{\text{th}}}{\rho_0 c_0}. \tag{C–9}$$

For an exponential horn the factor is, from Eq. C–4,

$$\text{TF}_{\text{exp}} = \sqrt{1 - (f^c/f)^2}, \tag{C–10}$$

whereas for a conical horn, from Eq. C–7,

$$\text{TF}_{\text{con}} = \frac{k_0^2 r_0^2}{1 + k_0^2 r_0^2}. \tag{C–11}$$

These quantities are plotted in Fig. 7.6. The exponential horn works better at high and medium frequencies, all the way down to near its cutoff frequency. But at low frequencies the conical horn is superior. Although it operates inefficiently, at least it does operate.

D. MORE GENERAL APPROACH: WKB METHOD

As has been demonstrated, the horn equation may be solved by elementary methods when the horn is exponential or conical. For other horn shapes more

[2]Morse (Ref. 6) uses the term "transmission coefficient" and the symbol τ. We use a different name and symbol to avoid confusion with the power transmission coefficient τ used in this book. Section 24 of Morse's book is devoted to horns and includes several topics not considered here.

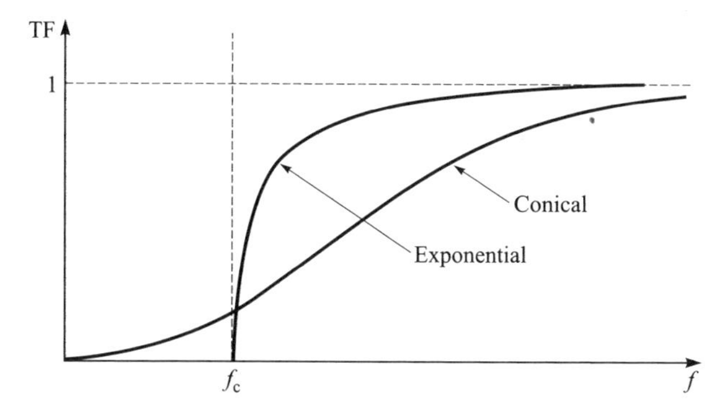

Figure 7.6 Transmission factor for exponential and conical horn.

advanced methods are usually necessary. The WKB method[3] is described here (see, for example, Ref. 5).

To solve ordinary differential equations of the form

$$P_{xx} + g(x)P_x + k_0^2 P = 0, \tag{D–1}$$

one may try a solution of the form

$$P = Ae^{-j\phi(x)},$$

where A is a constant and ϕ may be complex. (*Note*: In this section ϕ is not velocity potential.) Substitution in Eq. D–1 yields the following (nonlinear) equation in ϕ:

$$\phi'^2 + jg(x)\phi' - k_0^2 = -j\phi''. \tag{D–2}$$

This equation may be solved by the method of successive approximations. The basic assumption is that ϕ'' is small (note that if $g(x) = 0$, ϕ'' is identically zero). The first approximation, denoted by ϕ_1, is found by neglecting the ϕ'' term entirely, i.e.,

$$\phi_1'^2 + jg(x)\phi_1' - k_0^2 = 0.$$

Solve for ϕ_1',

$$\phi_1' = \tfrac{1}{2}\left[-jg \pm \sqrt{-g^2 + 4k_0^2}\right],$$

[3] W, K, and B stand for G. Wentzel, H. A. Kramers, and L. Brillouin, respectively, who used the method in quantum mechanics in the 1920s. Lord Rayleigh (Ref. 9) had, however, used the method earlier.

and integrate:

$$\phi_1 = \pm \int \sqrt{k_0^2 - (g/2)^2}\, dx - j\tfrac{1}{2} \int g(x)\, dx.$$

After the indicated integration is carried out, ϕ_1 is substituted for ϕ on the right-hand side of Eq. D–2 to get a second approximation, which is called ϕ_2:

$$\phi_2'^2 + jg(x)\phi_2' - k_0^2 = -j\phi_1'' = \left[\mp j\sqrt{k_0^2 - g^2/4} - g/2\right]'.$$

This equation may be solved for ϕ_2', which may in principle be integrated. One then proceeds to obtain a third approximation ϕ_3 if necessary, and so on.

1. Application to the Horn Equation: Direct Approach

The horn equation is put in the form of Eq. D–1 by taking the time dependence to be harmonic. Substitution of $p = e^{j\omega t} P(x)$ in Eq. A–7 yields

$$P_{xx} + (\ln S)_x P_x + k_0^2 P = 0. \tag{D–3}$$

The trial solution

$$P = A e^{-j\phi(x)} \tag{D–4}$$

then leads to

$$\phi'^2 + j(\ln S)_x \phi' - k_0^2 = -j\phi''. \tag{D–5}$$

For the first-order approximation, ignore the ϕ'' term entirely and solve for ϕ':

$$\phi_1' = \pm f_1(x) - j\tfrac{1}{2}(\ln S)_x, \tag{D–6}$$

where $f_1(x) = \sqrt{k_0^2 - [\tfrac{1}{2}(\ln S)_x]^2}$. The positive root is taken in order to represent an outgoing wave. Integration then yields

$$\phi_1 = \int f_1(x)\, dx - j \ln \sqrt{\frac{S}{S_0}},$$

where S_0 is a constant of integration (generally the value of S at the horn throat). Substitution in Eq. D–4 and addition of the time-harmonic dependence lead to

$$p = A \exp\left[j\left(\omega t - \int f_1\, dx\right)\right] \exp\left(-\ln\sqrt{\frac{S}{S_0}}\right)$$
$$= A\sqrt{\frac{S_0}{S}} \exp\left[j\left(\omega t - \int f_1\, dx\right)\right].$$

We thus obtain anew the result that the pressure amplitude varies as $S^{-1/2}$(see Eq. B–9). The second-order approximation is found by substituting ϕ_1'' for ϕ'' on the right-hand side of Eq. D–5. Rearrangement yields

$$\phi_2' = f_2(x) - j\tfrac{1}{2}(\ln S)_x,$$

where f_2 is the same as f_1 except that k_0^2 is replaced by $k_0^2 - j\phi_1''$. The last term on the right-hand side is the same as the corresponding term in the first-order expression; see Eq. D–6. Since it is this term that leads to the result that $p \propto 1/\sqrt{S}$, that result still holds, and in fact continues to hold as we proceed to higher and higher approximations. Under the circumstances, therefore, we may expect to simplify the algebra of the WKB approximation by assuming the appropriate dependence on S at the outset.

2. Modified Approach

Without going through the intermediate step of separating out the harmonic time dependence, we assume a solution of the form

$$\frac{A}{\sqrt{S}} e^{j[\omega t - \psi(x)]}.$$

Furthermore, let us for simplicity restrict our attention to horns of circular cross section, i.e., $S \propto y^2$, where $y(x)$ is the radius of the cross section. For this case the Webster horn equation reduces to

$$p_{xx} + \frac{2y'}{y} p_x - \frac{1}{c_0^2} p_{tt} = 0, \qquad \text{(D–7)}$$

and our trial solution becomes

$$p = \frac{A}{y} e^{j[\omega t - \psi(x)]}. \qquad \text{(D–8)}$$

The first and second derivatives of p with respect to x are

$$p_x = -A\left(\frac{\{y'}{y^2} + \frac{j\psi'}{y}\right)e^{j[\omega t-\psi(x)]} = -\left(\frac{y'}{y} + j\psi'\right)p, \tag{D–9}$$

$$p_{xx} = \left[-\frac{y''}{y} + 2\left(\frac{y'}{y}\right)^2 + 2j\frac{y'\psi'}{y} - j\psi'' - \psi'^2\right]p.$$

When these expressions and $p_{tt} = -\omega^2 p$ are substituted in Eq. D–7, many terms cancel. The final result is

$$\psi'^2 = k_0^2 - y''/y - j\psi''. \tag{D–10}$$

The first approximation is found by neglecting ψ'' entirely:

$$\psi_1' = \sqrt{k_0^2 - y''/y}, \tag{D–11}$$

or

$$\psi_1 = \int \sqrt{k_0^2 - y''/y}\, dx. \tag{D–12}$$

Similarly, the second approximation is given by

$$\psi_2' = \sqrt{k_0^2 - y''/y - j\psi_1''}, \tag{D–13}$$

or

$$\psi_2 = \int \sqrt{k_0^2 - y''/y - j\psi_1''}\, dx. \tag{D–14}$$

3. Impedance and Transmission Factor

If we are satisfied with the first-order approximation, the impedance and transmission factor may be computed without ever having to carry out the integration in Eq. D–12. First find the impedance. Substitution of Eq. D–9 in the momentum equation Eq. A–6 for time-harmonic waves yields

$$(y'/y + j\psi')p = j\omega\rho_0 u = jk_0\rho_0 c_0 u.$$

The impedance is therefore given by

$$Z \equiv \frac{p}{u} = \frac{k_0\rho_0 c_0}{\psi' - jy'/y},$$

or, on substitution of the first-order expression for ψ', Eq. D–11,

$$Z = \frac{\rho_0 c_0}{\sqrt{1 - y''/(yk_0^2) - jy'/(yk_0)}}. \tag{D–15a}$$

Separate into real and imaginary parts:

$$\frac{Z}{\rho_0 c_0} = \frac{\sqrt{1 - y''/(yk_0^2)}}{1 - k_0^{-2}(y'/y)'} + j\frac{y'/(yk_0)}{1 - k_0^{-2}(y'/y)'}. \tag{D–15b}$$

This is the expression for the impedance, relative to $\rho_0 c_0$, at any point in the horn. The real part, evaluated at the throat, is the power transmission factor TF (see Eq. C–9),

$$\mathrm{TM} = \left.\frac{\sqrt{1 - y''/(yk_0^2)}}{1 - k_0^{-2}(y'/y)'}\right|_{x=0}. \tag{D–16}$$

4. Examples

First consider the exponential horn, for which $y = y_0 \exp(mx/2)$, where y_0 is the throat radius. Since $y'/y = m/2$ for this case, Eq. D–16 gives

$$\mathrm{TF}_{\mathrm{exp}} = \sqrt{1 - (m/2k_0)^2} = \sqrt{1 - (mc_0/2\omega)^2},$$

which agrees with the result found previously, Eq. C–10. Why do the two calculations, one exact and one approximate, yield the same answer? The reason is that for the exponential horn the first approximation ψ_1' is a constant, namely

$$\psi_1' = \sqrt{k_0^2 - m^2/4}.$$

Consequently, $\psi_1'' = 0$, a result which means (see Eq. D–13) that $\psi_2 = \psi_1$, i.e., the first approximation is exact.

The catenoidal horn is a good example that is easily treated by the WKB method. The radius of the cross section of a catenoidal horn is

$$y = y_0 \cosh x/h = \tfrac{1}{2}y_0(e^{x/h} + e^{-x/h}). \tag{D–17}$$

This horn has the attractive feature that it may be joined very smoothly to a tube of radius y_0. Both radius and slope match at the junction. The details of the problem of propagation in a catenoidal horn are left as an exercise (see Problem 7–7). Here we compute only the power transmission factor. The real part of $Z/\rho_0 c_0$ (see Eq. D–15b) is readily calculated for arbitrary values of x; the result is

$$\mathrm{Re}\left(\frac{Z}{\rho_0 c_0}\right) = \frac{\sqrt{1-(k_0 h)^{-2}}}{1-[k_0 h \cosh(x/h)]^{-2}}.$$

The transmission factor, found by letting $x = 0$, is

$$\mathrm{TM}_{\mathrm{cat}} = \frac{\sqrt{1-(k_0 h)^{-2}}}{1-(k_0 h)^{-2}} = \frac{k_0 h}{\sqrt{k_0^2 h^2 - 1}}. \tag{D–18}$$

The cutoff frequency is seen to be

$$f^c = c_0/2\pi h. \tag{D–19}$$

As Fig. 7.7 shows, the behavior of the transmission factor is markedly different from that for the exponential horn.

E. HORN DUALS

If the pressure instead of the particle velocity is eliminated from the (linear) conservation equations (Eqs. A–4 to A–6), the result is

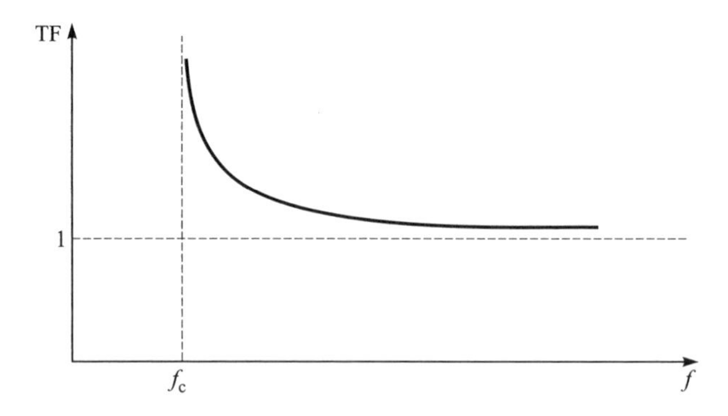

Figure 7.7 Transmission factor for a catenoidal horn.

$$u_{xx} + \frac{S_x}{S} u_x + u\left(\frac{S_x}{S}\right)_x - \frac{1}{c_0^2} u_{tt} = 0, \tag{E–1}$$

which does not have the form of the Webster horn equation, Eq. A–7. By suitable change of variables, however, Eq. E–1 may be made to take on the Webster form. The first three terms are equal to $[S^{-1}(uS)_x]_x$. Expressed in terms of the volume velocity $q = Su$, the equation is

$$S(S^{-1}q_x)_x - \frac{1}{c_0^2} q_{tt} = 0,$$

or, in terms of the inverse cross-sectional area of the horn $F = 1/S$,

$$q_{xx} + \frac{F_x}{F} q_x - \frac{1}{c_0^2} q_{tt} = 0. \tag{E–2}$$

Equation E–2 does have the form of the Webster horn equation.

It is seen from the identity of form of Eqs. A–7 and E–2 that the concept of duality from electrical network theory may be applied to horns: The pressure field in a horn of cross-sectional area $S(x)$ has the same functional form as the volume velocity field in a horn having a cross-sectional area proportional to $1/S(x)$ (Refs. 1, 8).

REFERENCES

1. M. B. C. Campos, "Some general properties of the exact acoustic fields in horns and baffles," *J. Sound Vib.* **95**, 177–201 (1984).
2. I. B. Crandall, *Theory of Vibrating Systems and Sound* (Van Nostrand, New York, 1926), pp. 152–153.
3. E. Eisner, "Complete solutions of the 'Webster' horn equation," *J. Acoust. Soc. Am.* **41**, 1126–1146 (1967).
4. F. V. Hunt, *Physical Acoustics*, unpublished draft.
5. J. Mathews and R. L. Walker, *Mathematical Methods of Physics* (W. A. Benjamin, New York, 1965), pp. 26–27.
6. P. M. Morse, *Vibration and Sound*, 2nd ed. (McGraw-Hill, New York, 1948), pp. 265–293. Reprinted by Acoustical Society of America, New York, 1981.
7. R. N. Nagarkar and R. D. Finch, "Sinusoidal horns," *J. Acoust. Soc. Am.* **50**, 23–31 (1971).
8. R. W. Pyle, "Duality principle for horns," *J. Acoust. Soc. Am.* **37**, 1178(A) (1965).
9. Lord Rayleigh, "On the propagation of waves through a stratified medium, with special reference to the question of reflection, " *Proc. Roy. Soc.* **A86**, 207–226 (1912).

PROBLEMS

7–1. An exponential horn for use in air is to be designed to

(i) Connect smoothly to the port, 50 mm in diameter, of a horn driver.

(ii) Have a cutoff frequency of 100 Hz.

(iii) Radiate effectively from the mouth into a 90° sector at a frequency of 200 Hz.

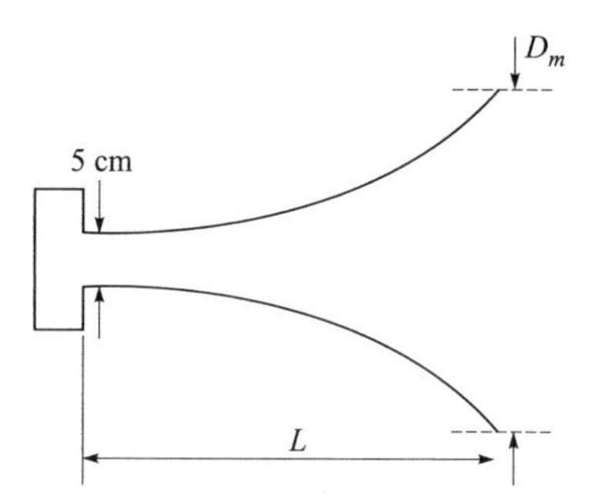

The third requirement may be met by specifying the horn mouth diameter D_m to be

$$D_m = \frac{3.232}{(\omega/c_0)\sin\theta_{\mathrm{HP}}}, \tag{P–1}$$

where θ_{HP} is half the desired sector angle (in this case $\theta_{\mathrm{HP}} = 45°$).[4] Neglecting reflections from the mouth, design a horn to meet these three requirements. In particular, specify the following:

(a) Flare constant m.

(b) Mouth diameter D_m.

(c) Length L of the horn.

(d) Color the outside of the horn is to be painted.

7–2. The length of an exponential horn is 10 m. The diameters of the mouth and throat are 10 m and 0.01 m, respectively. Neglect reflections from the mouth. A rigid piston just fitting the throat vibrates at 2000 Hz with a velocity amplitude $u_0 = 3$ m/s.

(a) Find the cutoff frequency for the horn.

[4]Equation P–1 comes from modeling the mouth of the horn as a vibrating piston of diameter D_m (the vibrating piston is discussed in Chap. 13). If one requires that the angular variation in SPL, relative to the axial value, be no more than 3 dB, Eq. P–1 follows.

(b) How much power does the horn radiate?

(c) How much power would a conical horn of the same overall dimensions (i.e., same length and same mouth and throat diameters) radiate?

7–3. Design an exponential horn to be attached to a 25-mm-diameter microphone so as to produce 26 dB of (pressure) gain at the microphone. The horn is to have a cutoff frequency of 200 Hz. The microphone diaphragm is relatively rigid so that pressure doubling takes place at the diaphragm. This provides 6 dB of the required 26 dB gain. The remaining 20 dB must come from the horn. Assume that negligible reflection of the incident wave takes place at the horn mouth. Find the required flare constant m, the length L, and the mouth diameter D. Note that in this case m is a negative number because the sound propagates from the mouth of the horn to the throat.

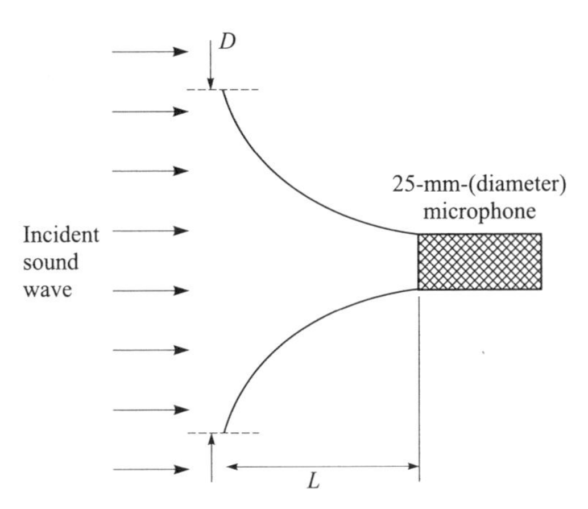

7–4. The cross-sectional area of a certain horn is

$$S = S_0(1 - x/a)^2,$$

where a is a positive constant.

(a) Draw a sketch of the horn.

(b) Find the most general solution of the Webster horn equation for this case for waves traveling in the forward ($+x$) direction (assume that absorbing material is used just before $x = a$ to prevent reflections from the apex). *Hint*: Try a solution of the form $p \propto S^{-1/2}F(x, t)$.

(c) The pressure-time waveform at the entrance to the horn ($x = 0$) is shown in the figure. Sketch the pressure-time waveform at $x = a/2$.

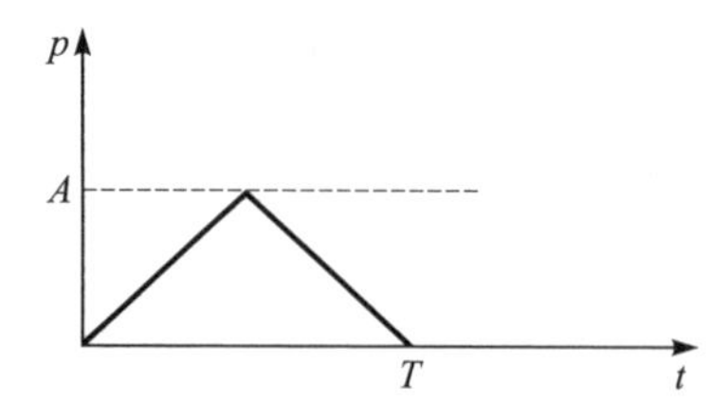

7–5. Radiation from an open tube may be enhanced by coupling the end of the tube smoothly to the throat of a horn, as shown in the sketch. Let the horn be exponential and neglect reflections from the horn mouth.

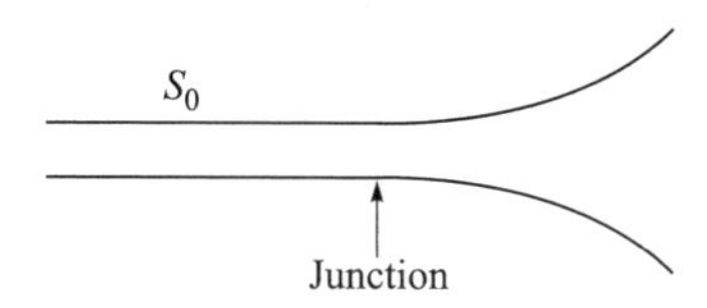

(a) Derive the following expression for the reflection coefficient R for waves incident on the junction from the tube:

$$R = \frac{jF}{1+\sqrt{1-F^2}} = jF^{-1}(1-\sqrt{1-F^2}),$$

where $F = f^c/f = mc_0/4\pi f$.

(b) How far must the frequency be above f^c in order that 91% of the incident power be transmitted into the horn?

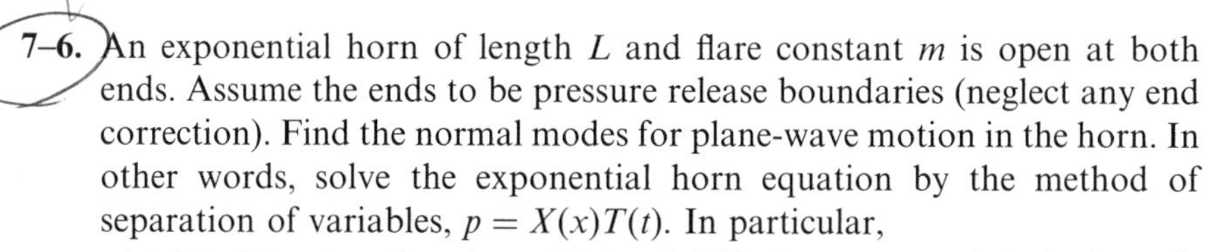

7–6. An exponential horn of length L and flare constant m is open at both ends. Assume the ends to be pressure release boundaries (neglect any end correction). Find the normal modes for plane-wave motion in the horn. In other words, solve the exponential horn equation by the method of separation of variables, $p = X(x)T(t)$. In particular,

(a) Find the eigenfunctions $X(x)$ and $T(t)$. Be sure to satisfy the boundary conditions at $x = 0, L$. Let n be the index denoting the nth eigenfunction. Your answer should be expressed as $p_n(x, t)$.

(b) Find the expression for the eigenfrequencies.

(c) What is the lowest (nontrivial) eigenfrequency? How does this compare (lower, higher, or the same?) with the lowest eigenfrequency of a straight tube of the same length, also open at both ends?

7–7. Use the WKB method to treat the problem of plane waves in an infinite catenoidal horn, for which the radius of the horn cross section is

$$y = y_0 \cosh x/h.$$

Assume that at the throat ($x = 0$) the pressure is $p = A_0 e^{j\omega t}$. In particular, find expressions for the following:

(a) Phase function ψ (this means finding $\psi_1, \psi_2, \ldots$).

(b) Pressure p at any point x.

(c) Phase velocity c^{ph} and the cutoff frequency, if any.

(d) Impedance Z at any point x.

7–8. A horn driver is capable of producing at its port a volume velocity amplitude of 0.004 m^3/s at a frequency of 125 Hz. The driver port is to be connected to the throat of a catenoidal horn that is 1 m in length, the power transmitted is to be 1 W, and the horn cutoff frequency is to be 100 Hz.

(a) Design a catenoidal horn to meet these requirements (for a driver frequency of 125 Hz). In particular, find the constant h, the throat diameter D_0, and the mouth diameter D_m.

(b) Find the sound pressure level of sound at the throat.

7–9. Consider the sinusoidal horn shown in the sketch (Ref. 7). The radius of the cross section is

$$y = y_0 \sin \frac{x + x_0}{h}.$$

[No practical horn has this form, of course, but some horns are shaped like a section (less than one cycle) of a sine curve, for example, the bell of the English horn.] For this horn determine

(s) The expression for the pressure p in the horn if the boundary condition (condition at the throat, $x = 0$) is $p(0, t) = A_0 e^{j\omega t}$.

(b) The phase velocity c^{ph} in the horn (include a sketch).

(c) The cutoff frequency f^c.

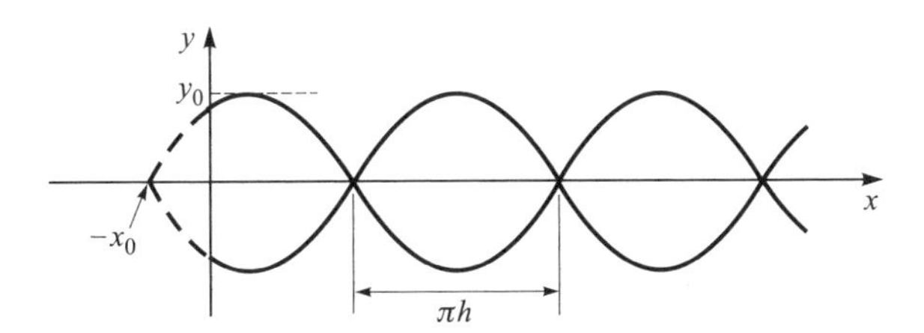

7–10. The Webster horn equation for a certain horn of circular cross section is

$$p_{xx} + \frac{2a}{1 + ax} p_x - \frac{1}{c_0^2} p_{tt} = 0,$$

where a is a positive constant. The source condition at the throat ($x = 0$) is

$$p|_{x=0} = A_0 e^{j\omega t}.$$

(a) Find the solution for this particular horn and source condition.
(b) Find the cutoff frequency.
(c) Draw a sketch of the horn.

7–11. The cross-sectional area of a circular duct is given by $S = S_0 e^{m|x|}$, where m is a positive constant. A time-harmonic wave incident from the left is partly reflected and partly transmitted when it reaches the point $x = 0$.

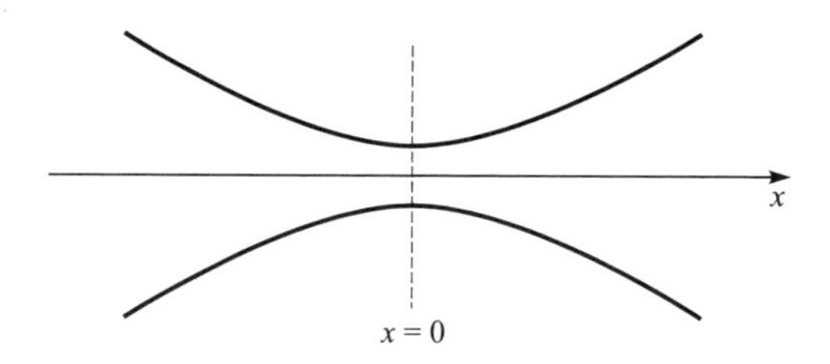

(a) Show that the reflection coefficient at $x = 0$ is

$$R = \frac{j(m/2k_0)}{\sqrt{1}} - \left(\frac{m}{2k_0}\right)^2,$$

where $k_0 = \omega/c_0$.

(b) The medium in the duct is air, and the frequency is $f = mc_0/2.4\pi$ Hz. The SPL of the incident wave is 80 dB (re 20 μPa) at the point $x = -1$ m.

(i) Find the SPL of the transmitted wave at $x = 1$ m.
(ii) Find the SPL of the reflected wave at $x = -1$ m.

8

PROPAGATION IN STRATIFIED MEDIA

In most of this book the ambient properties of the medium are assumed constant, that is, the medium is assumed homogeneous. In fact, most media are inhomogeneous to some degree. For example, turbulence causes the medium to be randomly inhomogeneous. A regular or a deterministic inhomogeneity is produced by gravity, which causes both the atmosphere and the ocean to be stratified. Propagation in a stratified medium is considered in this chapter.[1]

For both the ocean and the atmosphere, gravity causes static properties such as p_0 and ρ_0 to vary with depth (or height). Traveling in such a medium, an acoustic wave encounters, and is affected by, an ever-changing ambient environment. In Sec. A the static properties of the atmosphere and ocean are described. A wave equation for vertical propagation in a horizontally layered medium is derived in Sec. B. After propagation in an isothermal atmosphere is considered as an example, a more general solution is obtained by means of the WKB method. Ray theory is the subject of Sec. C. Continuous refraction of rays in the ocean and atmosphere can have a very significant effect on long-range propagation.

[1]Inhomogeneity is of course not restricted to the vertical dimension. In the atmosphere, for example, wind, temperature, and barometric pressure vary horizontally from one location on the earth to another. The inhomogeneity of the atmosphere is thus "range dependent" as well as height dependent. Horizontal dependence is, however, beyond the scope of this chapter.

A. STATIC PROPERTIES OF THE ATMOSPHERE AND THE OCEAN

1. Atmosphere

The temperature and sound speed profile for the U.S. Standard Atmosphere is shown in Fig. 8.1. In the troposphere (height 0–11 km) the temperature decreases linearly. It gradually increases through the stratosphere, decreases again through the mesosphere, and then increases rapidly in the thermosphere. If the region of interest is limited to the lower 120 km of the atmosphere, one might be tempted to ignore the temperature swings and instead use the average temperature for the region. That approach leads to the *isothermal* model of the atmosphere, which is often used and is described below.

Gravity is largely responsible for the stratification illustrated by Fig. 8.1. When gravity is included, the one-dimensional momentum equation for vertical motion in the atmosphere is (see Eq. 2A–11c)

$$\rho(u_t + uu_z) + P_z = -\rho g, \tag{A–1}$$

where g is the acceleration of gravity, u is vertical particle velocity, and z is height above sea level. If the fluid is motionless ($u = 0$, $P = p_0$, $\rho = \rho_0$), Eq. A–1 reduces to

$$\frac{dp_0}{dz} = -\rho_0 g, \tag{A–2}$$

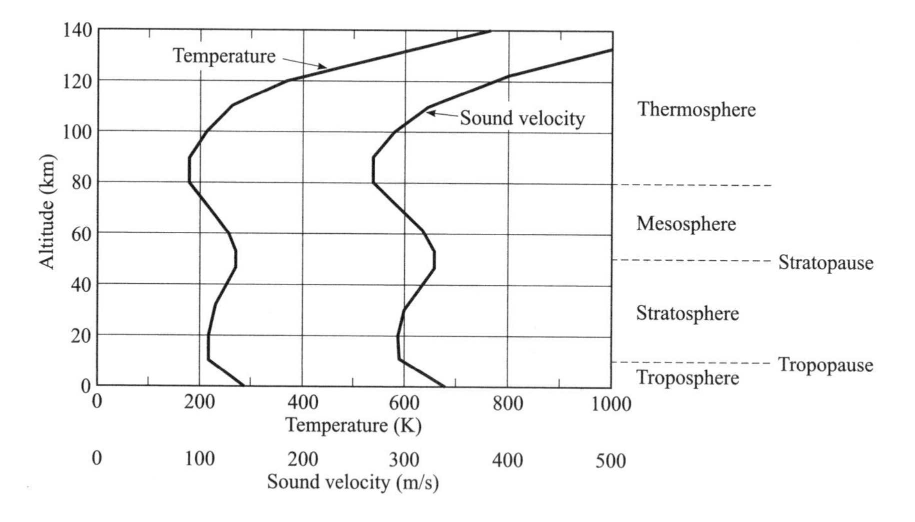

Figure 8.1 Temperature and sound velocity profiles for the 1962 U.S. Standard Atmosphere (from Ref. 2). Real atmosphere profiles vary with location on the earth and with the seasons.

the equation of static equilibrium, the solution of which depends on the relation between the static pressure and static density. The simplest approach is to assume ρ_0 constant, i.e., assume the atmosphere to be incompressible or uniform. For this case the solution of Eq. A–2 is

$$p_0 = \bar{p}_0 - \rho_0 g z, \tag{A–3}$$

where $\bar{p}_0$ is the sea level pressure. This is the familiar "$\Delta P = \rho g h$" law of hydrostatics. One major difficulty with the incompressible model is that the atmosphere is predicted to terminate (the pressure vanishes) at a height

$$H = \frac{\bar{p}_0}{\rho_0 g} = \frac{c_0^2}{\gamma g}. \tag{A–4}$$

Called the *scale height* of the atmosphere, H has a value of about 8.4 km (based on the sea level value $c_0 = 340.3$ m/s given by the 1962 U.S. Standard Atmosphere). Although the incompressible atmosphere is a poor model (except over short distances), H is a very convenient number and turns up as a characteristic distance in more realistic models of the atmosphere.

The next simplest model is the isothermal atmosphere mentioned in the introduction to this section.[2] Start with the ideal gas law (static conditions), $\rho_0 = p_0/RT_0 = \gamma p_0/c_0^2$, or if Eq. A–4 is used,

$$\rho_0 = \frac{p_0}{gH}. \tag{A–5}$$

When this relation is used to eliminate ρ_0 in Eq. A–2, the latter becomes

$$\frac{dp_0}{dz} = \frac{p_0}{H}, \tag{A–6}$$

integration of which yields

$$p_0 = \bar{p}_0 e^{-z/H}. \tag{A–7}$$

Moreover, application of Eq. A–5 yields the dependence of ρ_0 on altitude:

$$\rho_0 = \bar{\rho}_0 e^{-z/H}, \tag{A–8}$$

where $\bar{\rho}_0$ is the static density at sea level. In this model the scale height H represents the altitude at which the pressure and density drop to $1/e$ of their sea level values.[3]

[2]Isothermal here means only that the static temperature T_0 does not vary with altitude. Sound propagation itself remains isentropic as long as dissipation is not important.
[3]In his isothermal model of the atmosphere, Cook (Ref. 2) took c_0 to have its average value, 333 m/s, over the altitude range of interest. The implied average temperature is about 276 K (3°C). The scale height for this atmosphere is $H = 8.04$ km.

Another common model of the atmosphere is one in which T_0 varies linearly with altitude. As Fig. 8.1 shows, this model fits the $T_0(z)$ curve over short distances. Problems 8–1 and 8–2 illustrate the use of this model for the troposphere.

2. Ocean

The most important acoustical property affecting propagation in the ocean is the sound speed c_0, which is a function of static temperature, static density, static pressure, salinity, and even latitude. For gases c_0 is calculated analytically by starting with the equation of state and taking the isentropic derivative of P with respect to ρ. The same procedure for seawater is not practical, however, because of lack of knowledge about the equation of state of seawater. Consequently, empirical formulas for c_0 have been developed. See, for example, the extensive one by Del Grosso (Ref. 4). For purposes of this text, the following relatively simple formula given by Coppens (Ref. 3) is used:

$$\begin{aligned} c_0 =& 1449.05 + 45.7\left(\frac{t}{10}\right) - 5.21\left(\frac{t}{10}\right)^2 + 0.23\left(\frac{t}{10}\right)^3 \\ &+ \left[1.333 - 0.126\left(\frac{t}{10}\right) + 0.009\left(\frac{t}{10}\right)^2\right](S - 35) + \Delta(z), \end{aligned} \tag{A–9}$$

where the depth function $\Delta(z)$ is

$$\Delta(z) = 16.3\left(\frac{z}{1000}\right) + 0.18\left(\frac{z}{1000}\right)^2. \tag{A–10}$$

The units of c_0 are meters per second, t is temperature in degrees Celsius, S is salinity in parts per thousand (‰), and z is depth in meters. Although Eq. A–9 is valid only for latitude 45°, the actual dependence on latitude is so slight that we ignore it. Notice that static pressure and static density do not appear in the expression for c_0. Their effect is included implicitly in the depth function.

Example 8.1. Sound speed in isothermal seawater. Water at and near the surface is often so well stirred up because of surface wave action that it is isothermal and is called a "mixed layer." Let the temperature and salinity of a mixed layer be 13°C and 35‰, respectively. First use Eq. A–9 to find the sound speed at the surface, $z = 0$:

$$c_0 = 1449.05 + 45.7(1.3) - 5.21(1.3)^2 + 0.23(1.3)^3 = 1500.16 \text{ m/s}.$$

Next assume that the mixed layer continues to a depth of 100 m. Because T_0 and S do not vary, the only change in c_0 is that due to depth. From 0 to 100 m, however, the quadratic term in the depth function (Eq. A–10) is insignificant, and Eq. A–9 reduces to

$$c_0 = 1500.16 + 0.0163\,z \quad \text{m/s}, \tag{A–11}$$

which defines the linear sound speed profile shown in Fig. 8.2*a*. Although a 1.63-m/s change over 100 m may seem insignificant, it is enough to produce a very substantial refraction effect over the large distances that are typical in problems of underwater propagation.

A typical sound speed profile for the deep ocean is shown in Fig. 8.2*b*. The general shape of this curve may be understood by considering the temperature variation in the water column. At great depths the density has its maximum value (because of the enormous hydrostatic pressure). Since water is most dense at 4°C, this is the temperature of the deep ocean. At the surface, on the other hand, heating by the sun and atmosphere generally makes the water much warmer. The overall trend, therefore, is for water temperature to decrease with depth. For deep oceans (Fig. 8.2*b*) the 4°C value is reached well before the bottom.

However, the sound speed profile does not simply track the temperature profile. Begin at the top. As explained in Example 8.1, the surface layer is often mixed (therefore isothermal); the sound speed profile there has only the weak depth dependence indicated in Fig. 8.2*a*. Below the mixed layer is the thermocline, where the water is little influenced by surface heating or mixing. The drop in temperature is so rapid here that it dominates the sound speed profile. The negative sound speed gradient due to decreasing temperature overshadows the positive gradient due to increasing depth. Gradually, however, as the temperature drop slows down, the depth dependence begins to assert itself. Eventually, the two effects reach a balance (which is affected by the salinity), and a mini-

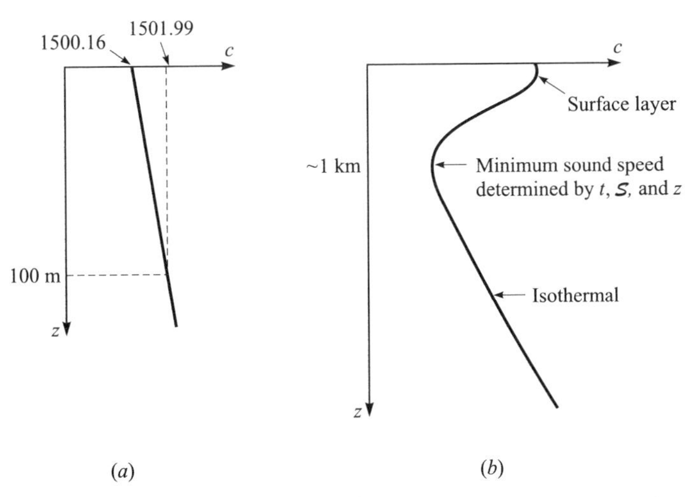

Figure 8.2 Ocean sound speed profiles. (*a*) Mixed layer near the surface. (*b*) Deep ocean.

mum sound speed is reached, as shown in Fig. 8.2*b*.[4] The minimum establishes the deep sound channel (sometimes called the SOFAR channel), which acts as a natural waveguide for sound propagation (see Sec. C below). Below the minimum the approach of the temperature to its asymptotic value 4°C leaves depth in control of the profile.

B. VERTICAL PROPAGATION OF PLANE WAVES

Vertical propagation of plane waves in a fluid stratified by gravity is considered here. After a wave equation for arbitrary stratification is derived, a solution is found for the particular case of upward propagation through an isothermal atmosphere. Sound radiated by an earthquake is taken as an example. Finally, a general solution is obtained by the WKB method.

1. One-Dimensional Wave Equation

In this section no specific model is assumed for the layering of the medium. The static properties p_0, ρ_0, and c_0 are simply taken to be functions of the propagation distance z. It is assumed, however, that the waves are plane and that they travel in the direction of the gradient of the properties, i.e., normal to the layers. In this case the layered medium cannot bend the rays, but it can cause dispersion and a change in amplitude.

To derive the appropriate wave equation, we start as usual with the conservation equations. The continuity equation is the same as that for a homogeneous fluid,

$$\rho_t + \rho_z u + \rho u_z = 0. \tag{B–1}$$

To linearize this equation, begin by expressing the density as a static part ρ_0, which in this case varies with distance, plus a small perturbation $\delta\rho$,

$$\rho = \rho_0(z) + \delta\rho. \tag{B–2}$$

Because the middle term in Eq. B–1 has a linear part, $(\rho_0)_z u$, linearization of Eq. B–1 yields

$$\delta\rho_t + (\rho_0)_z u + \rho_0 u_z = 0. \tag{B–3}$$

The momentum equation for a layered medium has already been given, Eq. A–1. The total pressure is also expressed as a static part (again a function of z) plus a small perturbation:

[4]The figure shows the minimum at $z \sim 1000$ m, but this is just a rough average value. For example, in the Arctic the minimum can be very close to the surface.

$$P = p_0(z) + p, \tag{B–4}$$

The result of substituting Eqs. B–2 and B–4 in Eq. A–1 is

$$(\rho_0 + \delta\rho)(u_t + uu_z) + (p_0)_z + p_z = -\rho_0 g - g\,\delta\rho.$$

By Eq. A–2, the two static terms cancel. Linearization of the remaining terms leads to

$$\rho_0 u_t + p_z = -g\delta\rho. \tag{B–5}$$

An equation of state must be added. A pressure-density-entropy relation is convenient; see Sec. 1C.3.c. Sound propagation generally takes place with little or no change in entropy. If the medium is homogeneous, the static entropy is also fixed throughout the fluid; as a result the linearized equation of state is $p = c_0^2\delta\rho$, Eq. 1C–56. However, the situation changes if the fluid is inhomogenous because the static entropy may vary with distance just as the static pressure and static density do. Even so, the assumption that sound propagation is lossless may be preserved by considering *local* pressure-density variations to take place without change in entropy. In other words, we assume that the entropy of a given fluid particle remains constant even though the static entropy may vary from particle to particle. To begin the mathematical expression of these ideas, take the material derivative of the relation $P = P(\rho, s)$, where s is the entropy per unit mass:

$$\frac{DP}{Dt} = \left(\frac{\partial P}{\partial \rho}\right)_s \frac{D\rho}{Dt} + \left(\frac{\partial P}{\partial s}\right)_\rho \frac{Ds}{Dt}.$$

Constancy of entropy for a fluid particle means that $Ds/Dt = 0$. Therefore [given the definition $(\partial P/\partial\rho)_s = c^2$], the equation for DP/Dt reduces to

$$\frac{DP}{Dt} = c^2 \frac{D\rho}{Dt}. \tag{B–6}$$

Linearization yields

$$p_t + u(p_0)_z = c_0^2[\delta\rho_t + u(\rho_0)_z]. \tag{B–7}$$

Equations B–3, B–5, B–7, and A–2 may now be combined to form a single wave equation in u. First use the continuity equation, Eq. B–3, to eliminate the density from the right-hand side of Eq. B–7. On the left-hand side, use Eq. A–2 to eliminate the static pressure gradient. The spatial derivative of the result is

$$p_{zt} - g(\rho_0 u)_z = -(\rho_0 c_0^2 u_z)_z. \tag{B–8}$$

Now take the time derivative of the momentum equation, Eq. B–5, and use the continuity equation to eliminate $\delta\rho_t$. The result is

$$p_{zt} - g(\rho_0 u)_z = -\rho_0 u_{tt}. \tag{B–9}$$

Subtraction of Eq. B–8 from Eq. B–9, expansion, and rearrangement lead to

$$u_{zz} + (\ln \rho_0 c_0^2)_z u_z - \frac{1}{c_0^2} u_{tt} = 0. \tag{B–10}$$

This is the wave equation for plane waves in an inhomogeneous medium.

Notice the resemblance of Eq. B–10 to Eq. 7A–7, the Webster horn equation. The adiabatic bulk modulus $\rho_0 c_0^2$ plays a role similar to that of the cross-sectional area S of a horn. On the other hand, Eq. B–10 is inherently more complicated than the Webster horn equation because while c_0^{-2}, the coefficient of the u_{tt} term, is always constant for propagation in a horn, it is generally not constant for an inhomogenous medium.

Equation B–10 is solved by an approximation method (WKB) in Sec. B.3 below. First, however, we consider the special case of an isothermal atmosphere, a medium for which an exact solution may be obtained.

2. Vertical Propagation through an Isothermal Atmosphere

For an isothermal atmosphere c_0 is constant and ρ_0 is given by Eq. A–8. Equation B–10 reduces to

$$u_{zz} - \frac{1}{H} u_z - \frac{1}{c_0^2} u_{tt} = 0. \tag{B–11}$$

In this case the analogy to propagation in a horn is complete. Equation B–11 has exactly the same form as the Webster horn equation for an exponential horn, Eq. 7B–2. In particular, $-1/H$ plays the role of the flare constant m. What does a negative flare constant mean? Simply that the horn cross section decreases exponentially with distance, as shown in Fig. 8.3. All the formulas for the exponential horn may be applied to vertical propagation in an isothermal atmosphere simply by replacing p by u and m by $-1/H$. For example, the particle velocity amplitude u_0 at any height is given by

$$u_0 = \bar{u}_0 e^{z/2H}, \tag{B–12}$$

where $\bar{u}_0$ is the value on the ground. The growth of the particle velocity signal, indeed *exponential* growth, does not violate conservation of energy. In fact, Eq. B–12 may be derived directly by recognizing that the intensity stays constant as the wave propagates upward. Since the intensity is given by

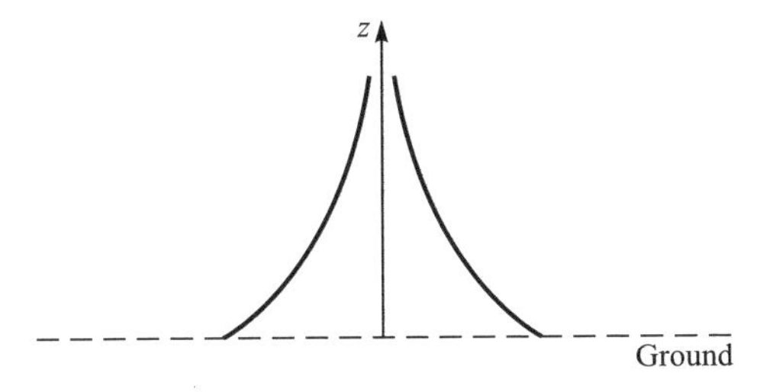

Figure 8.3 Exponential horn analog of vertical propagation through an isothermal atmosphere.

$$I = \rho_0 c_0 u_{\text{rms}}^2,$$

and since ρ_0 decreases as $\exp(-z/H)$, u_{rms} must vary as $e^{z/2H}$.

A spectacular example of the signal amplification indicated by Eq. B–12 was produced by the March 1964 Alaskan earthquake. Ground vibration in the form of Rayleigh waves as far away as Washington, D.C., and Boulder, Colorado, generated a low frequency (period about 25 s) plane wave at the ground that traveled upward through the atmosphere. Radiation produced by supersonic flexural waves on a panel is discussed in Sec. 5C. Here the panel is the ground, and the flexural waves are Rayleigh waves. Their velocity was about 3.5 km/s. Equation 5C–2 shows that the radiated plane wave entered the atmosphere at an angle of about 6° from the vertical. For purposes of the rest of the discussion, we approximate the direction as truly vertical. The particle velocity at the ground was about 5 mm/s. At the same time a layer in the ionosphere at about 240 km was being monitored in Boulder by means of radio wave reflections. About 9 min after the ground vibrations started, Doppler shift of the reflected radio waves indicated a strong vertical disturbance in the ionosphere.[5] The particle velocity associated with the Doppler shift was estimated to be more than 200 m/s. Although this represents a very substantial amplification, it is less by a considerable factor than one would predict by using Eq. B–12. The lower amplification was probably due to nonlinear effects in combination with atmospheric absorption. For a more complete account of this interesting natural experiment, see Ref. 1.

3. General Solution by Means of the WKB Method

Considering the similarity of Eq. B–10 to the Webster horn equation, Eq. 7A–7, it is not surprising that the WKB method (see Sec. 7D and Ref. 5) is useful

[5] If the value $c_0 = 333$ m/s is used, the predicted time for the sound wave to reach 240 km is 12 min. However, as Fig. 8.1 shows, c_0 increases rapidly beyond 110 km altitude. A rough calculation based on use of $c_0 = 333$ m/s for the first 110 km and the apparent slope of the sound speed profile above 110 km (Fig. 8.1) yields a total travel time estimate of 8.7 min, well in line with the reported time.

for solving Eq. B–10 in the general case. The primary assumption is that the inhomogeneity of the medium is small. If a solution of the form

$$u = Ae^{j[\omega t - \phi(z)]} \tag{B–13}$$

is substituted in Eq. B–10, the following differential equation for the phase factor ϕ is obtained:

$$-j\phi'' - \phi'^2 - j\phi'(\ln \rho_0 c_0^2)_z + \frac{\omega^2}{c_0^2} = 0, \tag{B–14}$$

where all derivatives are with respect to z. Note that if the medium is homogeneous, the first and third terms disappear. For a slightly inhomogeneous medium, therefore, it seems reasonable to use the second and fourth terms as the basis of a first approximation. Ignoring the first and third terms for the moment, we have $\phi_1'^2 = \omega^2/c_0^2$, or, for an outward traveling wave,

$$\phi_1' = \frac{\omega}{c_0}.$$

This equation does not necessarily imply a homogeneous medium, since c_0 may vary with z. Integration gives

$$\phi_1 = \omega \int \frac{dz}{c_0}$$

as the first approximation. To get a second approximation, substitute ϕ_1 for ϕ in the first and third terms of Eq. B–14 and solve again for ϕ':

$$\begin{aligned}(\phi_2')^2 &= \frac{\omega^2}{c_0^2} - j[\phi_1'' + \phi_1'(\ln \rho_0 c_0^2)_z] \\ &= \frac{\omega^2}{c_0^2} - j\left[-\frac{\omega}{c_0}(\ln c_0)_z + \frac{\omega}{c_0}(\ln \rho_0 c_0^2)_z\right].\end{aligned}$$

When the bracketed terms are combined, the result is

$$\phi_2' = \frac{\omega}{c_0}\left[1 - j\frac{c_0}{\omega}(\ln \rho_0 c_0)_z\right]^{1/2},$$

or, if the binomial expansion is used,

$$\phi_2' \doteq \frac{\omega}{c_0} - j\frac{1}{2}(ln\, \rho_0 c_0)_z. \tag{B–15}$$

Dropping the higher order terms in the binomial expansion is an approximation based on the smallness of the percentage change of $\rho_0 c_0$ over a wavelength. Equation B–15 may be integrated to give

$$\phi_2 = \omega \int \frac{dz}{c_0} - j\,\frac{1}{2}(\ln \rho_0 c_0)$$

as the second approximation. Substitution of this result in Eq. B–13 yields

$$u = \frac{A}{\sqrt{\rho_0 c_0}}\, e^{j\omega(t - \int dz/c_0)}. \tag{B–16}$$

To generalize this solution, we note that it satisfies the boundary condition $u = u_0 e^{j\omega t}$ at $z = 0$. If the excitation at $z = 0$ is some other time function, say $f(t)$, then by using Fourier transforms, we find that Eq. B–16 is replaced by

$$u = \sqrt{\frac{\bar{\rho}_0 \bar{c}_0}{\rho_0 c_0}}\, f\left[t - \int \frac{dz}{c_0}\right]. \tag{B–17}$$

Under what conditions is Eq. B–17 a valid approximate solution? If the second approximation is to be but a small improvement on the first approximation (and if it is, the implication is that no higher approximation is needed), then the percentage change in the impedance over a wavelength must be small (see Eq. B–15). In general, this means that the inequalities

$$\frac{\omega}{c_0} \gg (\ln \rho_0)_z, \tag{B–18}$$

$$\frac{\omega}{c_0} \gg (\ln c_0)_z \tag{B–19}$$

must hold.

Although the exponential signal amplification found in Sec. B.2 is dramatic, for most applications of sound propagation in the lower atmosphere the change in amplitude is only a minor effect. For example, the particle velocity amplitude increases by only 20% (a change in level of only 1.5 dB) as a plane sound wave travels vertically from the ground to 3 km (10,000 ft). Similarly, the change in $\rho_0 c_0^2$ in the ocean from the surface to even very great depths is seldom enough to produce more than a 1-dB change in signal level. A far greater effect of the inhomogeneity of the medium is refraction or bending of the sound rays when the propagation is not purely vertical. Omitting the effect of the bulk modulus variation on amplitude (Eq. B–10), and proceeding to three dimensions, we arrive again at the classical wave equation

$$\nabla^2 p - \frac{1}{c_0^2} p_{tt} = 0.$$

The inhomogeneity of the medium now makes itself felt only in the dependence of c_0 on position in the medium. Variation in sound speed causes refraction. Ray theory, to which we now turn, has long been used to describe wave propagation in a refracting medium.

C. RAY THEORY

The refraction of rays (that represent plane waves) at a plane interface separating two fluids of different sound speed is discussed in Chap. 5. The description of the sound field in terms of ray acoustics is exact in that case, since the wave equation is satisfied in each fluid and the boundary conditions at the interface are met. Indeed, it is the satisfying of the boundary conditions that leads to Snell's law. In this section we extend the ideas of Chap. 5 to media in which the sound speed varies continuously. A tacit assumption is that Snell's law continues to hold even though the wavefronts in an inhomogeneous fluid are generally not plane.

Ray theory is a high frequency approximation. In more advanced treatments of ray theory (see, for example, Ref. 6) the wave equation is shown to lead, at high frequencies, to two equations. One of these, called the eikonal equation, is used to find the ray paths. The other, called the transport equation, allows one to calculate the amplitude of the acoustic signal as it travels along the ray path. A small bundle of adjacent rays constitutes a ray tube. The physical interpretation of the transport equation is that the acoustic energy passing through a ray tube stays the same. This means that the intensity times the cross-sectional area S of the ray tube is a constant. Consequently the pressure amplitude varies as $1/\sqrt{S}$, just as it does for horns. Although these ideas are physically appealing, and a plot of the ray trajectories provides a very useful visualization of the sound field in a stratified medium, ray theory has certain deficiencies because of its limitation to high frequencies. One of the most glaring is that effects of diffraction are not taken into account. Diffraction is taken up in Chap. 14.

1. Ray Paths

As noted above, our treatment of ray theory is based on Snell's law. We begin with some terms and conventions. First, in this section the general symbol for sound speed is c, not c_0; when a subscript is used, it denotes sound speed at a particular place in the medium. Second, although a ray has heretofore been characterized by the angle it makes with the normal to the interface, the convention in ray acoustics is to use the grazing angle, that is, the angle the ray makes with the interface. The grazing angles in Fig. 8.4*a* are θ_1 and θ_2. Because of the change in angular description, Snell's law becomes

$$\frac{\cos\theta_1}{c_1} = \frac{\cos\theta_2}{c_2}. \tag{C–1}$$

Finally, the sign convention used here for the grazing angles is that downward angles are positive, upward angles negative.

As a primitive model, the ocean or the atmosphere may be thought of as a fluid made up of many discrete layers, as shown in Fig. 8.4*b*. Repeated application of Snell's law at each succeeding interface leads to $\cos\theta_1/c_1 = \cos\theta_2/c_2 = \cos\theta_3/c_3 = \cdots = \text{const}$. The "const" is sometimes called the Snell invariant; it is the same for any particular ray but varies from ray to ray. Since the invariant may often be identified from its value at the source, $\cos\theta_i/c_i$ (the subscript i stands for initial), a convenient way of expressing Snell's law for a multilayer fluid is

$$\frac{\cos\theta_1}{c_1} = \frac{\cos\theta_2}{c_2} = \cdots = \frac{\cos\theta_i}{c_i}. \tag{C–2}$$

A better representation of the ocean or the atmosphere is one in which the sound speed varies continuously with depth (or height). For this model a differential form of Snell's law is needed. Since $\cos\theta/c$ is constant, differentiation yields $(d/d\theta)(\cos\theta/c) = (\sin\theta/c) + (\cos\theta/c^2)(dc/d\theta) = 0$, or

$$\frac{dc}{d\theta} = -\frac{c\sin\theta}{\cos\theta}. \tag{C–3}$$

The general formula for the path of a ray may now be found. The path is usually expressed in terms of the radius of curvature R_c of the ray. Figure 8.5 shows an increment of arc Δs of a ray in the ocean,[6] the coordinates r (hor-

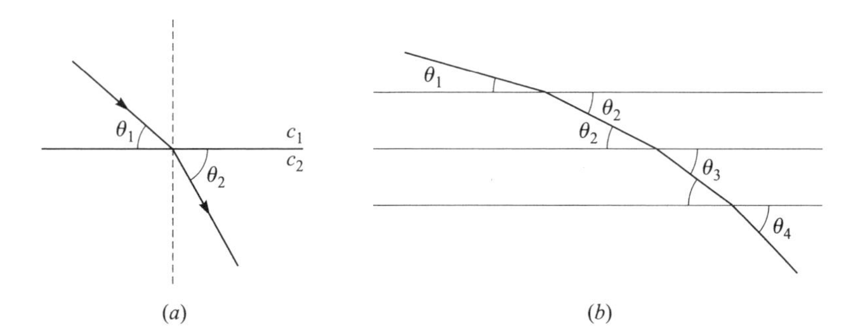

Figure 8.4 Refraction for layered fluids. (*a*) Two layers. (*b*) Multiple layers.

[6]Although the following analysis is particularized to bodies of water, the results also apply to the atmosphere.

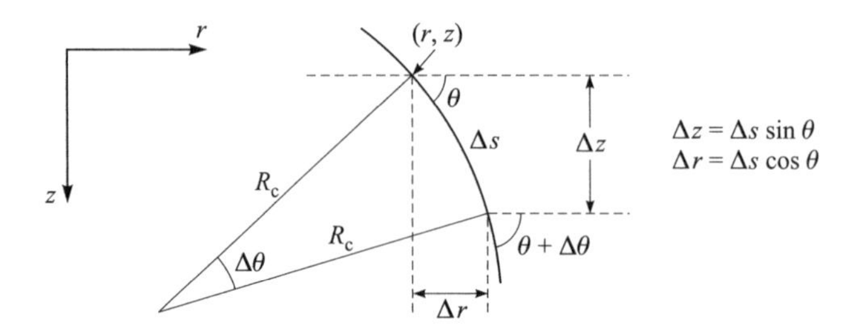

Figure 8.5 Radius of curvature of a small arc Δs of a ray.

izontal range) and z (depth), and the relation of Δs to Δr and Δz. The expression for R_c is

$$R_c = \frac{ds}{d\theta} = \frac{ds}{dz}\frac{dz}{dc}\frac{dc}{d\theta} = \left(\frac{1}{\sin\theta}\right)\left(\frac{1}{dc/dz}\right)\left(-\frac{c\sin\theta}{\cos\theta}\right),$$

where Eq. C–3 has been used. Since dc/dz is the sound speed gradient g, the expression for R_c reduces to

$$R_c = -\frac{1}{g}\frac{c}{\cos\theta}. \tag{C–4}$$

Although $c/\cos\theta$ is a constant, the gradient g generally varies with depth; see, for example, Fig. 8.2*b*. Consequently, the ray's radius of curvature also changes with depth. An important special case in which R_c stays constant is discussed below. Notice that R_c is positive when $dc/dz < 0$, negative when $dc/dz > 0$. A positive value of R_c signifies a ray path that is concave downward, as in Fig. 8.6*a*. The center of curvature of Δs is a point below the arc (frequently below the ocean bottom), as in Fig. 8.6*b*. A negative value of R_c means

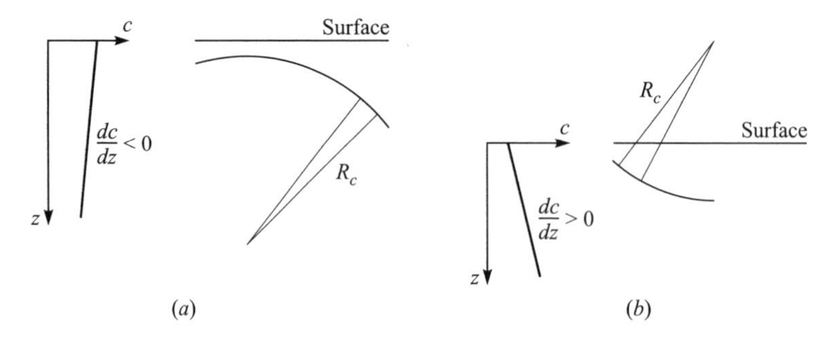

Figure 8.6 Gradient and radius of curvature. (*a*) R_c positive when $g < 0$. (*b*) R_c negative when $g > 0$.

that the ray is concave upward, i.e., the center of curvature is a point above, usually well beyond, the ocean surface.

Example 8.2. Radius of curvature of a ray. Find R_c for a ray starting downward at $\theta_i = 2^\circ$ from a source at the surface. Let the water be that described in Example 8.1 ($c_i = 1500.16$ m/s and $c = 1500.16 + 0.0163\,z$). Since the gradient is $g = 0.0163\,\text{s}^{-1}$, R_c is given by

$$R_c = -\frac{1}{0.0163}\frac{1500.16}{\cos 2^\circ} = -92,090 \text{ m}.$$

The arc is thus centered on a point up in the air at a distance of 92.1 km.

Two elementary applications are now discussed. First, it is often found that the continuous bending of a ray takes it through a maximum or a minimum, as shown in Fig. 8.6. The place where the ray turns to go down (or up) is called a vertex. Since the vertex angle θ_v is zero ($\cos\theta_v = 1$), Snell's law becomes

$$c_v = c_i / \cos\theta_i. \qquad \text{(C–5)}$$

The vertex of the ray in Example 8.2 is shown in Fig. 8.7*a*. Using Eq. C–5, we find the sound speed at the vertex to be $1500.16/\cos 2^\circ = 1501.07\text{m/s}$. Equation A–11 is then used to find the depth z_v at which the vertex is located:

$$z_v = \frac{1501.07 - 1500.16}{0.0163} = 56.1 \text{ m}.$$

Our second application is to limit rays and shadow zones. Figure 8.7*b* shows a family of rays from an omnidirectional source S in water for which the sound speed decreases with depth. Most of the upward traveling rays have no vertex because they hit the surface first, where they reflect and thereafter travel downward. One ray, however, has a vertex at the surface. It is known as the limit

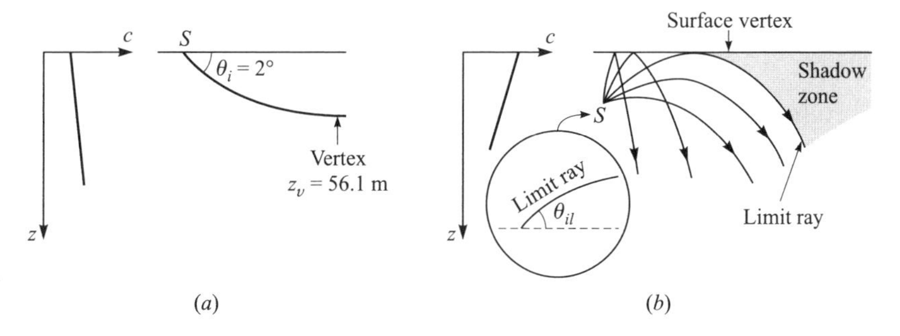

Figure 8.7 (*a*) Vertex for an upward bending ray, Example 8.2. (*b*) Limit ray and shadow zone.

ray; call its initial angle $\theta_{i\ell}$. All the remaining upward traveling rays leave the source at shallower angles, reach a vertex below the surface, and then turn downward. As can be seen from Fig. 8.7*b*, no rays penetrate the region between the limit ray (after its vertex) and the surface. This region is called a shadow zone. To find the initial angle of the limit ray, note that $c_v = c_s$, where c_s is the sound speed at the surface.

$$\therefore \quad \theta_{i\ell} = \cos^{-1}\frac{c_i}{c_s}.$$

Shadow zones are obviously of great importance in both underwater acoustics and atmospheric acoustics. Does the sound field shown in Fig. 8.7*a* have a shadow zone?

The section closes with the observation that rays always bend toward a region of lower sound speed. This handy rule of thumb is readily confirmed by sketches and examples throughout this and following sections.

2. Rays in a Fluid Having a Linear Sound Speed Profile

An important special case occurs when the sound speed profile is linear (the gradient g constant). For such a profile, as Eq. C–4 shows, the radius of curvature of any given ray is a constant. Rays are therefore circular arcs. For instance, since the profile for the seawater of Example 8.2 is linear, the ray shown in Fig. 8.7*a* follows a circular path. Calculation of circular ray trajectories is quite simple, so much so that even when the sound speed profile is curved, it is often approximated by a sequence of straight-line segments. The ray path is then made up of a sequence of circular arcs, one for each linear segment of the profile.

Example 8.3. Ray following a circular path. Isosalinity surface water often has a moderate temperature gradient. Let the salinity $S = 35‰$, and let the temperature have the linear profile

$$t = 20^\circ - 0.03z.$$

Substitution in Eq. A–9 leads, after much simplification, to

$$c = 1521.45 - 0.0666\ z + 0.0000343\ z^2 + \cdots.$$

For depths for which the quadratic and higher order terms are negligible (at what depth is the quadratic term 1% of the linear term?), the gradient has the constant value $g = -0.0666\ \mathrm{s}^{-1}$. Let a source be located at a depth $z = 20$ m ($c_i = 1519.45$ m/s). The radius of curvature for a ray traveling upward at an initial angle $\theta_i = -2^\circ$ is

$$R_c = -\frac{c_i}{g\cos\theta_i} = -\frac{1519.45}{-0.0666\ \cos(-2^\circ)} = 22.8\ \text{km}.$$

Will this ray reach the surface?

Formulas for change in depth Δz and change in horizontal range Δr are easily derived for a ray that follows a circular path. Refer to Fig. 8.8. As we move along the ray from point A (ray angle θ_i) to point B (ray angle θ), the change in depth Δz is $Z_2 - Z_1 = (z_A - z_C) - (z_B - z_C) = R_c \cos\theta_i - R_c \cos\theta$, or

$$\Delta z = R_c(\cos\theta_i - \cos\theta). \tag{C–6}$$

Similarly, the change in horizontal range Δr is $R_2 - R_1 = R_c \sin\theta - R_c \sin\theta_i$, or

$$\Delta r = R_c(\sin\theta - \sin\theta_i). \tag{C–7}$$

Division of Eq. C–6 by Eq. C–7 yields

$$\frac{\Delta z}{\Delta r} = \frac{\cos\theta_i - \cos\theta}{\sin\theta - \sin\theta_i} = \tan\left(\frac{\theta_i + \theta}{2}\right). \tag{C–8}$$

When the last three equations are used, checking signs may help to guard against errors. A positive value of Δz means an increase in depth, while a negative value means a decrease in depth. If the ray is traveling away from a source, Δr is always positive.

Equations C–7 and C–6 make it easy to calculate the horizontal and vertical distance from a source to a vertex, since $\theta_v = 0$. One obtains

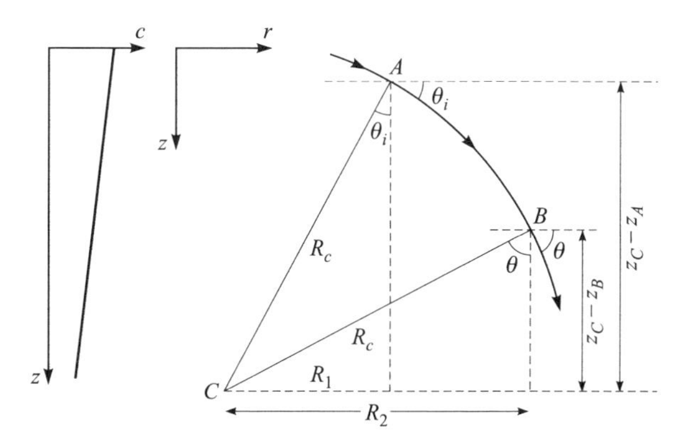

Figure 8.8 Change in range and depth along a ray in a medium having a linear sound speed profile.

$$\Delta r_v = -R_c \sin\theta_i = \frac{c_i}{g}\tan\theta_i, \qquad \text{(C–9)}$$

$$\Delta z_v = R_c(\cos\theta_i - 1) = \frac{c_i}{g}\,\frac{1-\cos\theta_i}{\cos\theta_i}, \qquad \text{(C–10)}$$

respectively. These results show that increasing the magnitude of the initial angle θ_i moves the vertex farther from the source. As an example, Fig 8.9 shows successive vertices for water in which sound speed increases linearly with depth.

Example 8.4. Cycle distance. Given a source at depth d in water for which sound speed increases linearly with depth, find the range r_0 at which the horizontal ray from the source strikes the surface. See Fig. 8.10. First note that in this problem $\theta_i = 0$ (the vertex is at the source!), $\Delta z = -d$, and $\Delta r = r_0$. Solve Eqs. C–6 and C–7 for $\cos\theta_s$ and $\sin\theta_s$, respectively, where θ_s is the angle of the ray at the surface, and square each result:

$$\cos^2\theta_s = \left(\frac{d}{R_c} + 1\right)^2,$$
$$\sin^2\theta_s = \left(\frac{r_0}{R_c}\right)^2.$$

Add these two equations and solve the result for r_0:

$$r_0 = \sqrt{-2R_c d - d^2}. \qquad \text{(C–11)}$$

Note that $R_c < 0$ for this ray. Since normally $d \ll |2R_c|$, Eq. C–11 is often approximated by

$$r_0 \doteq \sqrt{-2dR_c}. \qquad \text{(C–12)}$$

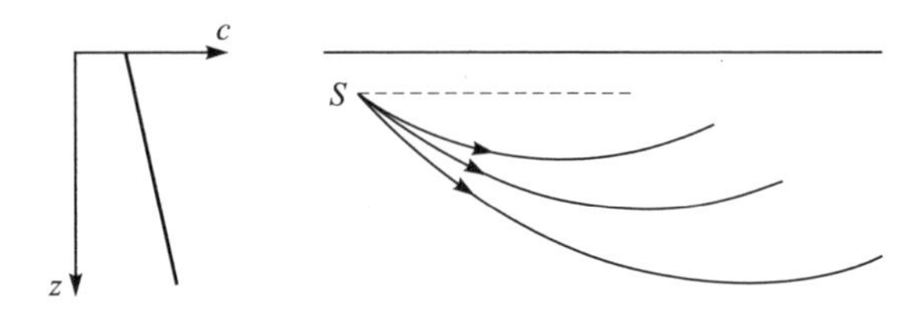

Figure 8.9 Family of rays with vertices, sound speed increasing linearly with depth.

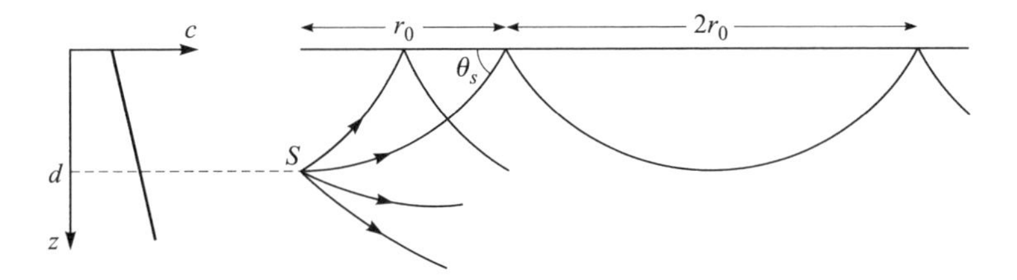

Figure 8.10 Example 8.4. Bouncing of rays from the surface.

Where does the ray hit the surface again? Clearly at $r_0 + 2r_0 = 3r_0 = 3\sqrt{-2dR_c}$. It is seen that the range to the nth bounce is

$$r_n = (2n+1)r_0 = (2n+1)\sqrt{-2dR_c}.$$

The *cycle distance* is the distance to the point at which the ray repeats itself. Here it is equal to $2r_0$.

Example 8.5. Bilinear sound speed profile; limit rays. The sound speed profile shown in Fig. 8.11 is bilinear. In particular, the gradient changes sign at depth h. The source is located at depth d in the upper layer, that is, between the surface and the change in gradient. Let $\theta_i = \theta_c$ denote the downward ray that has a vertex at depth $z = h$. Downward rays for which $\theta_i < \theta_c$ have vertices in the upper layer and are therefore trapped in the upper layer. If the angle θ_i exceeds θ_c, however, the ray does reach the level $z = h$. Since its direction is still downward at that point, the ray crosses over into the lower layer, whereupon it refracts further downward, never to return (unless it bounces off the bottom). The θ_c ray, called the critical ray, cannot make up its mind. When it reaches the change in gradient, it cannot decide which kind of vertex to form, a minimum or a maximum. So it splits and does both, in the process forming two limit rays. The region between the two limit rays cannot be reached by any direct ray and is therefore a shadow region (but it can be reached by rays that bounce off the

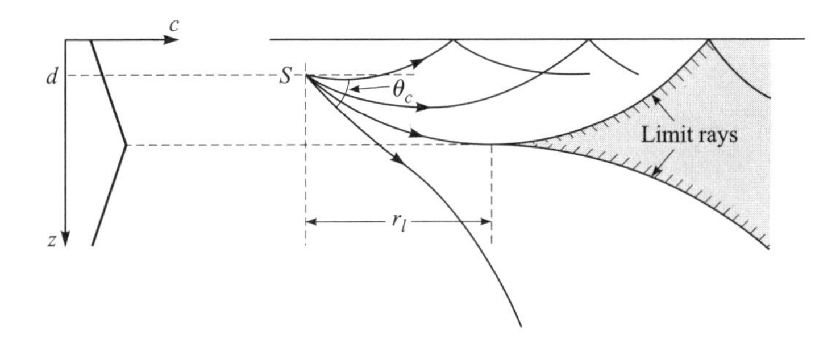

Figure 8.11 Example 8.5. Bilinear profile gives rise to two limit rays.

surface or the bottom). Our task is to find (1) the horizontal range r_ℓ to the point where the ray splits into two limit rays and (2) the critical angle θ_c. To find r_ℓ, use the same method that allowed us to find r_0 in Example 8.4. In this case the change in depth is $\Delta z = h - d$ and the change in range is $\Delta r = r_\ell$. The expression for r_ℓ is given by Eq. C–11, with d replaced by $h - d$:

$$r_\ell = \sqrt{-2R_c(h-d) - (h-d)^2},$$

where R_c here is the radius of curvature for the upper layer. To find θ_c, make use of Eq. C–6, which yields

$$\cos\theta_c = 1 + \frac{h-d}{R_c}.$$

In order for θ_c to be real, R_c must be negative, as it is in Fig. 8.11 (upper layer).

Of course a change in profile slope can be just the opposite of that shown in Fig. 8.11, that is, the sound speed can decrease through the upper layer and increase through the lower layer. The sound speed then has a minimum instead of a maximum at the depth where the slope changes. When this occurs, rays are attracted to the minimum layer. That is, after crossing the layer, they are bent back toward it, recross, are bent back again, and so on. Rays are thus trapped in the region of the minimum (just the opposite of the situation shown in Fig. 8.11). An important example is the minimum in the deep ocean profile shown in Fig. 8.2*b*. This minimum establishes the deep ocean sound channel, a natural waveguide of great importance. See Problems 8–8 and 8–9. As Fig. 8.1 shows, the atmosphere also has sound speed minima about which natural waveguides can develop.

3. Time of Travel along a Ray Path

What time t is required for a signal to travel from point A to point B along a ray? See Fig. 8.12. Since at any point along the path the travel speed is $ds/dt = c$, where s is the arc length, integration gives

$$t = \int_A^B \frac{ds}{c},$$

or, since $ds = R_c\, d\theta$,

$$t = \int_{\theta_i}^{\theta} \frac{R_c}{c}\, d\theta.$$

Substitution for c and R_c from Snell's law and Eq. C–4, respectively, leads to the following general result:

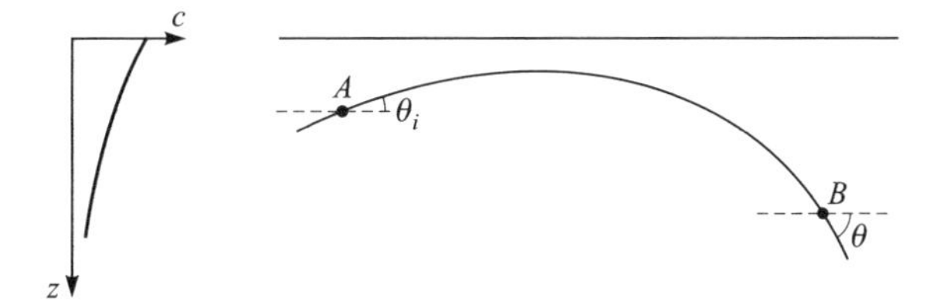

Figure 8.12 Calculation of time to travel between two points along a ray.

$$t = -\int_{\theta_i}^{\theta} \frac{d\theta}{g\cos\theta}. \tag{C–13}$$

To go further, we need to know how the gradient g varies along the path.

In the special case of a linear sound speed profile, constancy of g allows Eq. C–13 to be integrated immediately. The result is

$$t = \frac{1}{g}\left[\ln\tan\left(\frac{\theta_i}{2} + \frac{\pi}{4}\right) - \ln\tan\left(\frac{\theta}{2} + \frac{\pi}{4}\right)\right]. \tag{C–14}$$

Example 8.6. Travel time in a constant gradient medium. Refer to Fig. 8.7*a*. Calculate the time for a signal to travel from the source (on the surface) to the vertex and on back up to the surface along the ray path shown. A simple approach is to calculate the source-to-vertex time and double it. To calculate the time to the vertex, first note that since g is constant, Eq. C–14 applies. But because $\theta = \theta_v = 0$ in this case, the second ln tan term in the equation vanishes. Given $g = 0.0163\ \mathrm{s}^{-1}$ and $\theta_i = 2^\circ$, we obtain

$$t = 2t_v = \frac{2}{0.0163}\ln\tan 46^\circ = 4.2839\ \mathrm{s}.$$

For comparison calculate the time taken for a signal to travel the same range by the most direct route, i.e., by the horizontal path just below the surface. Since the speed is c_i over the entire path, the calculation is $t_{\mathrm{horizontal}} = 2r_v/c_i = \tan\theta_i/g$, where Eq. C–10 has been used, or $t_{\mathrm{horizontal}} = 4.2848$ s, or 0.9 ms longer than the time taken along the vertex ray, despite the fact that the vertex route is longer than the horizontal path. If other routes between the two points had been taken (involving two, three, and more vertices) we should have found that none require more travel time than the direct route: the direct ray always arrives last. The explanation is that travel is slowest along the direct path; higher sound speed along the vertex routes more than makes up for the greater travel distance.

REFERENCES

1. R. K. Cook, "Radiation of sound by earthquakes," *Proceedings of the Fifth International Congress on Acoustics*, Leige, Belgium (1965), Paper K19.
2. R. K. Cook, "Atmospheric sound propagation," in *Atmospheric Exploration by Remote Probes*, Proceedings of the Scientific Meetings of the Panel on Remote Atmospheric Probing (Committee on Atmospheric Sciences, National Academy of Sciences and National Research Council, Washington, DC, January 1969), Vol. 2, pp. 663–669.
3. A. B. Coppens, "Simple equations for the speed of sound in Neptunian waters," *J. Acoust. Soc. Am.* **69**, 862–863 (1981).
4. V. A. Del Grosso, "New equation for the speed of sound in natural waters (with comparisons to other equations)," *J. Acoust. Soc. Am.* **56**, 1084–1091 (1974).
5. J. Mathews and R. L Walker, *Mathematical Methods of Physics* (W.A. Benjamin, New York, 1965), pp. 26–27.
6. A. D. Pierce, *Acoustics: An Introduction to Its Physical Principles and Applications* (McGraw-Hill, New York, 1981). Reprinted by Acoustical Society of America, 1989, Chap. 8.

PROBLEMS

8–1. The decrease in temperature in the troposphere (see Fig. 8.1) is approximately linear (the gradient is about 6.5 K/km) for the first 11 km, i.e.,

$$T_0 = \bar{T}_0(1 - z/a),$$

where $\bar{T}_0 = 288.15$ K is the sea level temperature and $a = 44.4$ km is the height at which T_0 would become zero if the linear decrease continued past the tropopause. Show that the wave equation for vertical propagation through such an atmosphere is

$$u_{zz} - \frac{u_z}{H(1 - z/a)} - \frac{1}{c_0^2} u_{tt} = 0,$$

where $H = \bar{c}_0^2/\gamma g$ is the scale height based on the sea level value of the sound speed ($\bar{c}_0 = 340.3$ m/s).

8–2. A time-harmonic wave travels vertically upward through the troposphere. Using the model described in the previous problem:

(a) Calculate the increase in particle velocity amplitude u_0 as the wave travels from sea level to a height of 10 km. Put your answer in the form of a change in particle velocity level $20 \log_{10}[u_0(10 \text{ km})/u_0(0 \text{ km})]$.

(b) Find the time the wave takes to reach 10 km.

8–3. The radiation of sound into the atmosphere by ground vibration caused by the 1964 Alaskan earthquake is described in Sec. B.2. Assume that the SPL was 94 dB at the earth's surface near Washington, D.C.

(a) Use the isothermal model of the atmosphere to estimate the peak particle velocity u_0 of this wave when it reached a height of 200 km in the ionosphere. For your calculation assume $T_0 = \bar{T}_0 = 276$ K (3°C).

(b) Critically analyze your result. For example, is it consistent with assumptions on which the linear wave equation is based?

8–4. Consider the following first-order wave equation for propagation in an inhomogeneous fluid:

$$u_z + \frac{1}{c_0} u_t + \frac{(\rho_0 c_0)_z}{2\rho_0 c_0} u = 0. \qquad \text{(P–1)}$$

(a) Show that Eq. B–17 is an exact solution of Eq. P–1, which may thus be regarded as an approximate first integral of Eq. B–10.

(b) Neglecting the question of the approximation, what generality does Eq. B–10 have that Eq. P–1 does not? *Hint*: What type of wave does the solution Eq. B–17 represent?

8–5. A measured ocean temperature profile is given by

$$t = 13° + g_t z,$$

where t is temperature in degrees Celsius, z is depth, and g_t is the (constant) temperature gradient. The salinity is 35‰.

(a) Find an expression for the sound velocity in the form

$$c = c_0 + a_1 z + a_2 z^2 + a_3 z^3,$$

that is, find the values of c_0 and the coefficients a_1, a_2, and a_3 (the coefficients will be found to depend on g_t).

(b) If $g_t = 0.01$°C/m, over what range of z is c essentially linear in z? As a criterion, require that the magnitudes of the quadratic and cubic terms each be less than 1% of the magnitude of the linear term.

8–6. A source is located at a depth of 60 m in seawater in which the sound speed increases linearly with depth at a rate of 0.05 m/s per m. The sound speed at the surface is 1500 m/s. Consider the two rays that start out at angles of 3° (downward ray) and −3° (upward ray).

(a) Find the horizontal range and angle at which the upward ray strikes the surface.

(b) For the downward ray find the range to the vertex and the total depth at that point.

(c) Beyond the vertex the downward ray travels upward, reflects from the surface, and completes its cycle by crossing the 60-m depth plane at the same angle it began, 3°. What is the cycle distance, that is, the horizontal range between the two points at which $\theta = 3°$?

8–7. A surface ship uses its sonar to hunt for mines that are known to be resting on the bottom in 30 m of water. The sound speed increases linearly from 1500 m/s at the surface to 1503 m/s at the bottom. How close (in horizontal range) must the ship be to a mine for the mine to be detected? (*Hint*: Consider a shadow zone.)

8–8. A source S is located at the depth at which the sound speed stops decreasing and begins increasing. Instead of spreading uniformly through the ocean volume, the sound in this situation tends to be trapped in a relatively narrow channel, say roughly between depths d_1 and d_2. Give a brief qualitative explanation of this phenomenon. Sketch several sample ray paths to illustrate your argument.

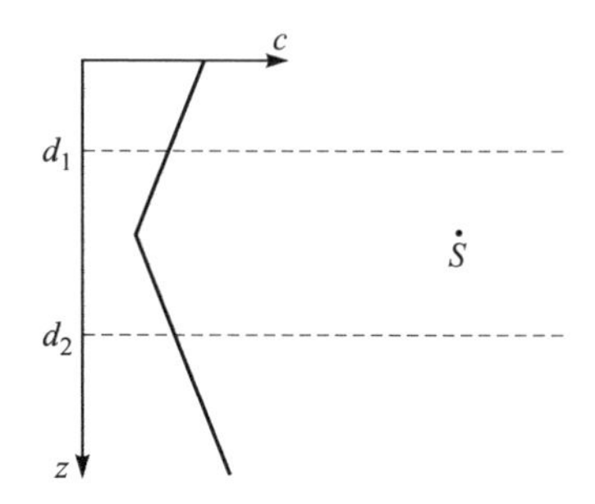

8–9. A source is located on the axis, depth 1200 m, of a deep sound channel (see the figure for the previous problem). The sound speed at that point is 1480 m/s. The sound speed increases linearly to a value of 1525 m/s at the surface. In the downward direction the sound speed increases linearly to a value of 1540 m/s at a depth of 4800 m.

(a) Consider the upward ray leaving the source at an angle $\theta_i = -6°$. Find the range to point P where the ray first returns to the channel axis.

(b) One of the downward rays from the source also has a first return to the channel axis at point P. What is the initial angle θ_i of this ray?

(c) Calculate the travel times to point P for the following rays:

(i) the ray in part (a),

(ii) the ray in part (b),

(iii) the direct ray (the one traveling straight down the axis).

8–10. A submarine S is located at a depth of 15 m, where the sound speed is 1500 m/s and the gradient has the constant value $g_s = 0.2\ \text{s}^{-1}$.

Waterborne helicopter noise is detected and found to be incident on the submarine at a grazing angle of 0°. Looking through his periscope, the captain sees the helicopter H at an angle of 45° relative to the horizon. Assume that the air temperature between the helicopter and the ocean surface is constant, 20°C.

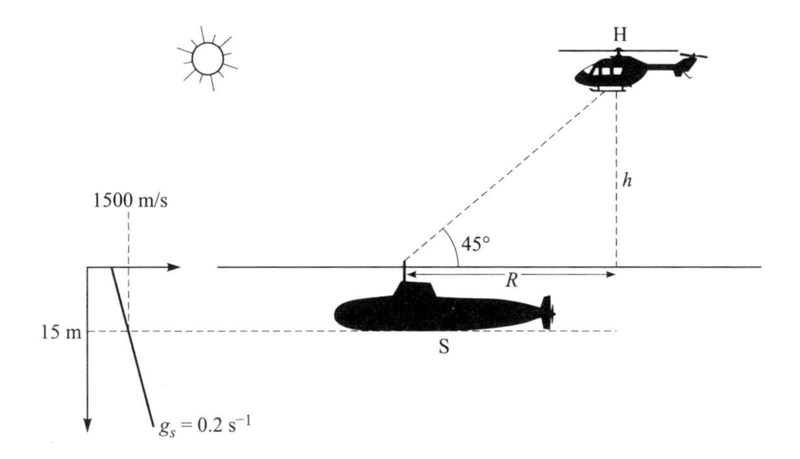

(a) Find the horizontal range R to the helicopter.

(b) Find the time for sound from the helicopter to reach the submarine.

8–11. Refer to the previous problem. Instead of assuming the atmosphere to be isothermal, allow for the slight decrease in sound speed with height that is characteristic of the lower atmosphere. For the U.S. Standard Atmosphere the gradient is approximately $g_{\text{atm}} = 0.004\ \text{s}^{-1}$ (positive in this case because the coordinate z is positive downward). Because the gradient is very small, it is suggested that you use the answer to the previous problem to calculate an approximate value of c_H, where c_H is the sound speed at the helicopter, and proceed from there. Reanswer parts (a) and (b) of the previous problem.

9

PROPAGATION IN DISSIPATIVE FLUIDS: ABSORPTION AND DISPERSION

The effect of dissipation in the medium is considered in this chapter. Viscosity, heat conduction, relaxation, and other loss mechanisms cause absorption of sound. The reduction in wave amplitude associated with absorption is different from the attenuation that takes place in a horn or in a diverging spherical or cylindrical wave field. The latter is strictly a geometrical effect. No energy is actually lost; it is simply spread over a larger area. In the case of viscosity and other true dissipation mechanisms, the sound field actually loses energy (which is transformed into heat). Absorption is generally accompanied by dispersion (dependence of sound speed on frequency).

Section A of the chapter is devoted to general aspects of absorption and dispersion of sound. Topics include the way dispersion relations arise, spatial and temporal modes of absorption, frequency dependence of the absorption coefficient α for various dissipation mechanisms, an overview of absorption in the atmosphere and the ocean, and methods of expressing absorption. Specific loss mechanisms are then taken up: viscosity and heat conduction in Sec. B, relaxation in Sec. C, and boundary layer losses in Sec. D. Section E is a summary.

A. INTRODUCTION

Taking the various mechanisms of dissipation one at a time, we determine their effect on sound propagation by following the procedure outlined in the flow diagram in Fig. 9.1. The physical effect of a given mechanism is first described

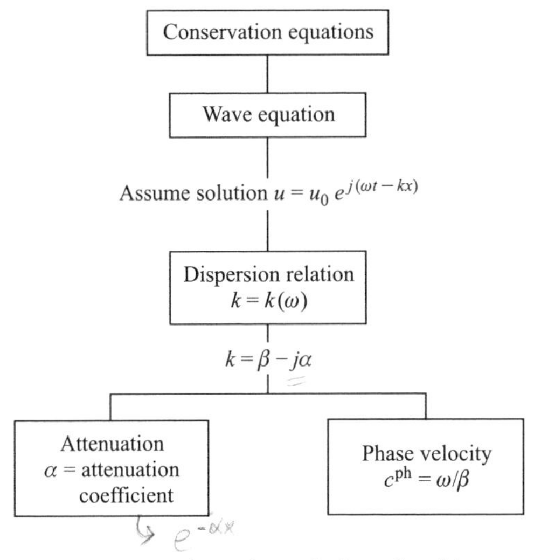

Figure 9.1 Dispersion relation algorithm.

by including it in the conservation equations. The conservation equations are then combined to obtain a wave equation, which is solved by assuming a time-harmonic solution

$$u = u_0 e^{j(\omega t - kx)}. \tag{A–1}$$

In this way the wave equation is transformed into a dispersion relation, that is, an algebraic equation in the wave number k, which has real and imaginary parts:

$$k = \beta - j\alpha. \tag{A–2}$$

Substitution of Eq. A–2 in Eq. A–1 yields

$$u = u_0 e^{-\alpha x} e^{j(\omega t - \beta x)} = u_0 e^{-\alpha x} \exp j\omega\left(t - \frac{x}{\omega/\beta}\right), \tag{A–3}$$

which shows that α is the absorption coefficient and β is related to the phase speed by $c^{ph} = \omega/\beta$.[1]

Although the derivations in this chapter are for simplicity done only for plane waves, the expressions obtained for α and c^{ph} also hold for spherical and other unidirectional waves. For example, the expression for the velocity poten-

[1] The quantities α and β are not independent. Application of the principles of causality and linearity leads to the Kramers-Kronig relations, which provide a functional relationship between the absorption coefficient and the phase velocity. See, for example, Ref. 13.

tial produced by a pulsating sphere in a lossless medium is given in Chap. 1; see Eq. 1D–16. When the medium is dissipative, the solution is

V

$$\phi = \frac{Ar_0}{r} e^{-\alpha(r-r_0)} e^{j\omega[t-(r-r_0)/c^{ph}]}.$$

Notice that the form of Eq. A–1 is ideal for time-harmonic forced waves. A source emitting a signal $u = u_0 e^{j\omega t}$ at $x = 0$ is implied. Because the wave generated decays with distance, α represents the *spatial* absorption coefficient. A slightly different approach is used for initial-value problems. Suppose a given wave field exists in space at time $t = 0$ and one wishes to know how the field decays with time. In this case the solution usually assumed is

$$u = u_0 e^{jk(x-ct)}. \tag{A–4}$$

Here the wave number k is regarded as known (and real) and c is to be determined from the appropriate dispersion relation. If c is separated into real and imaginary parts in the form

$$c = c^{ph} - jm/k,$$

the expression for u becomes

$$u = u_0 e^{-mt} e^{jk(x-c^{ph}t)},$$

i.e., c^{ph} is indeed the phase velocity and m is the *temporal* absorption coefficient. The initial-value problem is commonly encountered in room acoustics, where one wants to know how rapidly sound in a room decays with time; the decay is principally due to absorption by room surfaces. Our treatment in this chapter is restricted to forced waves.

A preview of the results developed in the ensuing sections of the chapter is now given. Table 9.1 shows how the spatial absorption coefficient α depends on frequency f for various loss mechanisms. For most media two or more mechanisms are important. The actual absorption coefficient is then a composite, usually obtained simply by adding the contributions due to the various mechanisms.[2] The sample absorption coefficient curve A shown in Fig. 9.2 is typical. Curve A is the sum of curves r (relaxation absorption) and tv (thermoviscous absorption). Relaxation dominates at low frequencies, thermoviscous effects at high frequencies.

[2]The validity of superposing the various contributions, e.g., putting $\alpha_{total} = \alpha_1 + \alpha_2 + \alpha_3 + \cdots$, is rarely discussed. In fact, superposition is in general not justified because although the various absorption mechanisms act mainly separately, they do have interactions with each other; see, for example, Sec. B.3. In practice, however, the effect of each mechanism is normally so small that the interactions may be neglected, in which case superposition is justified.

Table 9.1 Various Absorption Mechanisms

Mechanism	Section	Frequency Dependence of α
Viscosity	B	f^2
Heat conduction	B	f^2
Relaxation	C	$f^2/(f^2+f_r^2)$
Boundary-layer effects	D	$\sqrt{f}$

The symbol f_r appearing in the relaxation formula in Table 9.1 is called the relaxation frequency. It characterizes the transition in the relaxation absorption curve from the low frequency region ($\alpha \propto f^2$) to the high frequency plateau ($\alpha = \text{const}$); see Fig. 9.2. The value of f_r may be found from the intersection of the low frequency asymptote with the plateau asymptote.

Our two most common fluid media, the atmosphere and the ocean (see Appendix B and Refs. 2, 5), are blessed with *two* major relaxation processes, as indicated in Fig. 9.3. Part (*a*) of the figure (after Ref. 20) shows that the total atmospheric absorption (curve α_{air}) is the sum of the relaxation absorption associated with the vibration of nitrogen molecules (curve α_{N_2}), the relaxation absorption associated with the vibration of oxygen molecules (curve α_{O_2}), and the "frequency-squared" absorption due mainly to viscosity and heat conduction (curve α_{tv}). Because the two relaxation frequencies $f_{r,N}$ and $f_{r,O}$ are strongly dependent on the water vapor content of the air, the total absorption curve changes markedly with relative humidity (the curves shown are for 70% relative humidity). In part (*b*) of the figure, which is for seawater (after Ref. 7), individual relaxation absorption curves are not given. Instead the total absorption curves for temperatures 0, 10, and 20°C are shown. Nevertheless two relaxation processes are evident. The one at lower frequency ($f_r \approx 1$ kHz) is due to boric acid. The one at higher frequency ($f_r \approx 85$ kHz) is due to

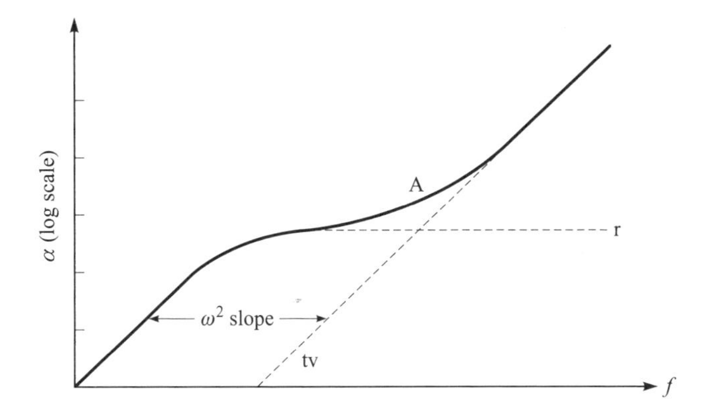

Figure 9.2 Absorption in a relaxing (r) and thermoviscous (tv) fluid.

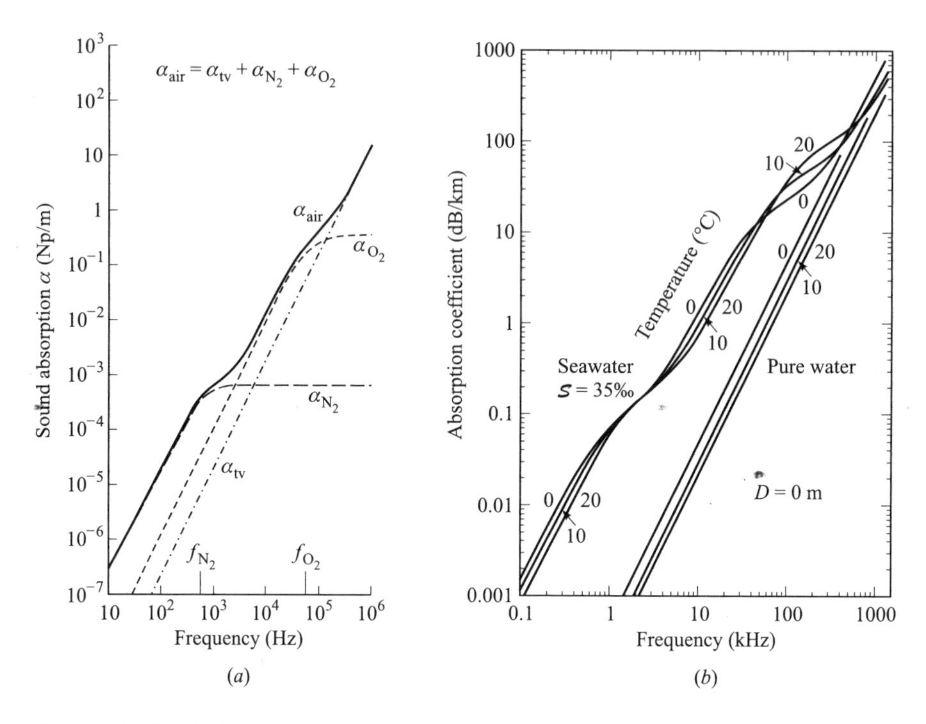

Figure 9.3 Typical absorption curves for (*a*) the atmosphere (Ref. 20), (*b*) the ocean (Ref. 7).

magnesium sulfate. Both relaxation frequencies are strong functions of temperature and, to a lesser extent, salinity. The curves labeled "pure water" represent the frequency-squared asymptotic absorption due primarily to viscosity. Formulas for the oceanic and atmospheric absorption coefficient are quite complicated and are given in Appendix B.

Our preliminary remarks close with a discussion of the units of the absorption coefficient. The pressure equivalent of Eq. A–1 (use the characteristic impedance relation to make the transformation) is

$$p = Ae^{-\alpha x}e^{j(\omega t - \beta x)}, \tag{A–5}$$

where A is the pressure amplitude at $x = 0$. The amplitude absorption coefficient α has units of nepers per unit distance ("neper," a unit without units, is sometimes omitted). The sound pressure level of the traveling wave is

$$\text{SPL} = 20\log_{10}\frac{A/\sqrt{2}}{p_{\text{ref}}}e^{-\alpha x} = 20\log_{10}\frac{A/\sqrt{2}}{p_{\text{ref}}} + 20\log_{10}e^{-\alpha x} = \text{SPL}_0 - 8.686\alpha x,$$

where SPL_0 is the level at the source ($x = 0$). The drop in sound pressure level is

$$\Delta SPL \equiv SPL_0 - SPL = \overline{\alpha}x, \qquad \text{(A–6)}$$

where

$$\overline{\alpha} = 8.686\alpha \qquad \text{(A–7)}$$

has units of decibels per unit distance, that is, $\overline{\alpha} = \Delta SPL/\Delta x$. In this text the overbar is used to distinguish the decibel absorption rate $\overline{\alpha}$ from the amplitude coefficient α. Often in the literature, unfortunately, no distinction is made between the two quantities, and one must read carefully to determine the author's intent. Notice, for example, that α is plotted in Fig. 9.3*a*, $\overline{\alpha}$ in Fig. 9.3*b*. Finally, since the intensity of a propagating plane or spherical wave is proportional to the square of the rms pressure, we have

$$I \propto e^{-2\alpha x}.$$

The intensity absorption coefficient is thus 2α.

B. VISCOSITY AND HEAT CONDUCTION

The first studies of sound absorption were for viscosity and heat conduction in gases. Stokes (1845, Ref. 17) investigated the loss due to viscosity. However, because he considered free waves, not forced waves, he found the temporal absorption coefficient m, not the spatial coefficient α.[3] Stefan (1866, Ref. 16) also considered viscosity and was the first to find the spatial absorption coefficient α. Shortly thereafter Kirchhoff (1868, Ref. 11) published what was to become a benchmark work on absorption in a thermoviscous gas. He obtained not only the absorption coefficient for propagation in an open fluid but also the coefficient for propagation in a tube, where absorption is due mainly to viscous and thermal boundary layers at the tube wall. Later Langevin (1923) generalized the analysis of free-medium absorption to include liquids; his work was reported by Biquard (1936, Ref. 3).[4]

This section has three parts. Viscosity by itself is treated in Sec. B.1, heat conduction by itself in Sec. B.2, and viscosity and heat conduction together in Sec. B.3. Boundary-layer absorption and dispersion, although due to viscosity and heat conduction, are postponed until Sec. D.

[3] Stokes later investigated absorption caused by heat radiation, for which case he found the spatial absorption coefficient (1851, Ref. 18). Absorption due to heat radiation is the subject of Problem 2–18.

[4] Truesdell has given a detailed discussion of thermoviscous absorption (Ref. 19).

1. Viscous Fluids

We begin with the linearized equations for planar motion in a viscous fluid:

$$\text{Continuity:} \quad \delta\rho_t + \rho_0 u_x = 0, \tag{B–1a}$$

$$\text{Momentum:} \quad \rho_0 u_t + p_x = (\lambda + 2\mu) u_{xx}, \tag{B–1b}$$

$$\text{State:} \quad p = c_0^2 \delta\rho. \tag{B–1c}$$

The continuity and state equations are the same as those for a lossless fluid, namely, Eqs. 1C–54 and 1C–56, respectively.[5] Only the momentum equation is different. It may be obtained by taking the spatial derivative of the planar form of Eq. 2A–27, substituting u for ϕ_x, and making use of Eq. 2A–21. The result is Eq. B–1b, where λ and μ are the dilatational and shear viscosity coefficients, respectively.

Although the shear viscosity coefficient μ has been measured for many fluids, tables of the dilatational viscosity coefficient λ will not be found. As explained in Sec. 2A.2.b, the use of Stokes's assumption ($\lambda = -2\mu/3$), which holds for some fluids, such as the noble gases, leads to replacement of $\lambda + 2\mu$ by $4\mu/3$. For fluids that do not follow Stokes's assumption, modern practice is to replace λ by $-2\mu/3 + \mu_B$, where μ_B is called the bulk viscosity coefficient. See Eqs. 2A–21 and 2A–22. The momentum equation then becomes

$$\rho_0 u_t + p_x = \mu \tilde{V} u_{xx}, \tag{B–2}$$

where $\tilde{V} = 4/3 + \mu_B/\mu$ is called the viscosity number. Of course nothing new has been added. We have simply replaced one coefficient that is hard to measure, λ, and added another that is equally hard to measure, μ_B (or $\tilde{V}$).

The equations of motion are now combined to form a wave equation. If $\delta\rho$ is eliminated between Eqs. B–1a and B–1c and the result used to eliminate p from Eq. B–2, the result is

$$\frac{\nu\tilde{V}}{c_0^2} u_{xxt} + u_{xx} - \frac{1}{c_0^2} u_{tt} = 0, \tag{B–3}$$

where $\nu = \mu/\rho_0$ is the kinematic viscosity coefficient. The three-dimensional version of this equation (in terms of pressure) is Eq. 2C–12.

[5] The use of Eq. B–1c, which appears to be the descendant of an isentropic equation of state, requires some explanation because entropy is not constant for fluid motion in a viscous medium. The entropy variation is, however, only second order, that is, it is described by a nonlinear (quadratic) term in the energy equation (see Eq. B–11 below). Since all nonlinear terms are negligible in the small-signal approximation, Eq. B–1c continues to hold, even for viscous fluids. On the other hand, Eq. B–1c may not be used for heat-conducting fluids because the term representing entropy change due to heat conduction is linear (see Sec. B.2 below).

The absorption and dispersion properties are found by substituting a time-harmonic, traveling-wave solution (Eq. A–1) in Eq. B–3. The resulting dispersion relation is

$$(1 + j\omega\nu\tilde{V}/c_0^2)k^2 - \omega^2/c_0^2 = 0,$$

or

$$k = \pm\frac{\omega/c_0}{\sqrt{1 + j\tilde{V}\delta_v}}, \tag{B–4}$$

where δ_v is the dimensionless coefficient

$$\delta_v = \frac{\omega\nu}{c_0^2}. \tag{B–5}$$

The plus sign in Eq. B–4 applies to outgoing traveling waves, the minus sign to incoming traveling waves; the plus sign is used hereafter.

To separate k into its real and imaginary parts, make use of the fact that δ_v is ordinarily very, very small, even for very high ultrasonic frequencies. For example, for air at 20°C and for fresh water at 20°C the values of ν are 1.50×10^{-5} and 1.004×10^{-6} m^2/s, respectively. The corresponding values of δ_v are

$$\delta_{v_{\mathrm{air}}} = 8.01 \times 10^{-10} f \tag{B–6a}$$

$$\delta_{v_{\mathrm{water}}} = 2.88 \times 10^{-12} f, \tag{B–6b}$$

where f is frequency in hertz. It is clear that even at frequencies as high as 10 MHz in air and 10 GHz in water the value of δ_v is still very small. Consequently, a simple binomial expansion of Eq. B–4, with only the first two terms retained, is sufficient,

$$k \doteq \frac{\omega}{c_0(1 - j\tilde{V}\delta_v/2)} = \frac{\omega}{c_0} - \frac{j\tilde{V}\delta_v\omega}{2c_0}. \tag{B–7}$$

Comparison with Eq. A–2 yields

$$\alpha_v = \frac{\tilde{V}\delta_v\omega}{2c_0} = \frac{\tilde{V}\nu\omega^2}{2c_0^3}, \tag{B–8a}$$

(the subscript v denotes absorption due to viscosity) and

$$c^{\mathrm{ph}} = c_0. \tag{B–8b}$$

Sound therefore propagates without dispersion in a viscous fluid but is attenuated. The absorption coefficient increases as the square of the frequency. Note that the absence of dispersion is due to our two-term approximation of Eq. B–4. If the third term had been included, we should have obtained

$$c^{\mathrm{ph}} = c_0/(1 - \tfrac{3}{8}\tilde{V}^2\delta_{\mathrm{v}}^2),$$

which shows that the phase velocity does after all depend on frequency. Because of the smallness of δ_{v}, however, the dispersion is usually too weak to be measurable.

What is the order of magnitude of the absorption caused by viscosity? Consider air at 20°C as an example. In Sec. B.3 the value of $\tilde{V}$ is estimated to be 1.94. Substitution in Eq. B–8a yields

$$\alpha_{\mathrm{v}} = 1.42 \times 10^{-11} f^2 \text{ nepers/m}, \tag{B–9a}$$

or

$$\bar{\alpha}_{\mathrm{v}} = 1.24 \times 10^{-10} f^2 \text{ dB/m}, \tag{B–9b}$$

where f is in hertz. At 1 kHz, for example, the absorption coefficient is only

$$\bar{\alpha}_{\mathrm{v}} = 1.24 \times 10^{-4} \text{ dB/m}.$$

Since at 1 kHz measured values of $\bar{\alpha}$ for air are of order 40 times larger (see Appendix B, Fig. B.1), it is clear that viscosity is not an important cause of absorption in air at audio frequencies. However, for fresh water, high viscosity liquids such as glycerine and castor oil, and other fluids at very high frequencies, viscosity is a major contributor to sound absorption.

2. Thermally Conducting Fluids

Up to now sound propagation has been assumed to be an adiabatic process, that is, one for which the compressions and expansions take place without the flow of heat. But all media conduct heat to some extent. The compressed portions (+) of the wave shown in Fig. 9.4 are regions where the medium is a little hotter, the expanded portions (−) where the medium is a little colder. If the fluid conducts heat, the variation in temperature gives rise to heat flow from hot regions to cold regions as the medium tries to reestablish temperature equilibrium. The flow of heat reduces the amount of energy available for the sound wave and thus represents an acoustical loss.

To include the effect of heat flow on the sound wave, we need to develop a proper conservation-of-energy relation. Although a general (three-dimensional) derivation of the energy equation is given in Chap. 2, for the benefit

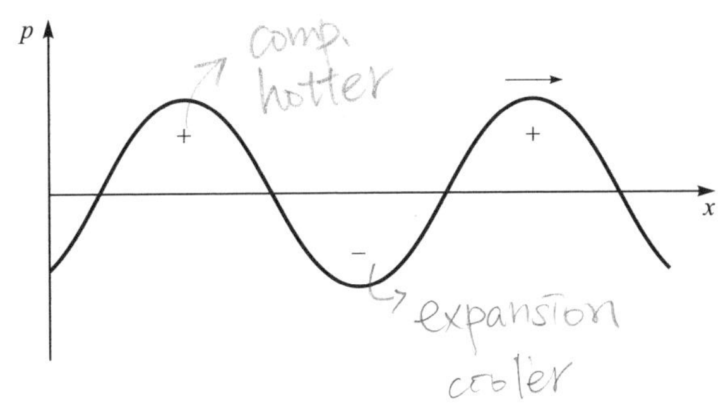

Figure 9.4 Propagating sound wave.

of readers who postponed Chap. 2 until later, a simple one-dimensional derivation is presented here. The approach follows that used in Chap. 1 to derive continuity and momentum equations for one-dimensional flow. For simplicity the fluid is assumed to be inviscid. The energy carried into and out of a small control volume is shown in Fig. 9.5. Here e is the total energy per unit mass, q is the heat energy per unit time per unit area, and S is the cross-sectional area. The time rate of increase of energy inside the control volume is equal to (i) the net inflow of energy carried by the fluid across the surfaces of the control volume, (ii) the work per unit time done on the fluid inside by the surface forces (restricted to the pressure for inviscid fluids), and (iii) the net heat per unit time added to the fluid. To calculate the work per unit time (power) done by the surface forces, note that power is force times velocity, or, on a per-unit-area basis, pressure times velocity Pu. The mathematical expression of conservation of energy is therefore

$$\frac{\partial}{\partial t}(\rho e S\,\Delta x) = \rho u e S|_x - \rho u e S|_{x+\Delta x} + PuS|_x - PuS|_{x+\Delta x} + qS|_x - qS|_{x+\Delta x}.$$

Dividing by $S\,\Delta x$ and taking the limit as $\Delta x \to 0$, we obtain

$$(\rho e)_t + (\rho u e)_x + (Pu)_x = -q_x,$$

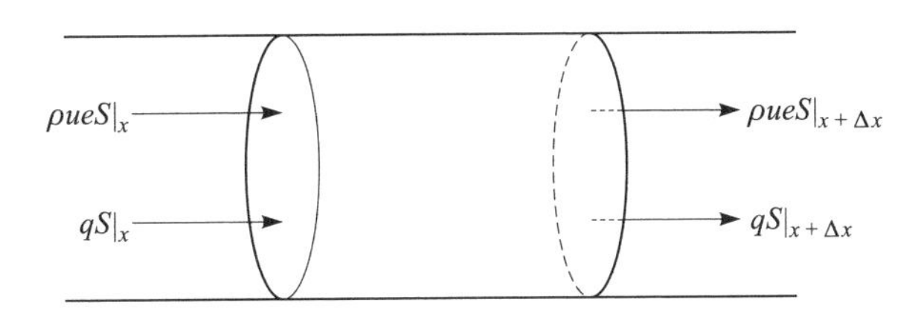

Figure 9.5 Incremental control volume for derivation of an energy conservation equation.

or, when the first two terms are expanded and the continuity equation Eq. 1C–28 is used,

$$\rho De/Dt + (Pu)_x = -q_x,$$

where De/Dt is the material derivative of the total energy, i.e., the time rate of energy change for a fluid particle. The total energy is the sum of the kinetic, potential, and internal energies. We assume the potential energy per unit mass to be constant (either the flow is horizontal or the change in altitude is small enough that gravity has little effect). The kinetic energy per unit mass is $u^2/2$. Letting the internal energy per unit mass be denoted ϵ, we have

$$\frac{De}{Dt} = \frac{D(u^2/2)}{Dt} + \frac{D\epsilon}{Dt} = u(u_t + uu_x) + \epsilon_t + u\epsilon_x,$$

and the energy equation becomes

$$\rho\epsilon_t + \rho u u_t + \rho u \epsilon_x + \rho u u u_x + uP_x + Pu_x = -q_x.$$

Application of the momentum equation, Eq. 1C–33, greatly simplifies this expression. The second, fourth, and fifth terms cancel, and all that remains is

$$\rho\epsilon_t + \rho u \epsilon_x + Pu_x = -q_x. \tag{B–10}$$

As an aside, note that if viscosity had been taken into account, the result would have been (Eq. 2A–34)

$$\rho\epsilon_t + \rho u \epsilon_x + Pu_x = (\lambda + 2\mu)u_x^2 - q_x. \tag{B–11}$$

To apply Eq. B–10 to the problem at hand, we need to linearize it, specify that the heat transfer is due to conduction, and select an equation of state. Linearization yields[6]

$$\rho_0\epsilon_t + P_0 u_x = -q_x. \tag{B–12}$$

When the heat flow is due to conduction, q is given by Fourier's law of heat conduction,

$$q = -\kappa T_x, \tag{B–13}$$

where κ is the heat conduction coefficient and T is the absolute temperature.

[6] Notice that linearization of Eq. B–11 also leads to Eq. B–12 because the viscous energy dissipation $(\lambda + 2\mu)u_x^2$ is a quadratic term. In the linearized conservation equations, therefore, viscosity appears only in the momentum equation.

Next take the fluid to be an ideal gas, for which the equation of state is $P = R\rho T$ (R is the gas constant), or, on linearization,

$$\frac{p}{p_0} = \frac{\delta\rho}{\rho_0} + \frac{T - T_0}{T_0}.$$

It is convenient to rearrange this equation as

$$T - T_0 = \frac{1}{R\rho_0}\left(p - \frac{c_0^2}{\gamma}\,\delta\rho\right). \tag{B–14}$$

Moreover, for an ideal gas the internal energy depends only on the temperature, in particular,

$$d\epsilon = C_v\,dT, \tag{B–15}$$

where C_v is the specific heat at constant volume. Substitution of Eqs. B–13, B–15, and B–1a (in the form $u_x = -\delta\rho_t/\rho_0$) into Eq. B–12 yields

$$\rho_0 C_v T_t - \frac{c_0^2}{\gamma}\,\delta\rho_t = \kappa T_{xx}. \tag{B–16}$$

With the help of Eq. B–14 and the relation $R = C_v(\gamma - 1)$, where $\gamma = C_p/C_v$ is the ratio of specific heats and C_p is the specific heat at constant pressure, we eliminate the temperature from Eq. B–16 and obtain

$$(p - c_0^2\delta\rho)_t = \frac{\kappa}{\rho_0 C_v}\left(p - \frac{c_0^2}{\gamma}\delta\rho\right)_{xx}. \tag{B–17}$$

Although derived here for an ideal gas, this form of the linear energy equation is valid for an inert fluid of arbitrary equation of state.

A wave equation is obtained by combining Eq. B–17 with the linear continuity equation Eq. B–1a and the linear momentum equation for an inviscid fluid (Eq. B–1b with λ, $\mu = 0$). The result is

$$\frac{\kappa}{\rho_0 C_p}\left[u_{xx} - \frac{\gamma}{c_0^2}u_{tt}\right]_{xx} - \left[u_{xx} - \frac{1}{c_0^2}u_{tt}\right]_t = 0. \tag{B–18}$$

This is the equation for plane sound waves in a thermally conducting but inviscid fluid.

Careful examination of the form of Eq. B–18 leads to considerable insight about the effect of heat conduction on sound propagation. The relative importance of the two bracketed pairs of terms depends on the magnitude of the heat conduction coefficient κ. Before the equation is analyzed on this basis, however, the coefficient in front of the first bracket should be made dimensionless. Otherwise, the coefficient may be made numerically small or large simply by choosing the right units. If dimensionless time and space variables

$$t^* = \omega t \qquad \text{and} \qquad x^* = \omega x / c_0$$

are introduced, Eq. B–18 becomes[7]

$$\delta_{\text{th}} \left[u_{xx} - \frac{\gamma}{c_0^2} u_{tt} \right]_{x^* x^*} - \left[u_{xx} - \frac{1}{c_0^2} u_{tt} \right]_{t^*} = 0, \tag{B–19}$$

where

$$\delta_{\text{th}} = \frac{\kappa \omega}{\rho_0 c_0^2 C_p} \tag{B–20}$$

is the dimensionless thermal conduction coefficient. The notation δ_{th} has been used to call attention to the fact that δ_{th} plays a role similar to that played by the dimensionless viscosity coefficient δ_{v} (see Eq. B–5). Inspection of Eq. B–19 shows that if $\delta_{\text{th}} \ll 1$ (poor heat conduction or low frequency), the first bracket of terms in Eq. B–19 is negligible. What remains (after an integration with respect to t) is

$$u_{xx} - \frac{1}{c_0^2} u_{tt} = 0,$$

which is the ordinary lossless wave equation. The propagation speed is the *adiabatic* value $c_0 = \sqrt{\gamma p_0 / \rho_0}$. On the other hand, if $\delta_{\text{th}} \gg 1$ (high heat conduction or very high frequency), the second bracket of terms in Eq. B–19 may be dropped. The result (after two integrations with respect to x) is

$$u_{xx} - \frac{\gamma}{c_0^2} u_{tt} = 0,$$

again a lossless wave equation, but now the propagation speed is $c_0/\sqrt{\gamma} = \sqrt{p_0/\rho_0}$. This is the isothermal speed of sound (see Sec. 1C.3.d).[8] It is physically plausible that heat conduction should cause sound to propagate isothermally at very high frequencies. When the wavelength is very short, if there were a variation in temperature from peak to trough (see Fig. 9.4), the

[7]Fully dimensionless form, where x^* and t^* would replace x and t inside the brackets as well as outside, is unnecessary. In order that the relative importance of the two bracketed expressions may be compared, all that is necessary is that their dimensions be the same.

[8]Notice that if the temperature is constant, $T = T_0$, differentiation of the ideal gas law yields the following expression for the sound speed: $dP/d\rho = RT_0 = p_0/\rho_0 = c_0^2/\gamma$.

temperature gradient T_x would be so great that the corresponding heat flow (see Eq. B–13) would immediately quench the temperature variation. At very high frequency, therefore, the effect of heat conduction is to prevent the temperature from changing, or in other words to impose isothermal conditions.

Although the previous reasoning seems plausible and provides a useful background, only the low frequency analysis is correct. The high frequency analysis turns out to be valid only for completely inviscid fluids. The presence of only the slightest amount of viscosity produces a drastic change in the high frequency behavior. The actual phase speed asymptote at high frequency is no longer the isothermal sound speed. A brief explanation is given at the end of the next section. However, the actual "high frequency" behavior is rarely an issue because, as we see next, most acoustical phenomena lie in the "low frequency" regime.

For most fluids δ_{th} is very small even for very high frequencies. For example, for 20°C air at 1 atm we find

$$\delta_{\text{th}_{\text{air}}} = 1.13 \times 10^{-9} f,$$

where f is in hertz. Thus even at frequencies up to the gigahertz range the second bracket of terms in Eq. B–19 is heavily dominant. The applicable sound speed for "ordinary acoustics" is therefore the adiabatic value $c_0 = \sqrt{\gamma p_0/\rho_0}$.

Let us now find the dispersion relation for a heat-conducting fluid. Substitution of a time-harmonic solution, Eq. A–1, in Eq. B–19 leads to

$$j\delta_{\text{th}} K^4 - (1 + j\delta_{\text{th}}\gamma)K^2 + 1 = 0, \tag{B–21}$$

where $K = kc_0/\omega$ is the dimensionless wave number. This is a quartic equation and has two pairs of roots. First solve for K^2:

$$K_1^2 = \frac{(1 + j\gamma\delta_{\text{th}}) - \sqrt{(1 + j\gamma\delta_{\text{th}})^2 - 4j\delta_{\text{th}}}}{j2\delta_{\text{th}}}, \tag{B–22a}$$

$$K_2^2 = \frac{(1 + j\gamma\delta_{\text{th}}) + \sqrt{(1 + j\gamma\delta_{\text{th}})^2 - 4j\delta_{\text{th}}}}{j2\delta_{\text{th}}}. \tag{B–22b}$$

Both K_1^2 and K_2^2 have positive and negative roots, which in each case signify outgoing and incoming waves, respectively. The two pairs of roots are associated with two entirely different types of waves. The roots $\pm K_1$ pertain to ordinary sound waves. The waves associated with the roots $\pm K_2$ are sometimes called thermal waves because they are very similar to wavelike solutions of the heat conduction equation $\kappa T_{xx} - T_t = 0$. At ordinary frequencies these waves damp out so rapidly as to be of little consequence (thermal waves are, however, important in liquid helium), and we shall not consider them further.[9] Taking the root $+K_1$, which corresponds to an outgoing wave, we have

$$K_1 = \sqrt{\frac{(1 + j\gamma\delta_{\mathrm{th}}) - \sqrt{(1 + j\gamma\delta_{\mathrm{th}})^2 - 4\delta_{\mathrm{th}}}}{j2\delta_{\mathrm{th}}}}.$$

For small values of δ_{th} (i.e., "low" frequencies), this complicated expression may be expanded as follows (much care must be used):

$$K_1 = 1 - j\tfrac{1}{2}(\gamma - 1)\delta_{\mathrm{th}} + O(\delta_{\mathrm{th}}^2),$$

where $O(\delta_{\mathrm{th}}^2)$ means "terms of order δ_{th}^2 and higher." If the higher-order terms are dropped, the (dimensional) expression for the wave number is

$$k_1 = \frac{\omega}{c_0} - j\frac{(\gamma - 1)\kappa\omega^2}{2\rho_0 c_0^3 C_p}. \qquad \text{(B–23)}$$

Comparison with Eq. A–2 shows that the wave is nondispersive ($c^{\mathrm{ph}} = c_0$), at least to order δ_{th}, and the absorption coefficient is

$$\alpha_{\mathrm{th}} = \frac{(\gamma - 1)\kappa\omega^2}{2\rho_0 c_0^3 C_p}, \qquad \text{(B–24)}$$

where the subscript "th" denotes thermal conduction.

Both the viscous and thermal absorption coefficients are seen to be proportional to f^2. For most gases the two coefficients are numerically similar as well. Introduction of the Prandtl number facilitates the comparison:

$$\mathrm{Pr} = \frac{\mu C_p}{\kappa}, \qquad \text{(B–25)}$$

which is a dimensionless number used to characterize the importance of viscosity to that of heat conduction. Because of the fact that

$$\delta_{\mathrm{th}} = \frac{\delta_{\mathrm{v}}}{\mathrm{Pr}},$$

α_{th} may be expressed as

[9]Thermal waves make it possible to satisfy temperature boundary conditions. For example, the face of a vibrating metal piston is not only a place where the particle velocity is specified but also a place where the temperature remains constant. Unless the thermal wave is included, the temperature boundary condition has to be ignored.

$$\alpha_{\text{th}} = \frac{\gamma - 1}{\text{Pr}} \frac{\omega^2 \nu}{2c_0^3}. \tag{B–26}$$

The ratio of α_{th} to α_{v} is

$$\frac{\alpha_{\text{th}}}{\alpha_{\text{v}}} = \frac{\gamma - 1}{\text{Pr}\, \tilde{V}}. \tag{B–27}$$

Since for most gases, Pr, $\tilde{V}$, and $\gamma - 1$ are of order unity, the absorptions due to heat conduction and viscosity are comparable. For most liquids (except mercury), on the other hand, the Prandtl number is large and γ is close to unity. Absorption due to heat conduction in liquids is therefore generally negligible compared to viscous absorption.

We now give a short-cut method for obtaining the absorption and dispersion due to heat conduction. Since δ_{th} is normally very small, the first bracket of terms in Eq. B–18 has only a small effect. Dropping the first bracket entirely yields a "zero-order approximation," namely, $u_{xx} - u_{tt}/c_0^2 = 0$. A better approximation is found by retaining the first bracket but modifying it according to the zero-order approximation. In other words, replace u_{xx} *in the first bracket only* by u_{tt}/c_0^2. Equation B–18 becomes

$$\frac{\kappa}{\rho_0 C_p} \left[\frac{\gamma - 1}{c_0^2} u_{tt} \right]_{xx} + \left[u_{xx} - \frac{1}{c_0^2} u_{tt} \right]_t = 0.$$

Integration with respect to t yields

$$\frac{(\gamma - 1)\kappa}{\rho_0 c_0^2 C_p} u_{xxt} + u_{xx} - \frac{1}{c_0^2} u_{tt} = 0, \tag{B–28}$$

which has the same form as the viscous wave equation, Eq. B–3. By inspection, therefore, Eq. B–24 follows.

Notice that the "short-cut" approximation reduces the order (meaning the number of differentiations with respect to x) of the wave equation from fourth (Eq. B–18) to second (Eq. B–28). The price paid for simplifying in this way is loss of the thermal waves (and any possibility of satisfying temperature boundary conditions). Because the thermal waves are usually not very important, however, the price is cheap.

3. Thermoviscous Fluids

Now consider the combined effects of heat condition and viscosity. The wave equation for thermoviscous fluids may be found by combining Eqs. B–1a and B–1b and Eq. B–17 (the latter holds even for viscous fluids because, as already pointed out, the viscosity term in the full-energy equation is nonlinear). The result is

$$\frac{\gamma \nu^2 \tilde{V}}{\mathrm{Pr}\, c_0^2} u_{xxxxt} + \nu \left[\frac{u_{xx}}{\mathrm{Pr}} - \left(\frac{\tilde{V} + \gamma}{\mathrm{Pr}} \right) \frac{u_{tt}}{c_0^2} \right]_{xx} - \left[u_{xx} - \frac{u_{tt}}{c_0^2} \right]_t = 0. \tag{B–29}$$

Although this equation appears formidable, it may easily be dealt with by the short-cut method described at the end of the previous section. First assess the relative importance of the various terms. The ν^2/Pr factor in front of the first term implies that the relative order of this term is $\delta_{\mathrm{v}}\delta_{\mathrm{th}}$. The factor ν in front of the first bracket implies that the terms inside the bracket are of relative order δ_{v} (if the Prandtl number does not appear as a coefficient) or δ_{th} (if the Prandtl number does appear). The second bracket of terms, having no small coefficient in front of it, is of relative order unity. The ordering scheme allows us to make the following physical interpretation: the first term represents the interaction of viscous and thermal effects, while the first set of bracketed terms represents their superposition. Since both δ_{v} and δ_{th} are usually very small, their product is so small that the interaction term may be dropped. Next, following the short-cut method described previously, replace u_{xx} by u_{tt}/c_0^2 in the first bracket of terms. Combining terms and integrating once with respect to time yield

$$\frac{\nu}{c_0^2} \left[\tilde{V} + \frac{\gamma - 1}{\mathrm{Pr}} \right] u_{xxt} + u_{xx} - \frac{1}{c_0^2} u_{tt} = 0. \tag{B–30}$$

Once again the form of the simple viscous wave equation, Eq. B–3, is recovered. It follows that the absorption coefficient for a thermoviscous fluid is

$$\alpha_{\mathrm{tv}} = \frac{\omega \delta_{\mathrm{v}}}{2c_0} \left[\tilde{V} + \frac{\gamma - 1}{\mathrm{Pr}} \right] = \frac{\omega^2 \nu}{2c_0^3} \left[\left(\frac{4}{3} + \frac{\mu_B}{\mu} \right) + \frac{\gamma - 1}{\mathrm{Pr}} \right]. \tag{B–31}$$

Again, the dispersion is negligible (at least to order δ_{v}). To a very good approximation, therefore, the effects of viscosity and heat conduction are simply additive in causing absorption of sound, that is, $\alpha_{\mathrm{tv}} = \alpha_{\mathrm{th}} + \alpha_{\mathrm{v}}$.

Finding a numerical value from Eq. B–31 requires that the term in brackets be known. Many writers refer to Eq. B–31 with μ_B set to zero as the "classical absorption,"

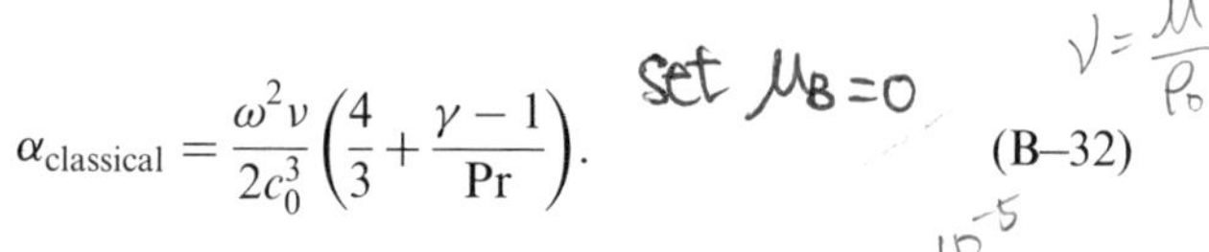

$$\alpha_{\mathrm{classical}} = \frac{\omega^2 \nu}{2c_0^3} \left(\frac{4}{3} + \frac{\gamma - 1}{\mathrm{Pr}} \right). \tag{B–32}$$

For 20°C air at 1 atm, for example, use of the values $\nu = 1.50 \times 10^{-6}$ m^2/s, $\mathrm{Pr} = 0.711$, $\gamma = 1.402$, and $c_0 = 343$ m/s yields $\alpha_{\mathrm{classical}} = 1.39 \times 10^{-11} f^2$, where f is in hertz. The accepted high frequency asymptotic expression for air absorption under these conditions (where relaxation is not important; see Appendix B,

Heat conduction 不重要(在 liquid) 對於 absorption

Eqs. B–1 and B–5c) is $\alpha = 1.84 \times 10^{-11} f^2$, a figure 32% higher. The discrepancy may be attributed to the fact that μ_B is not zero for air. To obtain an estimate of the true value of μ_B, match Eq. B–31 with the accepted high frequency asymptotic expression for air. The result is $\tilde{V} + (\gamma - 1)/\text{Pr} = 2.51$, or

$$\tilde{V}_{\text{air}} = 1.94 \qquad \text{and} \qquad \mu_B = 0.61\mu. \tag{B–33}$$

By analyzing Greenspan's data for absorption of air (Ref. 8), Pierce (Ref. 14) concludes that $\mu_B = 0.60\mu$. See the next section for an interpretation of bulk viscosity in terms of relaxation.

It has already been noted that heat conduction is not an important cause of absorption in liquids. For fresh water in particular the viscosity appears to be the principal mechanism. Experiments show that $\alpha_{\text{water}} = \text{const.} \times f^2$. The values of the constant are tabulated as a function of temperature (see, for example, Ref. 15). A value often used at 20°C is 25×10^{-15} nepers/m·Hz, in good agreement with the coefficient of the "pure water" component of absorption for seawater (see Appendix B, Eqs. B–7, B–9c, and B–10c). Use of this value to determine the bulk viscosity coefficient for water is left as an exercise (see Problem 9–5).

Finally, although the interaction term, i.e., the order $\delta_v \delta_{th}$ term in Eq. B–29, is not of much practical importance, it has some theoretical interest. First, if the effect of viscous-thermal interaction is to be considered, this term must be included. Second, the presence of the term voids the tentative conclusion reached in the previous section that sound propagation is isothermal in the high frequency limit. To investigate the behavior of Eq. B–29 at very high frequency, let both δ_v and δ_{th} be large compared to unity. In this case the second bracket of terms is negligible and may be dropped. Observe that if what remains is to reduce to the first bracket of terms in Eq. B–19 (from which we deduced that $c^{ph} = c_0/\sqrt{\gamma}$ at very high frequency), the viscosity must vanish entirely. Since no fluid is completely inviscid, the deduction that the sound speed at very high frequency is isothermal must be abandoned. A mathematical way to come to this conclusion is to solve the dispersion relation for Eq. B–29 for the case in which both δ_v and δ_{th} are very large.

C. RELAXATION

1. Introduction

One of our assumptions heretofore has been that the fluid is in local thermodynamic equilibrium. That is, when an equation of state, such as

$$P = P(\rho, T),$$

is used, an inherent assumption is that the pressure at any instant at any place in the fluid depends only on the local, instantaneous values of density and temperature. In particular, the pressure does not depend on the rate at which the density and temperature are changing. In a relaxing fluid, however, there is a rate dependence. Consider, for example, a chemically reacting fluid made up of two constituents A and B. In equilibrium, A constitutes a fraction ξ of the fluid, B the remainder. If conditions change, for example, if the pressure or temperature is increased, the equilibrium is upset and the fraction ξ tends to increase (or decrease) until a new equilibrium is reached. The quantity ξ is thus a new thermodynamic variable that must be included in the equation of state, for example,

$$P = P(\rho, T, \xi).$$

Associated with ξ is the relaxation time τ, which is the time needed for a new equilibrium to be established when a change occurs.

A rather clear physical picture of the effect of changing equilibrium on sound propagation was given in a 1920 article by Einstein (Ref. 4). Einstein, treating the problem of propagation in partially dissociated gases, made the following introductory remarks (p. 268 in English translation):

> If we change the volume of a partially dissociated gas adiabatically so fast that during the time of volume change practically no chemical transformation takes place, the gas behaves under these circumstances like an ordinary mixture. If, on the other hand, the volume is changed so slowly that the process consists practically entirely of chemical equilibrium states, the dependence of the pressure on the density will be a different one in such a way that the compressibility of the mixture is less than in the first case. The velocity of sound must therefore increase from an initial value to a limiting value as the frequency increases. For frequencies which lie between the two extremes, the reaction will lag behind the increase in density in such a fashion that there takes place a kind of temporal lag of the pressure curve with respect to the density curve with the concomitant transformation of mechanical energy into heat.

It hardly needs to be added that if the volume changes are due to passage of a sound wave, the "transformation of mechanical energy into heat" represents a loss to the sound wave.

Although the discussion above is for a chemically reacting fluid, it applies equally well to a fluid composed of molecules whose internal energy is associated with various modes of motion of the molecules. The following passage is taken from a review paper by Johannesen and Hodgson (Ref. 10, p. 634):

> When a pure diatomic or polyatomic gas or a mixture of gases is in thermodynamic equilibrium at meteorological temperatures the internal energy is distributed over the classical modes of translation and rotation, which are fully excited, and over the vibrational modes which have temperature-dependent specific heats.

> When the equilibrium is disturbed the various energy modes require vastly different times to readjust themselves to new equilibrium values. Under typical conditions for air these times may be of the order of 10^{-10} s, 10^{-9} s, 10^{-5} s and the 10^{-3}s for translation, rotation, oxygen vibration and nitrogen vibration, respectively, although the times for vibration adjustment are strongly humidity-dependent.

Translational and rotational molecular motion is often linked to bulk viscosity; see, for example, Ref 14. The relaxation times $\sim 10^{-5}$ s for oxygen vibration and $\sim 10^{-3}$ s for nitrogen vibration are the ones of most importance in determining sound absorption in air at audio and low ultrasonic frequencies. As noted in Sec. A above, the two important relaxation processes for seawater are those for magnesium sulfate (relaxation time $\sim 10^{-5}$ s) and boric acid (relaxation time $\sim 10^{-3}$ s); see, for example, Refs. 6 and 7.

In our analytical treatment here we shall for simplicity assume a fluid subject to a single relaxation process. Later we shall account for multiple relaxations by assuming that the absorption coefficients due to the several relaxations are simply additive.

2. Equation of State

The equation of state for a relaxing fluid (inviscid and thermally nonconducting) is given here without proof:

$$\tau(p - c_\infty^2 \delta\rho)_t + (p - c_0^2 \delta\rho) = 0, \tag{C–1}$$

where c_∞ and c_0 are the sound speeds at infinite frequency and zero frequency, respectively (notice that the symbol c_0 now has a special meaning not previously used). To explain the subscripts "∞" and "0," perform a simple analysis of Eq. C–1, rewritten in terms of a dimensionless time $t^* = \omega t$:

$$\omega\tau(p - c_\infty^2 \delta\rho)_{t^*} + (p - c_0^2 \delta\rho) = 0. \tag{C–2}$$

In the limit as $\omega\tau \to \infty$ (infinite frequency or infinite relaxation time), the last pair of terms of Eq. C–2 becomes negligible. Integration with respect to time yields

$$p - c_\infty^2 \delta\rho = 0,$$

which is the pressure-density relation for an acoustic wave having a propagation speed c_∞. Sound speed c_∞ is thus associated with propagation at infinitely high frequency. When, on the other hand, $\omega\tau \to 0$ (zero frequency or zero relaxation time), the first pair of terms in Eq. C–2 drops out. The remaining expression,

$$p - c_0^2 \delta \rho = 0,$$

is the relation for a wave that propagates with speed c_0.

Two other terms, *frozen sound speed* and *equilibrium sound speed*, are sometimes used for c_∞ and c_0, respectively. "Frozen" implies that, in terms of the chemically reacting fluid model initially described, the frequency of the sound wave is so high that pressure and density change too rapidly for the percentages of constituents A and B to change during a half period. The fraction ξ is frozen. The tendency is still for the amount of A to increase and that of B to decrease, or vice versa, but the half period is too short for any appreciable change to take place. At the opposite end of the scale, at very low frequencies, the change in pressure and density is so slow that the chemical reaction always carries to completion. Equilibrium is reestablished after each infinitesimal pressure change in the acoustical cycle. Because equilibrium always exists, c_0 is called the equilibrium sound speed.

3. Wave Equation

For simplicity assume that the fluid is inviscid and thermally nonconducting. In this case the continuity and momentum equations are Eqs. B–1a,b (with λ, $\mu = 0$), respectively. Elimination of the particle velocity between these two equations leads to

$$\delta \rho_{tt} = p_{xx}. \tag{C–3}$$

If Eq. C–1 is differentiated twice with respect to time and Eq. C–3 used to eliminate $\delta\rho$, one obtains

$$\tau\left(p_{tt} - c_\infty^2 p_{xx}\right)_t + \left(p_{tt} - c_0^2 p_{xx}\right) = 0. \tag{C–4}$$

This is the wave equation for a relaxing fluid.

4. Dispersion Relation

To find the absorption and dispersion, substitute $p = Ae^{j(\omega t - kx)}$ in Eq. C–4 and obtain $j\omega\tau(-\omega^2 + c_\infty^2 k^2) + (-\omega^2 + c_0^2 k^2) = 0$, or

$$K^2 = \frac{1 + j\omega\tau}{1 + j\omega\tau(c_\infty/c_0)^2}, \tag{C–5}$$

where $K = k/(\omega/c_0)$ is the dimensionless wave number based on the equilibrium sound speed. If a quantity called the *dispersion m* is introduced,

$$m \equiv \frac{c_\infty^2 - c_0^2}{c_0^2} = \frac{c_\infty^2}{c_0^2} - 1, \tag{C–6}$$

the expression for K becomes

$$K = \pm\sqrt{\frac{1 + j\omega\tau}{1 + j\omega\tau(1+m)}}. \tag{C–7}$$

The plus sign is used for outgoing waves. The task now is to separate K into its real and imaginary parts,

$$K = R - jI,$$

which are related to the absorption coefficient and phase velocity by

$$R = c_0/c^{\text{ph}}, \tag{C–8a}$$

$$I = \alpha\lambda/2\pi, \tag{C–8b}$$

where $\lambda = c_0/f$ is the wavelength based on the equilibrium sound speed. The dimensionless absorption coefficient $\alpha\lambda$ may be thought of as the absorption over a wavelength.

An exact analysis is first carried out. After introducing the expression for K in Eq. C–7, square the equation and rationalize the right-hand side:

$$(R^2 - I^2) - j2RI = \frac{[1 + \omega^2\tau^2(1+m)] - j\omega\tau m}{1 + \omega^2\tau^2(1+m)^2} = \frac{N_r - jN_i}{\Delta},$$

where the new symbols N_r, N_i, and Δ are defined by the equation. Next equate real and imaginary parts on either side of the equation,

$$R^2 - I^2 = N_r/\Delta, \qquad 2RI = N_i/\Delta,$$

and solve for I and R:

$$I = \left(\frac{\sqrt{N_r^2 + N_i^2} - N_r}{2\Delta}\right)^{1/2}, \tag{C–9a}$$

$$R = N_i/2\Delta I. \tag{C–9b}$$

Equations C–8a,b and C–9a,b constitute the formal solution of the problem. The actual expressions for α and c^{ph} obtained from these formulas are, however, quite complicated.

Fortunately, for most fluids c_∞ and c_0 are so close together, i.e., $m \ll 1$, that simpler but still very accurate expressions for α and c^{ph} may be obtained. First notice that N_i/N_r is of order m regardless of the value of $\omega\tau$. Thus for $m \ll 1$ an appropriate expansion of Eq. C–9a is

$$I = \sqrt{\frac{N_r}{2\Delta}}\left[1 + \frac{1}{2}\left(\frac{N_i}{N_r}\right)^2 - \frac{1}{8}\left(\frac{N_i}{N_r}\right)^4 \cdots - 1\right]^{1/2}.$$

If only the leading term is retained, the result is

$$I = N_i/2\sqrt{N_r\,\Delta}. \tag{C–10a}$$

The expression for R now reduces to

$$R = \sqrt{N_r/\Delta}. \tag{C–10b}$$

Application of Eqs. C–8a,b then leads to

$$\alpha\lambda = \frac{\pi\omega\tau m}{\sqrt{[1+\omega^2\tau^2(1+m)^2][1+\omega^2\tau^2(1+m)]}}, \tag{C–11a}$$

$$\frac{c^{\text{ph}}}{c_0} = \sqrt{\frac{1+\omega^2\tau^2(1+m)^2}{1+\omega^2\tau^2(1+m)}}. \tag{C–11b}$$

The expression for $\alpha\lambda$ may be further simplified. Because of the nature of the approximation used so far, the expression for $\alpha\lambda$ is valid only to $O(m)$. This means that it is consistent to ignore m completely in the denominator of Eq. C–11a. The final result is

$$\alpha\lambda = \frac{\pi\omega\tau m}{(1+\omega^2\tau^2)}. \tag{C–12}$$

It is not, however, appropriate to make the same approximation in Eq. C–11b because the result would be $c^{\text{ph}} = c_0$ independent of $\omega\tau$.

The first step in analyzing the results (Eqs. C–11b and C–12) is to consider the two frequency extremes, $\omega\tau \ll 1$ and $\omega\tau \gg 1$. At very low frequencies the expressions are

$$\alpha = m\omega^2\tau/2c_0 \qquad \text{and} \qquad c^{\text{ph}} = c_0.$$

The absorption coefficient thus varies as the square of the frequency, just as in the case of thermoviscous absorption, and the sound speed is the equilibrium value. At very high frequencies, on the other hand, we obtain

$$\alpha = \frac{m}{2c_0\tau(1+m)^{3/2}} = \frac{mc_0^2}{2c_\infty^3\tau} \qquad \text{and} \qquad c^{\text{ph}} = c_\infty.$$

The absorption coefficient is independent of frequency (see the horizontal asymptote of curve r in Fig. 9.2), and the sound speed has the frozen value.

A more detailed analysis is now given. Figure 9.6 shows a plot of Eq. C–12. Experimental data are often displayed in this manner, that is, $\alpha\lambda$ rather than α itself is plotted against frequency. The reason is that the $\alpha\lambda$ curve has a maximum, which may readily be identified. In fact, the location ($\omega = \omega_r$) of the maximum and the value there ($\alpha\lambda_{\text{max}}$) allows τ and m to be determined. Setting $d(\alpha\lambda)/d\omega = 0$, one finds

$$\omega_r = 1/\tau \qquad \text{and} \qquad \alpha\lambda_{\text{max}} = \pi m/2.$$

The expression for α may thus be written in the form

$$\alpha = \frac{\alpha\lambda_{\text{max}}}{\pi c_0}\,\frac{\omega_r\omega^2}{\omega^2+\omega_r^2},$$

or

$$\alpha = A\frac{f_r f^2}{f^2+f_r^2}, \tag{C–13}$$

where $A = 2\alpha\lambda_{\text{max}}/c_0$ and $f_r = (2\pi\tau)^{-1}$ is called the relaxation frequency. Equation C–13 is the form for α often seen in the literature on relaxation. See, for example, Appendix B.

Phase velocity vs. frequency, Eq. C–11b, is plotted in Fig. 9.7. At the relaxation frequency, $(c^{\text{ph}})^2$ is, for $m \ll 1$, the arithmetic mean of c_0^2 and c_∞^2.

As an example, consider air, which has two relaxation processes, curves for which are shown in Fig. 9.3*a*. At a temperature of 20°C the difference between c_∞ and c_0 is only 2.17 cm/s ($m = 1.26 \times 10^{-4}$) for nitrogen and only 11.5 cm/s ($m = 6.71 \times 10^{-4}$) for oxygen. Despite the small amount of dispersion, the two

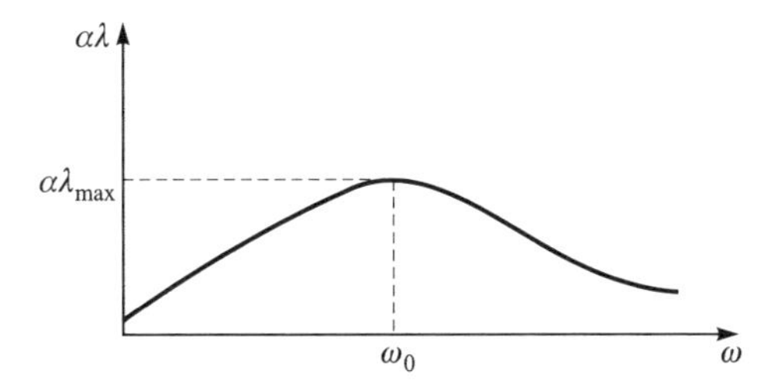

Figure 9.6 Dimensionless absorption coefficient curve.

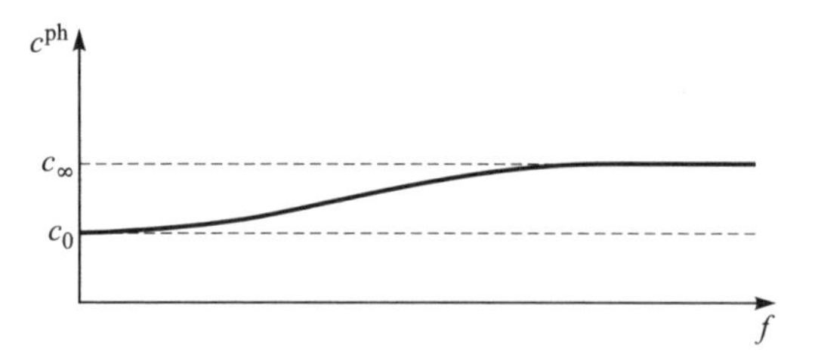

Figure 9.7 Phase velocity in a relaxing fluid.

relaxation processes are responsible for most of the sound absorption in the audio and low ultrasonic frequency region. As Appendix B shows, the shape of the absorption coefficient vs frequency curve depends heavily on the values of the two relaxation frequencies, $f_{r,\mathrm{N}}$ and $f_{r,\mathrm{O}}$, which in turn are strongly dependent on the water vapor content of the air. For example, at a relative humidity of 30% and a temperature of 20°C the relaxation frequencies are $f_{r,\mathrm{N}} \doteq 250$ Hz and $f_{r,\mathrm{O}} \doteq 21$ kHz; at 70% relative humidity the two values rise to about 575 Hz and 59 kHz, respectively (Refs. 1 and 9). Appendix B gives formulas for calculating $f_{r,\mathrm{N}}$, $f_{r,\mathrm{O}}$, and the total absorption coefficient.

D. BOUNDARY-LAYER ABSORPTION (AND DISPERSION)

A completely different mode of absorption, that due to passage of a wave over a surface boundary, is described here. Because of viscosity, the surface exerts a frictional shear force on the overlying fluid. At the same time, thermal conduction allows heat transfer to take place between the fluid and the surface.

1. Physical Phenomenon: Viscous Boundary Layer

Consider propagation of time-harmonic sound waves (angular frequency ω) through a pipe. Although fluid particles in the mainstream (i.e., away from the pipe wall) oscillate back and forth in response to passage of the sound wave, the layer of particles adjacent to the wall adheres to the wall. A transition must therefore take place where the particle oscillation amplitude decreases from its nominal value in the mainstream to zero at the wall, as shown in Fig. 9.8. The transition region is called the *acoustic boundary layer*. If the fluid were inviscid, it could "slip" past the wall perfectly, i.e., the flow would be frictionless and the boundary layer thickness would be zero. The presence of viscosity causes the boundary layer to be finite though nevertheless quite thin. The thickness of the viscous acoustic boundary layer is (Appendix C)

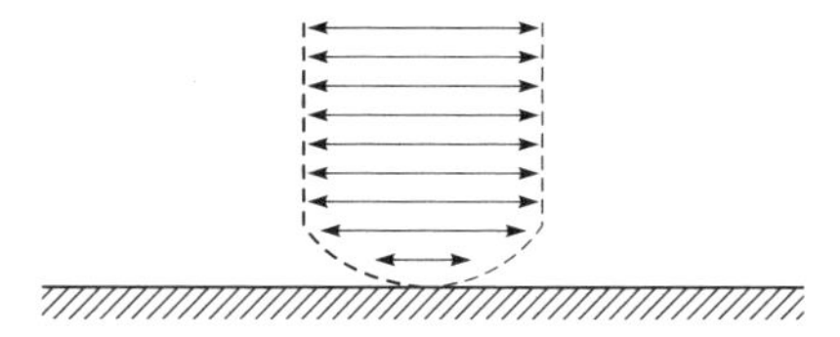

Figure 9.8 Viscous boundary layer for oscillating flow.

$$\delta_{\text{visc}}^{\text{BL}} = \sqrt{\frac{2\mu}{\omega\rho_0}} = \sqrt{\frac{2\nu}{\omega}}, \tag{D–1}$$

which is the "$1/e$" distance from the wall, that is, the distance required for the particle velocity amplitude to increase from zero at the wall to $1/e$ of its mainstream value. Notice the dependence on frequency. At 100 Hz in air, for example, the thickness is about 0.2 mm. In pipes of ordinary size, therefore, the acoustic boundary layer takes up very little of the cross section of the pipe.[10]

The viscous acoustic boundary layer may be visualized. In an experiment reported by Meyer and Güth (Ref. 12), electrically charged oil drops were used as tracer particles to image the motion in the boundary layer. As the droplets oscillated back and forth in a horizontal sound field above a rigid surface, at the same time they were swept down to the surface by an electric field. The zigzag path traced out by the particles provided a graphic picture of the boundary layer. The experiment may be thought of as the Millikan oil drop experiment with a sound field added.

2. Thermal Boundary Layer

Coexisting with the viscous boundary layer is a thermal boundary layer. In the mainstream the compressions and expansions take place adiabatically. Near the pipe wall, however, the flow tends to be isothermal, not adiabatic. Because the wall is effectively an infinite heat source or sink, adjacent fluid particles stay at the wall temperature (any temperature change in the adjacent fluid would immediately be quenched by heat flow into or out of the wall). The region near the wall where the flow changes character from adiabatic to isothermal is called the *thermal* boundary layer. The thickness $\delta_{\text{therm}}^{\text{BL}}$ of this layer is comparable to that of the viscous boundary layer:

[10]Notice that the acoustic boundary layer is qualitatively different from the ordinary viscous boundary layer that develops in a pipe when the flow through the pipe is steady (unidirectional). The thickness of the steady flow boundary layer grows with the distance, eventually filling the entire cross section of the pipe. The acoustic boundary layer, on the other hand, cannot grow very much because the fluid never flows in one direction for more than half a period.

$$\delta_{\text{therm}}^{\text{BL}} = \sqrt{\frac{2\kappa}{\rho_0 \omega C_p}} = \sqrt{\frac{2\mu}{\omega \rho_0 \, \text{Pr}}} = \frac{\delta_{\text{visc}}^{\text{BL}}}{\sqrt{\text{Pr}}}. \tag{D–2}$$

3. Effect of the Two Boundary Layers

The shear and thermal losses associated with the boundary layers give rise to absorption and dispersion of sound. Their effects on the sound wave in the mainstream may be modeled in one-dimensional fashion by letting the control volume (see, for example, Fig. 1.21) contain just the fluid in the mainstream, that is, outside the boundary layer. The effect of the viscous boundary layer is to exert a shear force on the side surface of the control volume. Similarly, the effect of the thermal boundary layer (see Fig. 9.5) is to transfer heat to the control volume through the side surface. In this way one-dimensional models of the momentum and energy equations may be developed. The following wave equation may then be derived (see Appendix C):

$$u_{xx} - \frac{1}{c_0^2} u_{tt} = B \int_0^\infty u_{xx}(x, t - \tau) \frac{d\tau}{\sqrt{\tau}}, \tag{D–3}$$

where

$$B = \frac{4}{\text{HD}} \sqrt{\frac{\mu}{\pi \rho_0}} \left(1 + \frac{\gamma - 1}{\sqrt{\text{Pr}}} \right). \tag{D–4}$$

Here $\text{HD} = 4S/C$ is the hydraulic diameter, and S and C are the area and perimeter, respectively, of the cross section. The hydraulic diameter is introduced in order that Eq. D–3 apply to ducts of arbitrary cross section, e.g., round, square, or rectangular. For a round pipe of radius a, for example, $\text{HD} = 2a$, and B is given by

$$B = \frac{2}{a} \sqrt{\frac{\mu}{\pi \rho_0}} \left(1 + \frac{\gamma - 1}{\sqrt{\text{Pr}}} \right), \tag{D–5}$$

The derivation of Eq. D–3 is based on the assumption that the boundary layer is (1) small compared to the effective radius of the duct (HD/2), but (2) not so small that mainstream thermoviscous losses are important. In quantitative terms the restriction is

$$\delta_{\text{visc}}^{\text{BL}} \ll \frac{\text{HD}}{2} \ll \frac{c_0^2}{\omega^2 \delta_{\text{visc}}^{\text{BL}}}. \tag{D–6}$$

Substitution of a time-harmonic solution in Eq. D–3 leads to the following dispersion relation:

$$(-jk)^2 + \frac{\omega^2}{c_0^2} = B(-jk)^2 \underbrace{\int_0^\infty e^{-j\omega\tau} \frac{d\tau}{\sqrt{\tau}}}_{\sqrt{\pi/j\omega}},$$

which may be solved for k,

$$k = \pm \frac{\omega/c_0}{\sqrt{1 - B\sqrt{\pi/j\omega}}}. \tag{D–7}$$

Expand and use the plus sign for outgoing waves:

$$k \doteq \frac{\omega}{c_0}\left[1 + \frac{B}{2}\sqrt{\frac{\pi}{j\omega}} + \cdots\right]. \tag{D–8}$$

For a round tube (B given by Eq. D–5) the expression is

$$k \doteq \frac{\omega}{c_0}\left[1 + (1-j)\,\frac{1}{a}\,\sqrt{\frac{\mu}{2\rho_0\omega}}\left(1 + \frac{\gamma - 1}{\sqrt{\mathrm{Pr}}}\right)\right], \tag{D–9a}$$

from which we deduce

$$\alpha = \frac{1}{a}\sqrt{\frac{\omega\mu}{2\rho_0 c_0^2}}\left(1 + \frac{\gamma - 1}{\sqrt{\mathrm{Pr}}}\right) \tag{D–9b}$$

and

$$c^{\mathrm{ph}} = \frac{\omega}{\beta} = \frac{c_0}{1 + \alpha c_0/\omega} \doteq c_0\left(1 - \frac{\alpha c_0}{\omega}\right). \tag{D–9c}$$

Thus α varies as $\sqrt{\omega}$, and $c^{\mathrm{ph}} \to c_0$ at high frequency but is less than c_0 at low frequencies.

E. SUMMARY OF SOUND ABSORPTION IN FLUIDS

1. Viscous Fluids

(a) Wave equation:

$$\frac{\tilde{V}\mu}{\rho_0 c_0^2} u_{xxt} + u_{xx} - \frac{1}{c_0^2} u_{tt} = 0\,, \quad \tilde{V} = \frac{\lambda + 2\mu}{\mu} = \frac{4}{3} + \frac{\mu_B}{\mu}.$$

(b) Absorption: $\alpha = \tilde{V}\mu\omega^2/2\rho_0 c_0^3$ (for "low" frequencies).[11]
(c) Dispersion: $c^{\text{ph}} = c_0$ (for "low" frequencies).

2. Thermally Conducting Fluids

(a) Wave equation:

$$\frac{\kappa}{\rho_0 C_p}\left(u_{xx} - \frac{\gamma}{c_0^2} u_{tt}\right)_{xx} - \left(u_{xx} - \frac{1}{c_0^2} u_{tt}\right)_t = 0.$$

Approximate wave equation:

$$\frac{(\gamma - 1)\kappa}{\rho_0 c_0^2 C_p}\, u_{xxt} + u_{xx} - \frac{1}{c_0^2}\, u_{tt} = 0.$$

(b) Absorption: $\alpha = (\gamma - 1)\kappa\omega^2/2\rho_0 c_0^3 C_p$ (for "low" frequencies).
(c) Dispersion: $c^{\text{ph}} = c_0$ (for "low" frequencies).

3. Thermoviscous Fluids

(a) Wave equation:

$$\frac{\tilde{V}\mu}{\rho_0 c_0^2}\frac{\gamma\kappa}{C_p}\, u_{xxxxt} + \frac{\kappa}{\rho_0 C_p}\, u_{xxxx} - \left(\frac{\tilde{V}\mu}{\rho_0 c_0^2} + \frac{\gamma\kappa}{c_0^2 C_p}\right) u_{xxtt} - \left(u_{xx} - \frac{1}{c_0^2} u_{tt}\right)_t = 0.$$

Approximate wave equation:

$$\left(\tilde{V} + \frac{\gamma - 1}{\text{Pr}}\right)\frac{\mu}{\rho_0 c_0^2} u_{xxt} + u_{xx} - \frac{1}{c_0^2} u_{tt} = 0, \quad \text{Pr} = \frac{\mu C_p}{\kappa}.$$

(b) Absorption: $\alpha = \left[(\mu\omega^2)/(2\rho_0 c_0^3)\right]\left(\tilde{V} + (\gamma - 1)/\text{Pr}\right)$ (for "low" frequencies).
(c) Dispersion: $c^{\text{ph}} = c_0$ (for "low" frequencies).

4. Relaxing Fluids

(a) Wave equation: $\tau\left(u_{tt} - c_\infty^2 u_{xx}\right)_t + \left(u_{tt} - c_0^2 u_{xx}\right) = 0$.
(b) Absorption: $\alpha = m\omega^2\tau/2c_0(1 + \omega^2\tau^2)$ (for $m \ll 1$), $m = (c_\infty^2 - c_0^2)/c_0^2$.
(c) Dispersion: $c^{\text{ph}} = c_0\sqrt{[1 + \omega^2\tau^2(1 + m)^2]/[1 + \omega^2\tau^2(1 + m)]}$.

[11] By "low" frequency is meant $f \ll 50$ MHz for air, $f \ll 10^4$ MHz for water.

5. Boundary-Layer Absorption: Thermoviscous Fluids

(a) Wave equation:

$$u_{xx} - \frac{1}{c_0^2} u_{tt} = B \int_0^\infty u_{xx}(x, t-\tau) \frac{d\tau}{\sqrt{\tau}}, \quad B = \frac{4}{\text{HD}} \sqrt{\frac{\mu}{\pi \rho_0}} \left(1 + \frac{\gamma - 1}{\sqrt{\text{Pr}}}\right).$$

Validity: $\delta_{\text{visc}}^{\text{BL}} \ll \text{HD}/2 \ll (c_0^2/\omega^2)\delta_{\text{visc}}^{\text{BL}}, \quad \delta_{\text{visc}}^{\text{BL}} = \sqrt{2\mu/\rho_0 \omega}.$

(b) Absorption (round tube, radius a):

$$\alpha = a^{-1} \sqrt{\omega\mu/2\rho_0 c_0^2} \left[1 + (\gamma - 1)/\sqrt{\text{Pr}}\right].$$

(c) Dispersion: $c^{\text{ph}} = c_0/(1 + \alpha c_0/\omega) \doteq c_0(1 - \alpha c_0/\omega).$

REFERENCES

1. ANSI S1.26-1995, "American National Standard Method for Calculation of the Absorption of Sound by the Atmosphere" (Acoustical Society of America, New York, 1995).
2. H. E. Bass, L. C. Sutherland, A. J. Zuckerwar, D. T. Blackstock, and D. M. Hester, "Atmospheric absorption in air: Further developments," *J. Acoust. Soc. Am.* **97**, 680–683 (1995); "Erratum," *J. Acoust. Soc. Am.* **99**, 1259 (1996).
3. P. Biquard, "Sur l'absorption des ondes ultra-sonores par les liquides." *Ann. Phys. (Paris)* (Ser. 11) **6**, 195–304 (1936).
4. A. Einstein, "Schallausbreitung in teilweise dissozierten Gasen," *Sitzber. preussischem Akad. Wiss. Berlin*, **24**, 380–385 (1920). English translation: "Sound propagation in partially dissociated gases," in R. B. Lindsay, Ed., *Physical Acoustics*, Vol. 4 in Benchmark Papers in Acoustics series (Dowden, Hutchinson, and Ross, Stroudsberg, PA, 1974), pp. 268–272.
5. L. B. Evans, H. E. Bass, and L. C. Sutherland, "Atmospheric absorption of sound: Theoretical predictions," *J. Acoust. Soc. Am.* **51**, 1565–1575 (1972).
6. R. E. François and G. R. Garrison, "Sound absorption based on ocean measurements: Part I: Pure water and magnesium sulfate contributions," *J. Acoust. Soc. Am.* **72**, 896–907 (1982).
7. R. E. François and G. R. Garrison, "Sound absorption based on ocean measurements: Part II. Boric acid contribution and equation for total absorption," *J. Acoust. Soc. Am.* **72**, 1879–1890 (1982).
8. M. Greenspan, "Rotational relaxation in nitrogen, oxygen, and air," *J. Acoust. Soc. Am.* **31**, 155–160 (1959).

9. ISO 9613-1:1993, "Acoustics—Attenuation of sound during propagation outdoors—part 1: Calculation of the absorption of sound by the atmosphere" (International Organization for Standardization, Geneva, Switzerland, 1993).
10. N. H. Johannesen and J. P. Hodgson, "The physics of weak waves in gases," *Rep. Prog. Phys.* **42**, 629–676 (1979).
11. G. Kirchhoff, "Über den Einfluss der Wärmeleitung in einem Gase auf die Schallbewegung," *Ann. Phys. Chem.* **134**, 177–193 (1868). English translation: "On the influence of thermal conduction in a gas on sound propagation," in R. B. Lindsay, Ed., *Physical Acoustics*, Vol. 4 in Benchmark Papers in Acoustics series (Dowden, Hutchinson, and Ross, Stroudsberg, PA, 1974), pp. 7–19.
12. E. Meyer and W. Güth, "Zur akustischen Zähigkeitsgrenzschicht," *Acustica* **3**, 185–187 (1953).
13. M. O'Donnell, E. T. Jaynes, and J. G. Miller, "Kramers–Kronig relationship between ultrasonic attenuation and phase velocity," *J. Acoust. Soc. Am.* **69**, 696–701 (1981).
14. A. D. Pierce, *Acoustics: An Introduction to Its Physical Principles and Applications* (McGraw-Hill, New York, 1981), Sec. 10–7. Reprinted by Acoustical Society of America, New York, 1989.
15. J. M. M. Pinkerton, "A pulse method for the measurement of ultrasonic absorption in liquids: Results for water," *Nature* **160**, 128–129 (July 26, 1947).
16. M. J. Stefan, "Über den Einfluss der inneren Reibung in der Luft auf die Schallbewegung," *Sitzber. Akad. Wiss. Wien* (Mathematisch-Naturwissenschaftliche Klasse) **53**, 529–537 (1866).
17. G. G. Stokes, "On the theories of the internal friction of fluids in motion, and of the equilibrium and motion of elastic solids," *Trans. Cambr. Phil. Soc.* **8**, 287–319 (1845).
18. G. G. Stokes, "An examination of the possible effect of radiation of heat on the propagation of sound," *Phil. Mag.* (Ser. 4) **1**, 305–317 (1851).
19. C. Truesdell, "Precise theory of the absorption and dispersion of forced plane infinitesimal waves according to the Navier-Stokes equations," *J. Ration. Mech. Anal.* **2**, 643–730 (1053).
20. A. J. Zuckerwar and R. W. Meredith, "Low-frequency sound absorption measurements in air," NASA Ref. Publ. 1128 (November 1984).

PROBLEMS

When numerical values of absorption in the atmosphere or ocean are needed for these problems, see Appendix B.

9–1. A standing wave pattern is developed in a closed pipe. The pipe is many wavelengths long. The absorption of the sound as it travels through the pipe is small but not negligible. The particular mechanism of the absorption is not specified here. Simply denote the wave number by Eq. A–2 and for numerical calculations take $\alpha/\beta = 0.02$. Assume that dispersion is negligible, i.e., $c^{\mathrm{ph}} = c_0$.

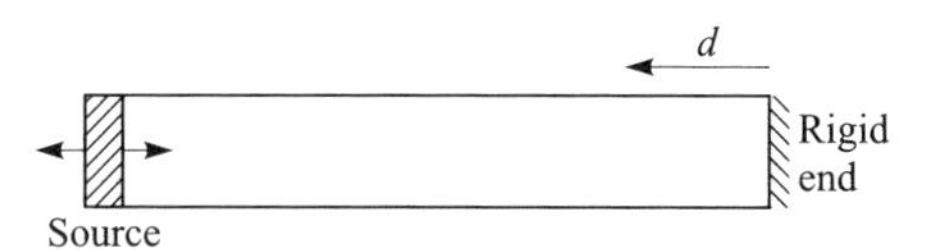

(a) First plot the pressure standing wave pattern for the lossless case ($\alpha = 0$). This will serve as the reference pattern.

(b) Next, use rotating phasor diagrams to develop the standing wave pattern for the case in which α is small but nonzero. Sketch the pattern. In particular, show the envelope of the peaks ($\beta d = 0, \pi, 2\pi, \ldots$) and the envelope of the troughs ($\beta d = \pi/2, 3\pi/2, 5\pi/2, \ldots$). Show numerical values at the first few peaks and troughs.

(c) Show that the pressure standing wave ratio, which may be defined here as the ratio of the pressure amplitude at the nth peak to that at the nth trough, decreases as d increases. What is the limiting value of SWR as n becomes very large (assume that the pipe length becomes correspondingly large)?

9–2. A civil defense siren at A (see sketch) is to be designed to operate at 2 kHz in a residential neighborhood and produce a sound pressure level of at least 74 dB at a distance of 1 mile. Because a horn is used, the sound from the siren spreads uniformly over a 15° (elevation) segment of a hemisphere (show that the area S of the segment is $2\pi r^2 \sin 15°$). Ignore reflection and attenuation from the ground itself and from trees, bushes, and houses. Find the power output required from the siren for the following cases:

(a) Atmospheric absorption neglected.

(b) Atmospheric absorption taken into account. Make calculations for the following values of relative humidity:

(i) RH = 0%

(ii) RH = 30%

(iii) RH = 60%

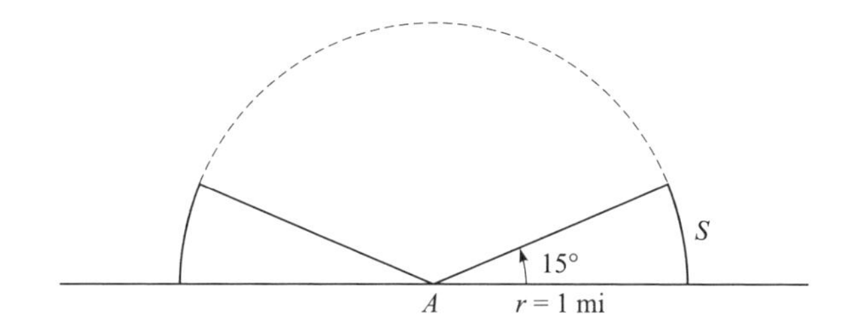

You should find that atmospheric absorption causes the power requirements for the siren to be rather high. Consider lowering the operating frequency to 600 Hz. Reanswer parts (a) and (b) and draw a conclusion about which frequency would be preferable.

9–3. An underwater research vessel has a sonar that can transmit a beam of 50-kHz sound waves in any direction. The sound waves spread spherically from the source. The sonar also acts as a receiver to detect and measure the reflected waves. The vessel is located 50 m below the surface in 13°C seawater that is 550 m deep. First the sonar is pointed directly upward and sends out a tone burst that is reflected from the ocean surface (assume the surface is perfectly flat). The sound pressure level of the surface reflected signal, when it returns to the research vessel, is $\mathrm{SPL_{SR}} = 137$ dB re 1 μPa. The sonar is then pointed downward and sends out another tone burst. The purpose is to measure the absolute value of the reflection coefficient $|R_B|$ of the bottom. The signal received from the bottom is found to have a sound pressure level $\mathrm{SPL_{BR}} = 100$ dB.

(a) If the seawater is assumed lossless, what value do you calculate for $|R_B|$?

(b) By taking absorption into account, find the true value of $|R_B|$. Assume a salinity of 35‰, a pH value of 8.0, and values $P_2 = 1$ and $P_3 = 1$ for the pressure coefficients (i.e., ignore the depth dependence of P_2 and P_3). Use the value $c_0 = 1500$ m/s; show that this is very close to the sound speed at the mean depth (275 m) of the water column.

9–4. Derive formulas for the normal-incidence reflection and transmission coefficients R and T for the interface between two sound absorbing fluids. First, show that Eqs. 3A–8 and 3A–9 continue to hold even for dissipative fluids. Into these equations substitute the appropriate expressions for Z_1 and Z_2. To find the latter, it is suggested that you assume the dispersion is very small, i.e., $k_1 = \omega/c_1 - j\alpha_1$ and $k_2 = \omega/c_2 - j\alpha_2$, so that the continuity equation in the form $p_t + \rho_0 c_0^2 u_x = 0$ is applicable. To show the relation of your expression for R to that for a lossless fluid, put your expression in the form

$$R = \frac{\rho_2 c_2 - \rho_1 c_1 + A}{\rho_2 c_2 + \rho_1 c_1 + B},$$

i.e., find A and B. Put your result for T in a similar form.

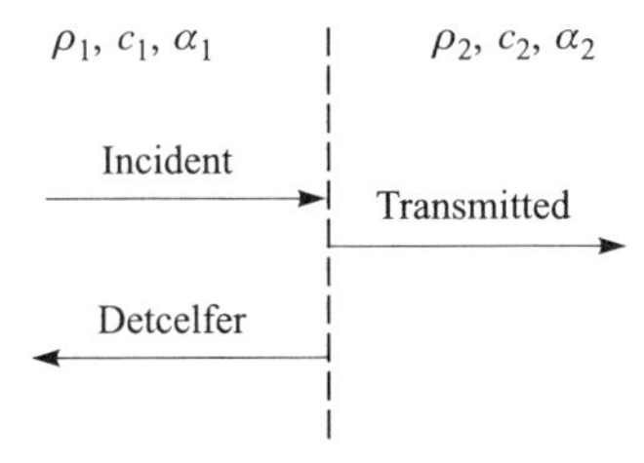

9–5. A widely accepted formula, based on experiments, for sound absorption in fresh water at 20°C is

$$\alpha = 25 \times 10^{-15} f^2 \text{ nepers/m},$$

where f is frequency in hertz. By assuming that the absorption is entirely due to viscosity, calculate $\tilde{V}$ and μ_B for water.

9–6. Find the expression for the characteristic impedance of a viscous medium. (Remember that Z_0 is defined as p/u for an outgoing plane wave.) Express your answer as a real plus an imaginary part.

$$\text{ANSWER: } Z_0 = \rho_0 c_0 / \sqrt{1 + j\tilde{V}\delta_{\mathrm{v}}} \doteq \rho_0 c_0 [1 - j\alpha_{\mathrm{v}} c_0/\omega].$$

9–7. A source generates a sawtooth wave of amplitude P_0. The Fourier series for the sawtooth at the source is

$$p = \frac{2}{\pi} P_0 \sum_{n=1} \frac{1}{n} \sin n\omega t.$$

Let $\mathrm{SPL}_n(x)$ stand for the SPL of the nth frequency component of the wave at a distance x from the source.

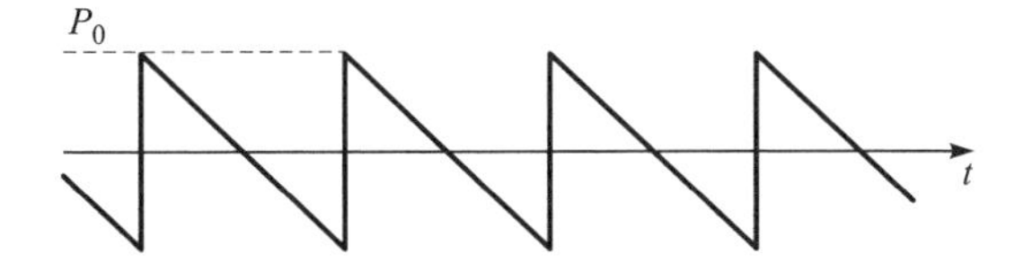

(a) Show that $\mathrm{SPL}_2(0) = \mathrm{SPL}_1(0) - 6$ dB.

The fundamental frequency of the sawtooth wave is $f = 10$ kHz, and the wave propagates through castor oil.

(b) At what distance is $\mathrm{SPL}_2(x)$ 20 dB below $\mathrm{SPL}_1(x)$?

9–8. Derive the wave equation for sound in a thermoviscous fluid, Eq. B–29.

9–9. Derive the wave equation for sound in a viscous, relaxing fluid. For the equation of state, use Eq. C–1.

Answer: $\tau(u_{tt} - c_\infty^2 u_{xx} - \nu\tilde{V}u_{xxt})_t + (u_{tt} - c_0^2 u_{xx} - \nu\tilde{V}u_{xxt}) = 0.$

9–10. Find the absorption coefficient, both α and $\overline{\alpha}$, due to tube wall effects in a 5-cm (inside diameter) tube containing air at 20°C. Give your result in the form $\overline{\alpha} = \text{const} \times \sqrt{f}$. Plot $\overline{\alpha}$ (in dB/100 m) on the copy of Fig. B.1 given below. From the plot, determine the frequency at which atmospheric absorption appears to be about the same as absorption due to tube wall effects. Is the formula for tube wall absorption valid at this frequency?

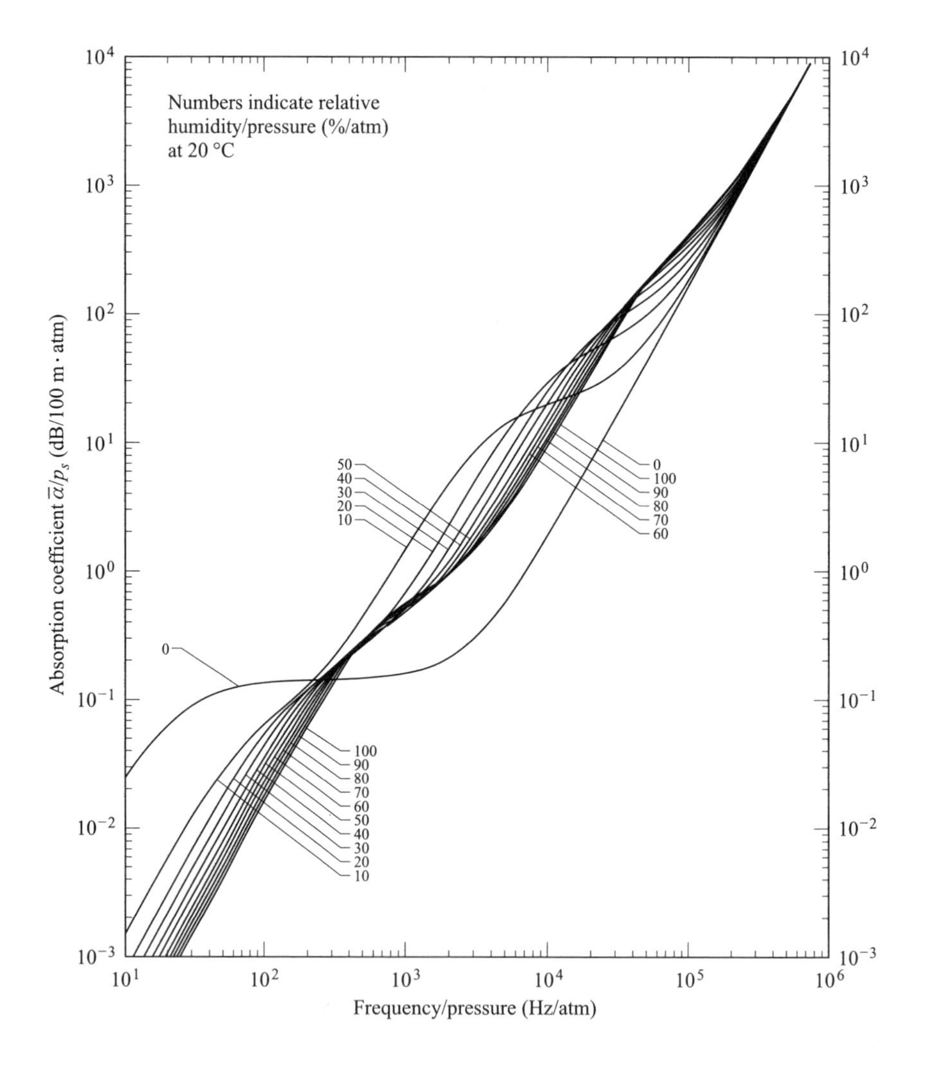

9–11. For practical application the formula for the tube wall absorption coefficient may be expressed as follows:

$$\overline{\alpha} = (K/a)\sqrt{f} \quad \text{dB/m},$$

where a is in meters and f is in hertz. Find the value of K for air at 20°C.

9–12. A long, air-filled steel tube, 3 cm inside diameter, is terminated at one end by a material of unknown acoustical reflection coefficient R. The temperature is 20°C. Tone bursts of 850-Hz sound are sent down the pipe and reflected by the material. A microphone located 10 m from the material is used to measure the SPL of the incident signal and that of reflected signal. The difference in levels is 16 dB. Find $|R|$.

9–13. A rigid tube 5 m long and 20 cm in diameter is closed at one end by a very long plug of super light cork and at the other end by a rigid piston. The characteristic impedance of the cork is $Z_0 = 30,000$ MKS rayls (one-quarter that of ordinary cork). The fluid in the tube is air. A microphone is located at the midpoint of the tube. The piston vibrates and sends out a 10-cycle tone burst (pulse) of 1000-Hz sound in the form of a plane wave. After emitting the pulse, the piston comes to rest and may be considered rigidly fixed thereafter. The pulse bounces back and forth between the cork and the piston. When the pulse first passes by the microphone, the recorded SPL is 80 dB. Twenty passes later (i.e., after 10 round trips, where a round trip is a 10-m path that includes one reflection at each end), what is the SPL recorded by the microphone? Be sure to consider all acoustical effects.

9–14. An acoustical transmitter in air generates a collimated beam of plane waves in the form of a 50-cycle tone burst of 50-kHz sound. The transmitter is located 10 m above the surface of a fresh water lake and is pointed straight down toward the water surface. In the water at a depth of 10 m (directly below the transmitter) is a hydrophone. Both air and water are at a temperature of 20°C, and a heavy fog lies over the lake. The sound pressure level of the tone burst from the transmitter is $\text{SPL}_{\text{air}} = 120$ dB. Find the sound pressure level of the tone burst received by the hydrophone ($\text{SPL}_{\text{water}}$).

9–15. A 5-cm (inside diameter) pipe is used to carry methane (natural gas) over a long distance. The acoustical properties of methane at 0°C are as follows: $\rho_0 = 0.7168\ \text{kg/m}^3$, $c_0 = 430$ m/s, $\mu = 1.04 \times 10^{-5}\ \text{kg/m} \cdot \text{s}$, $\text{Pr} = 1.30$, $\gamma = 1.307$. The pipe is buried in the ground, where the temperature is 0°C (it is winter). However, the pipe comes above ground in order to cross over a stream. On a sunny day the temperature of the methane in the above-ground part of the pipe is 100°C. Consider the place where the 0°C methane meets the 100°C methane to be a plane interface. The ambient pressure is the same on both sides of the interface. A sound pulse of frequency 2600 Hz is incident on the interface from the underground side. Before reaching the interface, the pulse passes by a microphone located 20 m in front of the interface. The SPL of the incident signal is 100 dB when it passes by the microphone.

(a) Show that the reflection coefficient associated with the temperature jump across the interface is

$$R = \frac{\sqrt{T_1} - \sqrt{T_2}}{\sqrt{T_1} + \sqrt{T_2}},$$

where T_1 and T_2 are absolute temperatures before and after the interface, respectively.

(b) Find the SPL of the reflected signal when it passes by the microphone.

9–16. A rigid pipe (length L) with rigid ends contains a viscous fluid, for which the wave equation is (boundary-layer dissipation excluded)

$$\frac{\tilde{V}\nu}{c_0^2}\, p_{xxt} + p_{xx} - \frac{1}{c_0^2}\, p_{tt} = 0.$$

(a) Use the method of separation of variables to obtain the solution

$$p = e^{-mt}\begin{Bmatrix} e^{j\Omega t} \\ e^{-j\Omega t} \end{Bmatrix}\begin{Bmatrix} \cos kx \\ \sin kx \end{Bmatrix} \quad \text{or} \quad e^{-mt}\begin{Bmatrix} \cos \Omega t \\ \sin \Omega t \end{Bmatrix}\begin{Bmatrix} \cos kx \\ \sin kx \end{Bmatrix}.$$

You will have to supply the expressions for k, Ω, and m. (*Suggestion*: to solve the T equation, use a trial solution $T = T_0 e^{jbt}$, where b is to be determined.)

(b) The fluid is castor oil and the length of the pipe is $L = 0.1$ m. Consider the lowest mode (not counting the "dc mode" $n = 0$).

(i) What is the eigenfrequency of this mode?

(ii) Assume that the mode is excited by some means and that its pressure amplitude is A_0 at time $t = 0$. How long does it take the mode to decay by 20 dB?

10

SPHERICAL WAVES

A. INTRODUCTION

Up to this point our treatment of spherical waves has been limited to omnidirectional radiation; see, for example, Sec. 1D. In this chapter we consider more complicated problems involving spherical waves. In terms of the velocity potential ϕ the (lossless) wave equation, Eq. 1D–7, in spherical coordinates r, θ, ψ is

$$\frac{1}{r^2}\left[(r^2\phi_r)_r + \frac{1}{\sin\theta}(\sin\theta\phi_\theta)_\theta + \frac{1}{\sin^2\theta}\phi_{\psi\psi}\right] - \frac{1}{c_0^2}\phi_{tt} = 0. \tag{A–1}$$

Figure 10.1 shows the relation of the spherical coordinate system to the rectangular coordinate system. The functional relationships are

$$x = r\sin\theta\cos\psi, \tag{A–2a}$$

$$y = r\sin\theta\sin\psi, \tag{A–2b}$$

$$z = r\cos\theta. \tag{A–2c}$$

The z axis is called the polar axis, θ the polar angle, and ψ the axial, or azimuthal, angle. Let $\mathbf{r}_1$, $\boldsymbol{\theta}_1$, and $\boldsymbol{\psi}_1$ be unit vectors in the r, θ, and ψ directions, respectively. The expression for the particle velocity $\mathbf{u} = \nabla\phi$ in spherical coordinates is

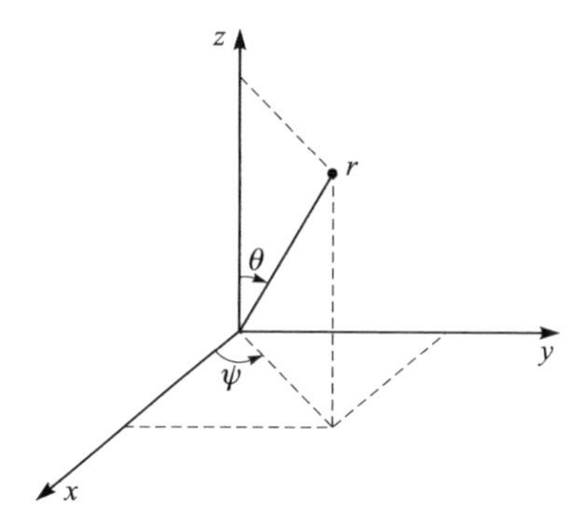

Figure 10.1 Spherical coordinate system.

$$\mathbf{u} = \mathbf{r}_1\phi_r + \boldsymbol{\theta}_1\frac{1}{r}\phi_\theta + \boldsymbol{\psi}_1\frac{1}{r\sin\theta}\phi_\psi, \tag{A–3}$$

from which the components $u^{(r)}$, $u^{(\theta)}$, and $u^{(\psi)}$ may be identified. The pressure may, as usual, be found from the relation

$$p = -\rho_0\phi_t. \tag{A–4}$$

The simplest kind of spherical wave is the pure radial disturbance, treated briefly in Sec. 1D. In this case Eq. A–1 reduces to Eq. 1D–12, repeated here for convenience,

$$(r\phi)_{rr} - \frac{1}{c_0^2}(r\phi)_{tt} = 0. \tag{A–5}$$

The general solution is Eq. 1D–13, also repeated here for convenience,

$$\phi = \frac{f(t - r/c_0)}{r} + \frac{g(t + r/c_0)}{r}. \tag{A–6}$$

The $1/r$ dependence shows that an outgoing wave diminishes in amplitude as it propagates whereas an incoming wave grows in amplitude. Each wave preserves its shape, however, as it travels. The outgoing wave described by the first term in Eq. A–6 is sometimes called monopole radiation. Monopoles are discussed in detail in Sec. D.2 below. A more complicated type of spherical wave motion is one in which the field depends on the polar angle θ as well as on r. The most complicated fields also depend on the axial angle ψ. Figure 10.2 illustrates the progression of complexity. If a function is described as being constant on the surface of a sphere, as in Fig. 10.2*a*, no dependence on either θ or ψ exists. If the function is constant on a general surface of revolution, as in Fig. 10.2*b*, we take the z axis to be the axis of revolution, in which case the function depends on the polar angle θ but not on ψ. Finally, in the most complicated case, for example, Fig. 10.2*c*, the function depends on both angles as well as on r.

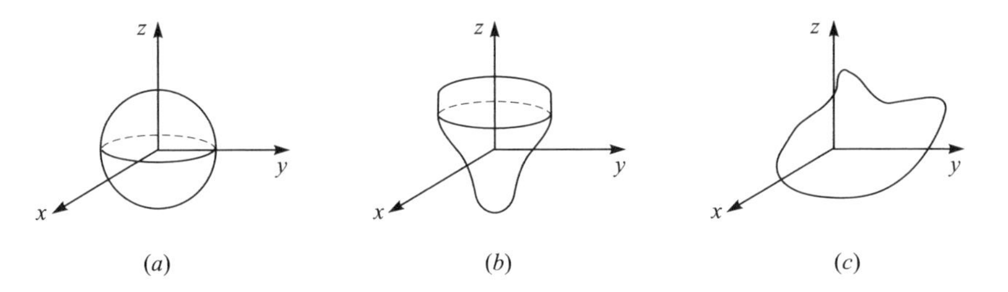

Figure 10.2 Symmetries in spherical wave motion: (*a*) radial symmetry, (*b*) axial symmetry, and (*c*) no symmetry.

B. SOLUTION BY SEPARATION OF VARIABLES

In this section we first solve the wave equation for fields not dependent on the axial angle ψ. Separation of variables leads to a solution in terms of spherical Bessel and Neumann functions $j_n(kr)$ and $n_n(kr)$, respectively, and Legendre polynomials $P_n(\cos\theta)$. Later, when the dependence on ψ is included, the solution is found to include functions called spherical harmonics (see, for example, Ref. 13).

When $\phi = \phi(r, \theta, t)$ only, the wave equation is

$$\frac{1}{r^2}(r^2\phi_r)_r + \frac{1}{r^2 \sin\theta}(\sin\theta\phi_\theta)_\theta - \frac{1}{c_0^2}\phi_{tt} = 0. \tag{B–1}$$

This equation may be solved by the method of separation of variables (see Chap. 6 for a detailed exposition of the method). Assume a product solution

$$\phi = R(r)\Theta(\theta)T(t), \tag{B–2}$$

substitute this expression into the wave equation, divide through by $R\Theta T$, and put the time-dependent part on the right-hand side:

$$\frac{(r^2 R_r)_r}{r^2 R} + \frac{(\sin\theta\Theta_\theta)_\theta}{r^2 \sin\theta\Theta} = \frac{1}{c_0^2}\frac{T_{tt}}{T}. \tag{B–3}$$

The left-hand side is a function of r, θ only while the right-hand side is a function of t only. The equality can hold only if both sides are equal to a constant. It is convenient to choose $-k^2$ as the symbol for the constant (because k turns out to be the wave number):

$$\frac{1}{c_0^2}\frac{T''}{T} = -k^2,$$

or

$$T'' + k^2 c_0^2 T = 0. \tag{B–4}$$

Solutions are, with $k^2 c_0^2 = \omega^2$,

$$T = \begin{Bmatrix} \cos \omega t \\ \sin \omega t \end{Bmatrix} \tag{B–5a}$$

or

$$T = \begin{Bmatrix} e^{j\omega t} \\ e^{-j\omega t} \end{Bmatrix}. \tag{B–5b}$$

As in Chap. 6, the curly brackets mean any linear combination of the two functions. Whether to use the trigonometric functions or the complex exponentials depends on the particular problem being solved. The trigonometric functions are generally more useful for describing wave motion in enclosures, where standing waves are encountered. The exponential functions find widespread use in radiation problems. These are not hard and fast rules, however, only guidelines.

The r, θ part of the equation is now dealt with. Multiply the equation through by r^2 and again separate variables. Let the separation constant be called C:

$$\frac{1}{R}(r^2 R_r)_r + k^2 r^2 = -\frac{(\sin\theta\Theta_\theta)_\theta}{\Theta\sin\theta} = C. \tag{B–6}$$

This relation yields two ordinary differential equations, one in θ and one in r. Take the θ equation first.

1. Legendre Polynomials

The θ equation obtained from Eq. B–6 is

$$\frac{1}{\sin\theta}\frac{d}{d\theta}\left(\sin\theta\frac{d\Theta}{d\theta}\right) + C\Theta = 0. \tag{B–7}$$

As Fig. 10.1 shows, $\cos\theta$ is the projection on the z axis of a point on the unit sphere $r = 1$. Using the relation $z = \cos\theta$, so that $d/d\theta = -\sin\theta d/dz$, we transform Eq. B–7 into[1]

$$\frac{d}{dz}\left[(1 - z^2)\frac{d\Theta}{dz}\right] + C\Theta = 0,$$

or

[1]Note carefully the difference between the general relation $z = r\cos\theta$ and the special relation $z = \cos\theta$ used in this section. The latter is valid only for the unit sphere.

$$(1 - z^2)\frac{d^2\Theta}{dz^2} - 2z\frac{d\Theta}{dz} + C\Theta = 0. \tag{B–8}$$

This is Legendre's differential equation, and its solution is given in Appendix A. The important point here is that the solution is unbounded at $z = \pm 1$, i.e., at the poles $\theta = 0, \pi$, unless C is chosen to have one of the special values

$$C = n(n+1), \tag{B–9}$$

where n is an integer. In physical problems involving sound waves we do not expect unbounded solutions, even if the unboundedness occurs only at the poles. We are therefore forced to pick from all the possible solutions of Eq. B–8 only those that are bounded. In other words, C must have the value given by Eq. B–9. Note the difference between this way of selecting the value of the separation constant and the way k is chosen in the vibrating string problem. In the string problem k is chosen to satisfy boundary conditions. Here C is chosen to assure a bounded solution.

The solutions of Legendre's equation when C is given by Eq. B–9 are the Legendre polynomials (Ref. 1, Chap. 8; Ref. 2, Chap. 12),

$$\Theta = P_n(z) = P_n(\cos\theta). \tag{B–10}$$

(The notation used here agrees with that in most standard texts. Morse (Ref. 7) and Morse and Ingard (Ref. 9) use m where we have used n.) The Legendre polynomials may be defined in terms of their power series,

$$P_n(z) = \sum_{r=0}^{N} (-1)^r \frac{(2n-2r)!}{2^n r!(n-r)!(n-2r)!} z^{n-2r}, \tag{B–11}$$

where $N = n/2$ if n is even, $N = (n-1)/2$ if n is odd. An alternative is Rodrigues's formula,

$$P_n(z) = \frac{1}{2^n n!}\frac{d^n}{dz^n}(z^2 - 1)^n. \tag{B–12}$$

A few of the lowest order Legendre polynomials are as follows:

$$\begin{aligned}
P_0(z) &= 1, & P_0(\cos\theta) &= 1,\\
P_1(z) &= z, & P_1(\cos\theta) &= \cos\theta,\\
P_2(z) &= (3z^2 - 1)/2, & P_2(\cos\theta) &= (3\cos 2\theta + 1)/4,\\
P_3(z) &= (5z^3 - 3z)/2, & P_3(\cos\theta) &= (5\cos 3\theta + 3\cos\theta)/8,\\
P_4(z) &= (35z^4 - 30z^2 + 3)/8, & P_4(\cos\theta) &= (35\cos 4\theta + 20\cos 2\theta + 9)/64.
\end{aligned} \tag{B–13}$$

Figure 10.3 shows graphs of the Legendre polynomials P_0 through P_5. Note that the even-numbered polynomials (Fig. 10.3*a*) are symmetric, while the odd-numbered polynomials (Fig. 10.3*b*) are anti-symmetric, about the equatorial plane $z = 0$. Also notice that the amplitudes of all the P_n functions are such that they become unity at the pole ($\theta = 0$), i.e., $P_n = 1$ at $z = 1$. This is the traditional choice of normalization for the Legendre polynomials. Some useful properties of the P_n follow.

a. Recurrence Relations

Two useful recurrence formulas are

$$(2n + 1)zP_n(z) = (n + 1)P_{n+1}(z) + nP_{n-1}(z), \tag{B–14a}$$

$$(2n + 1)P_n(z) = P'_{n+1}(z) - P'_{n-1}(z). \tag{B–14b}$$

b. Orthogonality Relation

The Legendre polynomials are orthogonal over the interval $[-1, 1]$, in particular,

$$\int_{-1}^{1} P_m(z)\, P_n(z)\, dz = \frac{2}{2n+1}\, \delta_{mn}, \tag{B–15}$$

where δ_{mn} is the Kronecker delta ($= 1$ if $m = n$; $= 0$ otherwise). Because they are orthogonal and form a complete set, the Legendre polynomials may be used to form a series expansion of a function $f(z)$,

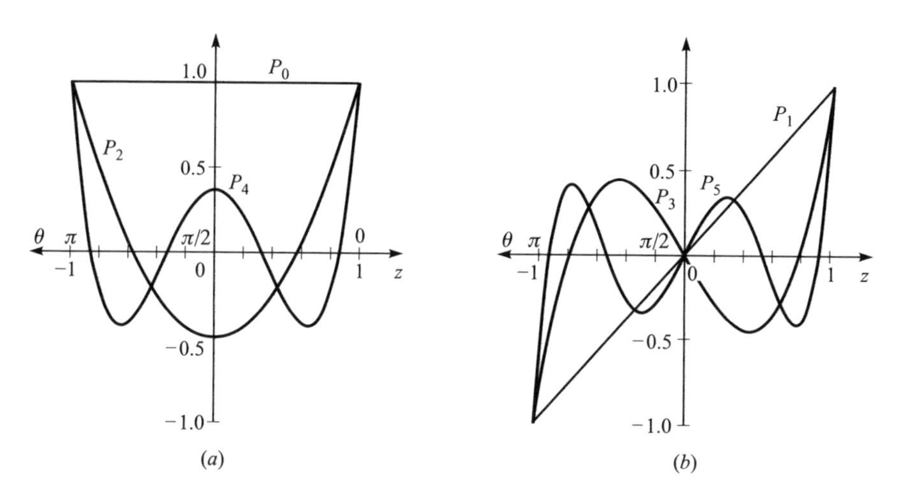

Figure 10.3 First six Legendre polynomials P_n, plotted as functions of z and θ: (*a*) n even; (*b*) n odd.

$$f(z) = \sum a_n P_n(z),$$

over the interval $[-1, 1]$, just as the sine and cosine functions are used to form a Fourier series expansion over the interval $[-\pi, \pi]$. Note, however, that although orthogonal, the P_n are not orthonormal (the right-hand side of Eq. B–15 is not simply δ_{mn}). The traditional normalization of the P_n, i.e., $P_n(1) = 1$, precludes orthonormality. Finally, the following result is sometimes useful in finding the coefficients a_n:

$$\int_{-1}^{+1} z^n P_n(z)\, dz = \frac{2^{n+1}(n!)^2}{(2n+1)!}. \tag{B–16}$$

In certain problems involving hemispheres, the domain of z is only $[0, 1]$, i.e., $0 \leq \theta \leq \pi/2$. What orthogonality relation may be used for this interval? It turns out that if $m + n$ is even, that is, if m and n are both odd or both even, Eq. B–15 reduces to

$$\int_0^1 P_m(z)P_n(z)\, dz = \frac{1}{2n+1}\delta_{mn}. \tag{B–17}$$

2. Spherical Bessel Functions

With C given by Eq. B–9, the radial equation obtained from Eq. B–6 is, when expanded,

$$\frac{d^2R}{dr^2} + \frac{2}{r}\frac{dR}{dr} + \left[k^2 - \frac{n(n+1)}{r^2}\right]R = 0. \tag{B–18}$$

This equation resembles Bessel's equation,

$$\frac{d^2y}{dr^2} + \frac{1}{r}\frac{dy}{dr} + \left(k^2 - \frac{n^2}{r^2}\right)y = 0.$$

To determine whether Eq. B–18 can be transformed into Bessel's equation, try the substitution $R = r^\alpha Z(r)$. It is found that if $\alpha = -\frac{1}{2}$, Eq. B–18 becomes

$$\frac{d^2Z}{dr^2} + \frac{1}{r}\frac{dZ}{dr} + \left[k^2 - \frac{(n+1/2)^2}{r^2}\right]Z = 0,$$

which is Bessel's equation of order $\mu = n + \frac{1}{2}$. The solution of the radial equation is therefore

$$R = \frac{1}{\sqrt{kr}} \left\{ \begin{array}{c} J_{n+1/2}(kr) \\ N_{n+1/2}(kr) \end{array} \right\}, \tag{B–19}$$

where J_μ and N_μ are the ordinary Bessel and Neumann functions, respectively.

Although Eq. B–19 appears formidable, the solution is actually rather simple because Bessel functions of half-integer order are closely related to the ordinary sine and cosine functions. In particular,

$$x^{-1/2} J_{n+1/2}(x) = (2/\pi)^{1/2} j_n(x), \tag{B–20a}$$

$$x^{-1/2} N_{n+1/2}(x) = (2/\pi)^{1/2} n_n(x), \tag{B–20b}$$

where $j_n(x)$ and $n_n(x)$ are the so-called spherical Bessel and Neumann functions, respectively (Ref. 1, Chap. 10; Ref. 2, Sec. 11.7). Thus the simpler way of writing the solution of the radial equation is

$$R = \left\{ \begin{array}{c} j_n(kr) \\ n_n(kr) \end{array} \right\}. \tag{B–21}$$

Some properties of the spherical Bessel and Neumann functions are as follows. Their Maclaurin series are

$$j_n(x) = (2x)^n \sum_{s=0}^{\infty} \frac{(-1)^s (s+n)! x^{2s}}{s!(2s+2n+1)!}, \tag{B–22a}$$

$$n_n(x) = \frac{(-1)^{n+1}}{x(2x)^n} \sum_{s=0}^{\infty} \frac{(-1)^s (s-n)!}{s!(2s-2n)!} x^{2s}. \tag{B–22b}$$

More commonly used expressions are in terms of sines and cosines:

$$\begin{aligned}
j_o(x) &= \frac{\sin x}{x}, & n_0(x) &= -\frac{\cos x}{x}, \\
j_1(x) &= \frac{\sin x}{x^2} - \frac{\cos x}{x} = -j_0'(x), & n_1(x) &= -\frac{\cos x}{x^2} - \frac{\sin x}{x} = -n_0'(x), \\
j_2(x) &= \left(\frac{3}{x^3} - \frac{1}{x}\right) \sin x - \frac{3}{x^2} \cos x, & n^2(x) &= -\left(\frac{3}{x^3} - \frac{1}{x}\right) \cos x - \frac{3}{x^2} \sin x, \\
&\vdots & &\vdots \\
j_n(x) &= (-x)^n \left(\frac{1}{x}\frac{d}{dx}\right)^n j_0(x), & n_n(x) &= (-x)^n \left(\frac{1}{x}\frac{d}{dx}\right)^n n_0(x).
\end{aligned} \tag{B–23}$$

Here the operator $[(1/x)\,d/dx]^n$ means $(1/x)\,d/dx$ applied n times, not $(1/x)^n\, d^n/dx^n$. Two recursion relations for the spherical Bessel functions are as follows:

$$j_{n+1}(x) = \frac{2n+1}{x} j_n(x) - j_{n-1}(x), \tag{B–24a}$$

$$j_n'(x) = \frac{1}{2n+1}\left[n j_{n-1}(x) - (n+1) j_{n+1}(x)\right]. \tag{B–24b}$$

The same relations apply to the spherical Neumann functions $n_n(x)$, to the spherical Hankel functions $h_n^{(1)}(x)$ and $h_n^{(2)}(x)$ (which are discussed below), and to any linear combination of the four functions provided the multiplying coefficients are independent of n and x.

The first few j_n and n_n are shown in Fig. 10.4. Note that the spherical Neumann functions, like the ordinary Neumann functions N_n, are singular at the origin. Thus the n_n functions may not be used to represent spherical wave motion that includes the origin, such as the wave motion inside a spherical enclosure.

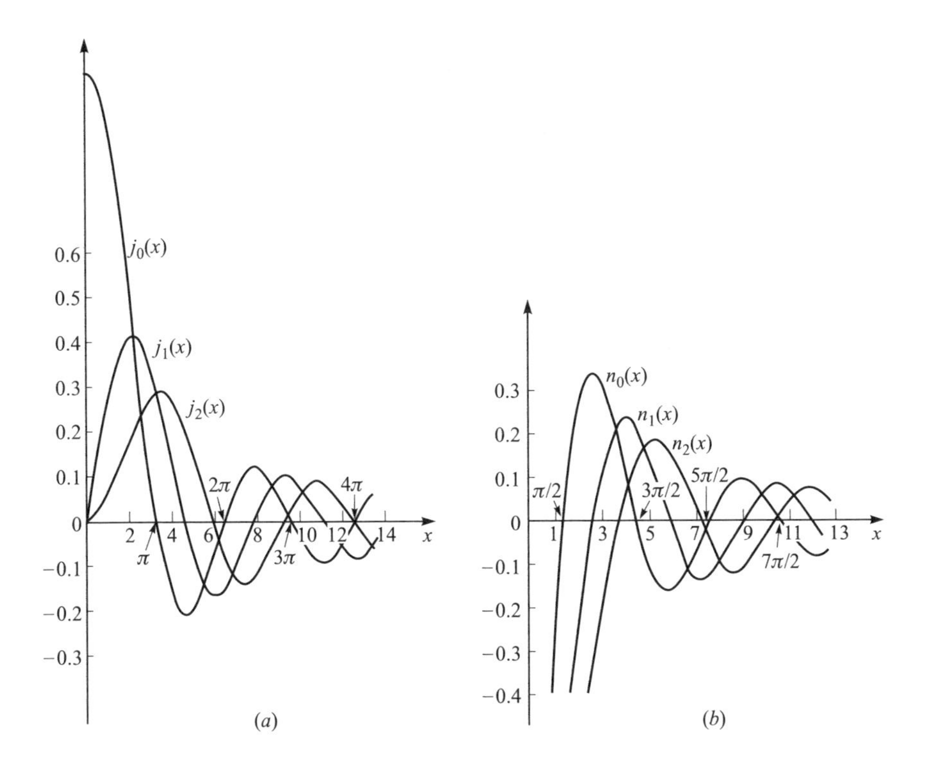

Figure 10.4 (*a*) Spherical Bessel functions. (*b*) Spherical Neumann functions.

Zeros of spherical Bessel functions and derivatives of spherical Bessel functions are given in Table 10.1. The indexing notation is made clear by example: $x_{23} = 12.323$ is the third zero of j_2. It is customary to regard $x = 0$ as the "zeroth" zero of $j_1, j_2, j_3, \ldots$. For other tables, see Ref. 8, p. 1576.

3. Spherical Hankel Functions

Recall that the exponential functions e^{jkx}, e^{-jkx} are useful alternatives to the trigonometric functions $\sin kx$, $\cos kx$ as solutions of the equation for time-harmonic plane waves. Analogous alternatives to $j_n(kr)$, $n_n(kr)$ are the spherical Hankel functions

$$R(r) = \begin{Bmatrix} h_n^{(1)}(kr) \\ h_n^{(2)}(kr) \end{Bmatrix}, \tag{B–25}$$

where

$$h_n^{(1)}(x) = j_n(x) + jn_n(x) = (-x)^n \left(\frac{1}{x}\frac{d}{dx}\right)^n h_0^{(1)}(x), \tag{B–26a}$$

$$h_n^{(2)}(x) = j_n(x) - jn_n(x) = (-x)^n \left(\frac{1}{x}\frac{d}{dx}\right)^n h_0^{(2)}(x). \tag{B–26b}$$

For example, the zero-order spherical Hankel functions are

$$h_0^{(1)}(x) = j_0(x) + jn_0(x) = \frac{\sin x}{x} - j\,\frac{\cos x}{x} = -j\,\frac{e^{jx}}{x},$$

$$h_0^{(2)}(x) = j_0(x) - jn_0(x) = \frac{\sin x}{x} + j\,\frac{\cos x}{x} = j\,\frac{e^{-jx}}{x}.$$

Table 10.1 Tables of Zeros of Spherical Bessel Functions

	Roots $x_{n\ell}$ of $j_n(x) = 0$					Roots $x'_{n\ell}$ of $j'_n(x) = 0$				
ℓ	$n = 0$	$n = 1$	$n = 2$	$n = 3$	$n = 4$	$n = 0$	$n = 1$	$n = 2$	$n = 3$	$n = 4$
1	π	4.493	5.763	6.988	8.183	0	2.082	3.342	4.514	5.647
2	2π	7.725	9.095	10.417	11.705	4.493	5.940	7.290	8.578	9.840
3	3π	10.904	12.323	13.698	15.040	7.725	9.206	10.614	11.973	13.296
4	4π	14.066	15.515	16.924	18.301	10.904	12.405	13.846	15.245	16.609
5	5π	17.221	18.689	20.122	21.525	14.066	15.579	17.043	18.468	19.862

The spherical Hankel functions are generally used to describe progressive waves. To see this, combine the T and R functions for the case $n = 0$ [recall that in this case $\Theta = P_0(\cos\theta) = 1$]:

$$\phi = \begin{Bmatrix} h_0^{(1)}(kr) \\ h_0^{(2)}(kr) \end{Bmatrix} \begin{Bmatrix} e^{j\omega t} \\ e^{-j\omega t} \end{Bmatrix}.$$

For instance, let $h_0^{(2)}$ be combined with $e^{j\omega t}$. The result is

$$h_0^{(2)}(kr)e^{j\omega t} = j\frac{e^{j(\omega t - kr)}}{kr},$$

which represents an outward traveling spherical wave. To represent an inward traveling wave, use $h_0^{(1)}$ in place of $h_0^{(2)}$.[2]

On the other hand, when a standing wave field is to be described, the spherical Bessel and Neumann functions are usually preferred. They are combined with the trigonometric time functions as follows, again for the case $n = 0$:

$$\phi = \begin{Bmatrix} j_0(kr) \\ n_0(kr) \end{Bmatrix} \begin{Bmatrix} \cos\omega t \\ \sin\omega t \end{Bmatrix}.$$

For example, to find the field inside an enclosure when the origin lies within the domain of interest, one would discard the Neumann functions and use

$$j_0(kr)\begin{Bmatrix} \cos\omega t \\ \sin\omega t \end{Bmatrix} = \frac{\sin kr}{kr}[A\cos\omega t + B\sin\omega t].$$

This combination of functions closely resembles that used for a vibrating string with fixed ends.

4. Summary of Solutions for Axially Symmetric Wave Motion

If $\phi \neq \phi(\psi)$, spherical waves are conveniently described by

[2]The combinations given here are not the only ones in common use. Those who prefer to use the function $e^{-j\omega t}$ to express harmonic time dependence represent an outgoing wave by

$$h_0^{(1)}(kr)e^{-j\omega t} = -j\frac{e^{j(kr-\omega t)}}{kr};$$

$h_0^{(2)}$ is then used to represent an incoming wave. In this book, however, the time function is always $e^{j\omega t}$; accordingly, $h_n^{(2)}$ represents an outgoing wave.

$$\phi = \begin{Bmatrix} h_n^{(1)}(kr) \\ h_n^{(2)}(kr) \end{Bmatrix} P_n(\cos\theta) \begin{Bmatrix} e^{j\omega t} \\ e^{-j\omega t} \end{Bmatrix}, \tag{B–27a}$$

or, alternatively,

$$\phi = \begin{Bmatrix} j_n(kr) \\ n_n(kr) \end{Bmatrix} P_n(\cos\theta) \begin{Bmatrix} \cos\omega t \\ \sin\omega t \end{Bmatrix}. \tag{B–27b}$$

The first form is generally used to represent the field in an open medium (radiation problem), the second to represent the field in an enclosure (standing wave problem).

5. Most General Spherical Waves; Spherical Harmonics

In the most general type of spherical wave motion the acoustic field depends on the axial angle ψ as well as on r, θ, t; see Fig. 10.2*c*. Let the velocity potential be represented by the product of a spatial function $\Phi(r, \theta, \psi)$ and a time function $T(t)$,

$$\phi = \Phi(r, \theta, \psi)T(t). \tag{B–28}$$

Substitution in Eq. A–1 and separation of variables lead to the same time equation—and its solution, Eqs. B–5a,b—as before. The remaining (spatial) part is

$$\nabla^2\Phi + k^2\Phi = 0, \tag{B–29}$$

which is known as the Helmholtz equation. Express Φ as the product function

$$\Phi = R(r)\,\Theta(\theta)\,\Psi(\psi)$$

and separate variables again:

$$\sin^2\theta\left[\frac{1}{R}(r^2R_r)_r + k^2r^2 + \frac{1}{\sin\theta}(\sin\theta\Theta_\theta)_\theta\right] = -\frac{\Psi''}{\Psi}.$$

Let the separation constant be m^2. The Ψ equation is then

$$\Psi'' + m^2\Psi = 0,$$

and its solution is

$$\Psi = \begin{Bmatrix} e^{jm\psi} \\ e^{-jm\psi} \end{Bmatrix} \tag{B–30a}$$

or

$$\Psi = \begin{Bmatrix} \cos m\psi \\ \sin m\psi \end{Bmatrix}. \tag{B–30b}$$

Because Ψ must be periodic with period 2π, m must be an integer. Note that the choice $m = 0$ leads to the ψ-independent wave motion already studied, $\phi = \phi(r, \theta, t)$.

Separation of the R, Θ equation proceeds in much the same way as before. We obtain

$$\frac{1}{R}(r^2 R_r)_r + k^2 r^2 = -\frac{(\sin\theta\Theta_\theta)_\theta}{\Theta\sin\theta} + \frac{m^2}{\sin^2\theta} = C.$$

Again, to assure boundedness of the Θ equation at the poles, C must be chosen equal to $n(n+1)$. Because this choice yields the same radial equation as before, Eq. B–18, the solutions are again the spherical Bessel and Neumann functions or the spherical Hankel functions. The Θ equation is a little more complicated, in particular,

$$\frac{1}{\sin\theta}\frac{d}{d\theta}\left(\sin\theta\frac{d\Theta}{d\theta}\right) + \left[n(n+1) - \frac{m^2}{\sin^2\theta}\right]\Theta = 0.$$

Use of the transformation $z = \cos\theta$ leads to

$$(1 - z^2)\frac{d^2\Theta}{dz^2} - 2z\frac{d\Theta}{dz} + \left[n(n+1) - \frac{m^2}{1 - z^2}\right]\Theta = 0, \tag{B–31}$$

which is called the associated Legendre equation. Its solutions are the associated Legendre polynomials $P_n^m(z)$,[3]

$$\Theta = P_n^m(z) = (1 - z^2)^{m/2}\frac{d^m}{dz^m}P_n(z), \tag{B–32}$$

or if Rodrigues's formula for P_n, Eq. B–12, is used,

$$P_n^m(z) = \frac{1}{2^n n!}(1 - z^2)^{m/2}\frac{d^{m+n}}{dz^{m+n}}(z^2 - 1)^n. \tag{B–33}$$

[3]Because in our case z is a real number in the range $-1 \le z \le 1$, the factor $(1 - z^2)^{m/2}$ is real. When it is desirable to treat z as a complex variable, some authors define the associated Legendre functions by writing $(z^2 - 1)^{m/2}$ in place of $(1 - z^2)^{m/2}$. We shall not have occasion to use the alternative definition.

This equation shows that if $m + n > 2n$, i.e., if $m > n$, the quantity $(z^2 - 1)^n$ is differentiated so many times that the result is zero. Therefore, $P_n^m = 0$ if $m > n$. Furthermore, in order that the formula not call for differentiating a negative number of times, it is necessary that $m + n \geq 0$. The index m is therefore limited to the range

$$-n \leq m \leq n. \tag{B–34}$$

When $m = 0$, Eq. B–32 gives the ordinary Legendre polynomial

$$P_n^0(z) = P_n(z).$$

A few of the associated Legendre polynomials for $m \neq 0$ are as follows[4]:

$$\begin{aligned}
P_1^1(z) &= \sqrt{1 - z^2} = \sin\theta, \\
P_2^1(z) &= 3z\sqrt{1 - z^2} = 3\sin\theta\cos\theta = \tfrac{3}{2}\sin 2\theta, \\
P_2^2(z) &= 3(1 - z^2) = 3\sin^2\theta = \tfrac{3}{2}(1 - \cos 2\theta), \\
P_3^1(z) &= \tfrac{3}{2}(5z^2 - 1)\sqrt{1 - z^2} = \tfrac{3}{2}(5\cos^2\theta - 1)\sin\theta, \\
P_3^2(z) &= 15z(1 - z^2) = 15\cos\theta\sin^2\theta, \\
P_3^3(z) &= 15(1 - z^2)^{3/2} = 15\sin^3\theta.
\end{aligned} \tag{B–35}$$

The product $Y(\theta, \psi) = \Theta\Psi$ is called the surface spherical harmonic

$$Y_{mn} = \begin{Bmatrix} e^{jm\psi} \\ e^{-jm\psi} \end{Bmatrix} P_n^m(\cos\theta) \tag{B–36a}$$

or

$$Y_{mn} = \begin{Bmatrix} \cos m\psi \\ \sin m\psi \end{Bmatrix} P_n^m(\cos\theta). \tag{B–36b}$$

A number of conventions are in use about which functions to associate with each other; see, for example, the books by Arfken (Ref. 2) or Morse and Ingard (Ref. 9). We shall not need to go into details here.

In summary, when the wave motion depends on the axial angle ψ as well as on the polar angle θ, the solution of the wave equation has the functional form

[4]The odd, even properties of $P_n^m(z)$ are worth noting. Equation B–32 shows that P_n^m is an odd function of z if $m + n$ is an odd number, an even function if $m + n$ is an even number. This information is useful, for example, when one analyzes the sound field in a hemispherical enclosure.

$$\phi = \begin{Bmatrix} j_n(kr) \\ n_n(kr) \end{Bmatrix} \begin{Bmatrix} \cos m\psi \\ \sin m\psi \end{Bmatrix} P_n^m(\cos\theta) \begin{Bmatrix} \cos\omega t \\ \sin\omega t \end{Bmatrix}. \tag{B–37a}$$

This form is usually convenient for standing wave problems. For radiation problems the formulation

$$\phi + \begin{Bmatrix} h_n^{(1)}(kr) \\ h_n^{(2)}(kr) \end{Bmatrix} \begin{Bmatrix} e^{jm\psi} \\ e^{-jm\psi} \end{Bmatrix} P_n^m(\cos\theta) \begin{Bmatrix} e^{j\omega t} \\ e^{-j\omega t} \end{Bmatrix} \tag{B–37b}$$

is generally preferable. Other combinations are of course possible. For example, the time function in Eq. B–37a may be switched with that in Eq. B–37b. The choice should always be made on the basis of the problem at hand. In any case it should be remembered that the *complete* solution is a sum of all possible solutions of this type, in particular a sum over at least all values of the indices m and n. In standing wave problems summation over a third index, say ℓ, is in general also necessary.

6. Example: Bipolar Pulsating Sphere

Before studying classical standing wave and radiation problems in detail, let us consider the following example, which illustrates the use and manipulation of some of the functions that have been introduced. Radial pulsations of a sphere of radius a generate sound in the surrounding fluid $r > a$. The upper and lower halves of the sphere pulsate out of phase; see Fig. 10.5. The source may be described as a bipolar spherical radiator. The boundary condition at the (exterior) surface of the sphere is

$$u^{(r)}(a, \theta, t) = u_0 e^{j\omega t}, \qquad 0 \leq \theta < \pi/2, \tag{B–38a}$$

$$= -u_0 e^{j\omega t}, \qquad \pi/2 < \theta \leq \pi. \tag{B–38b}$$

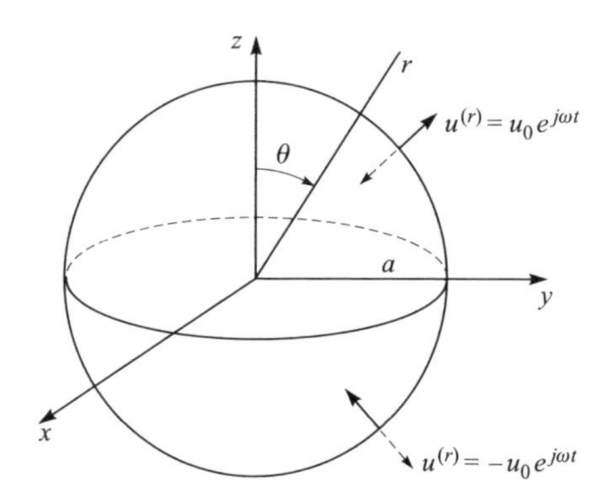

Figure 10.5 Radiation by a bipolar pulsating sphere.

After the sound field radiated by the sphere is found, the solution is analyzed to determine what sort of condition is satisfied on the dividing plane $\theta = \pi/2$. We are led to deduce the solution of a related hemispherical radiation problem.

To solve the radiation problem, first note that since the sound field produced is progressive, not standing (the field outside, not inside, the sphere is desired), we should begin with Eq. B–37b, not Eq. B–37a, as the general solution. The time dependence specified in the boundary condition, Eqs. B–38, calls for $e^{j\omega t}$ to be chosen as the appropriate time function. In turn, if the solution is to represent an outgoing wave, $h_n^{(2)}(kr)$ must be picked as the companion radial function. Next, because the pulsation is axially symmetric, the radiated field cannot depend on ψ. To meet this condition, set m equal to zero. In other words, the ordinary Legendre polynomials $P_n(\cos\theta)$ are sufficient to describe the angular dependence of the field. The solution therefore takes the form of a summation over the single index n:

$$\phi = \sum_n A_n h_n^{(2)}(kr) P_n(\cos\theta) e^{j\omega t}. \tag{B–39}$$

Next apply the boundary condition at the surface of the sphere. The radial particle velocity is

$$u^{(r)} = \phi_r = \sum_n k A_n h_n^{(2)\prime}(kr) P_n(\cos\theta) e^{j\omega t}. \tag{B–40}$$

Matching this expression to Eqs. B–38 yields

$$\sum_n k A_n h_n^{(2)\prime}(ka) P_n(\cos\theta) = \begin{cases} u_0 & \text{for } 0 \le \theta < \pi/2, \\ -u_0 & \text{for } \pi/2 < \theta \le \pi. \end{cases} \tag{B–41}$$

To find the coefficients A_n, use the orthogonality of the Legendre polynomials. We anticipate that since the right-hand side of Eq. B–41 is odd about $z = 0$ (i.e., about $\theta = \pi/2$), only the odd Legendre polynomials will be needed. Multiply Eq. B–41 by $P_m(z)\,dz$ and integrate over the domain $-1 \le z \le 1$:

$$\sum_{n=0}^{\infty} k A_n h_n^{(2)\prime}(ka) \int_{-1}^{1} P_n(z) P_m(z)\,dz = \int_0^1 u_0 P_m(z)\,dz - \int_{-1}^{0} u_0 P_m(z)\,dz. \tag{B–42}$$

Application of Eq. B–15 reduces the left-hand side to

$$k A_m h_m^{(2)\prime}(ka) \frac{2}{2m+1},$$

whereas the right-hand side simplifies to

$$u_0 \int_0^1 [P_m(z) - P_m(-z)]dz.$$

Because $P_m(-z) = (-1)^m P_m(z)$, the right-hand side vanishes for even values of m and equals $2u_0 \int_0^1 P_m(z)\,dz$ for odd values. Solving for A_n (the index reverts from m to n), we obtain

$$A_n = \begin{cases} \dfrac{(2n+1)u_0}{kh_n^{(2)\prime}(ka)} \displaystyle\int_0^1 P_n(z)\,dz & \text{for } n \text{ odd}, \quad \text{(B–43a)} \\ 0 & \text{for } n \text{ even}. \quad \text{(B–43b)} \end{cases}$$

To evaluate the integral in Eq. B–43a, make use of Eq. B–14b:

$$\int_0^1 P_n(z)\,dz = \frac{1}{2n+1} \int_0^1 [P'_{n+1}(z) - P'_{n-1}(z)]\,dz$$
$$= \frac{1}{2n+1} [P_{n+1}(1) - P_{n-1}(1) - P_{n+1}(0) + P_{n-1}(0)].$$

The first two terms on the right-hand side cancel (why?). Because $n+1$ and $n-1$ are even numbers, the last two terms do not vanish. They may be combined by invoking Eq. B–14a, which yields $P_{n+1}(0) = -n(n+1)^{-1}P_{n-1}(0)$. The integral thus reduces to $(n+1)^{-1}P_{n-1}(0)$, and Eq. B–43a becomes

$$A_n = \frac{(2n+1)u_0}{(n+1)kh_n^{(2)\prime}(ka)} P_{n-1}(0) \qquad (\text{for } n \text{ odd}). \tag{B–44}$$

This result is sufficient for our purposes,[5] and we obtain

$$\phi = \sum_{n \text{ odd}} \frac{u_0(2n+1)P_{n-1}(0)}{(n+1)kh_n^{(2)\prime}(ka)} h_n^{(2)}(kr)P_n(z)e^{j\omega t} \tag{B–45}$$

as the final expression for the velocity potential.

What hemispherical radiation problem is also satisfied by Eq. B–45? Inspection of Eq. B–45 reveals an interesting property about the field on the equatorial plane $\theta = \pi/2$. Since only the odd Legendre polynomials participate in the motion, and since all of them are zero on this plane, ϕ vanishes there and the pressure as well (why?). The equatorial plane could therefore be replaced by an infinite pressure release surface without disturbing the sound field. Consider, for example, the problem of downward radiation in water from a

[5]For computation one may wish to use the following relation, obtained by using Eq. B–11:

$$P_{n-1}(0) = (-1)^{(n-1)/2} \frac{(n-1)!}{2^{n-1}\{[(n-1)/2]!\}^2} \quad \text{for } n \text{ odd}.$$

pulsating hemisphere immersed with its plane at the air-water interface. The solution is given by Eq. B–45; in this case the polar angle is restricted to $\theta \geq \pi/2$.

C. STANDING SPHERICAL WAVES: ENCLOSURE PROBLEMS

Standing waves in spherical enclosures are treated in this section. The simplest spherical enclosures are the sphere with a pressure release surface and the sphere with a rigid surface, sometimes called a hollow sphere. An example of the pressure release sphere is a spherical glass vessel filled with a liquid such as water. If the vessel wall is thin enough to be acoustically transparent, the sound field in the liquid "sees" only the surrounding air. Since the impedance of the air is so much less than the impedance of the liquid, the boundary condition at the wall is approximately $p = 0$, or, as Eqs. A–4 and B–5 imply, $\phi = 0$. An acoustical system of this sort has been used to measure the absorption of sound in liquids. After a resonance mode has been excited in the liquid sphere, the source is turned off. The time decay of the modal signal then provides a measure of the absorption of sound in the liquid. The method and some results are described in a paper by Wilson and Leonard (Ref. 15). Gas-filled spheres, "hollow" in the sense that the vessel wall is taken to be rigid, have also been used to measure fluid properties, mainly the speed of sound in gases. See, for example, the review by Moldover, Mehl, and Greenspan (Ref. 6).

Another example of the hollow sphere is of course a spherical room. Not many spherical rooms are built, however. One is at the entrance to the Christian Science Reading Room in Boston, Massachusetts. The spherical "room" is really an overlarge glass globe, about 15 ft in diameter, with a map of the world on the surface. By walking through the inside of the globe on a catwalk, the visitor gets a worm's eye view of the apple earth. The acoustical effects, principally focusing, are awesome. Another example is the hemispherical theater in the Bell Telephone System Building built for the 1965–1966 New York World's Fair. For an analysis of the modes of the room before sound-absorbing materials were installed, see Ref. 4.

Before the analysis of specific examples is undertaken, a few remarks about boundary conditions are in order. Spheres, cones, planes, and wedges have shapes well suited to spherical coordinates (see Fig. 10.6). First consider rigid surfaces, for which the boundary condition is that the normal component of particle velocity $u^{(n)} = \partial\phi/\partial n$ must vanish. See Eq. A–3. If the surface is spherical (Fig. 10.6*a*), the condition is $\phi_r = 0$. If the surface is conical (Fig. 10.6*b*), orient the cone so that its axis coincides with the z axis. At the half-angle θ_0 of the cone, the vanishing of $u^{(\theta)}$ implies $\phi_\theta = 0$. If the surface is a major plane through a sphere, orient the sphere so that the plane is either the equatorial plane $\theta = \pi/2$, in which case $\phi_\theta = 0$, or a $\psi = \text{const}$ plane, in which case the vanishing of $u^{(\psi)}$ implies $\phi_\psi = 0$. A wedge is a special case of two intersecting planes. If the coordinate system is oriented as in Fig. 10.6*c* (the

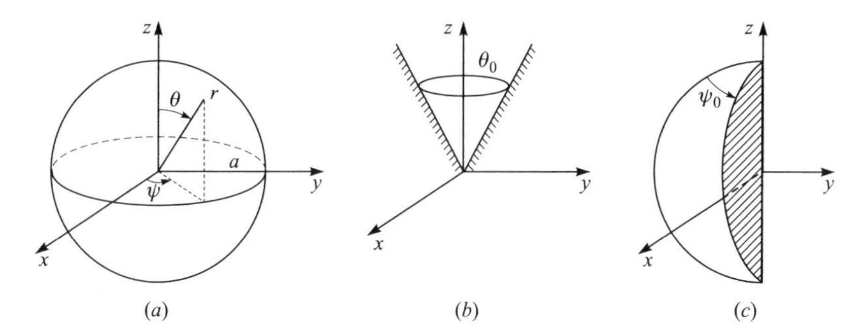

Figure 10.6 Geometry for spherical wave boundary conditions: (*a*) sphere, (*b*) cone, and (*c*) wedge.

z axis is along the intersection and the wedge angle is ψ_0), the boundary conditions are that $\phi_\psi = 0$ at $\psi = 0,\ \psi_0$. For a pressure release surface, the velocity potential itself vanishes on the surface.

In the case of plane waves, we found (Chap. 6) an especially simple relation between p and $u^{(n)}$ at plane boundaries: a zero in pressure is always accompanied by an extremum in normal particle velocity and vice-versa. This principle does not always apply in spherical wave fields. Consider, for example, a simple time-harmonic radial wave incident on a spherical surface. The continuity equation, Eq. 2C–4a, with $\delta\rho$ replaced by p/c_0^2 is

$$j\omega p + \rho_0 c_0^2\left[u_r^{(r)} + \frac{2}{r}u^{(r)}\right] = 0.$$

It is clear that the vanishing of p does not imply the vanishing of $u_r^{(r)}$ (unless r is very large). A particle velocity maximum does not, therefore, occur at a spherical pressure release surface. If, on the other hand, the surface is rigid, the vanishing of $u^{(r)}$ *is* accompanied by a pressure extremum, just as it is in the case of plane waves. To see this, refer to the momentum equation, Eq. 2C–4b, for a radial wave:

$$j\omega\rho_0 u^{(r)} + p_r = 0.$$

1. Pressure Release Sphere

A time-harmonic sound field within a pressure release sphere of radius a is assumed. The general solution is Eq. B–37a. The vanishing of the pressure at $r = a$ implies that ϕ also vanishes at $r = a$. Denote the spatial part of the velocity potential by $\Phi(r, \theta, \psi)$:

$$\Phi(r, \theta, \psi) = \begin{Bmatrix} \cos m\psi \\ \sin m\psi \end{Bmatrix} P_n^m(\cos\theta) \begin{Bmatrix} j_n(kr) \\ n_n(kr) \end{Bmatrix}. \tag{C–1}$$

First, since all the spherical Neumann functions $n_n(kr)$ are unbounded at the origin, they must be discarded. Next, the requirement that $\Phi = 0$ at $r = a$ means that $j_n(ka) = 0$, i.e.,

$$ka = x_{n\ell} \tag{C–2}$$

or

$$f_{n\ell} = x_{n\ell} c_0 / 2\pi a, \tag{C–3}$$

where $x_{n\ell}$ is the ℓth zero of j_n; see Sec. B.2. Let the indices ℓ, m, n denote the (ℓmn)th mode in the sphere,

$$\Phi_{\ell mn} = \begin{Bmatrix} \cos m\psi \\ \sin m\psi \end{Bmatrix} P_n^m(\cos\theta) j_n\left(x_{n\ell} \frac{r}{a}\right), \quad -n \le m \le n. \tag{C–4}$$

The complete solution is of course a sum of all the modes and includes the time function:

$$\phi = \sum_{\ell,m,n} \Phi_{\ell mn}(A_{\ell mn} \cos \omega_{\ell mn} t + B_{\ell mn} \sin \omega_{\ell mn} t). \tag{C–5}$$

A few of the simple modes of vibration in the sphere are now discussed.

a. $\boldsymbol{n = 0}$

When $n = 0$, m is restricted to be zero as well. The roots are therefore $x_{n\ell} = x_{0\ell} = \ell\pi$.

$$\therefore \quad \Phi_{\ell 00} \propto j_0\left(\frac{\ell\pi r}{a}\right) = \frac{\sin(\ell\pi r/a)}{\ell\pi r/a}. \tag{C–6}$$

Furthermore, since $k = \ell\pi/a$, the eigenfrequencies are

$$f_{\ell 00} = \ell c_0 / 2a. \tag{C–7}$$

The pressure vanishes at the surface of the sphere, as required by the boundary condition, and is a maximum at the center. Let $U^{(r)}$ stand for the radial particle velocity with the time dependence removed. Since $U^{(r)} = \Phi_r$, we have

$$U_{\ell 00}^{(r)} \propto \frac{d}{dr} j_0(kr) = k j_0'(kr) = -(\ell\pi/a) j_1(\ell\pi r/a). \tag{C–8}$$

As the curve of j_1 in Fig. 10.4 shows, the radial particle velocity vanishes at the center of the sphere, regardless of the value of ℓ. Given the symmetry of the field, this is a physically reasonable result. Next we notice an example of the fact that the pressure zeros do not always coincide with the particle velocity extrema: $U^{(r)}$ has an extremum, not at the surface of the sphere $r = a$, but at a point in front of the surface, $r < a$. For example, when $\ell = 1$, the particle velocity amplitude has a single maximum, which occurs at about $r = 0.66a$. As ℓ increases (ka also increases, as Eq. C–2 shows), more particle velocity extrema develop, and the one nearest the surface approaches the surface. Finally, as $\ell \to \infty$, the nearest maximum arrives at the surface. The fact that the limiting behavior is the same as that for plane waves is not surprising: at very high frequencies the surface appears locally plane.

b. $\boldsymbol{n = 1}$

Now m may take on values $-1, 0, +1$. Since it turns out that $P_n^{(-m)}$ is related to P_n^m by a constant, it is sufficient to consider the cases $m = 0, +1$. Figure 10.6*a* may be used to visualize geometrical relations. First let $m = 0$. The velocity potential is

$$\Phi_{\ell 01} \propto P_1(\cos\theta) j_1(x_{1\ell} r/a) = \cos\theta j_1(x_{1\ell} r/a). \tag{C–9}$$

In this case the pressure vanishes at the center as well as at the surface. The equatorial plane ($\theta = \frac{1}{2}\pi$) is also a pressure node. For the $m = 1$ mode we have

$$\begin{aligned} \Phi_{\ell 11} &= \begin{Bmatrix} \cos\psi \\ \sin\psi \end{Bmatrix} P_1^1(\cos\theta)\, j_1\left(x_{1\ell}\frac{r}{a}\right) \\ &= \begin{Bmatrix} \cos\psi \\ \sin\psi \end{Bmatrix} \sin\theta\, j_1\left(x_{1\ell}\frac{r}{a}\right). \end{aligned} \tag{C–10}$$

The pressure signal still has nodes at $r = 0$ and $r = a$, but the z axis is now nodal instead of the equatorial plane. One of the $\psi = \text{const}$ planes is also nodal; its position depends on initial conditions.

It will be seen that the eigenfrequency is the same for both the $m = 0$ and $m = 1$ modes. This is a general result: once the indices n and ℓ have been selected, the eigenfrequency is determined; it is not further affected by the value of m. Note also that the *lowest* eigenfrequency is that associated with the simple radial mode $n = 0$.

2. Hollow Sphere

Only a few remarks are made here about the standing wave field inside a hollow sphere. For more discussion the reader is referred to the books by Lord Rayleigh (Ref. 11) and Rschevkin (Ref. 12) or to the article by Flanagan (Ref. 4).

When the wall of the sphere is rigid, the velocity potential must satisfy the condition $\phi_r = 0$ at $r = a$, which is fulfilled if $j'_n(ka) = 0$, i.e., the eigenvalues are now $ka = x'_{n\ell}$. The modes are still given by Eq. C–4 but $x'_{n\ell}$ replaces $x_{n\ell}$. Examine Table 10.1 for the lowest values of $x'_{n\ell}$. The null eigenvalue $x'_{0\,0}$ implies a constant pressure throughout the sphere, since $j_0(0) = 1$ (the 0,0,0 mode is sometimes called the "dc mode"). What is the lowest non-dc mode, i.e., what mode has the lowest (nonzero) eigenfrequency? The answer may prove surprising. Further study of the hollow-sphere problem is left to Problems 10–7 to 10–9, 10–11, and 10–13.

D. RADIATION PROBLEMS

1. Introduction: Multipole Expansion

To describe the radiation of spherical waves, one generally uses complex exponentials for the time function and spherical Hankel functions for the radial function. In particular, our choice for outgoing waves is the combination

$$\phi_{mn} = \begin{Bmatrix} \cos m\psi \\ \sin m\psi \end{Bmatrix} P_n^m(\cos\theta)e^{j\omega t}h_n^{(2)}(kr). \tag{D–1}$$

For many sources in actual practice, for example, a circular-cone loudspeaker, the sound field is symmetric about a main axis of propagation. If this axis is chosen as the z axis of the coordinate system, the solution does not depend on the axial angle ψ. In mathematical terms, independence of ψ is achieved by choosing $m = 0$. The expression for the velocity potential then simplifies to

$$\phi_n \propto P_n(\cos\theta)e^{j\omega t}h_n^{(2)}(kr).$$

The most general solution is the linear combination

$$\phi = \sum_{n=0}^{\infty} A_n P_n(\cos\theta)e^{j\omega t}h_n^{(2)}(kr). \tag{D–2}$$

Let us now explore the physical meaning of the various terms in this solution.

First suppose the source emits sound equally in all directions. In this case the field is independent of the angle θ. Because only one term in the entire infinite series is not a function of θ, namely the $n = 0$ term [recall that $P_0(\cos\theta) = 1$], all the coefficients A_n must be set to zero except for A_0. Equation D–2 reduces to

$$\phi = \phi_0 = A_0 e^{j\omega t}h_0^{(2)}(kr) = A_0\,\frac{e^{j(\omega t - kr)}}{r}, \tag{D–3}$$

where the factor j/k has been absorbed in the coefficient A_0. As we know, this solution describes the sound field radiated by a pulsating sphere. A source that radiates uniformly in all directions is called a monopole.

The next simplest term in Eq. D–2 is the one for $n = 1$:

$$\phi_1 = A_1 \cos\theta e^{j\omega t} h_1^{(2)}(kr),$$

or, since $h_1^{(2)}(x) = -d/dx[h_0^{(2)}(x)]$,

$$\phi_1 = A_1 \cos\theta \frac{\partial}{\partial r}\left[\frac{e^{j(\omega t - kr)}}{r}\right]. \tag{D–4}$$

(Again, various constants have been absorbed in the coefficient.) Sound described by this equation is called dipole radiation. The directivity factor

$$D(\theta) = \cos\theta$$

prescribes a figure-eight pattern of radiation, as shown in Fig. 10.7*a*. Although no proof is given at this point, one can see that two oppositely phased monopoles placed close together along the z axis, as in Fig. 10.7*b*, should produce the same radiation pattern. The sound emitted by the two sources cancels in the x, y plane (the plane bisecting and perpendicular to the line between the two sources) and is strongest along the z axis. Such a pair of monopoles, called a dipole source, is a good physical model of the $n = 1$ term in Eq. D–2.

Two oppositely phased dipoles placed close together constitute a quadrupole. An equivalent model for the quadrupole is four monopoles appropriately phased and grouped. Two basic groupings are possible. Four monopoles arranged on a line, as in Fig. 10.8*a*, make up a *longitudinal* quadrupole, which has a directivity factor proportional to $\cos^2\theta$ (see Sec. D.3.e below,

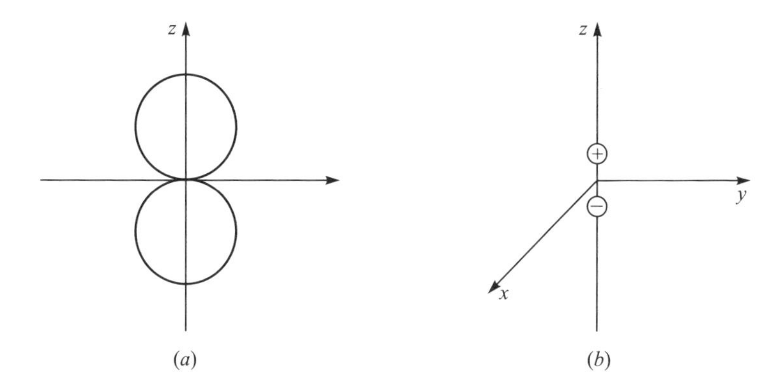

Figure 10.7 Dipole sound: (*a*) radiation pattern; (*b*) two-monopole model.

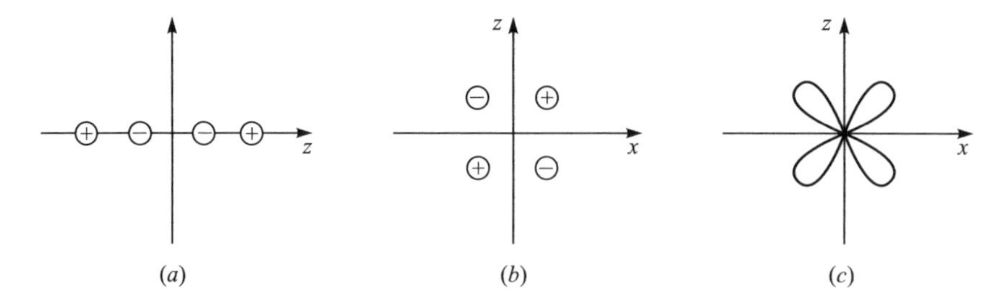

Figure 10.8 Quadrupoles: (*a*) longitudinal quadrupole, (*b*) lateral quadrupole, and (*c*) radiation pattern for lateral quadrupole.

particularly Eq. D–35). If the four monopoles are arranged in a plane, however, as in Fig. 10.8*b*, the result is a *lateral* quadrupole. The cloverleaf radiation pattern for the lateral quadrupole is shown in Fig. 10.8*c*.

It should be mentioned that quadrupoles are sometimes defined mathematically by the $n = 2$ terms of expansions like Eq. D–2. In this interpretation the longitudinal quadrupole is associated with $P_2(\cos\theta)$ and two lateral quadrupoles (one in the x, z plane, the other in the x, y plane) with $P_2^1(\cos\theta)$ and $P_2^2(\cos\theta)$. We shall, however, stick with the more physical model based on combinations of two dipoles or four monopoles.

2. Monopoles

The most general expression for monopole radiation is

$$\phi = \frac{f(t - r/c_0)}{r} \tag{D–5a}$$

(the special function given by Eq. D–3 describes a time-harmonic monopole). The radial component of the particle velocity $u^{(r)}$ (the only component in this case) is

$$u^{(r)} = \phi_r = -\frac{f(t - r/c_0)}{r^2} - \frac{f'(t - r/c_0)}{rc_0}. \tag{D–5b}$$

The pressure is given by

$$p = -\rho_0\phi_t = -\rho_0\frac{f'(t - r/c_0)}{r}. \tag{D–5c}$$

Next let us find a physical attribute of the source that is associated with the function f.

a. Simple Source

A point monopole is called a simple source. Consider the radiation from a pulsating sphere of small but finite radius a. The volume flow of fluid $Q(t)$ from the source is

$$Q(t) = 4\pi a^2 u^{(r)}(a, t), \tag{D–6}$$

or, if Eq. D–5b is used,

$$Q(t) = -4\pi\left[f\left(t - \frac{a}{c_0}\right) + \frac{a}{c_0} f'\left(t - \frac{a}{c_0}\right)\right].$$

The volume flow for a simple source is found by taking the limit as the source radius $a \to 0$, more precisely, as $a/c_0\tau \to 0$, where τ is a characteristic time for the source motion. We obtain

$$\lim_{a \to 0} Q(t) = -4\pi f(t),$$

and Eq. D–5a becomes

$$\phi = -\frac{Q(t - r/c_0)}{4\pi r}.$$

The expressions for f and its derivative are

$$f(t) = -\frac{Q(t)}{4\pi} \quad \text{and} \quad f'(t) = -\frac{\dot{Q}(t)}{4\pi}.$$

These relations show that f is proportional to the volume flow from the source.

Next consider the pressure radiated by the simple source. From Eq. D–5c we have

$$p = \frac{\rho_0 \dot{Q}(t - r/c_0)}{4\pi r}. \tag{D–7}$$

Since $\rho_0 \dot{Q}$ is the mass flow per unit time, or mass acceleration, we see that the pressure at any point r in the field is determined by the mass acceleration of the simple source. The time delay r/c_0 represents the time required for the wave to travel from the source to the field point r. The amplitude diminution factor $1/r$ is due to spherical spreading of the wave. The quantity Q (or sometimes $\dot{Q}$) is called the *strength* of the simple source. The simple source plays an important role in practical applications. For example, at low enough frequency a loudspeaker mounted in a rigid closed cabinet may act as a simple source.

If an actual source is not small, we may divide it into elements each of which acts individually as a simple source. See Fig. 10.9. The total radiation received

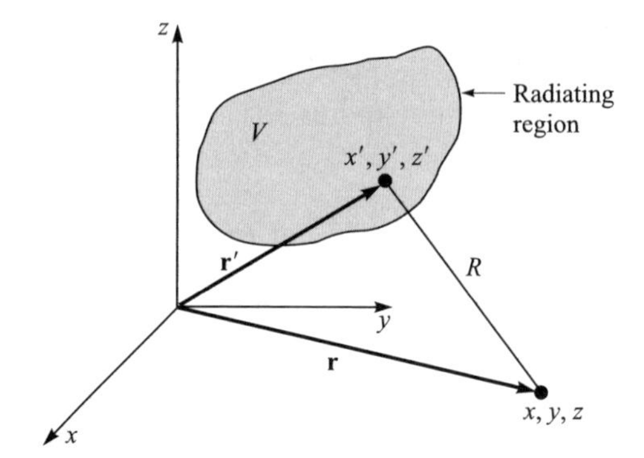

Figure 10.9 Radiation from a region of sources.

at the field point x, y, z is then the sum of the radiations from the individual simple sources, that is,

$$p(x, y, z, t) = \rho_0 \int_V \frac{\dot{q}_v(x', y', z'; t - R/c_0)}{4\pi R} \, dx' \, dy' \, dz', \tag{D–8}$$

where $\dot{q}_v$ is the source strength per unit volume, V is the volume of the source region, and R is the distance between an individual source point (x', y', z') and the field point (x, y, z):

$$R = |\mathbf{r} - \mathbf{r}'| = \sqrt{(x - x')^2 + (y - y')^2 + (z - z')^2}. \tag{D–9}$$

Although arrived at somewhat intuitively, Eq. D–8, called the Green function solution of the wave equation, may be shown to be a solution of the inhomogenous wave equation,

$$\nabla^2 p - \frac{1}{c_0^2} p_{tt} = -\rho_0 \dot{q}_v, \tag{D–10}$$

where $\rho_0 \dot{q}_v$ is the mass acceleration per unit volume of the sources. To derive this wave equation (Problem 10–15), start with the continuity equation for a fluid in which mass sources are present (see Problem 2–1).

If the sources are distributed over an area S instead of the volume V, the volume integral in Eq. D–8 is replaced by a surface integral. Let the coordinate system be chosen so that the source surface S is on the x, y plane, that is, $z' = 0$. In this case the pressure is given by

$$p(x, y, z, t) = \rho_0 \int_S \frac{\dot{q}_a(x', y'; t - R/c_0)}{4\pi R} \, dx' \, dy', \tag{D–11}$$

where $\dot{q}_a$ is the source strength per unit area, and R is given by Eq. D–9 except that $z' = 0$. The radiation is often restricted to the hemisphere in front of the source plane, for example, when the radiation is from a piston set in a plane rigid baffle. In this case each simple source radiates into a hemispherical space (not a full spherical space). The factor 4π in Eq. D–6, which gives the volume flow from the simple source, is therefore replaced by 2π. As a consequence, Eq. D–11 is also modified by having 4π replaced with 2π. The modified version of Eq. D–11 is used in Chap. 13.

b. Time-Harmonic Finite Monopole

Suppose the radius a of the pulsating sphere is not vanishingly small. The analysis is simplest if the pulsation is time harmonic, that is, if the boundary condition is

$$u^{(r)}(a, t) = u_0 \sin \omega t = u_0 \operatorname{Im}(e^{j\omega t}).$$

(More complicated pulsations may be represented by their Fourier series.) Starting with the monopole solution in the form of Eq. D–3, we find $u^{(r)}$ and match the boundary condition by choosing

$$A_0 = -\frac{u_0 a^2}{1 + jka}\, e^{jka}.$$

The potential, particle velocity, and pressure are therefore given by

$$\phi = -\frac{u_0 a^2}{1 + jka} \frac{e^{j[\omega t - k(r-a)]}}{r}, \tag{D–12a}$$

$$u^{(r)} = u_0 \frac{a^2(1 + jkr)}{r(1 + jka)} \frac{e^{j[\omega t - k(r-a)]}}{r}, \tag{D–12b}$$

$$p = \rho_0 c_0 u_0 \frac{jka^2}{1 + jka} \frac{e^{j[\omega t - k(r-a)]}}{r}, \tag{D–12c}$$

respectively. The impedance at any point $r \geq a$ is

$$Z \equiv \frac{p}{u^{(r)}} = \rho_0 c_0 \frac{jkr}{1 + jkr} = \rho_0 c_0 \left[\frac{k^2 r^2}{1 + k^2 r^2} + j \frac{kr}{1 + k^2 r^2} \right]. \tag{D–13}$$

A graph of the real and imaginary parts of $Z/\rho_0 c_0$ is shown in Fig. 10.10.

An analysis of the impedance presented to the source by the medium leads to useful conclusions about efficiency of spherical radiators. The mechanical

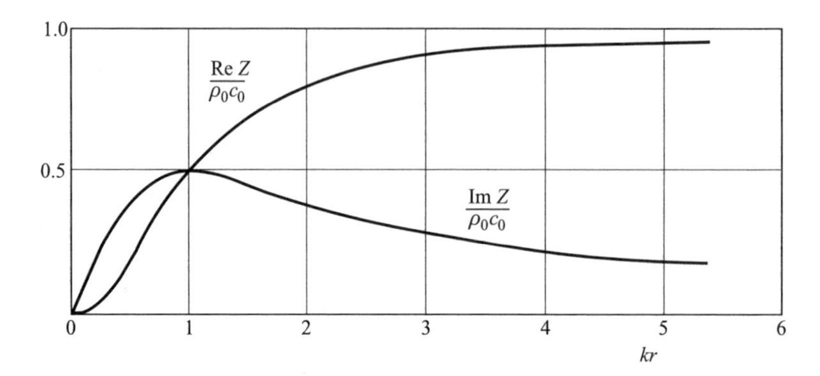

Figure 10.10 Impedance of a diverging radial wave (from Ref. 12).

impedance Z_{mech} seen by the source may be found by evaluating Eq. D–13 at the source and then multiplying by the source surface area:

$$Z_{\text{mech}}|_{r=a} = \frac{Sp}{u^{(r)}}\bigg|_{r=a} = 4\pi a^2 \rho_0 c_0 \left[\frac{k^2a^2}{1+k^2a^2} + j\,\frac{ka}{1+k^2a^2}\right].$$

For values of $ka \gg 1$, i.e., for very large sources and/or very high frequencies, this expression reduces to

$$Z_{\text{mech}}|_{r=a} = 4\pi a^2 \rho_0 c_0,$$

a pure real number. A large source (more precisely, a source that is large compared with a wavelength) therefore sees a purely resistive load, ideal for efficient energy transfer, as it radiates sound into the surrounding fluid. At the other extreme are very small sources or very low frequencies ($ka \ll 1$), for which

$$Z_{\text{mech}}|_{r=a} = jk4\pi a^3 \rho_0 c_0 = j\omega 3 \underbrace{\left(\tfrac{4}{3}\pi a^3 \rho_0\right)}_{m_d}, \tag{D–14a}$$

where m_d is the mass of the fluid displaced by the sphere. Since the load in this case is purely reactive, no acoustic power is transferred to the surrounding fluid. In the general case (ka not restricted to small values), the mechanical reactance is

$$X_{\text{mech}}|_{r=a} = \omega \frac{3m_d}{1+k^2a^2}. \tag{D–14b}$$

The effective mass $3m_d/(1+k^2a^2)$ is called the *accession to inertia*. At low frequencies the accession to inertia is simply three times the mass of fluid displaced by the sphere.

Example 10.1. Resonance frequency of a bubble in a liquid. When a small bubble of radius a is excited radially, the natural frequency of the pulsation is readily found by a mass-spring model. The mass is the accession to inertia produced by the bubble monopole as it tries to radiate sound into the liquid. The stiffness is provided by the gas volume inside the bubble (see Chap. 4) as the oscillating bubble surface attempts to radiate sound into the interior. Since most bubbles are small enough that $ka \ll 1$, the acoustic mass is (see Eq. D–14a) $M_{\text{ac}} = 4\pi a^3 \rho_0/S^2 = \rho_0/4\pi a$, where ρ_0 is the density of the liquid. The acoustic compliance of the gas volume is, from either Problem 10–12 or Eq. 4C–8, $C_{\text{ac}} = V/\gamma p_{0\text{g}} = 4\pi a^3/3\gamma p_{0\text{g}}$, where $p_{0\text{g}}$ is the static pressure of the gas inside the bubble. The impedance of the system is therefore

$$Z_{\text{ac}} = j\omega M_{\text{ac}} + \frac{1}{j\omega C_{\text{ac}}} = \frac{j\omega\rho_0}{4\pi a} + \frac{3\gamma p_{0\text{g}}}{j\omega 4\pi a^3}.$$

Setting the impedance equal to zero yields the resonance frequency f_0:

$$f_0 = \frac{1}{2\pi}\sqrt{\frac{3\gamma p_{0\text{g}}}{\rho_0 a^2}}. \tag{D–15}$$

Checking to determine whether $ka \ll 1$ is a good assumption is left as an exercise for the reader.[6]

c. Intensity of Monopole Radiation

It will be recalled that the acoustic intensity I is the time average of the power per unit area. Equation 1E–15, repeated here for convenience, is

$$I = \frac{1}{t_{\text{av}}}\int_0^{t_{\text{av}}} pu\,dt.$$

The choice of t_{av} depends on the type of signal. The following cases are identified:

1. For time-harmonic signals, let t_{av} be the period of the wave.
2. For continuous, aperiodic signals, such as steady noise, let t_{av} become very large.

[6] In more detailed treatments of bubble resonance, account is taken of the effect of surface tension; see Ref. 10. Still more elaborate effects occur if the oscillations are strong enough to elicit nonlinear behavior on the part of the bubble. See, for example, Ref. 5.

3. For transients, let t_{av} be the duration of the wave.

To compute the intensity of monopole radiation, start with Eqs. D–5b and D–5c and arrange the product pu as follows:

$$pu = \frac{\rho_0}{r^3} f\left(t - \frac{r}{c_0}\right) f'\left(t - \frac{r}{c_0}\right) + \frac{1}{\rho_0 c_0}\left[\frac{\rho_0 f'(t - r/c_0)}{r}\right]^2$$
$$= \frac{\rho_0}{r}\phi\phi_t + \frac{1}{\rho_0 c_0} p^2.$$

Substitution in Eq. 1E–15 yields

$$I = \frac{\rho_0}{2r}\frac{\phi^2(t_{av}) - \phi^2(0)}{t_{av}} + \frac{p_{rms}^2}{\rho_0 c_0}.$$

The first set of terms vanishes for cases 1 and 2 above. It also vanishes in the third case provided the transient is zero at its beginning and end. Even if the transient is one with nonzero extremes, such as an N wave, however, we can make the first term vanish simply by redefining the limits of integration to be $0-$ and $t_{av}+$. In all three cases, therefore, the final result is

$$I = p_{rms}^2 / \rho_0 c_0, \tag{D–16}$$

which is the same result found for plane waves. For example, Eq. D–12c shows that for a time-harmonic monopole of finite radius, $p_{rms} = P_s a / r\sqrt{2}$, where $P_s = \rho_0 c_0 u_0 ka / \sqrt{1 + k^2 a^2}$ is the pressure amplitude at the source. The intensity in this case is

$$I = \frac{P_s^2 a^2}{2\rho_0 c_0 r^2}.$$

d. Applications

First consider the sound field produced in water by a small time-harmonic, omnidirectional radiator located a short distance h below the water surface. Let us model the radiator as a simple source and the air-water interface as a pressure release boundary. Figure 10.11 shows the x axis along the water surface and the source S a short distance h below the surface. The origin O is chosen to be the point on the surface immediately above the source S. We wish to calculate the pressure and intensity at any point L in the farfield, here meaning $r \gg h$. An expression for the total power radiated is also desired.

The problem is easily worked by the image method. An image source S' is placed a distance h above the boundary, and the boundary is removed, that is, the air is replaced by water. If S' is driven out of phase with respect to S, the

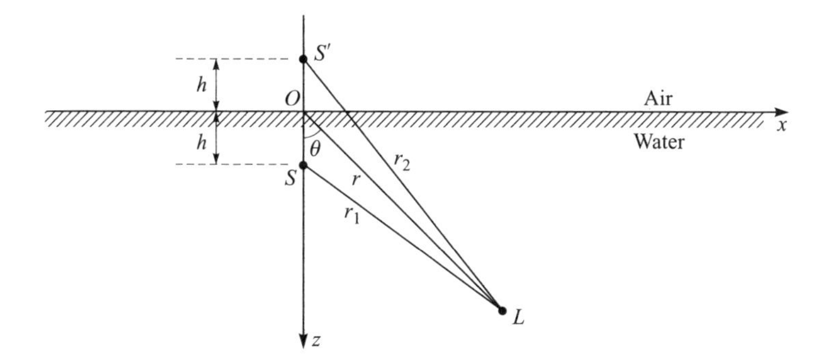

Figure 10.11 Underwater monopole near the water surface.

original boundary condition, that p vanish at $z = 0$, is satisfied. The field at L is now the sum of the fields due to S and S',

$$p = \frac{A}{r_1} e^{j(\omega t - kr_1)} - \frac{A}{r_2} e^{j(\omega t - kr_2)}. \tag{D–17}$$

Because L has been assumed to be in the farfield, lines SL, OL, and $S'L$ are approximately parallel, and we have

$$r_1 \doteq r - h\cos\theta \qquad r_2 \doteq r + h\cos\theta.$$

These expressions are used in the exponents (phase factors) in Eq. D–17. Although the extra terms $\pm h\cos\theta$ are small compared with r, they can be significant because of their effect on the respective phases of the two signals. The extra terms can, for example, determine whether the two signals arrive in phase and thus reinforce each other or arrive out of phase and therefore cancel. In the amplitude factors $1/r_1$ and $1/r_2$, however, it is sufficient to replace r_1 and r_2 by r because the extra distance $\pm h\cos\theta$ has little effect on the amplitude, that is, L is so far away that the difference between $1/r$ and $1/(r \pm h\cos\theta)$ is insignificant. With these approximations in place, Eq. D–17 becomes

$$\begin{aligned} p &= \frac{A}{r} e^{j(\omega t - kr)}[e^{jkh\cos\theta} - e^{-jkh\cos\theta}] \\ &= \frac{A}{r} e^{j(\omega t - kr)}[2j\sin(kh\cos\theta)]. \end{aligned} \tag{D–18}$$

The expression in front of the brackets represents the free-field pressure p_{free} at L, that is, the pressure signal if no boundary were present. The ratio p/p_{free} thus characterizes the effect of the boundary:

$$\frac{p}{p_{\text{free}}} = j2\sin(kh\cos\theta) = jF(\theta).$$

If $kh \ll 1$, F reduces to $2kh\cos\theta$. The field is that of a dipole, and the radiation is very weak. This result shows that an omnidirectional underwater sound source is very inefficient if it is located within a fraction of a wavelength of the water surface. If $kh = \pi/2$ ($h = \lambda/4$), however, the pressure on the z axis ($\theta = 0$) is twice the free-field value.

Now find the intensity and power radiated. Equation D–18 may be used to compute p_{rms}. Substitution in Eq. D–16 (why is Eq. D–16 applicable in this case?) yields

$$I = I_{\text{free}}F^2(\theta) = I_{\text{free}}4\sin^2(kh\cos\theta), \tag{D–19}$$

where $I_{\text{free}} = A^2/2r^2\rho_0 c_0$ is the intensity of a single simple source in a free field. Integration of the intensity over a hemisphere centered at O (see Fig. 10.12) gives the total power radiated:

$$W = \int I\, dS = 8\pi r^2 I_{\text{free}} \int_0^{\pi/2} \sin^2(kh\cos\theta)\sin\theta\, d\theta,$$

which is easily integrated by letting $q = \cos\theta$ so that $dq = -\sin\theta\, d\theta$. The result is

$$W = W_{\text{free}}\left(1 - \frac{\sin 2kh}{2kh}\right). \tag{D–20}$$

Here $W_{\text{free}} = 4\pi r^2 I_{\text{free}}$ is the power radiated by a simple source in a free field. The ratio W/W_{free} is plotted in Fig. 10.13. It is seen that sources very close to the surface are most affected by it.

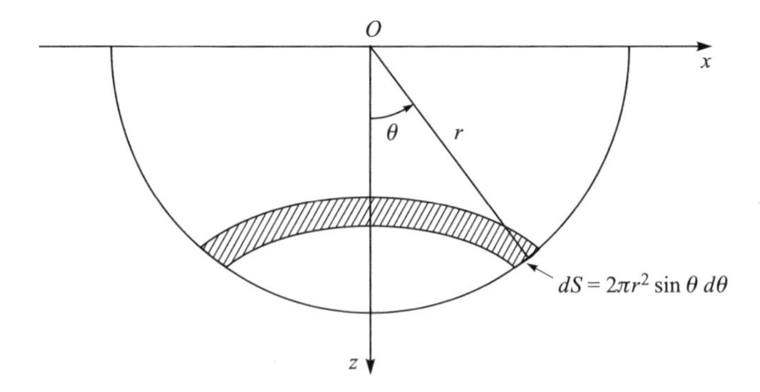

Figure 10.12 Calculation of total power output from a time-harmonic monopole located near a pressure release surface.

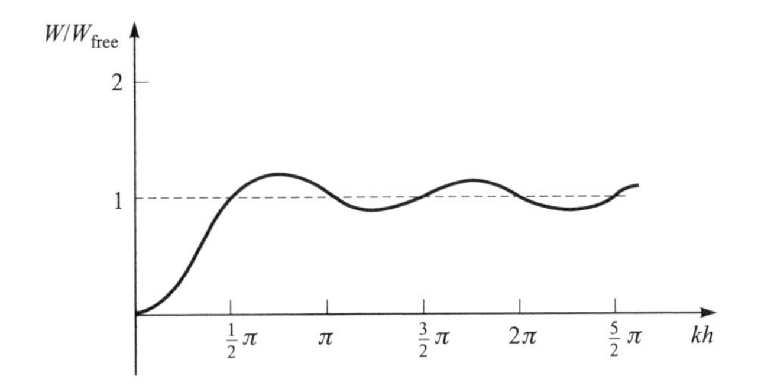

Figure 10.13 Relative power output of a monopole near a pressure release surface.

Another important example is that of a monopole near a rigid plane. In this case the total power radiated may be shown to be (see Problem 10–20)

$$W = W_{\text{free}}\left(1 + \frac{\sin 2kh}{2kh}\right). \tag{D–21}$$

It is seen that the total power radiated is doubled for sources close to the surface. Additional gains may be realized by putting the source in a two-surface or three-surface corner. The reader may qualitatively verify these results by listening to a speaker who stands first in the middle of a room, then as close as possible to a single wall, and finally in a corner.

3. Dipoles

Recall that for the monopole the time-harmonic solution is given by Eq. D–3 and the more general solution by Eq. D–5a. This suggests that for the dipole, for which the time-harmonic solution is Eq. D–4, a more general solution might be

$$\phi = \cos\theta \frac{\partial}{\partial r}\left[\frac{F(t - r/c_0)}{r}\right]. \tag{D–22a}$$

Indeed, this expression is found to satisfy the ψ-independent wave equation, Eq. B–1. The corresponding pressure is

$$p = -\rho_0 \phi_t = -\rho_0 \cos\theta \frac{\partial}{\partial r}\left[\frac{F'(t - r/c_0)}{r}\right], \tag{D–22b}$$

and the components of particle velocity are

$$u^{(r)} = \phi_r = \cos\theta \frac{\partial^2}{\partial r^2}\left[\frac{F(t - r/c_0)}{r}\right], \tag{D–22c}$$

$$u^{(\theta)} = \frac{1}{r}\phi_\theta = -\frac{\sin\theta}{r}\frac{\partial}{\partial r}\left[\frac{F(t - r/c_0)}{r}\right]. \tag{D–22d}$$

Equations D–22 are the formal mathematical expressions for the acoustic dipole. Next we consider two actual physical sources that, in the limit, produce dipole radiation.

a. Model of a Dipole as Two Simple Sources of Opposite Phase

Two simple sources a distance a apart are of equal but opposite strength, as shown in Fig. 10.14. What are the characteristics of the radiation from the pair? Assume for the moment that the observation point L is far enough away that both sources are at the same angle θ from the observer. To obtain the pressure field at L, add the contributions from the two sources:

$$\begin{aligned} p &= \sum_{i=1}^{2} \frac{\rho_0 \dot{Q}_i(t - r_i/c_0)}{4\pi r_i} \\ &= -\frac{\rho_0}{4\pi}\left[\frac{\dot{Q}[t - (r + \Delta r)/c_0]}{r + \Delta r} - \frac{\dot{Q}[t - r/c_0]}{r}\right] \\ &= -\frac{\rho_0 a \cos\theta}{4\pi}\left\{\frac{1}{\Delta r}\left[\frac{\dot{Q}[t - (r + \Delta r)/c_0]}{r + \Delta r} - \frac{\dot{Q}[t - r/c_0]}{r}\right]\right\}. \end{aligned}$$

In the limit as $\Delta r \to 0$, the quantity in curly brackets becomes $(\partial/\partial r)[\dot{Q}(t - r/c_0)/r]$.

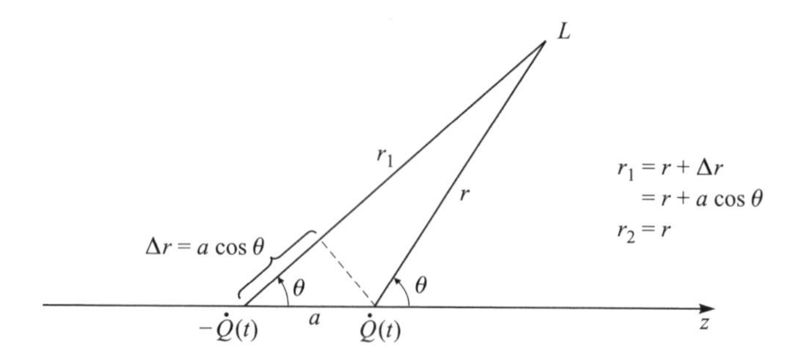

Figure 10.14 Dipole composed of two oppositely phased simple sources.

$$\therefore \lim_{\Delta r \to 0} p = -\cos\theta \frac{\partial}{\partial r}\left[\frac{\rho_0 a\dot{Q}(t - r/c_0)}{4\pi r}\right].$$

Now, since we have let $\Delta r \to 0$, we must also let $a \to 0$. Ordinarily this means canceling out the two sources completely, leaving zero for the pressure field. To avoid such a trivial result, we let the source strength become infinite while the separation becomes zero, the limit being taken so that the product $a\dot{Q}$ stays finite. The pair of monopoles has now become a point dipole. For the time being, refer to $\rho_0 a\dot{Q}(t)$ as the *dipole source strength*. Since this quantity has units of force, denote it by $f_d(t)$. The pressure due to a point dipole is therefore

$$p = -\cos\theta \frac{\partial}{\partial r}\left[\frac{f_d(t - r/c_0)}{4\pi r}\right]. \tag{D–23}$$

Notice that this expression is valid everywhere. Although we started by assuming the observer far away, the assumption became irrelevant when the limit $a \to 0$ was taken.

b. Radiation Characteristics of a Dipole

Perhaps the most striking property of dipole radiation is its figure-eight directivity pattern, which has already been noted; see Fig. 10.7*a*. The range and frequency dependence of the dipole field is conveniently analyzed by assuming a time-harmonic source,

$$f_d(t) = F_0 e <^{j\omega t} .$$

The differentiation indicated in Eq. D–23 yields

$$p = \cos\theta \frac{1 + jkr}{4\pi r^2} F_0 e^{j(\omega t - kr)}. \tag{D–24}$$

At low frequencies and/or short ranges ($kr \ll 1$), the pressure amplitude falls as $1/r^2$, that is, the sound pressure level drops 12 dB every time the distance is doubled. The region in which this occurs is called the nearfield of the dipole. See Fig. 10.15. Notice that monopole radiation does not have this characteristic; pressure due to a monopole decreases as $1/r$ regardless of distance or frequency. In the dipole farfield ($kr \gg 1$) ordinary spherical spreading sets in, and the drop in sound pressure level is only 6 dB for each doubling of distance. Because the farfield amplitude is proportional to k, however, radiation at low frequencies is poor.

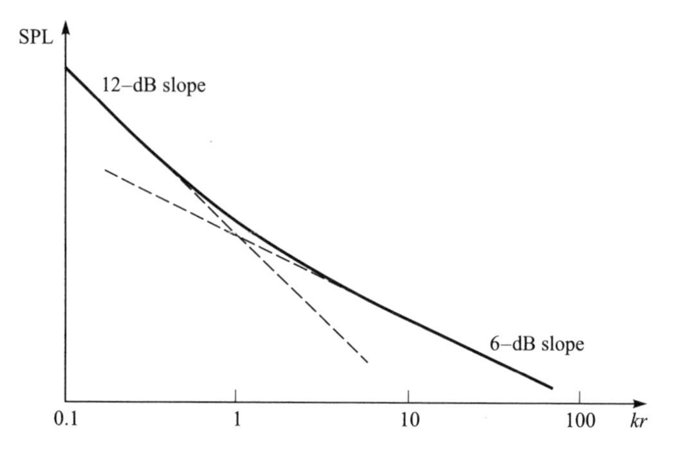

Figure 10.15 Sound pressure level of dipole radiation as a function of kr.

c. *Model of a Dipole as a Translating Sphere*

Can the dipole solution satisfy the boundary condition at the surface of a translating sphere? As Fig. 10.16 shows, the radial velocity at the surface of a sphere restricted to move along the z axis with velocity $v_a(t)$ is[7]

$$u^{(r)}|_{r=a} = v_a(t)\cos\theta. \tag{D–25}$$

Requiring that Eq. D–22c satisfy this condition yields

$$v_a(t) = \frac{2F(t - a/c_0)}{a^3} + \frac{2F'(t - a/c_0)}{a^2 c_0} + \frac{F''(t - a/c_0)}{a c_0^2}. \tag{D–26}$$

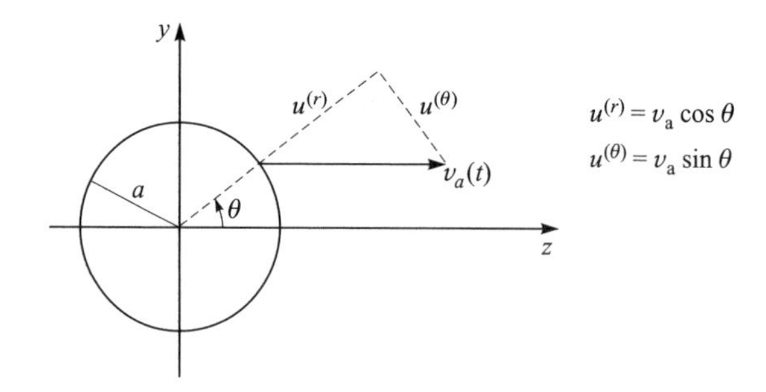

Figure 10.16 Radiation from a translating sphere.

[7]We assume that the movement of the sphere is vibratory and small enough that specifying the boundary condition at the fixed position a, θ rather than at the surface, which is moving, is a reasonably good approximation.

The function F may be obtained by solving this equation (Ref. 3). The pressure at any point may then be computed by means of Eq. D–22b:

$$p = \rho_0 \cos\theta \left[\frac{F'(t - r/c_0)}{r^2} + \frac{F''(t - r/c_0)}{rc_0} \right]. \tag{D–27}$$

Thus the dipole solution does indeed represent the acoustic radiation from a translating sphere, although in practice the relation between the translational velocity v_a and the function F may be somewhat complicated. Incidentally, why does the tangential component of the sphere velocity $u^{(\theta)} = v_a \sin\theta$ not play a role in the problem?

Next consider the force exerted on the fluid by the translating sphere. First, using Eq. D–27, calculate the force in the z direction $f^{(z)}$ (see Fig. 10.17):

$$\begin{aligned} f^{(z)} &= \int_0^\pi p(a, t)\, dS \cos\theta \\ &= 2\pi\rho_0 \left[F'\left(t - \frac{a}{c_0}\right) + \frac{a}{c_0} F''\left(t - \frac{a}{c_0}\right) \right] \int_0^\pi \sin\theta \cos^2\theta \, d\theta \\ &= \frac{4}{3} \pi\rho_0 \left[F'\left(t - \frac{a}{c_0}\right) + \frac{a}{c_0} F''\left(t - \frac{a}{c_0}\right) \right]. \end{aligned} \tag{D–28}$$

The force in the y direction $f^{(y)}$ is proportional to the integral

$$\int_0^\pi \sin^2\theta \cos\theta \, d\theta = \left. \frac{\sin^3\theta}{3} \right|_0^\pi = 0.$$

$$\therefore\ f^{(y)} = 0.$$

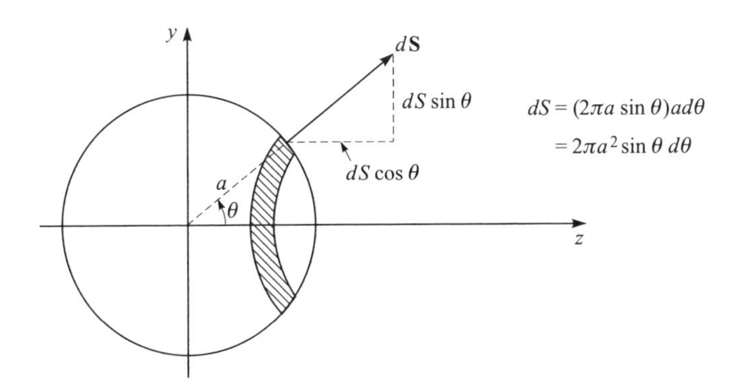

Figure 10.17 Force exerted by a translating sphere on the surrounding fluid.

Finally, calculate the total force needed to move the sphere. The total force is $f^{(z)}$ plus the inertial force of the sphere itself. Let the inertial force be that of the fluid displaced by the sphere,

$$f^{\text{inertial}} = \text{mass} \times \text{acceleration}$$
$$= \frac{4}{3}\pi a^3 \rho_0 \dot{v}_a,$$

or, if Eq. D–26 is used,

$$f^{\text{inertial}} = \frac{8}{3}\pi\rho_0\left[F'\left(t - \frac{a}{c_0}\right) + \frac{a}{c_0}F''\left(t - \frac{a}{c_0}\right) + \frac{a^2}{2c_0^2}F'''\left(t - \frac{a}{c_0}\right)\right].$$

The total force $f_d = f^{(z)} + f^{\text{inertial}}$ is therefore

$$f_d = 4\pi\rho_0\left[F'\left(t - \frac{a}{c_0}\right) + \frac{a}{c_0}F''\left(t - \frac{a}{c_0}\right) + \frac{a^2}{3c_0^2}F'''\left(t - \frac{a}{c_0}\right)\right]. \quad \text{(D–29)}$$

Analysis of Eq. D–29 leads to a new physical interpretation of dipole radiation. For a point dipole ($a \to 0$), Eq. D–29 reduces to

$$f_d = 4\pi\rho_0 F'(t), \quad \text{or} \quad F'(t) = \frac{f_d(t)}{4\pi\rho_0},$$

whence Eq. D–22b reduces to Eq. D–23. Both the two-monopole model and the translating sphere model therefore lead to the same formulas for a point dipole. The translating sphere model does, however, provide a simpler physical explanation of dipole source strength: f_d is the total force needed to move the point sphere in translatory fashion. A monopole is a *mass source*, a dipole a *force source*.

Example 10.2. Unbaffled loudspeaker as an example of a dipole. The cone of a loudspeaker may be modeled as a disk that vibrates back and forth. The following qualitative analysis for the disk is due to Stokes (Ref. 14). An unbaffled disk, Fig. 10.18*a*, is a very inefficient radiator of sound at low frequencies. As the disk moves to the right, the fluid pushed out of the way is simply sucked around the edge to fill in at the back side of the disk. The movement of the disk to the right does push the fluid on the right and pull the fluid on the left, but these forces give rise just to local hydraulic, that is, incompressible, flow of the fluid around the edge of the disk. Stokes called this reciprocating flow. Since the medium is compressed almost not at all in this case, very little sound is radiated. At high frequencies, however, time is too short for the reciprocating flow to equalize the pressure fore and aft before the

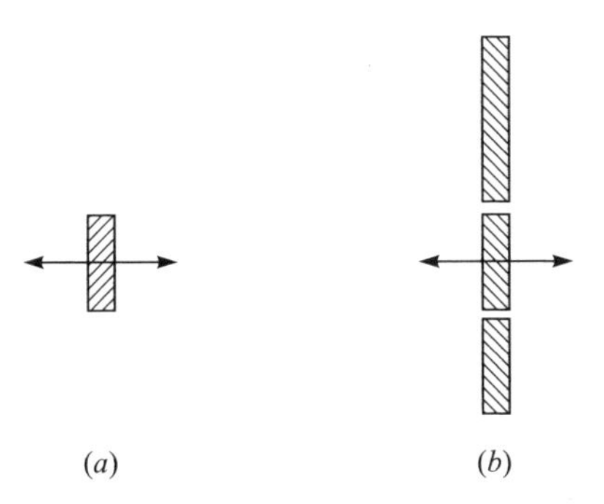

Figure 10.18 Vibration of (*a*) unbaffled and (*b*) baffled disks.

disk turns around and moves the other way. The medium does compress in this case, therefore, and as a result sound is radiated. How does one improve the sound radiation at low frequencies? By introducing a baffle, which prevents the reciprocating flow. See Fig. 10.18*b*. Since the pressure cannot now be equalized by flow from front to back, the medium compresses and sound is radiated. Baffling the disk effectively changes its radiation from dipole into monopole. Radiation from a baffled disk (or piston) is the subject of Chap. 13.

d. Dipole Applications

The radiation of a time-harmonic dipole in the presence of a reflecting plane is considered here. The pressure due to a single dipole in a free field is given by Eq. D–24.

First consider the field produced by a dipole with its axis parallel to the plane. See Fig. 10.19*a*. Let the observation point L be in the farfield in the sense that $r \gg h$. Furthermore, to keep the problem simple, (1) let L be in the y, z plane and (2) take $kr \gg 1$, i.e., let L also be in the farfield in terms of wavelengths. Given the last assumption, the pressure p_1 due to the direct radiation from the dipole is

$$\begin{aligned} p_1 &= \cos\theta' \frac{jk}{4\pi r_1} F_0 e^{j(\omega t - kr_1)} \\ &= \sin\theta \frac{jkF_0}{4\pi} \frac{e^{j[(\omega t - kr) + kh\cos\theta]}}{r - h\cos\theta}. \end{aligned}$$

In the farfield ignore $h\cos\theta$ in the amplitude factor but not in the phase factor. Thus

$$\therefore\ p_1 = \sin\theta \frac{jkF_0}{4\pi r} e^{j(\omega t - kr)} e^{jkh\cos\theta}. \tag{D–30}$$

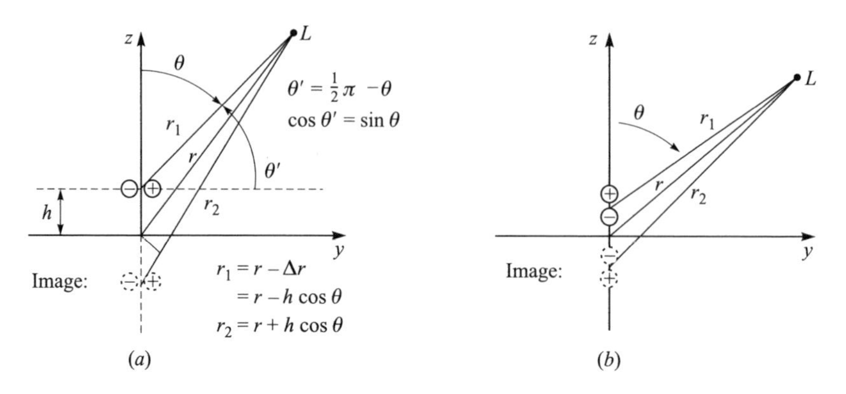

Figure 10.19 Dipole in front of a rigid plane. Dipole axis (*a*) parallel to the plane and (*b*) perpendicular to the plane.

Now use the image method to find the reflected field p_2. Because the plane is rigid, the image must have the same phase as the source, i.e.,

$$p_2 = \sin\theta \frac{jk}{4\pi r_2} F_0 e^{j(\omega t - kr_2)} = \sin\theta \frac{jk}{4\pi r} F_0 e^{j(\omega t - kr)} e^{-jkh\cos\theta}.$$

The total field is

$$p_{\text{tot}} = p_1 + p_2 = 2\sin\theta\cos(kh\cos\theta)\frac{jkF_0}{4\pi r} e^{j(\omega t - kr)}. \qquad \text{(D–31)}$$

In particular, for $kh \ll 1$ (the dipole very close to the plane), the expression reduces to

$$p_{\text{tot}} = 2\sin\theta \frac{jkF_0}{4\pi r} e^{j(\omega t - kr)} = 2p_1, \qquad \text{(D–32)}$$

where p_1 is given by Eq. D–30. This result is just what one would expect, since the real and image dipoles become superposed. Finally, it may be seen that the restriction $kr \gg 1$ may be removed simply by replacing the approximate factor jk/r in the amplitude in Eq. D–31 by its exact version $(1 + jkr)/r^2$.

Next let the dipole axis be perpendicular to the plane. See Fig. 10.19*b*. Again the observation point L is the farfield, but because of the symmetry, it is not necessary to restrict L to the y, z plane. The pressure due to the direct radiation is

$$p_1 = \cos\theta \frac{jkF_0}{4\pi r} e^{j(\omega t - kr)} e^{jkh\cos\theta}. \qquad \text{(D–33)}$$

Because in this case the orientation of the image dipole is opposite to that of the actual dipole (see Fig. 10.19*b*), the image pressure is

$$p_2 = -\cos\theta \frac{jkF_0}{4\pi r} e^{j(\omega t - kr)} e^{-jkh\cos\theta}.$$

$$\therefore\ p_{\text{tot}} = p_1 + p_2 = -2\cos\theta \sin(kh\cos\theta) \frac{kF_0}{4\pi r} e^{j(\omega t - kr)}. \quad \text{(D–34)}$$

If $kh \ll 1$, the total pressure is

$$p_{\text{tot}} = -2kh\cos^2\theta \frac{kF_0}{4\pi r} e^{j(\omega t - kr)}. \quad \text{(D–35)}$$

The farfield is extremely weak because, as the dipole and its image approach each other, they tend to cancel out. An example is an unbaffled loudspeaker placed close to a wall with its face parallel to the wall. Comparison of Figs. 10.8*a* and 10.19*b* indicates that the field is that of a longitudinal quadrupole; notice the $\cos^2\theta$ directivity.

REFERENCES

1. M. Abramowitz and I. A. Stegun (Eds.), *Handbook of Mathematical Functions*, National Bureau of Standards, Applied Mathematics Series 55 (U.S. Government Printing Office, Washington, DC,1964).
2. G. Arfken, *Mathematical Methods for Physicists*, 3rd ed. (Academic Press, New York, 1985), Chap. 12 and Art. 11.6.
3. P. E. Doak, "An introduction to sound radiation and its sources," in E. J. Richards and D. J. Mead (Eds.), *Noise and Fatigue in Aeronautics* (John Wiley and Sons, New York, 1968), Chap. 1, pp. 1–42.
4. J. L. Flanagan, "Acoustic modes of a hemispherical room," *J. Acoust. Soc. Am.* **37**, 616–618 (1965).
5. H. G. Flynn, "Physics of acoustic cavitation in liquids," in W. P. Mason, Ed., *Physical Acoustics*, Vol. IB (Academic Press, New York, 1964), pp. 57–152.
6. M. B. Moldover, J. B. Mehl, and M. Greenspan, "Gas-filled spherical radiators: Theory and experiment," *J. Acoust. Soc. Am.* **79**, 253–272 (1986).
7. P. M. Morse, *Vibration and Sound*, 2nd ed. (McGraw-Hill Book Co., New York, 1948), pp. 314–319. Reprinted by Acoustical Society of America, New York, 1981.
8. P. M. Morse and H. Feshbach, *Methods of Theoretical Physics* (McGraw-Hill Book Co., New York, 1953).
9. P. M. Morse and K. U. Ingard, *Theoretical Acoustics* (McGraw-Hill Book Co., New York, 1968), pp. 332–338. Reprinted by Princeton University Press, Princeton, NJ, 1986.

10. A. D. Pierce, *Acoustics: An Introduction to Its Physical Principles and Applications* (McGraw-Hill, New York, 1981, p. 438). Reprinted by Acoustical Society of America, New York, 1989.
11. Lord Rayleigh, *Theory of Sound*, Vol. II (Dover Publications, New York, 1945), Chap. 17.
12. S. N. Rschevkin, *A Course of Lectures on the Theory of Sound* (MacMillan, New York, 1963), Chaps. 4 and 8.
13. A. Sommerfeld, *Partial Differential Equations* (Academic Press, New York, 1949), pp. 124–134, 143–152.
14. G. G. Stokes, "On the communication of vibration from a vibrating body to a surrounding gas," *Phil. Trans. Roy. Soc. London* **158**, 447–456 (1868). Reprinted in R. B. Lindsay (Ed.), *Physical Acoustics* (Dowden, Hutchinson, & Ross, Stroudsberg, PA, 1974), pp. 22–31.
15. O. B. Wilson and R. W. Leonard, "Measurements of sound absorption in aqueous salt solutions by a resonator method," *J. Acoust. Soc. Am.* **26**, 223–225 (1954).

PROBLEMS

10–1. Uniform pulsation of a sphere of radius r_0 sends sound into the surrounding fluid $r > r_0$. The boundary condition at the source is

$$u(r_0, t) = u_0 \sin \omega t = u_0 \operatorname{Im} e^{j\omega t}.$$

The surrounding fluid contains no other bodies and extends to $r = \infty$.

(a) By evaluating the functions f and g in Eq. A–6, find the solution in terms of the velocity potential ϕ (do not use separation of variables to solve the problem). You should find that your solution is made up of a steady-state part and a transient part.

(b) Show that the transient part dies out as $t \to \infty$.

The rest of the problem has to do with the steady-state solution only.

(c) Find the pressure p and the radial particle velocity u that correspond to the steady-state part of ϕ. Check whether your expression for u satisfies the boundary condition.

(d) Find the impedance $Z = p/u$.

10–2. A source S of spherical waves of angular frequency ω is located at the center of a sphere of radius a (see sketch). The values of density and sound speed are ρ_1 and c_1, respectively, inside the sphere and ρ_2 and c_2 outside. The surface of the sphere is simply the interface between the two fluids. Reflection and transmission of sound at the interface is the subject of this problem. The incident, reflected, and transmitted waves may be represented, respectively, by

$$p_{\text{inc}} = A\frac{a}{r}\exp j[\omega t - k_1(r - a)],$$

$$p_{\text{refl}} = B\frac{a}{r}\exp j[\omega t + k_1(r - a)],$$

$$p_{\text{tr}} = C\frac{a}{r}\exp j[\omega t - k_2(r - a)].$$

(What then are the values of p^+, p^-, and p^{tr}?) The impedance for an outgoing spherical wave is given by Eq. 1D–19 or Eq. D–13; you will have to derive the analogous expression for an incoming spherical wave. In what follows, it simplifies the algebra to let $\alpha_1 = 1/k_1 a$ and $\alpha_2 = 1/k_2 a$.

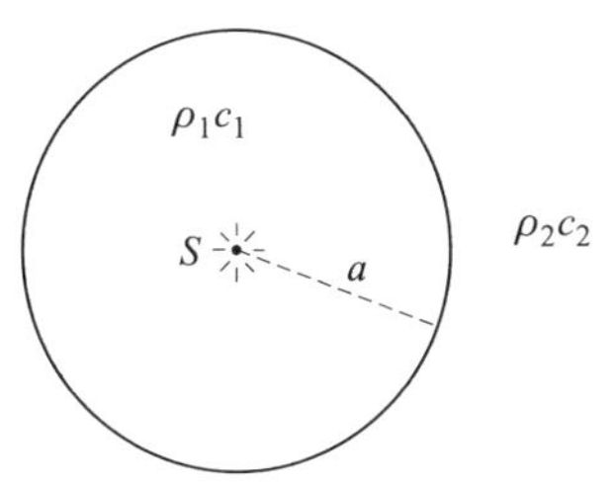

(a) Derive expressions for the pressure reflection and transmission coefficients R and T, respectively, for the interface.

PARTIAL ANSWER: $$R = \frac{(1 - Z_1/Z_2) - j\alpha_1(1 - \rho_1/\rho_2)}{(1 + Z_1/Z_2) + j\alpha_1(1 - \rho_1/\rho_2)},$$

where $Z_1 = \rho_1 c_1$ and $Z_2 = \rho_2 c_2$.

(b) Find limiting forms for R and T for high frequencies ($k_1 a$, $k_2 a \to \infty$) and for low frequencies ($k_1 a$, $k_2 a \to 0$).

(c) Find the conditions under which perfect transmission ($R = 0$) occurs. Compare with the conditions for perfect transmission of plane waves through a plane interface.

10–3. Using results from the previous problem, find the values of R and T for a helium-filled balloon, air on the outside, for a frequency such that $k_1 a = 2$, where k_1 is the wave number in the helium. The temperature is 20°C, at which point helium has sound speed 1000 m/s and density 0.166 kg/m^3. How do the values you find compare with those for plane waves normally incident on a plane interface between helium and air?

10–4. When one wishes to compute the diffraction of a plane wave by a sphere, it is convenient to express the incident plane wave as an infinite series of spherical waves, as follows (Ref. 2, p. 665):

$$e^{j(\omega t-kz)}=e^{j(\omega t-kr\cos\theta)}=e^{j\omega t}\sum_{n=0}^{\infty}a_n j_n(kr)P_n(\cos\theta).$$

In other words, put

$$e^{-jkr\cos\theta}=\sum_{n=0}^{\infty}a_n j_n(kr)P_n(\cos\theta).$$

Show that $a_n=(-j)^n(2n+1)$. It is suggested that you follow these steps:

(a) Use the orthogonality of the Legendre polynomials to obtain an expression for the coefficients $a_n j_n(kr)$ in terms of an integral.

(b) To simplify the integral, eliminate the r-dependence by differentiating n times with respect to kr and then set $kr=0$.

(c) Use Eq. B–16 to evaluate the remaining integral.

10–5. The sonic boom is an example of an acoustical transient called an N wave. The waveform is defined by $N(t)=-At/T$ for $|t|<T$, $N(t)=0$ otherwise.

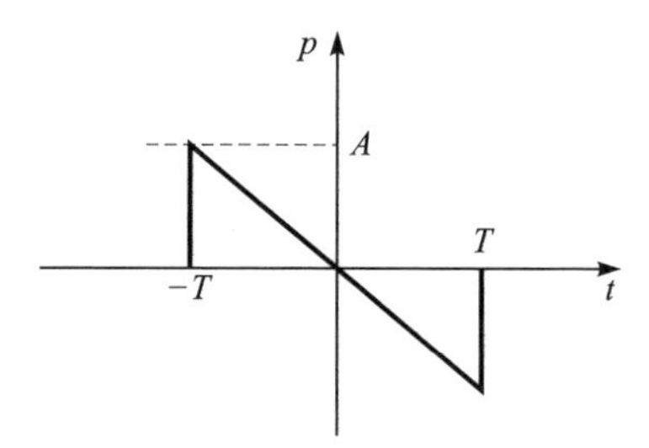

(a) Find the amplitude spectrum $|\overline{N}(\omega)|$ of the N wave, where $\overline{N}(\omega)$ is the Fourier transform

$$\overline{N}(\omega)=\int_{-\infty}^{\infty}N(t)e^{-j\omega t}\,dt.$$

You should be able to express $\overline{N}(\omega)$ in terms of a first-order spherical Bessel function.

(b) Plot the spectrum in the form $20\log_{10}|\overline{N}(\omega)|$ vs. ωt (log scale).

(c) Obtain the formula for the frequency f_{max} of the most prominent peak in the spectrum. Why does this differ from $1/T$?

(d) Calculate the numerical value of f_{max} for a supersonic passenger aircraft for which $T \doteq 0.15$ s.

10–6. The spherical-glass-vessel method is to be used to measure the attenuation of sound in seawater at $f = 1$ kHz. Assume that the glass is acoustically transparent to the waterborne sound waves and that the medium outside is air.

(a) What must be the radius of the vessel? Assume the eigenfrequency that gives the smallest possible radius.

(b) How many gallons of seawater are required, and what is the weight in metric tons of the seawater?

10–7. Replace the seawater in the glass vessel (previous problem) with air. Now what is the lowest eigenfrequency? *Hint*: What sort of boundary does the glass represent in this case?

10–8. Show that the eigenfrequencies of the hollow sphere (rigid wall) are determined from the roots of

(a) $\tan x = x$ for $n = 0$

(b) $\tan x = 2x/(2 - x^2)$ for $n = 1$,

where $x = ka$ and a is the radius of the sphere.

10–9. Two enclosures, one spherical and one cubical, have the same volume and contain the same fluid. Which has the lower lowest eigenfrequency

(a) If the walls are rigid?

(b) If the walls are pressure release?

In each case give the percentage difference between the two frequencies.

10–10. A fluid is confined between two concentric pressure release spherical surfaces, inner radius a, outer radius b.

(a) Find the general equation (motion not restricted to being radial) that determines the eigenfrequencies.

(b) For the case of pure radial motion, simplify the general equation and solve it explicitly for the eigenfrequencies.

10–11. A rigid hemispherical bowl of radius a is filled with water (see sketch). The boundary conditions are

(i) $p = 0$ at the top surface ($\theta = \pi/2$),

(ii) $u^{(r)} = 0$ at the surface of the bowl ($r = a$).

A small bubble is located at S on the z axis. A violent collapse of the bubble at time $t = 0$ produces a broadband sound field in the water.

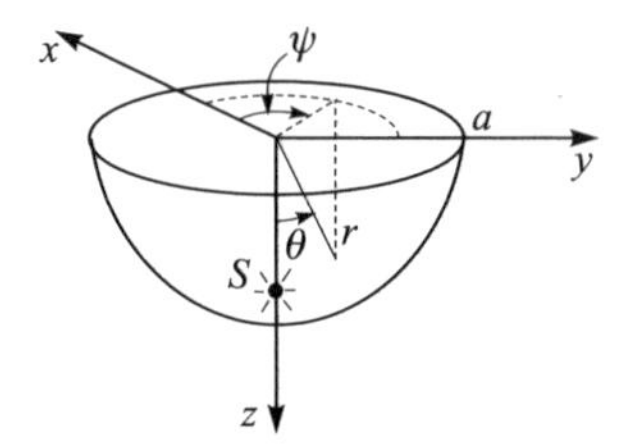

(a) Starting with the most general solution of the wave equation, e.g., Eq. B–37a or b (explain your choice), specialize this solution so that it applies to the problem at hand. Use all of the symmetry conditions and boundary conditions available to you.

(b) What is the lowest eigenfrequency in this case?

(c) Give an expression for the complete solution (sum of all the appropriate modal solutions) in terms of coefficients that could be determined if the specific initial conditions were known.

10–12. A spherical enclosure is completely filled with fluid. Radial pulsation of the enclosure wall produces a sound field in the fluid. The boundary condition is $u^{(r)} = u_0 e^{j\omega t}$ *at* $r = a$.

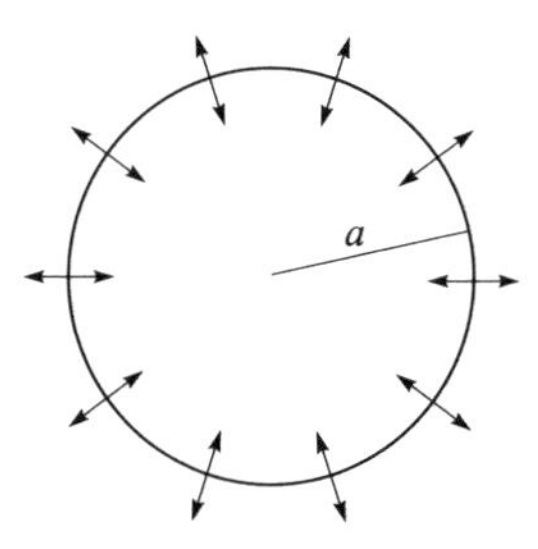

(a) Find the pressure $p(r, t)$ and radial particle velocity $u^{(r)}(r, t)$ (region $r \leq a$).

(b) Obtain an expression for the resonance frequencies (at resonance the pressure amplitude is unbounded).

(c) Find the impedance seen by the source $Z|_{r=a}$. In this case, since a positive pressure produces a negative particle velocity, define the impedance as $Z = p/(-u^{(r)})$.

(d) By taking the low frequency limit ($ka \ll 1$) of the expression found in part (c), determine the lumped-element impedance of the spherical volume. In particular, does the volume behave as a spring, mass, or resistance? Does your analysis support or disagree with the results found in Chap. 4 for short closed cavities? Discuss.

10–13. A certain sound field inside a hollow hemisphere (radius a, all surfaces rigid) has the property that the pressure is zero along the z axis. Find the lowest two eigenfrequencies for this field, and identify their corresponding eigenfunctions ϕ_{lmn}.

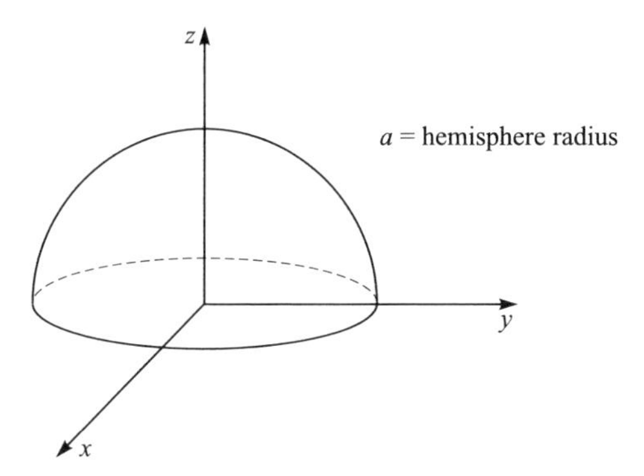

10–14. A pulsating sphere of radius $1/2\pi$ m (about 16 cm) is required to produce a sound pressure level of 63.33 dB at a distance of $1000/2\pi$ m. The medium is air at 20°C. Determine the velocity amplitude u_0 of the sphere required at the following frequencies:

(a) 3430 Hz

(b) 343 Hz

(c) 34.3 Hz

10–15. Derive the inhomogenous wave equation, Eq. D–10, for a fluid having mass sources. Start with the exact continuity equation for this case (see Problem 2–1), linearize the equation, and combine it with the linearized momentum and state equations for a lossless fluid.

10–16. Small explosive charges are often used as omnidirectional sound sources in underwater acoustics. The waveform of the pressure pulse produced is an abrupt rise (a shock wave) followed by a tail that resembles a decaying exponential function. Assume the charge is detonated at time $t = 0$ and the following data are measured in the free field at a distance r_0 from the charge:

$$p(r_0, t) = \begin{cases} 0, & t < r_0/c_0, \\ Ae^{-(t-r_0/c_0)/t_0}, & t > r_0/c_0. \end{cases}$$

(a) Find the source strength $\dot{Q}(t)$ (assume the charge to be a point source).

(b) Obtain the expression for the pressure at any distance r in an infinite ocean.

10–17. The explosive charge described in the previous problem is located a short distance h below the ocean surface, as shown in the sketch. The receiver is in the farfield ($r \gg h$).

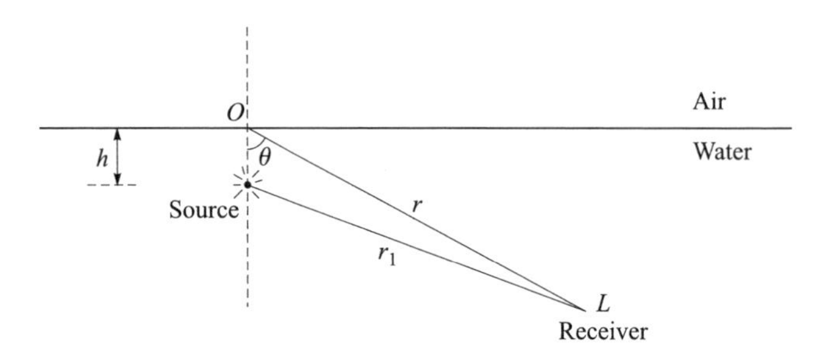

(a) Find the expression for the pressure measured by a receiver. Suggestion: This is a good problem in which to make use of the step function $H(t)$.

(b) Sketch the time waveform of the received signal for a case in which $t_0 = 50$ μs, $h = 0.43$ m, and $\theta = 88°$. Suggestion: Let $\tau = t - r_1/c_0$ (time starting when the first signal arrives at the receiver), and plot pr_1/Ar_0 vs. τ. Show relative amplitudes and times at important points on the waveform.

10–18. In underwater acoustics the relative positions of source S and receiver L (see sketch) are often denoted by the horizontal separation, called the range R, and the depths h of the source and D of the receiver. Find the expression for p_L for the case in which $R \gg h, D$. In particular, show that the received pressure signal varies inversely as R^2. Consider two cases:

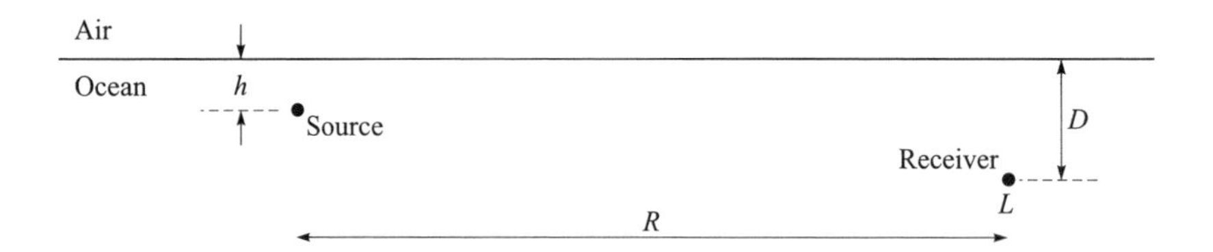

(a) The source is a time-harmonic monopole, $p = (A/r)e^{j(\omega t - kr)}$.

(b) The source is a general monopole, $p = (1/r)f(t - r/c_0)$.

10–19. A time-harmonic simple source is located a distance h above a rigid plane (see sketch). Let the origin be the point O on the plane just below the simple source.

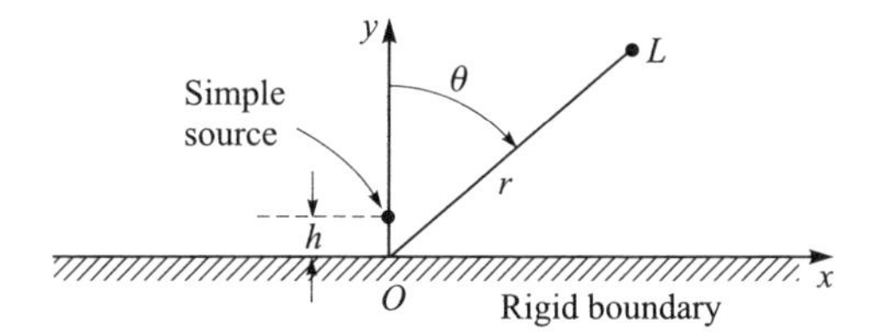

(a) Derive the expression for the signal $p(r, \theta)$ received at a point L in the farfield ($r \gg h$). How does your signal differ from the signal expected when no rigid plane is present?

(b) Determine whether your expression confirms the notion that pressure doubling occurs at a rigid plane.

(c) To obtain the pressure doubling result, is it necessary:

(i) That the restriction $kh \ll 1$ hold?

(ii) That L be in the farfield?

10–20. For the simple source near a rigid plane (previous problem), show that the total power radiated W is given by Eq. D–21. Sketch W/W_{free} as a function of kh.

10–21. Analyze the sound field produced by a sphere in translational oscillation. In other words, take the function $v_a(t)$ in Eq. D–25 to be $v_0 \exp(j\omega t)$.

(a) Solve Eq. D–26 for this case. You should obtain $F(t) = F_0 e^{j(\omega t + ka)}$, where

$$F_0 = \frac{v_0 a^3}{2 + j2ka - k^2a^2}.$$

(b) Find the expressions for the velocity potential ϕ, the pressure p, and the radial particle velocity $u^{(r)}$.

(c) Obtain an expression for the radial specific acoustic impedance

$$Z = p/u^{(r)}$$

and compare it with the impedance for a pulsating sphere. The comparison may be aided by expressing both impedances in terms of their real and imaginary parts, i.e., $Z = \rho_0 c_0 (r_1 + jx_1)$. In particular, consider the two cases $kr \ll 1$ and $kr \gg 1$.

10–22. A time-harmonic dipole is placed a distance h above a pressure release surface. The axis of the dipole is parallel to the surface (see Fig. 10.19*a*).

(a) Find the general expression for the pressure in the y, z plane in the farfield, i.e., $r \gg h$ and $kr \gg 1$, but no restriction on the value of kh.

(b) Let $kh \ll 1$. Simplify the expression obtained in part (a), and sketch a polar plot of the radiation pattern. Does your result imply quadrupole radiation? If so, what kind of quadrupole?

10–23. Repeat the previous problem for a dipole whose axis is perpendicular to the pressure release surface (see Fig. 10.19*b*).

10–24. Consider the sound made by a mosquito in your bedroom at night (acknowledgments to C. L. Morfey, who first formulated the problem). Assume that your floor has a rug, which you take to be a perfect sound absorber, and that the walls and ceiling are hard.

(a) How would you model the free-field sound radiation of the mosquito? That is, would you characterize the mosquito as a monopole, dipole, quadrupole, or what? Given your answer, what sort of directional characteristics do you expect the sound to have? Defend your answers. (For purposes of discussion, assume that the two wings of the mosquito go up and down as a unit, that is, do not be concerned with the fact that they are actually hinged at the body and move separately.)

(b) To swat the mosquito, you first have to locate it, primarily by your sense of hearing. The mosquito likes to fly close to surfaces, that is, wall, ceiling, and floor. Assume that the mosquito always maintains a horizontal attitude regardless of the surface. Discuss your detection chances for each of these surfaces. Assume that for each surface you are at the same distance from the surface but can move parallel to the surface so that the angle you make with the mosquito (angle of incidence with the normal to the surface) can vary from 0° to 45°. It is suggested that for each surface you examine the function p/p_{free}, where p is the actual farfield pressure and p_{free} is the free-field value, i.e., the pressure if the surface were not present.

10–25. A conical horn of infinite length and of half angle θ_0 is shown in the sketch. The horn is truncated at the small end by a spherical cap at $r = r_0$. Uniform vibration of the spherical cap gives rise to a wave that travels outward through the horn. Can the traveling wave be a simple radial wave, that is, a wave of the form

$$p = p(r, t) = \frac{Ar_0}{r} e^{j[\omega t - k(r - r_0)]},$$

where A is the pressure amplitude at the cap? Consider two cases:

(a) Rigid horn, i.e., normal component of particle velocity zero at the horn surface $\theta = \theta_0$.

(b) Pressure release horn, i.e., acoustic pressure is zero at $\theta = \theta_0$.
Defend your answers.

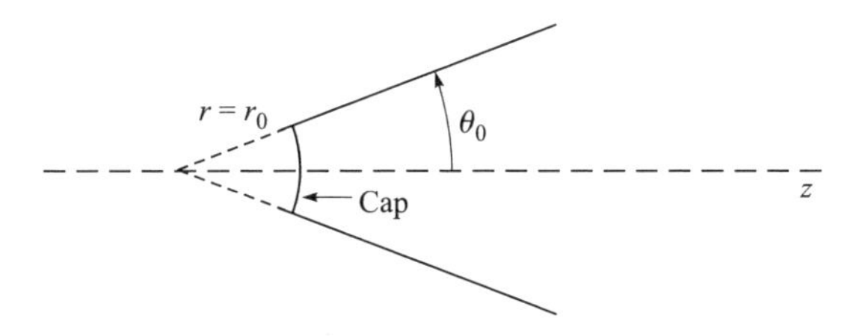

10–26. A hollow sphere (rigid wall) of radius a contains a viscous gas. Use the method of separation of variables to solve the wave equation for this case,

$$2\delta\,\nabla^2 p_t + \nabla^2 p - c_0^{-2} p_{tt} = 0,$$

where $2\delta = (\lambda + 2\mu)/\rho_0 c_0^2$. Start by assuming a product solution

$$p(r, \theta, \psi, t) = P(r, \theta, \psi)T(t).$$

You should find that $P(r, \theta, \psi)$ must satisfy the ordinary Helmholtz equation, Eq. B–29. The solution for P should thus be the same as that for an inviscid gas. The T equation is, however, different. One way to solve it is to assume a solution

$$T = e^{jmt}$$

and solve for m. Your solution should show that because of viscosity, the wave motion decays with time.

(a) Find the expression for the eigenvalue k.

(b) Give the expression for the temporal damping coefficient α_t, which is the imaginary part of m.

(c) Give the expression for the eigenfrequencies, which is found from the real part of m.

(d) How (quantitatively) does the formula for α_t differ from the spatial attenuation coefficient α for a progressive wave in a viscous fluid (Chap. 9)? What physical reason can you give for the difference between the two coefficients?

11

CYLINDRICAL WAVES

Having investigated plane and spherical waves in some detail, we now turn to wave motion in cylindrical coordinates. In Sec. A the wave equation in the cylindrical coordinate system is solved by the method of separation of variables, and the properties of Bessel functions are discussed. Circular membrane problems are treated in Sec. B. Section C is devoted to three-dimensional cylindrical waves, first standing waves inside cylindrical enclosures and then radiation from cylindrical sources.

A. SOLUTION OF THE WAVE EQUATION IN CYLINDRICAL COORDINATES

The cylindrical coordinate system is shown in Fig. 11.1. The wave equation, $\nabla^2\phi - (1/c_0^2)\,\phi_{tt} = 0$, in this system is

$$\phi_{rr} + \frac{1}{r}\,\phi_r + \frac{1}{r^2}\,\phi_{\theta\theta} + \phi_{zz} - \frac{1}{c_0^2}\,\phi_{tt} = 0, \tag{A–1}$$

where the relations between the x, y, z and r, θ, z coordinate systems are

$$x = r\cos\theta \tag{A–2a}$$

$$y = r\sin\theta \tag{A–2b}$$

$$z = z. \tag{A–2c}$$

The expression for the particle velocity $\mathbf{u} = \nabla\phi$ in cylindrical coordinates is

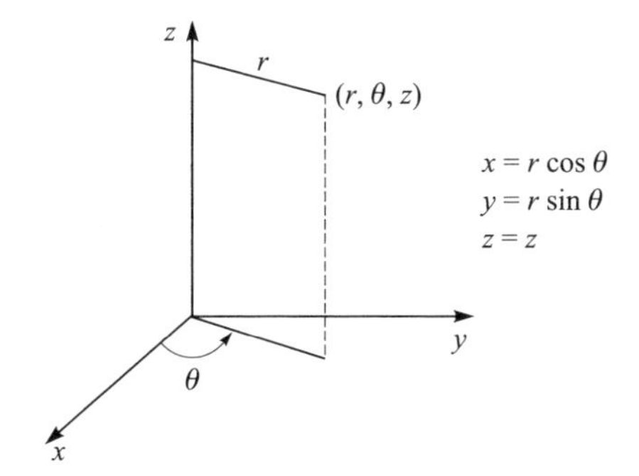

Figure 11.1 Cylindrical coordinate system.

$$\mathbf{u} = \mathbf{r}_1 \phi_r + \boldsymbol{\theta}_1 \frac{1}{r} \phi_\theta + \mathbf{z}_1 \phi_z, \tag{A–3}$$

where $\mathbf{r}_1$, $\boldsymbol{\theta}_1$, and $\mathbf{z}_1$ are unit vectors in the r, θ, and z directions, respectively. The components $u^{(r)}$, $u^{(\theta)}$, and $u^{(z)}$ may be identified from Eq. A–3.

Before proceeding with the general solution of this equation, we remind the reader that pure radial cylindrical waves (no dependence on θ or z) cannot be expressed in elementary form. No simple solution comparable to the one for radial spherical waves, Eq. 10A–6, is known. This point is discussed briefly in Sec. 1D.2.

1. Solution by Separation of Variables

If a product solution $\phi = R(r)\Theta(\theta)\mathcal{Z}(z)T(t)$ is assumed and substituted in Eq. A–1, the equation may be separated into space- and time-dependent parts,

$$\frac{R''}{R} + \frac{R'}{rR} + \frac{\Theta''}{r^2\Theta} + \frac{\mathcal{Z}''}{\mathcal{Z}} = \frac{1}{c_0^2}\frac{T''}{T}. \tag{A–4}$$

Setting both sides equal to a constant, say $-k^2$, leads to the familiar equation for the time function,

$$T'' + k^2 c_0^2 T = 0,$$

which has the equally familiar solution

$$T = \begin{Bmatrix} \cos \omega t \\ \sin \omega t \end{Bmatrix} \text{ or } \begin{Bmatrix} e^{j\omega t} \\ e^{-j\omega t} \end{Bmatrix}, \tag{A–5}$$

where $\omega^2 = k^2 c_0^2$. Next separate the spatial part of Eq. A–4 as follows:

$$\frac{R''}{R} + \frac{R'}{rR} + \frac{\Theta''}{r^2\Theta} + k^2 = -\frac{\mathcal{Z}''}{\mathcal{Z}} = k_z^2. \tag{A–6}$$

The $\mathcal{Z}$ equation

$$\mathcal{Z}'' + k_z^2 \mathcal{Z} = 0$$

has the solution

$$\mathcal{Z} = \begin{Bmatrix} \cos k_z z \\ \sin k_z z \end{Bmatrix} \text{ or } \begin{Bmatrix} e^{jk_z z} \\ e^{-jk_z z} \end{Bmatrix}. \tag{A–7}$$

Further manipulation of Eq. A–6 allows separation of the r and θ parts,

$$r^2 \frac{R''}{R} + r \frac{R'}{R} + k_r^2 r^2 = -\frac{\Theta''}{\Theta} = m^2, \tag{A–8}$$

where k_r is defined by

$$k_r^2 + k_z^2 = k^2 = \omega^2/c_0^2. \tag{A–9}$$

The θ-dependent part of Eq. A–8 is

$$\Theta'' + m^2\Theta = 0,$$

the solution of which is

$$\Theta = \begin{Bmatrix} \cos m\theta \\ \sin m\theta \end{Bmatrix} \text{ or } \begin{Bmatrix} e^{jm\theta} \\ e^{-jm\theta} \end{Bmatrix}. \tag{A–10}$$

The usual domain is $0 \leq \theta \leq 2\pi$. In this case m must be an integer because the field must be the same at $\theta + 2\pi$ as at θ. The r-dependent part of Eq. A–8 is

$$R'' + \frac{1}{r} R' + (k_r^2 - m^2/r^2)R = 0, \tag{A–11}$$

which is Bessel's equation of order m. It has solutions

$$R = \begin{Bmatrix} J_m(k_r r) \\ N_m(k_r r) \end{Bmatrix} \text{ or } \begin{Bmatrix} H_m^{(1)}(k_r r) \\ H_m^{(2)}(k_r r) \end{Bmatrix}, \tag{A–12}$$

where J_m and N_m are the ordinary Bessel and Neumann functions, respectively, and $H_m^{(1)}$ and $H_m^{(2)}$ are the two Hankel functions. The properties of these functions are discussed in the next section.

In summary a solution of the wave equation in cylindrical coordinates has been developed. In assembling the results given by Eqs. A–5, A–7, A–10, and A–12, we have several choices. The one normally most convenient for interior domains, for example, a cylindrical cavity, is

$$\phi = \begin{Bmatrix} J_m(k_r r) \\ N_m(k_r r) \end{Bmatrix} \begin{Bmatrix} \cos m\theta \\ \sin m\theta \end{Bmatrix} \begin{Bmatrix} \cos k_z z \\ \sin k_z z \end{Bmatrix} \begin{Bmatrix} \cos \omega t \\ \sin \omega t \end{Bmatrix}. \qquad \text{(A–13)}$$

For exterior domain problems, for example, for radiation into free space, we shall find that the combination

$$\phi = \begin{Bmatrix} H_m^{(1)}(k_r r) \\ H_m^{(2)}(k_r r) \end{Bmatrix} \begin{Bmatrix} e^{jm\theta} \\ e^{-jm\theta} \end{Bmatrix} \begin{Bmatrix} e^{jk_z z} \\ e^{-jk_z z} \end{Bmatrix} \begin{Bmatrix} e^{j\omega t} \\ e^{-j\omega t} \end{Bmatrix} \qquad \text{(A–14)}$$

is usually preferable. Still other combinations are called for in some problems. In all cases, however, Eq. A–9 gives the relation that k_r, k_z, and ω must satisfy. Before exploring the various solutions, we need to review Bessel functions and their properties.

2. Properties of Bessel Functions

See Refs. 1, 2, 3, and 9 for collections of formulas and relations about Bessel functions. Our survey is mainly limited to material of direct use in the text.

a. Bessel's Equation

Bessel's equation,[1]

$$y'' + \frac{1}{x}y' + \left(1 - \frac{m^2}{x^2}\right)y = 0, \qquad \text{(A–15)}$$

has a regular singular point at $x = 0$. The rules about singular points of differential equations are as follows:

1. Given $y'' + P(x)y' + Q(x)y = 0$, if P and Q are regular everywhere, then all points x are ordinary points of the differential equation.
2. Points at which P and Q are singular are called singular points of the differential equation. In general, no Taylor series solution exists about these points.

[1]Equation A–15 is the dimensionless form of Bessel's equation, obtained from Eq. A–11 by putting $x = k_r r$.

3. If x_0 is a singular point but $(x - x_0)P(x_0)$ and $(x - x_0)^2 Q(x_0)$ exist, then x_0 is a "regular singular point," and at least one solution in Taylor series about x_0 does exist.

It is seen that Bessel's equation fits rule 3, with $x_0 = 0$. Accordingly, we expect at least one solution that may be expressed as a Maclaurin series.

b. Bessel Functions of the First Kind: $J_m(x)$

Bessel's equation has one solution that is finite everywhere. It is called the Bessel function of the first kind and is denoted $J_m(x)$. The Maclaurin series for J_m are now given. First consider m an integer. The series for $m = 0$ is

$$J_0(x) = \sum_{n=0}^{\infty} \frac{(-1)^n (x/2)^{2n}}{(n!)^2}. \tag{A–16}$$

Note that $J_0(x)$ is an even function of x, and its value at the origin is $J_0(0) = 1$. The general Maclaurin series formula, valid for all integral values of m, is

$$J_m(x) = \left(\frac{x}{2}\right)^m \sum_{n=0}^{\infty} (-1)^n \frac{(x/2)^{2n}}{n!(n+m)!}. \tag{A–17}$$

It is clear that J_m is an even function if m is even, an odd function if m is odd, that is,

$$J_m(-x) = (-1)^m J_m(x). \tag{A–18}$$

The first four Bessel functions J_0, J_1, J_2, and J_3 are sketched in Fig. 11.2. Notice that $J_0(x)$ resembles a damped cosine function and $J_1(x)$ a damped sine function.

Next consider nonintegral values of m, say $m = \mu$. The power series for $J_\mu(x)$ is Eq. A–17 with $(n+m)!$ replaced by $\Gamma(n + \mu + 1)$, where Γ is the gamma function (Refs. 1 and 2). Of special interest is the gamma function of argument $m + \frac{1}{2}$ (m an integer):

$$\Gamma\left(m + \frac{1}{2}\right) = \frac{1 \cdot 3 \cdot 5 \cdot 7 \cdots (2m-1)}{2^m} \sqrt{\pi}. \tag{A–19}$$

Since Bessel's equation is second order, it has a second solution that is linearly independent of the first one. When m is not an integer, the second solution is $J_{-\mu}(x)$. The general solution of Eq. A–15 in this case is

$$y = AJ_\mu(x) + BJ_{-\mu}(x). \tag{A–20}$$

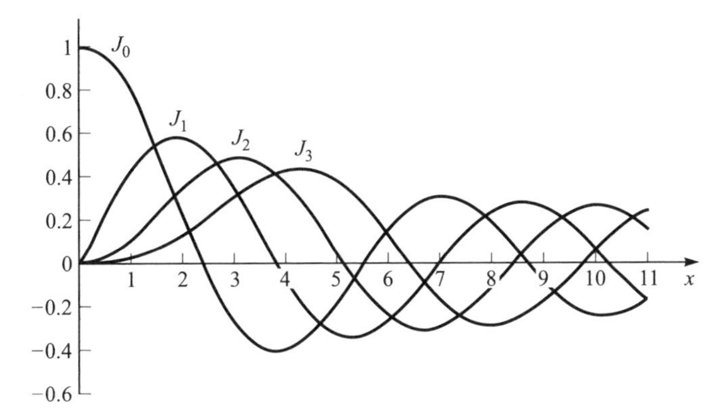

Figure 11.2 $J_m(x)$ for $m = 0, 1, 2$.

It may be seen from Eq. A–17 that $J_{-\mu}(x)$ is unbounded at the origin. If a bounded solution is required and if the origin is included in the domain of interest, one must put $B = 0$.

c. *Bessel Functions of the Second Kind: $N_m(x)$*

When m is an integer, $J_{-m}(x)$ is not linearly independent of the first solution $J_m(x)$. Some other second solution must therefore be sought. It turns out to be the Neumann function $N_m(x)$, sometimes called the Bessel function of the second kind. The Neumann function is singular at $x = 0$. For $m = 0$, a series expression is

$$N_0(x) = \frac{2}{\pi}\left\{\ln\left|\frac{x}{2}\right| + \gamma\right\}J_0(x) + \sum_{n=1}^{\infty}(-1)^{n+1}\frac{(x/2)^{2n}}{(n!)^2}\left(1 + \frac{1}{2} + \frac{1}{3} + \cdots + \frac{1}{n}\right), \tag{A–21}$$

where $\gamma = 0.5772$ is Euler's constant. Series expressions for other Neumann functions may be found in Refs. 3, 7, 2, and 1. The behavior of N_m $(m > 0)$ near the origin is given by

$$N_m(x) = -\frac{(m-1)!}{\pi}\left(\frac{2}{x}\right)^m \quad \text{for } |x| \ll 1, \tag{A–22}$$

an expression that holds for nonintegral as well as integral values of m. Sketches of N_0, N_1, N_2, and N_3 are shown in Fig. 11.3.

Finally, it should be noted that J_μ and N_μ are independent solutions for all values of μ, including nonintegral values. Therefore, N_μ may be used in place of $J_{-\mu}$ in Eq. A–20.

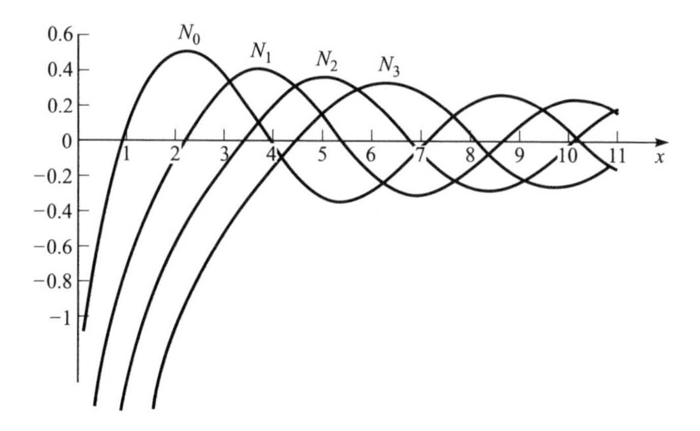

Figure 11.3 $N_m(x)$ for $m = 0, 1, 2$.

d. Hankel Functions

The Hankel functions are to the Bessel and Neumann functions as the complex exponentials are to the sine and cosine functions:

$$H_m^{(1)}(z) = J_m(z) + jN_m(z), \tag{A–23a}$$

$$H_m^{(2)}(z) = J_m(z) - jN_m(z). \tag{A–23b}$$

These relations are analogous to those that relate spherical Hankel functions to spherical Bessel and Neumann functions; see Sec. 10B.3. Equations A–23a and A–23b also define the Hankel functions for nonintegral order μ.

e. Recursion Relations

Some of the more useful recursion relations are as follows:

$$J_{m+1}(x) = \frac{2m}{x} J_m(x) - J_{m-1}(x), \tag{A–24a}$$

$$J_m'(x) = \tfrac{1}{2}\left[J_{m-1}(x) - J_{m+1}(x)\right], \tag{A–24b}$$

$$\frac{d}{dx}[x^m J_m(x)] = x^m J_{m-1}(x). \tag{A–24c}$$

The same relations hold for the Neumann functions $N_m(x)$, the Hankel functions $H_m^{(1)}(x)$ and $H_m^{(2)}(x)$, and any linear combinations of the four functions provided the multiplying coefficients are independent of m and x. Another useful relation is

$$J_{m+1}(x)N_m(x) - J_m(x)N_{m+1}(x) = \frac{2}{\pi x}. \tag{A–25}$$

The recursion relations for the spherical Bessel functions, Eqs. 10B–24a,b, follow from Eqs. A–24a and A–24b, respectively, simply by letting $m = n + \frac{1}{2}$ and employing Eqs. 10B–20.

f. Roots of $J_m(x)$ and $J'_m(x)$

Table 11.1 gives the roots of $J_m(x)$ and its first derivative $J'_m(x)$ for the lower values of m (Ref. 1, Table 9.5). The notation is as follows: If $J_m(x) = 0$, then $x = \alpha_{mn}$ is the nth root of J_m. Similarly, α'_{mn} is the nth root of the equation $J'_m(x) = 0$. Note that although $x = 0$ is a root of $J_m(x)$ for $m > 0$, it does not appear in the table because it is a trivial root, i.e., $J_m(0) = 0$ for $m > 0$.

g. Integral Relations

Several useful expressions for J_m in terms of integrals are now given. First, J_m may be defined as follows:

$$J_m(x) = \frac{2(x/2)^m}{\sqrt{\pi}\,\Gamma(m+1/2)} \int_0^1 (1-t^2)^{m-1/2} \cos xt\, dt, \qquad \mathrm{Re}(m) > -\frac{1}{2}, \tag{A–26}$$

where the gamma function is given by Eq. A–19. Another definition of J_m as an integral is

$$J_m(x) = \frac{1}{\pi} \int_0^{\pi} \cos(m\theta - x\sin\theta)\, d\theta, \tag{A–27a}$$

or an alternative,

Table 11.1 Zeros of Bessel Functions

	Root α_{mn} of $J_m(x) = 0$					Root α'_{mn} of $J'_m(x) = 0$				
n	$m = 0$	$m = 1$	$m = 2$	$m = 3$	$m = 4$	$m = 0$	$m = 1$	$m = 2$	$m = 3$	$m = 4$
1	2.405	3.832	5.136	6.380	7.588	0	1.841	3.054	4.201	5.318
2	5.520	7.016	8.417	9.761	11.065	3.832	5.331	6.706	8.015	9.282
3	8.654	10.173	11.620	13.015	14.373	7.016	8.536	9.969	11.346	12.682
4	11.792	13.324	14.796	16.223	17.616	10.173	11.706	13.170	14.586	15.964
5	14.931	16.471	17.960	19.409	20.827	13.324	14.864	16.348	17.789	19.196

$$J_m(x) = \frac{j^{-m}}{\pi}\int_0^{\pi} e^{jx\cos\theta}\cos m\theta\, d\theta. \tag{A–27b}$$

The usefulness of Eq. A–27a may be demonstrated by using it to find an expression for Bessel functions of negative order:

$$\begin{aligned} J_{-m}(x) &= \frac{1}{\pi}\int_0^{\pi} \cos(m\theta + x\sin\theta)\, d\theta \\ &= J_m(-x) \\ &= (-1)^m J_m(x). \end{aligned} \tag{A–28}$$

Combining Eqs. A–24b and A–28, we obtain the following useful formula for the derivative of J_0:

$$J_0'(x) = -J_1(x). \tag{A–29}$$

The same relation also applies to the Neumann and Hankel functions. Finally, the following formula, called Sonine's first finite integral (Ref. 9), is found to be useful in Chap. 13:

$$J_{m+n+1}(x) = \frac{x^{n+1}}{2^n n!}\int_0^{\pi/2} J_m(x\sin\theta)\sin^{m+1}\theta\cos^{2n+1}\theta\, d\theta. \tag{A–30}$$

Frequency modulation (FM) offers a good example of the use of Eq. A–27a. Given a time-varying signal, say for a traveling plane wave of pressure,

$$p = Ae^{j\psi\tau},$$

where $\tau = t - x/c_0$, we may define the instantaneous angular frequency ω_i by $\omega_i = d\psi/d\tau$. A frequency-modulated signal is one for which

$$\psi = \omega_c\tau + \mu\sin\omega_m\tau,$$

where ω_c is called the (angular) carrier frequency, ω_m the (angular) modulation frequency, and μ the modulation index. The instantaneous (angular) frequency for this case is $\omega_i = \omega_c + \mu\omega_m\cos\omega_m\tau$, which varies between $\omega_c - \mu\omega_m$ and $\omega_c + \mu\omega_m$. An FM pressure signal may therefore be represented by

$$p = Ae^{j\omega_c\tau}e^{j\mu\sin\omega_m\tau}. \tag{A–31}$$

The second exponential may be expressed as a complex Fourier series,

$$e^{j\mu\sin\theta} = \sum_{n=-\infty}^{\infty} C_n e^{jn\theta},$$

where $\theta = \omega_m \tau$. To find the coefficients C_n, use the orthogonality of the functions $e^{jn\theta}$. Multiply both sides of the equation by $e^{jm\theta}\, d\theta$ and integrate from $-\pi$ to π. On the right-hand side all the terms vanish except for the one for which $n = -m$. Integration for that term, and replacement of m with $-n$, yields

$$C_n = \frac{1}{2\pi} \int_{-\pi}^{\pi} e^{-j(n\theta - \mu \sin\theta)}\, d\theta.$$

The imaginary part of the integral is zero (why?). The real part is, by application of Eq. A–27a,

$$C_n = \frac{1}{2\pi} \int_{-\pi}^{\pi} \cos(n\theta - \mu \sin\theta)\, d\theta = J_n(\mu).$$

Equation A–31 now becomes

$$p = A \sum_{n=-\infty}^{\infty} J_n(\mu) e^{j(\omega_c + n\omega_m)\tau}. \tag{A–32}$$

The FM pressure signal therefore has a spectrum composed of the carrier signal (angular frequency ω_c) and pairs of sidebands at angular frequencies $\omega_c \pm \omega_m$, $\omega_c \pm 2\omega_m$, $\omega_c \pm 3\omega_m$, and so on. The amplitudes of the various components depend on the size of the modulation index μ. If the modulation is very weak ($\mu \ll 1$), the sidebands are very small and the effective bandwidth of the FM signal is narrow. As μ increases, the sidebands grow in importance and the effective bandwidth broadens.

A special case occurs when $\mu = \alpha_{01}$ ($= 2.405$), for then the component at the carrier frequency vanishes; only the sidebands remain. An interesting application of this property occurs in nonlinear acoustics. When an intense low frequency wave (ω_{LF}) and a weak high frequency wave (ω_{HF}) propagate together in the same direction, they interact nonlinearly in such a fashion that the low frequency wave modulates the high frequency wave ($\omega_c = \omega_{\mathrm{HF}}$ and $\omega_m = \omega_{\mathrm{LF}}$). The modulation index μ depends on the distance traveled. At a distance such that $\mu = 2.405$, the carrier vanishes; at this point all the carrier's energy has been shifted to the sidebands ($\omega_{\mathrm{HF}} \pm n\omega_{\mathrm{LF}}$). The use of a low frequency wave to nullify a high frequency wave is known as the suppression of sound by sound (see, for example, Ref. 6).

h. Orthogonality and Fourier-Bessel Series

Like the sine and cosine functions, the Bessel functions have useful orthogonality properties. First suppose that $m = 0$. The functions $J_0(\alpha_{0n} x)$ constitute a complete orthogonal set of functions. An arbitrary function $f(x)$ may therefore be expanded in terms of them,

$$f(x) = \sum_{n=1}^{\infty} A_n J_0(\alpha_{0n} x). \tag{A–33}$$

The expansion is called a "Fourier-Bessel" series. The orthogonality relation[2]

$$\int_0^1 x J_0(\alpha_{0n} x) J_0(\alpha_{0n'} x)\, dx = \frac{1}{2}[J_1(\alpha_{0n})]^2 \delta_{nn'}, \tag{A–34}$$

where $\delta_{nn'}$ is the Kronecker delta function, is used to find the coefficients A_n. Multiply Eq. A–33 by $xJ_0(\alpha_{0n'}x)\,dx$, and integrate from 0 to 1:

$$\begin{aligned}\int_0^1 f(x) x J_0(\alpha_{0n'} x)\, dx &= \sum_{n=0}^{\infty} A_n \int_0^1 x J_0(\alpha_{0n} x) J_0(\alpha_{0n'} x)\, dx \\ &= A_{n'} \frac{1}{2} [J_1(\alpha_{0n'})]^2\end{aligned}$$

Thus

$$A_n = \frac{2}{[J_1(\alpha_{0n})]^2} \int_0^1 x f(x) J_0(\alpha_{0n} x)\, dx. \tag{A–35}$$

Expansions in J_m, where $m > 0$, may also be made. The general orthogonality relation is

$$\int_0^1 x J_m(\alpha_{mn} x) J_m(\alpha_{mn'} x)\, dx = \frac{1}{2}\left[J'_m(\alpha_{mn})\right]^2 \delta_{nn'}. \tag{A–36}$$

Another useful Fourier-Bessel expansion of $f(x)$ is one in terms of the functions $J_0(\alpha'_{0n}x)$, which also constitute a complete orthogonal set,

$$f(x) = \sum_{n=1}^{\infty} A_n J_0(\alpha'_{0n} x). \tag{A–37}$$

In this case the orthogonality relation is Eq. A–34 with the roots α replaced by the roots α' on the left-hand side and $J_1(\alpha_{0n})$ replaced by $J_0(\alpha'_{0n})$ on the right-hand side. The expression for the coefficients A_n is, in place of Eq. A–35,

$$A_n = \frac{2}{[J_0(\alpha'_{0n})]^2} \int_0^1 x f(x) J_0(\alpha'_{0n} x)\, dx. \tag{A–38}$$

[2]Notice the presence of the weighting factor x in the integrand. In ordinary Fourier series the weighting factor is unity.

A similar generalization applies when the order of the Bessel function is m rather than zero. See Ref. 2 for details.

i. Other Series of Bessel Functions

Another common expansion is a series in the order m. Equation A–32, the series representing an FM signal, is an example. Another example is the representation of a plane wave as a sum of cylindrical waves. A time-harmonic plane wave propagating in the x direction may be expressed as $e^{j(\omega t - kx)} = e^{j\omega t}e^{-jkr\cos\theta}$ (see Eq. A–2a). The second exponential may be expanded in a series of Bessel functions of order m:

$$e^{-jkr\cos\theta} = \sum_{m=0}^{\infty} A_m J_m(kr). \tag{A–39}$$

It turns out (see Problem 11–3) that the coefficients are given by

$$A_0 = 1, \qquad A_m = 2(-j)^m \cos m\theta. \tag{A–40}$$

j. Asymptotic Forms

For large argument z (and $z \gg m$)

$$J_m(z) \sim \sqrt{\frac{2}{\pi z}} \cos\left[z - \left(m + \frac{1}{2}\right)\frac{\pi}{2}\right], \tag{A–41}$$

$$N_m(z) \sim \sqrt{\frac{2}{\pi z}} \sin\left[z - \left(m + \frac{1}{2}\right)\frac{\pi}{2}\right], \tag{A–42}$$

$$H_m^{(1)}(z) \sim \sqrt{\frac{2}{\pi z}} \exp j\left[z - \left(m + \frac{1}{2}\right)\frac{\pi}{2}\right], \tag{A–43}$$

$$H_m^{(2)}(z) \sim \sqrt{\frac{2}{\pi z}} \exp -j\left[z - \left(m + \frac{1}{2}\right)\frac{\pi}{2}\right]. \tag{A–44}$$

Thus as z becomes large, J_m behaves as a slowly decaying cosine function and N_m as a slowly decaying sine function. From the asymptotic form of the Hankel functions, it is clear that the product $e^{j\omega t}H_m^{(2)}(kr)$ represents an outward traveling wave, $e^{j\omega t}H_m^{(1)}(kr)$ an inward traveling wave.

B. CIRCULAR MEMBRANE

1. Introduction

The circular membrane is a good example of a physical system for which plane polar coordinates (no dependence on the axial coordinate z) are appropriate. The wave equation for the membrane displacement η is

$$\eta_{rr} + \frac{1}{r}\,\eta_r + \frac{1}{r^2}\,\eta_{\theta\theta} - \frac{1}{c^2}\,\eta_{tt} = 0. \tag{B–1}$$

The propagation speed c is given by

$$c = \sqrt{\mathcal{T}_\ell / m}, \tag{B–2}$$

where $\mathcal{T}_\ell$ is the tension force per unit length and m is the surface density (mass per unit area) of the membrane. Solutions of Eq. B–1 have the form of Eq. A–13 or A–14, with $k_z = 0$. Equation A–14 is appropriate for exterior domain wave problems, such as the radiation from a source on an infinite membrane. See, for example, Problem 11–7. For wave motion on finite membranes, however, such as the circular membranes considered here, Eq. A–13 is a more suitable starting point. Dropping the z-dependence, we have

$$\eta = \begin{Bmatrix} J_m(kr) \\ N_m(kr) \end{Bmatrix} \begin{Bmatrix} \cos m\theta \\ \sin m\theta \end{Bmatrix} \begin{Bmatrix} \cos \omega t \\ \sin \omega t \end{Bmatrix}, \tag{B–3}$$

where $k^2 = \omega^2/c^2$. If the membrane includes the origin ($r = 0$), the Neumann function must be discarded because of its singularity there. (In the case of an annular membrane, however, the Neumann function must be retained; see Problem 11–9.)

The modes of vibration of the membrane are determined by the boundary conditions. If the membrane is fixed, or clamped, at its edge, the displacement η must vanish there. In particular, for a full circular membrane clamped at $r = a$, one must set

$$J_m(ka) = 0,$$

or

$$ka = \alpha_{mn}. \tag{B–4}$$

The expression for the (mn)th mode is therefore

$$\eta_{mn} = J_m\!\left(\frac{\alpha_{mn} r}{a}\right) \begin{Bmatrix} \cos m\theta \\ \sin m\theta \end{Bmatrix} \begin{Bmatrix} \cos \omega_{mn} t \\ \sin \omega_{mn} t \end{Bmatrix}, \tag{B–5}$$

and the eigenfrequency of the mode is

$$f_{mn} = \frac{\omega_{mn}}{2\pi} = \frac{\alpha_{mn} c}{2\pi a}. \tag{B–6}$$

The displacement patterns of some of the lowest modes are shown in Fig. 11.4.

The other common boundary condition is that for a free, or unrestrained, edge. At a free edge the restoring force normal to the edge vanishes. Because the restoring force is proportional to the gradient of the displacement, the mathematical restriction is that $\eta_r = 0$ at $r = a$, that is,

$$J_m'(ka) = 0.$$

$$\therefore\ ka = \alpha_{mn}'. \tag{B–7}$$

The eigenvalue α_{mn}' now replaces α_{mn} in the expression for the normal mode, Eq. B–5, and in the formula for the eigenfrequency, Eq. B–6.

The general solution is the sum of all possible modes. For example, the general solution for a full circular membrane clamped at its edge is

$$\eta = \sum_{m,n} J_m\left(\frac{\alpha_{mn} r}{a}\right) \cos(m\theta - \theta_{0m})\,[A_{mn} \cos \omega_{mn} t + B_{mn} \sin \omega_{mn} t], \tag{B–8}$$

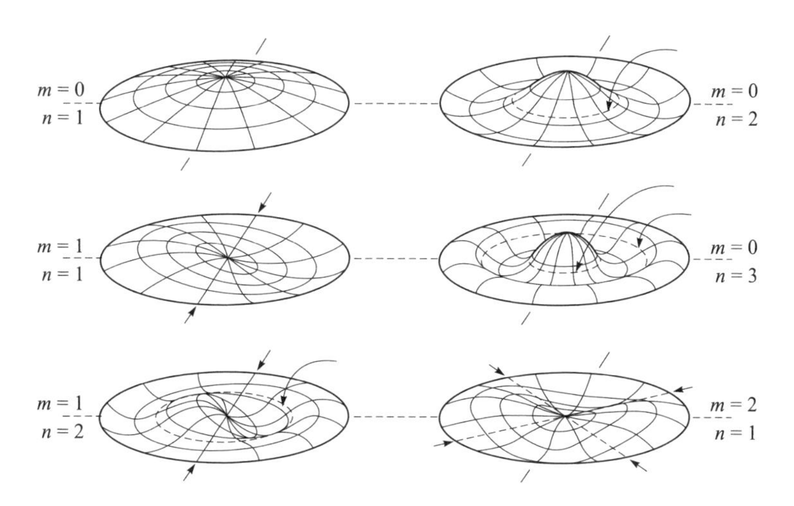

Figure 11.4 Shapes of lower order normal modes of a vibrating circular membrane clamped at its edge. Arrows point to nodal lines. (From Morse, Ref. 7, Fig. 42; also Ref. 8, Fig. 5.12.)

where the linear combination of $\cos m\theta$ and $\sin m\theta$ has been written as a cosine function with a phase angle θ_{0m}. The coefficients A_{mn} and B_{mn} are to be determined by applying initial conditions. The method is demonstrated in the example below.

2. Example: Membrane with Uniform Initial Displacement

Figure 11.5*a* shows a circular membrane clamped at its edge $r = a$. The membrane is initially displaced a uniform amount η_0, as shown in Fig. 11.5*b*.[3] At time $t = 0$ the membrane is released from rest. The initial conditions are therefore

$$\eta(r, 0) = \eta_0, \qquad 0 \le r < a, \tag{B–9a}$$

$$\dot{\eta}(r, 0) = 0, \qquad 0 \le r \le a. \tag{B–9b}$$

Starting with the general solution, Eq. B–3, we first discard the Neumann function because the origin is in the domain of interest. Second, m is restricted to the single value $m = 0$ because the initial conditions are independent of θ. Third, the clamped-edge boundary condition leads to $ka = \alpha_{0n}$. At this stage, therefore, the solution is reduced to

$$\eta = \sum_n J_0\left(\frac{\alpha_{0n} r}{a}\right)[A_n \cos \omega_n t + B_n \sin \omega_n t], \tag{B–10}$$

where the index $m = 0$ has been dropped.

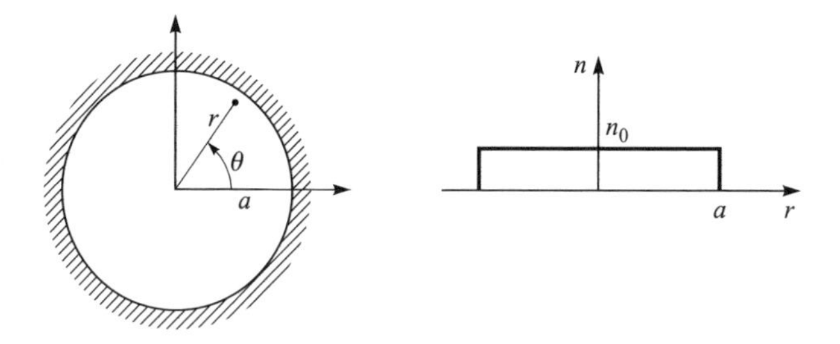

Figure 11.5 (*a*) Clamped-edge circular membrane. (*b*) Initial displacement.

[3] At $r = a$ the initial condition seems to violate the clamped-edge boundary condition. The apparent contradiction could be removed by letting the uniform displacement extend to just $r = a - \epsilon$ and then drop steeply to zero at the actual edge. After obtaining the solution of the modified problem, one would take the limit as $\epsilon \to 0$. However, since solving the problem in this fashion would be cumbersome, we assume that the method of solution described in the example gives the same result as the more cumbersome method.

Next find the coefficients A_n and B_n. Application of the second initial condition, Eq. B–9b, requires that $B_n = 0$. Application of the first initial condition, Eq. B–9a, yields

$$\eta_0 = \sum_n A_n J_0\left(\frac{\alpha_{0n} r}{a}\right). \tag{B–11}$$

For convenience, let y stand for r/a. To find the coefficients A_n, use the orthogonality of the Bessel functions $J_0(\alpha_{0n} y)$. Multiply both sides of Eq. B–11 by $y\, J_0(\alpha_{0s} y)\, dy$ and integrate over the region of the membrane $(0 \leq y \leq 1)$:

$$\int_0^1 \eta_0 y J_0(\alpha_{0s} y)\, dy = \sum_n A_n \int_0^1 y J_0(\alpha_{0n} y)\, J_0(\alpha_{0s} y)\, dy. \tag{B–12}$$

Because of the orthogonality of the Bessel functions, only the $n=s$ term of the series on the right-hand side survives the integration; see Eq. A–34. The integral on the left-hand side may be evaluated by making use of Eq. A–24c. Equation B–12 thus reduces to

$$\frac{\eta_0}{\alpha_{0s}} J_1(\alpha_{0s}) = \frac{A_s}{2} [J_1(\alpha_{0s})]^2.$$

Change s to n and solve for A_n:

$$A_n = \frac{2\eta_0}{\alpha_{0n} J_1(\alpha_{0n})}. \tag{B–13}$$

Substitution in Eq. B–10 yields the final form of the solution,

$$\eta = 2\eta_0 \sum_n \frac{J_0(\alpha_{0n} r/a)}{\alpha_{0n} J_1(\alpha_{0n})} \cos \omega_n t, \tag{B–14}$$

where $\omega_n = \alpha_{0n} c/a$.

The shapes of the first three modes ($m = 0$; $n = 1, 2, 3$) are shown in Fig. 11.4. It can be seen that if the three modes are to add to give a good approximation of the initial displacement (Fig. 11.5*b*), the coefficients A_1 and A_3 must be positive and A_2 negative. That the signs of A_1, A_2, and A_3 are indeed ordered in this fashion may be seen by inspecting Fig. 11.2. For quantitative evaluation of the coefficients, one may use tables of $J_1(\alpha_{0n})$; see, for example, Ref. 1.

3. Variations

In this section two variations on the standard membrane problem are treated.

a. Wedge-Shaped Membrane

Consider the pie-shaped section of a circular membrane shown in Fig. 11.6. The membrane is clamped at its edges $r = a$, $\theta = 0$, and $\theta = \theta_0$.

Although the angular function Θ for this problem may still be represented by

$$\Theta = \begin{Bmatrix} \cos m\theta \\ \sin m\theta \end{Bmatrix},$$

Θ is no longer constrained to be periodic with period 2π. This means that m is not necessarily an integer. The vanishing of η at $\theta = 0$ eliminates $\cos m\theta$ from consideration. The vanishing of η at $\theta = \theta_0$ requires that $\sin m\theta_0 = 0$, that is, m must be given by

$$m = \ell\pi/\theta_0, \qquad \ell = 1, 2, 3, \ldots. \tag{B–15}$$

(Why is the value $\ell = 0$ not considered?) For nonintegral values of m, say $m = \mu$, the radial function R is made up of a linear combination of J_μ and $J_{-\mu}$, but since $J_{-\mu}$ is singular at the origin, it must be deleted. In several important cases, however, m is still an integer. For example, for a semicircle ($\theta_0 = \pi$) m is any integer, and for a quarter circle ($\theta_0 = \pi/2$) m is any even integer. In such instances the ordinary Bessel function $J_m(kr)$ is used, and the final result is

$$\eta = \sum_{m,n} J_m\left(\frac{\alpha_{mn} r}{a}\right) \sin m\theta\,[A_{mn} \cos \omega_{mn} t + B_{mn} \sin \omega_{mn} t], \tag{B–16}$$

where the allowed values of m are set by Eq. B–15.

b. Driven Membrane

A second problem of interest is that of a circular membrane driven by an external force, for example, the pressure p_s due to a sound field. The wave equation for a driven membrane may be shown to be[4]

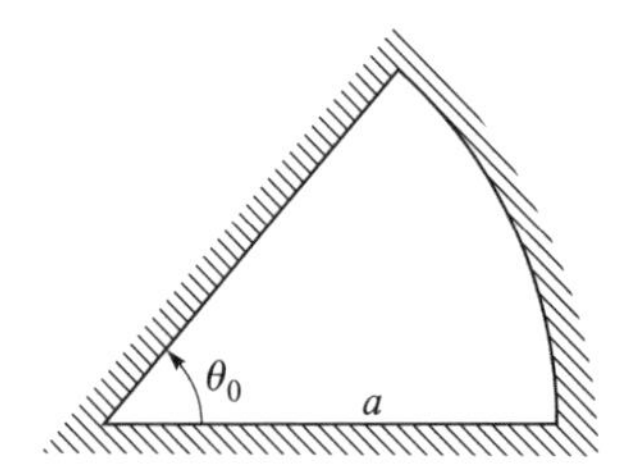

Figure 11.6 Pie-shaped membrane.

[4]The derivation is quite similar to that carried out in Sec. 1C.2.f for forced vibration of a string.

$$\nabla^2\eta - \frac{1}{c^2}\,\eta_{tt} = -\frac{p_s}{\mathcal{T}_\ell}. \tag{B–17}$$

If the wavelength of the sound is much larger than the diameter of the membrane, p_s is approximately uniform over the membrane and may be taken to be a function of time only. If the sound field is time harmonic, i.e.,

$$p_s = p_{s0}e^{j\omega t},$$

we expect a steady-state solution of the form $\eta = R(r)e^{j\omega t}$ (why no dependence on θ?). In this case Eq. B–17 reduces to

$$R'' + \frac{1}{r}\,R' + k^2R = -\frac{p_{s0}}{\mathcal{T}_\ell}, \tag{B–18}$$

where $k = \omega/c$. The solution of the homogeneous equation is

$$R_h = AJ_0(kr)$$

(as usual, the Neumann function has been rejected). Next find the particular solution R_p. Because the driving term on the right-hand side of Eq. B–18 is simply a constant, try $R_p = K$. By inspection, the value of K is $-p_{s0}/k^2\mathcal{T}_\ell$, i.e.,

$$R_p = -p_{s0}/k^2\mathcal{T}_\ell.$$

The complete solution of Eq. B–18 is therefore

$$R = R_h + R_p = A\,J_0(kr) - p_{s0}/k^2\mathcal{T}_\ell.$$

The boundary condition at the edge $r = a$ must now be invoked. Assume that the edge is clamped. In order that the motion vanish there, we must put

$$A = \frac{p_{s0}}{k^2\mathcal{T}_\ell\,J_0(ka)}.$$

The final expression for η is therefore

$$\eta = \frac{p_{s0}}{k^2\mathcal{T}_\ell}\left[\frac{J_0(kr)}{J_0(ka)} - 1\right]e^{j\omega t}. \tag{B–19}$$

The solution demonstrates the resonance behavior of the membrane. At driving frequencies such that $ka = \alpha_{0n}$, that is, $\omega = \omega_{0n}$, the membrane responds with infinite excursions (in practice, of course, damping keeps the excursion finite). This example illustrates the membrane's preference to be driven at one of its eigenfrequencies. See Ref. 4 for an application to the condenser microphone.

C. THREE-DIMENSIONAL CYLINDRICAL COORDINATES

Passing from problems in the plane to problems in space, we again make use of solutions in the form of Eqs. A–13 and A–14. The former is used primarily for enclosure problems, the latter for radiation problems.

1. Enclosure Problems

As in Chap. 10, wave fields constrained by enclosures are considered first.

a. Hollow Cylinder

In the hollow-cylinder problem "hollow" implies that the wall and ends of the cylinder are rigid. If the axis of the cylinder is the z axis (Fig. 11.7), the boundary conditions are

$$u^{(r)} = \phi_r = 0 \quad \text{at} \quad r = a,$$
$$u^{(z)} = \phi_z = 0 \quad \text{at} \quad z = 0, L.$$

Equation A–13 is taken as the solution and is fitted to the problem as follows:

(a) Discard the function $N_m(k_r r)$ because the origin is included.
(b) Set $J'_m(k_r a) = 0$ to satisfy the boundary condition at $r = a$.

$$\therefore\ k_r a = \alpha'_{mn}.$$

(c) Discard the function $\sin k_z z$ to satisfy the boundary condition at $z = 0$.
(d) To satisfy the boundary condition at $z = L$, set $\sin k_z L = 0$, i.e.,

$$k_z L = N\pi, \quad N = 0, 1, 2, 3, \ldots. \tag{C–1}$$

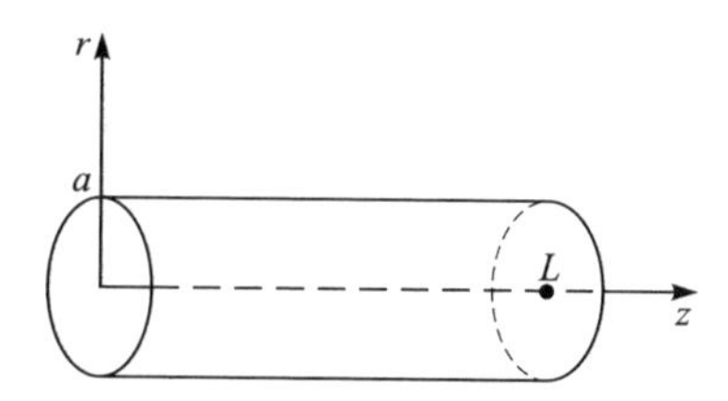

Figure 11.7 Hollow cylinder containing a sound field.

(e) For convenience, express the θ function as $\cos(m\theta - \theta_{0m})$, where θ_{0m} is a phase angle.

The (time-independent) velocity potential Φ_{mnN} is thus

$$\Phi_{mnN} = A_{mnN} J_m\left(\alpha'_{mn} \frac{r}{a}\right) \cos(m\theta - \theta_{0m}) \cos \frac{N\pi z}{L}, \tag{C–2}$$

which is a description of the (mnN)th mode of the cylinder. The modal frequency is, from Eq. A–9,

$$f_{mnN} = \frac{c_0}{2} \sqrt{\left(\frac{\alpha'_{mn}}{\pi a}\right)^2 + \left(\frac{N}{L}\right)^2}. \tag{C–3}$$

Several important families of modes may be identified. The pure longitudinal modes, ones for which the field is independent of r and θ, are found by setting $m = 0$ and $n = 1$,

$$\Phi_{01N} = A_N \cos N\pi z/L. \tag{C–4}$$

These modes are the familiar ones for ordinary plane waves in a pipe with closed ends. The eigenfrequencies are

$$f_{01N} = \frac{Nc_0}{2L}. \tag{C–5}$$

Exactly N half wavelengths fit lengthwise in the cylinder. To identify the pure radial modes, one must find indices that make Φ independent of θ and z. The required values are $m = 0$ and $N = 0$, i.e.,

$$\Phi_{0n0} = A_n J_0\left(\alpha'_{0n} \frac{r}{a}\right), \tag{C–6}$$

and the corresponding eigenfrequencies are

$$f_{0n0} = \frac{c_0 \alpha'_{0n}}{2\pi a}. \tag{C–7}$$

It might seem that a third simple family of modes, pure angular modes, having no dependence on r and z, could be identified. No such family exists, however. No combination of the indices m, n, N can be chosen to make $\Phi = \Phi(\theta)$ alone (except for the trivial case $\Phi_{010} = \text{const}$). The simplest θ-dependent modes are in the family of mixed radial, angular modes, found by setting $N = 0$,

$$\Phi_{mn0} = A_{mn} J_m\left(\alpha'_{mn} \frac{r}{a}\right) \cos(m\theta - \theta_{0m}), \tag{C–8}$$

for which the eigenfrequencies are

$$f_{mn0} = \frac{c_0 \alpha'_{mn}}{2\pi a}. \tag{C–9}$$

The pure radial modes are thus a special subgroup ($m = 0$) of this family. Which member of the r, θ family has the lowest eigenfrequency?

b. Other Cylinder Problems

First consider a rigid wall cylinder having pressure release ends. An example is a pipe open at both ends, sometimes called an open-open pipe (each open end is idealized as a place where p, and therefore ϕ, vanishes). In this case the $\mathcal{Z}$ eigenfunction must be

$$\mathcal{Z} = A_N \sin k_z z,$$

and k_z must be given by

$$k_z = \frac{N\pi}{L}, \qquad N = 1, 2, 3, \ldots.$$

(Why is the value $N = 0$ not included?) Since the expression for k_z is the same as that for a closed-closed pipe (Eq. C–1), the eigenfrequency formula is also the same (Eq. C–3). But are pure longitudinal modes possible in this case? Pure radial modes? What is the frequency of the lowest nontrivial mode?

Next consider the complementary problem: the ends are rigid but the cylinder wall is pressure release. Because in this case ϕ must be zero at the wall, the Bessel function $J_m(k_r a)$, not its derivative, must vanish.

$$\therefore\ k_r a = \alpha_{mn}.$$

The restriction on k_r is just like that on k for a vibrating membrane clamped at its edge. The expression for the resonance frequencies is obtained by replacing α'_{mn} in Eq. C–3 with α_{mn}. What simple families of modes are possible in this case? With what modes are the lowest eigenfrequencies associated?

Finally, let the enclosure be the space between two concentric cylinders. Annular sound fields are encountered in a number of industrial processes. So-called double-pipe heat exchangers (Fig. 11.8) carry one fluid inside the small pipe ($r < a$) and another fluid in the annular region between the two pipes ($a < r < b$). Each fluid may be a liquid, each a gas, or one may be a liquid and the other a gas. Depending on the nature of the fluids and the thickness of the walls, the boundaries at $r = a, b$ may act as rigid surfaces, pressure release surfaces, or impedance surfaces. Another example is the annu-

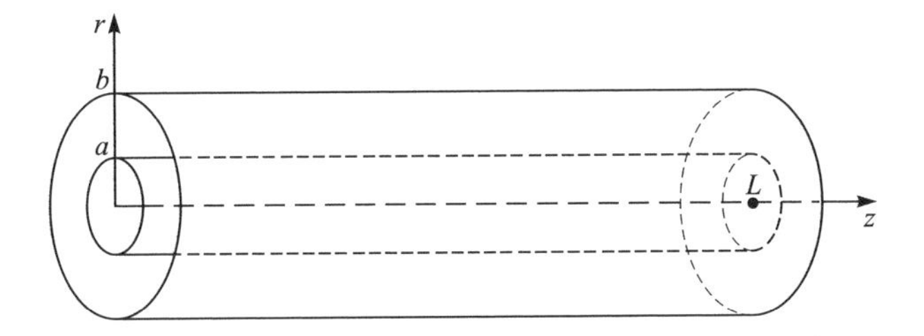

Figure 11.8 Double-pipe heat exchanger as an example of sound in an annular cylinder.

lar cylinder formed by a nuclear reactor fuel rod and its surrounding jacket. A gas such as helium is introduced in the annulus for mechanical balance. In this case both the inner and outer walls are rigid surfaces.

Because the origin falls outside the annular region, the Neumann function cannot be discarded. Equation A–13 in full is therefore the expression for ϕ. In the case of pressure release walls at $r = a, b$, application of the boundary conditions there yields

$$AJ_m(k_r a) + BN_m(k_r a) = 0, \tag{C–10a}$$

$$AJ_m(k_r b) + BN_m(k_r b) = 0. \tag{C–10b}$$

Move the Neumann function terms to the right-hand side, divide one equation by the other, and simplify (or require that the determinant of the coefficients vanish because the equations are homogeneous). The result is

$$J_m(k_r a)N_m(k_r b) - J_m(k_r b)N_m(k_r a) = 0. \tag{C–11}$$

The roots of this equation, which have been tabulated for lower values of m and for specified ratios b/a (see, for example, Ref. 1, Table 9.7), yield the allowed values of k_r. In the case of rigid walls at $r = a, b$, the relation is the same except that J_m and N_m are replaced by their first derivatives. Once k_r has been found, the rest of the analysis is the same as that for the hollow cylinder. The eigenfrequencies are then found as usual by applying Eq. A–9.

2. Radiation Problems

Traveling cylindrical waves are relatively common in nature and technology. An ordinary example is thunder, which is generated by sudden heating of the medium by lightning. Because the sound is produced along a line, albeit a crooked one, the spreading of the wave is two dimensional rather than three dimensional, at least in the early stages of the propagation. A truly straight-line source that generates sound by thermal effects is the laser. Examples of mechanical sources that produce cylindrical waves are the strings of stringed instruments, particularly the harp,[5] steam and water pipes that "sing," and vibrating wires, such as telephone or electrical power lines. Steady automobile noise from a straight roadway densely packed with traffic is frequently modeled as a cylindrical wave.

Infinitely long cylinders that produce sound (1) by radial pulsation and (2) by translational vibration are considered in this section and are analogous to pulsating and vibrating spheres, respectively. The section concludes with a brief analysis of cylindrical sources of finite length.

a. Infinite Pulsating Cylinder

The first problem to be considered is the sound radiated by a uniformly pulsating cylinder of infinite length in an open medium. Here we solve for the external field $r > a$; see Fig. 11.9. The boundary condition at the surface of the cylinder is

$$u^{(r)}(a, \theta, z; t) = u_0 \cos \omega t = u_0 \operatorname{Re}(e^{j\omega t}). \tag{C–12}$$

The sound field must also satisfy a radiation condition: because the medium is infinite, no incoming wave can be generated.

Equation A–14 is simplified for this problem as follows: First, because the cylinder is infinitely long and the boundary condition is independent of z, the field can have no dependence on z. To make the function $\mathcal{Z}$ a constant, choose $k_z = 0$ (the problem is thus similar to radiation on an infinite membrane). Second, because the cylinder pulsates uniformly, the field must be independent of θ. Therefore choose $m = 0$. Finally, because the time function must be $e^{j\omega t}$ to match the source condition, choose $H_0^{(2)}$ in order to represent an outgoing wave.

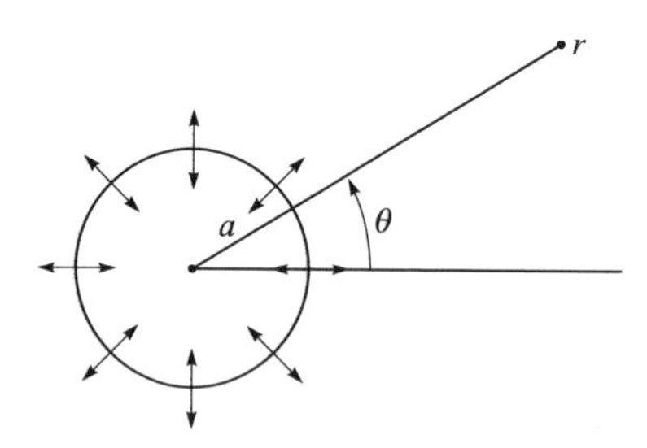

Figure 11.9 Uniformly pulsating cylinder.

[5]But the principal sound one hears from a violin, say, is not directly from the strings themselves but from the violin body, which vibrates in response to the string motion.

$$\therefore \quad \phi = AH_0^{(2)}(kr)e^{j\omega t}. \tag{C–13}$$

The constant A is determined by the boundary condition, C–12. First, the radial particle velocity is found from Eq. C–13,

$$u^{(r)} = \phi_r = AkH_0^{(2)\prime}(kr)e^{j\omega t} = -AkH_1^{(2)}(kr)e^{j\omega t},$$

which at $r = a$ reduces to $u_0 = -AkH_1^{(2)}(ka)$, or

$$A = -\frac{u_0}{kH_1^{(2)}(ka)}. \tag{C–14}$$

The solution is therefore

$$\phi = -\frac{u_0 H_0^{(2)}(kr)}{kH_1^{(2)}(ka)} e^{j\omega t}, \tag{C–15}$$

and the particle velocity and pressure are given respectively by

$$u^{(r)} = \phi_r = \frac{u_0 H_1^{(2)}(kr)}{H_1^{(2)}(ka)} e^{j\omega t} \tag{C–16}$$

and

$$p = -j\omega\rho_0\phi = \frac{j\rho_0 c_0 u_0 H_0^{(2)}(kr)}{H_1^{(2)}(ka)} e^{j\omega t}. \tag{C–17}$$

Next consider the impedance. At any point r the impedance is

$$Z = \frac{p}{u^{(r)}} = j\rho_0 c_0 \frac{H_0^{(2)}(kr)}{H_1^{(2)}(kr)}. \tag{C–18}$$

Two limiting forms of this expression are now developed. At great distances ($kr \gg 1$), the asymptotic expression for the Hankel functions, Eq. A–44, may be used. The result is

$$Z \sim j\rho_0 c_0 \frac{\sqrt{2/\pi kr}\, e^{-j(kr-\pi/4)}}{\sqrt{2/\pi kr}\, e^{-j(kr-3\pi/4)}} = \rho_0 c_0, \tag{C–19}$$

which shows that, as expected, the plane-wave impedance relation holds when the radius of curvature of the phase fronts becomes large. The other extreme, $kr \ll 1$, is of interest for the field near a cylinder of small radius, i.e., $ka \ll 1$.

Using the small-argument approximations of the Hankel functions (see Eqs. A–23b, A–17, A–21, and A–22), we find

$$\begin{aligned} Z &= j\rho_0 c_0 \frac{1 - j2[\ln kr - 0.116]/\pi}{kr/2 + j2/\pi kr} \\ &= \rho_0 c_0 \frac{kr[\pi/2 + j(0.116)] + jkr \ln(1/kr)}{1 - j\pi(kr/2)^2}. \end{aligned}$$

The leading term in the numerator is the last one. The behavior near the origin is thus given by

$$Z = j\omega\rho_0 r \ln(1/kr), \tag{C–20}$$

i.e., the impedance is small and pure imaginary. In particular, the impedance seen by the source is

$$Z|_{r=a} = j\omega\rho_0 a \ln(1/ka). \tag{C–21}$$

The positive reactance implies masslike behavior. The effective mass per unit length, $\rho_\ell = 2\pi a^2 \rho_0 \ln(1/ka)$, suggests an accession to inertia similar to that associated with small spherical radiators. However, the similarity is not close because the effective mass for the pulsating cylinder is frequency dependent. In any case, because its acoustical load is reactive, a cylinder for which $ka \ll 1$ is a poor radiator of sound.

Despite its low radiation efficiency, the line source deserves attention because of its frequent occurrence in noise control and other acoustical problems. Here we consider the farfield radiation. For $ka \ll 1$ and $kr \gg 1$, Eq. C–16 reduces to

$$u^{(r)} \sim u_0 \sqrt{\frac{\pi ka}{2}} \sqrt{\frac{a}{r}} e^{j(\omega t - kr + \pi/4)}. \tag{C–22}$$

Thus

$$\text{Re}\, u^{(r)} = u_0 \sqrt{\frac{\pi ka}{2}} \sqrt{\frac{a}{r}} \cos\left(\omega t - kr + \frac{\pi}{4}\right). \tag{C–23}$$

The pressure and intensity are found to be

$$\text{Re}\, p = \rho_0 c_0 u_0 \sqrt{\frac{\pi ka}{2}} \sqrt{\frac{a}{r}} \cos\left(\omega t - kr + \frac{\pi}{4}\right), \tag{C–24}$$

$$I = \frac{1}{T} \int_0^T pu\, dt = \frac{\rho_0 c_0 u_0^2 \pi k a^2}{4r}, \tag{C–25}$$

respectively. Note that p, $u^{(r)}$, and I all increase with frequency, provided u_0 is constant. The pressure and particle velocity vary as $1/\sqrt{r}$ and the intensity as $1/r$. This is the expected effect of cylindrical spreading. How do the pressure and particle velocity vary with distance near the source?

b. Infinite Vibrating Wire

How much sound does a vibrating string or wire make? The simplest analysis is for a wire of infinite length ($k_z = 0$ in Eq. A–14). As indicated by Fig. 11.10, the boundary condition is

$$u^{(r)}(a, \theta, z; t) = v_0 \cos\theta e^{j\omega t}.$$

Because the angle function Θ has to be $\cos\theta$ in order to match the boundary condition, the choice for m in Eq. A–14 must be $m = 1$.

$$\therefore \ \phi = A \cos\theta e^{j\omega t} H_1^{(2)}(kr),$$

$$u^{(r)} = Ak \cos\theta \, e^{j\omega t} H_1^{(2)\prime}(kr).$$

Satisfaction of the boundary condition requires that

$$v_0 = AkH_1^{(2)\prime}(ka).$$

The expressions for ϕ, $u^{(r)}$, and p are therefore, respectively,

$$\phi = \frac{v_0 \cos\theta e^{j\omega t} H_1^{(2)}(kr)}{kH_1^{(2)\prime}(ka)}, \tag{C–26}$$

$$u^{(r)} = \frac{v_0 \cos\theta \, e^{j\omega t} H_1^{(2)\prime}(kr)}{H_1^{(2)\prime}(ka)}, \tag{C–27}$$

$$p = -j\omega\rho_0\phi = -j\rho_0 c_0 v_0 \cos\theta \frac{H_1^{(2)}(kr)}{H_1^{(2)\prime}(ka)} e^{j\omega t}. \tag{C–28}$$

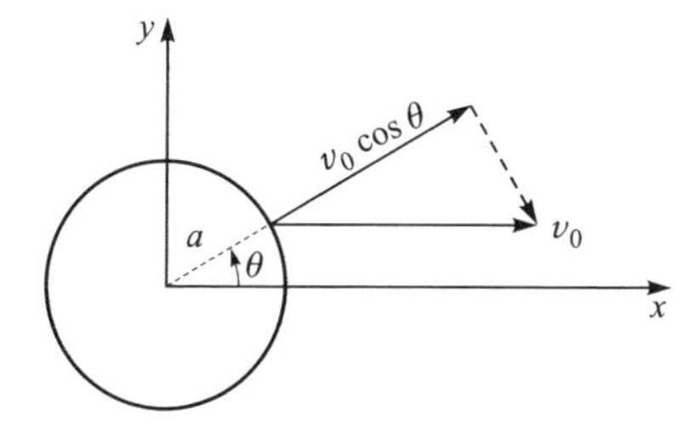

Figure 11.10 Vibrating wire.

For $ka \ll 1$ (the usual condition for vibrating wires), the farfield expressions are

$$u^{(r)} = v_0 \cos\theta\, ka \sqrt{\frac{\pi ka}{2}} \sqrt{\frac{a}{r}} e^{j(\omega t - kr + 3\pi/4)}, \tag{C–29}$$

$$p = \rho_0 c_0 u^{(r)}, \tag{C–30}$$

$$I = \frac{\pi}{4} \rho_0 c_0 v_0^2 (ka)^3 \frac{a}{r} \cos^2\theta. \tag{C–31}$$

The vibrating wire is obviously a *very* poor radiator at low frequencies. If this is so, why is the violin such a successful musical instrument? For further analysis of the vibrating wire, see, for example, Ref. 7 pp. 299–300, or Ref. 8, pp. 358–359.

c. Finite Line Source

Here we consider uniform pulsation of a very small cylinder, or line source, of finite length 2ℓ; see Fig. 11.11. Because m must be zero and the plane $z = 0$ bisects the source, the radiation could be described by using the functions $H_0^{(2)}$ $(k_r r)$ and $\cos k_z a$. An alternative approach is as follows. Consider a source element of length dz (see Fig. 11.11). This element radiates a spherical wave

$$d\phi = \frac{f(t - R/c_0)}{R}\, dz$$

that is received at the observation point L. In terms of our discussion of monopoles in Chap. 10, the product $f(t)\, dz$ is equal to $-Q(t)/4\pi$, where Q is the volume flow of fluid out of the element. To obtain contributions from all the elements, we simply integrate over the length of the source:

$$\phi = \int_{-\ell}^{\ell} \frac{f(t - R/c_0)}{R}\, dz.$$

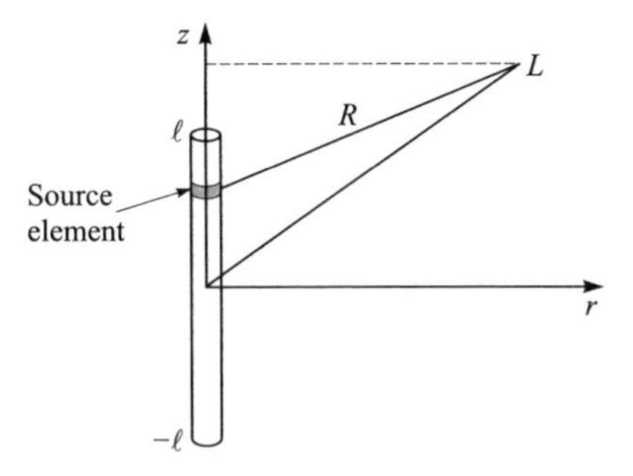

Figure 11.11 Radiation from a line source of finite length.

For an example of this approach see Ref. 10, which is about acoustic radiation from an electric spark. The directivity effects are quite interesting. The description given by Landau and Lifshitz (Ref. 5) is similar but more general in that an integral representing an incoming cylindrical wave is included.

REFERENCES

1. M. Abramowitz and I. A. Stegun (Eds.), *Handbook of Mathematical Functions*, National Bureau of Standards, Applied Mathematics Series 55 (U.S. Government Printing Office, Washington, DC, 1964), Chap. 9.
2. G. Arfken, *Mathematical Methods for Physicists*, 3rd ed. (Academic Press, New York, 1985), Chap. 11.
3. I. S. Gradshteyn and I. M. Ryzhik, *Table of Integrals, Series, and Products*, 4th ed. (Academic Press, New York, 1980). English translation by Scripto Technica, Inc., A. Jeffrey (Ed.), Secs. 8.4–8.5, pp. 951–991.
4. L. E. Kinsler, A. R. Frey, A. B. Coppens, and J. V. Sanders, *Fundamentals of Acoustics*, 3rd ed. (John Wiley and Sons, New York, 1982), Chap. 4.
5. L. D. Landau and E. M. Lifshitz, *Fluid Mechanics* (Addison-Wesley, Reading, MA, 1959), Sec. 70.
6. M. B. Moffett, W. L. Konrad, and L. F. Carlton, "Experimental demonstration of the absorption of sound by sound in water," *J. Acoust. Soc. Am.* **63**, 1048–1051 (1978).
7. P. M. Morse, *Vibration and Sound*, 2nd ed. (McGraw-Hill Book Co., New York, 1948), Secs. 2, 19, and 26. Reprinted by Acoustical Society of America, New York, 1981.
8. P. M. Morse and K. U. Ingard, *Theoretical Acoustics* (McGraw-Hill Book Co., New York, 1968), Secs. 1.2, 5.2, 7.3. Reprinted by Princeton University Press, NJ, 1986.
9. G. B. Watson, *A Treatise on the Theory of Bessel Functions* (Cambridge University Press, Cambridge, England, 1944), p. 373.
10. W. M. Wright and N. W. Medendorp, "Acoustic radiation from a finite line source with N-wave excitation," *J. Acoust. Soc. Am.* **43**, 966–971 (1968).

PROBLEMS

11–1. Use the relation between $j_n(x)$ and $J_m(x)$ (Eq. 10B–20a) to derive the recursion relations for $j_n(x)$, Eqs. 10B–24a,b, from those for $J_m(x)$, Eqs. 11Aa,b.

11–2. A Bessel function identity is

$$\frac{d}{dx}\left[\frac{J_m(x)}{x^m}\right] = -\frac{J_{m+1}(x)}{x^m}.$$

(a) Derive the identity.

(b) As an application, find the extrema of the function $J_1(x)/x$.

11–3. Find the coefficients A_m in the expansion of the function $e^{-jkx} = e^{-jkr\cos\theta}$ as a Fourier cosine series

$$e^{-jkr\cos\theta} = \sum_{m=0}^{\infty} A_m \cos m\theta.$$

In other words derive Eq. A–40. Why is a sine series not needed?

11–4. Show that the following integral is equal to $J_0(x)$:

$$I = \frac{1}{2\pi}\int_0^{2\pi} e^{jx\cos\theta}\, d\theta.$$

11–5. Make use of Eq. A–30 to show that

$$J_{n+1}(x) = \frac{x^{n+1}}{2^n n!}\int_0^1 J_0(x\mu)\mu\left(1-\mu^2\right)^n d\mu.$$

11–6. Investigate the normal modes of oscillation of a hanging chain. The chain, of length L and mass per unit length ρ_ℓ, hangs free from the ceiling. Let distance x be measured from the free end. The tension at any point on the chain is determined by the weight of the chain below that point, i.e., $\mathcal{T} = \rho_\ell g x$. The wave equation for a string under variable tension $\mathcal{T} = \mathcal{T}(x)$ may be found by generalizing the derivation in Sec. 1C.2.a. The right-hand side of Eq. 1C–11 is replaced by $(\mathcal{T}\sin x)_x$, or $\approx (\mathcal{T}\xi_x)_x$ for small angles.

(a) Show that the wave equation for the hanging chain is

$$\xi_{tt} - g(x\xi_x)_x = 0.$$

(b) Solve this equation by the method of separation of variables. Hint for solving the x-dependent equation: the substitution $x = z^2$ should lead to a form of Bessel's equation.

(c) Specialize the general solution found in part (b) so as to satisfy the requirements that (i) the solution be finite everywhere and (ii) the chain displacement vanish at the ceiling, $x = L$.

(d) Find the formula for the eigenfrequencies.

(e) As shown in Sec. 1C.2.d, the slope ξ_x vanishes at a free end of an ordinary string. Is the slope zero at the free end, $x = 0$, of the hanging chain? If not, why not? Give a physical argument.

11–7. A rigid metal disk of radius r_0 is glued to a membrane as shown in the sketch. The membrane is of infinite extent. Sinusoidal vibration of the disk (in the direction perpendicular to the membrane) generates an outgoing progressive wave on the membrane. Find the expression for the membrane displacement $\eta(r, \theta)$, $r > r_0$, for the following two cases:

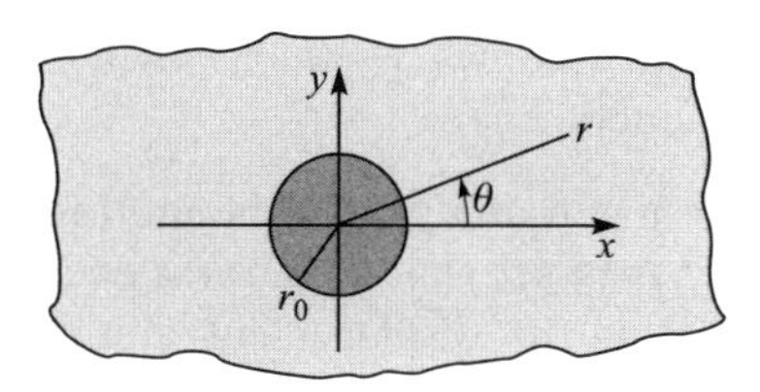

(a) The disk moves up and down as a unit. The displacement amplitude is η_0 and the time dependence is $\exp(j\omega t)$.

(b) Fixed along the y axis, the disk vibrates seesaw fashion. That is, as the right side comes up, the left side goes down, and vice-versa. The maximum displacement, at the points $x = \pm r_0$, $y = 0$, is η_0, but assume that $\eta_0 \ll r_0$. Again the time dependence is $\exp(j\omega t)$.

11-8. Given a square membrane and a circular membrane, of the same area and material and under the same tension, each clamped at its edge(s), which has the lower lowest eigenfrequency? What is the percentage difference between the two frequencies?

11–9. The radii of an annular membrane (see sketch) are a (inner) and b (outer).

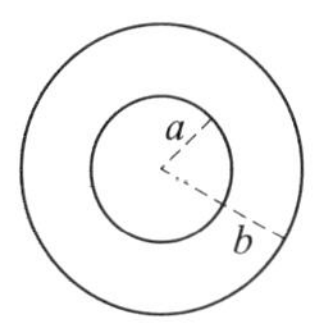

(a) Find the equation that determines the eigenfrequencies of the membrane. Consider two cases:

 (i) Membrane clamped at $r = a, b$

 (ii) Membrane free at $r = a, b$

(b) Let the vibration of the membrane be independent of the angle θ. Find the ratio of the lowest eigenfrequency for case (i) to that for case (ii) for a value $a/b = 0.2$. (To find the roots of the transcendental equations from which the eigenfrequencies are determined, consult Ref. 1.)

11–10. Design a circular vibrating membrane for which the eigenfrequencies are harmonically related (for at least one family of modes). *Hint*: If you can restrict the membrane so that the angle index m is equal to $\frac{1}{2}$, i.e.,

$$\Theta = \begin{Bmatrix} \cos\theta/2 \\ \sin\theta/2 \end{Bmatrix},$$

the radial function becomes $J_{1/2}(kr)$, which is closely related to the spherical Bessel function j_0.

11–11. Section B.3 contains an analysis of the motion of a membrane driven by a time-varying pressure that is uniform over the membrane surface. The function $R(r)$ is the displacement amplitude. Find the average displacement amplitude of the membrane

$$R_{\text{ave}} = \frac{1}{\pi a^2}\int_0^a R\, 2\pi r\, dr.$$

11–12. A flexible membrane of radius a is secured to a semicircular frame.

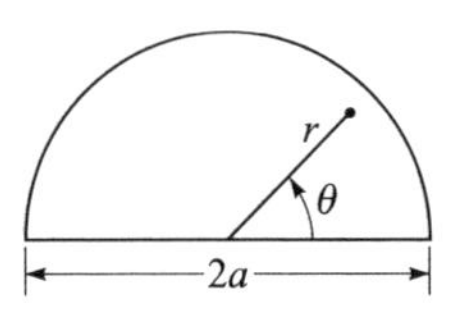

(a) Particularize the general solution of the wave equation in polar coordinates so that it applies to this problem, i.e., pick appropriate eigenfunctions and eigenvalues. Then give the complete solution.

(b) Find the lowest nontrivial eigenfrequency of a membrane for which $c = 10$ m/s and $a = 0.05$ m.

11–13. A flexible membrane fills the first quadrant of a circle of radius a. The membrane is clamped at its boundaries $\theta = 0$, $\theta = \pi/2$, and $r = a$.

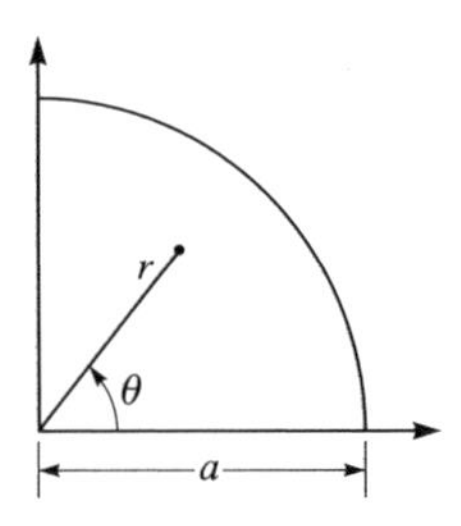

(a) Find the solution of the wave equation for this membrane.

(b) Find the expression for the eigenfrequencies. What is the lowest nontrivial eigenfrequency?

11–14. A membrane is stretched between two rigidly fixed hoops, like a lamp shade. This problem is the same as Problem 6–7. Here you are to use cylindrical coordinates to investigate the wave motion.

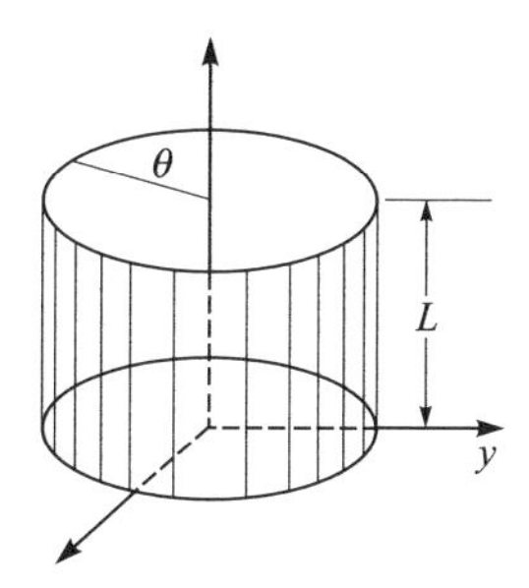

(a) Starting with the wave equation in cylindrical coordinates r, θ, and z, find the expression for the membrane displacement η that is appropriate for this problem.

(b) Deduce the expression for the eigenfrequencies.

11–15. A cylinder of radius 10 cm and length 17.06 cm contains air. The cylinder wall and ends are rigid. The enclosure is found to resonate at 1005 Hz. What mode(s) is (are) being excited?

11–16. A 5-cm-i.d. pipe is 12.7 cm long and capped at both ends. The medium inside the pipe is air at 20°C. List, in order of increasing frequency, all the modes and their frequencies that can be excited by a white noise source having an upper frequency limit of 7 kHz. The source is positioned so that it can excite all modes.

11–17. Radial pulsations of a 5-cm-o.d. pipe produce sound in the air outside the pipe. The frequency is 100 Hz. Consider the pipe length to be infinite. What is the particle velocity amplitude u_0 and the displacement amplitude ξ_0 of the pipe wall required to produce a sound pressure level of 95 dB at a distance of 8 m? Check before you use the farfield, small-source formulas to determine whether they are applicable.

11–18. If the pipe in the previous problem vibrates laterally instead of radially, i.e., if the boundary condition is

$$u^{(r)}(a, \theta, t) = v_0 \cos\theta e^{j\omega t},$$

what must be the vibration amplitude v_0 and displacement amplitude ξ_0 in order to realize the same SPL at 8 m at the angle $\theta = 0$?

11–19. Consider a pulsating vertical line source of sound in the ocean. The line source extends from the ocean surface ($z = 0$), which is treated as a pressure release boundary, to the bottom ($z = D$), which is treated as a rigid boundary. The particular boundary condition at the line source is not specified except that the signal there is proportional to $e^{j\omega t}$ and is the same in all directions.

(a) Specialize the solution of the wave equation in 3-D cylindrical coordinates so that it describes the acoustic pressure field in the water. Explain your choices. Each possible solution of the type you have found is called a mode.

(b) Give the total solution $p(r, z; t)$ as an infinite sum of the modes found in part (a). Each mode should have an amplitude coefficient which could be determined by applying the boundary condition at the surface of the line source. Because the specific boundary condition is not given in this problem, it is not necessary to calculate the coefficients.

(c) By examining the relationship between k_r, k_z, and ω, determine whether a cutoff phenomenon is involved. If so, give the formula for the cutoff frequency for each mode.

11–20. The phase velocity of a time-harmonic traveling wave may be defined in the following way. Let the pressure be expressed in the form

$$p = A(r)e^{j\psi(r,t)},$$

where the amplitude A and the phase ψ are real functions. Differentiation of the phase function,

$$d\psi = \psi_r\, dr + \psi_t dt,$$

shows that the velocity with which a constant phase surface is forwarded is

$$c^{\text{ph}} \equiv \left(\frac{dr}{dt}\right)_{\psi=\text{const}} = -\frac{\psi_t}{\psi_r}.$$

Use this formula to calculate the phase velocity for a pure radial cylindrical wave, i.e., one described by Eq. C–13.

(a) Obtain the general formula for c^{ph} (valid for all distances).

(b) Using the formula found in part (a), find the asymptotic value of c^{ph} for $kr \gg 1$.

11–21. How much sound does a vibrating string make? Assume that the vibration is due to a progressive wave traveling along an infinite string. Let the wave on the string be described by its displacement

$$\xi = \xi_0 e^{j(\omega t - Kz)},$$

where $K = \omega / c_{\text{string}}$ and c_{string} is the wave speed on the string. Take the z axis to be at the center of the string and the string vibration to be in the x, z plane (see sketch). Assume that $\xi_0 \ll a$ so that the boundary condition is that $u^{(r)}|_{r=a}$ is the radial component of the string velocity $v = v_0 e^{j(\omega t - Kz)}$, where $v_0 = j\omega\xi_0$.

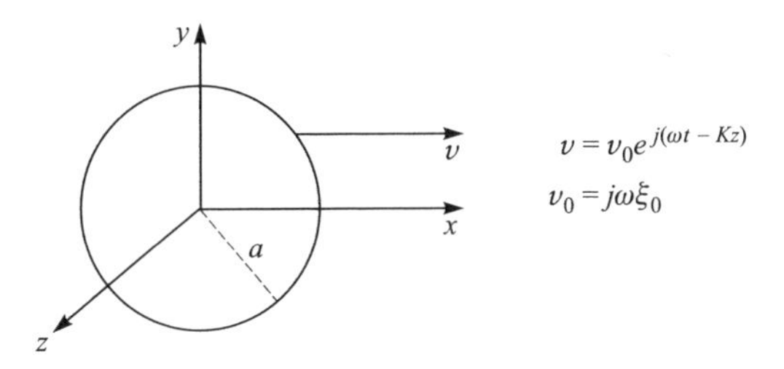

(a) Find the expression for the pressure $p(r, \theta, z; t)$ in the surrounding fluid. Explain or justify your choice of functions.

(b) What condition must be satisfied for the string vibration to produce sound in the surrounding fluid? Show what happens to the sound wave if the condition is not met.

11–22. A paraboloid of revolution defined by the function $y^2 = a(x_0 + x)$ is to be used as a horn. The designer cuts off the small end at the plane $x = 0$ and locates a plane time-harmonic source $p = Ae^{j\omega t}$ there.

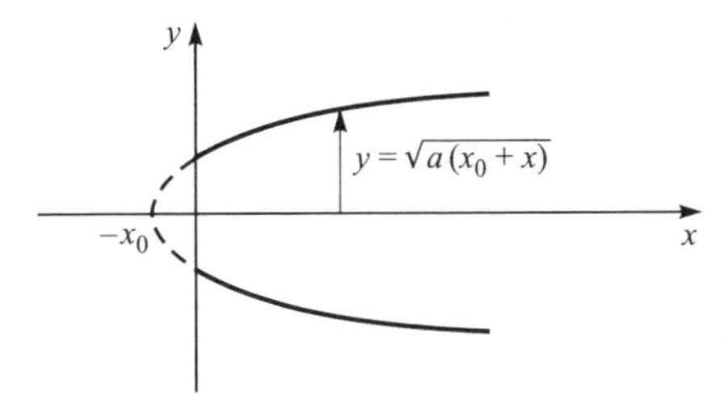

(a) Show that the horn equation for this case is

$$p_{xx} + \frac{1}{x_0 + x} p_x - \frac{1}{c_0^2} p_{tt} = 0.$$

(b) Find the solution for the wave generated by the source.

12

WAVEGUIDES

A. INTRODUCTION

As the name suggests, a waveguide is a device for guiding the propagation of waves. The wave motion in the guide is usually partly progressive and partly standing. The progressive wave motion is in the direction of the desired transmission. The standing waves are caused by the presence of the confining surfaces that do the guiding. In the typical waveguide problem a source at a given location emits a known acoustical signal, usually time harmonic, and one wishes to calculate the sound field produced in the waveguide.

An analysis of an elementary, two-dimensional waveguide appears in Sec. 6.D. Refer to that section for a discussion of waveguide modes, phase velocity, cutoff (now often called "cuton" in the field of duct acoustics), the physical interpretation of a mode as a combination of two traveling waves, and the role of a source in exciting modes. In the present chapter waveguides for rectilinear and cylindrical waves are considered.

In Chaps. 10 and 11 the wave equation is solved for the velocity potential ϕ. The relations $\mathbf{u} = \nabla\phi$ and $p = -\rho_0\phi_t$ are then used to find the particle velocity and pressure, respectively. An alternative method is used here, not for any fundamental reason but simply to demonstrate another approach. We solve the wave equation in p,

$$\nabla^2 p - \frac{1}{c_0^2} p_{tt} = 0, \qquad \text{(A–1)}$$

and employ the momentum equation,

$$\rho_0 \mathbf{u}_t = -\nabla p, \tag{A–2}$$

to find the components of particle velocity. For time-harmonic waves, Eq. A–2 becomes

$$\mathbf{u} = -\frac{1}{j\omega\rho_0} \nabla p. \tag{A–3}$$

B. RECTANGULAR WAVEGUIDE[1]

1. General Solution

A rigid-wall duct of rectangular cross section, height a and width b, is shown in Fig. 12.1. As Eq. A–3 shows, the vanishing of the normal component of particle velocity at each wall requires that

$$p_x = 0 \quad \text{at} \quad x = 0, a, \tag{B–1}$$

$$p_y = 0 \quad \text{at} \quad y = 0, b. \tag{B–2}$$

At $z = 0$ is a source (not shown in the figure but described in detail in the next section). The source emits a signal that is proportional to $e^{j\omega t}$.

For this problem the most convenient form of the general solution of Eq. A–1 is[2]

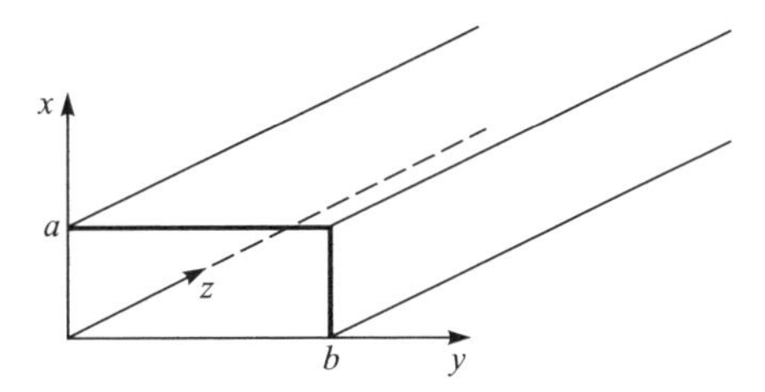

Figure 12.1 Rectangular waveguide.

[1]Waveguides with rigid or pressure release walls are considered here. Various aspects of rigid-wall rectangular waveguides are discussed in Ref. 3. For walls of finite, complex impedance see Refs. 8 and 9.

[2]Throughout this chapter β is used to represent the wave number of the traveling wave. In this case β replaces the symbol s used in Eq. 6C–5 to denote the wave number in the z direction.

$$p = \begin{Bmatrix} \cos qx \\ \sin qx \end{Bmatrix} \begin{Bmatrix} \cos ry \\ \sin ry \end{Bmatrix} \begin{Bmatrix} e^{j\beta z} \\ e^{-j\beta z} \end{Bmatrix} \begin{Bmatrix} e^{j\omega t} \\ e^{-j\omega t} \end{Bmatrix}, \tag{B–3}$$

where

$$q^2 + r^2 + \beta^2 = (\omega/c_0)^2. \tag{B–4}$$

Complex exponentials have been chosen for the $\mathcal{Z}(z)$ and $T(t)$ functions in order that they may be easily combined to represent a time-harmonic progressive wave in the z direction. In particular, the form $\exp j(\omega t - \beta z)$ is consistent with the source time function. To satisfy Eqs. B–1 and B–2, pick the cosine rather than the sine functions, and take $q = m\pi/a$ and $r = n\pi/b$ (m and n integers) as the eigenvalues. In this way we arrive at the following expression for the (mn)th traveling-wave mode of the waveguide:

$$p_{mn} = A_{mn} \cos \frac{m\pi x}{a} \cos \frac{n\pi y}{b} e^{j(\omega t - \beta_{mn} z)}, \tag{B–5}$$

where the formula for the wave number β_{mn} is found by using Eq. B–4:

$$\beta_{mn} = \sqrt{(\omega/c_0)^2 - (m\pi/a)^2 - (n\pi/b)^2}. \tag{B–6}$$

Notice that the choice $m = 0$, $n = 0$ does not lead to a trivial result. In fact, as Eq. B–5 shows, p_{00} represents an ordinary traveling plane wave and for this reason is frequently called the plane-wave mode (sometimes the term "principal mode" is used).

The character of the wave motion depends markedly on the value of the drive frequency $f = \omega/2\pi$. If f is sufficiently high, β_{mn} is real, and true wave propagation down the guide takes place. If the frequency is sufficiently low, however, β_{mn} is imaginary. In this case the exponential function in Eq. B–5 becomes

$$e^{j(\omega t - \beta_{mn} z)} = e^{j\omega t} e^{-|\beta_{mn}| z},$$

that is, the "wave" is evanescent. The transition from propagating to evanescent behavior occurs at the cutoff frequency, found by setting $\beta_{mn} = 0$,

$$f_{mn}^c = \tfrac{1}{2} c_0 \sqrt{(m/a)^2 + (n/b)^2}. \tag{B–7}$$

A convenient way of writing Eq. B–6 is therefore

$$\beta_{mn} = (\omega/c_0)\sqrt{1 - (f_{mn}^c/f)^2}.$$

The phase velocity c_{mn}^{ph} for the (mn)th mode is

$$c_{mn}^{\text{ph}} = (\omega/\beta_{mn}) = \frac{c_0}{\sqrt{1 - [(m/a)^2 + (n/b)^2](c_0/2f)^2}}, \tag{B–8}$$

or, in terms of the cutoff frequency,

$$c_{mn}^{\text{ph}} = c_0/\sqrt{1 - (f_{mn}^c/f)^2}. \tag{B–9}$$

Although each mode has its own phase velocity, at high enough frequency $(f \gg f_{mn}^c)$, the phase velocity approaches the free-medium sound speed. As the frequency is decreased, the mode's phase velocity increases. It becomes infinite at the cutoff frequency, as shown in Fig. 6.13.

The complete solution is the sum of all the modes,

$$p = \sum_{m,n} A_{mn} \cos\frac{m\pi x}{a} \cos\frac{n\pi y}{b}\, e^{j(\omega t - \beta_{mn} z)}, \tag{B–10}$$

where the coefficients A_{mn} are to be determined from source conditions.

2. Source Conditions and Mode Excitation

The general solution is now applied to the problem of wave motion generated by a time-harmonic source (what if the source is not time harmonic?) of arbitrary cross section located at the entrance to the waveguide. The source excitation may be expressed as follows:

$$p(x, y, 0; t) = F(x, y)e^{j\omega t}. \tag{B–11}$$

The amplitude $F(x, y)$ is assumed to be known at all points in the plane $z = 0$. Notice that the wave motion in the guide is forced, not free. The frequency is determined by the source, not by the geometry of the guide.

The coefficients A_{mn} are found by requiring that the solution Eq. B–10 satisfy the source condition, Eq. B–11. One obtains

$$F(x, y) = \sum_{m,n} A_{mn} \cos\frac{m\pi x}{a} \cos\frac{n\pi y}{b},$$

which is a double Fourier series. To solve for A_{mn}, use the orthogonality of the trigonometric functions. Multiply both sides by $\cos(m'\pi x/a)\cos(n'\pi y/b)\,dx\,dy$ and integrate over the domain $0 \le x \le a$, $0 \le y \le b$. Every term on the right-hand side vanishes with the integration except the one for which $m = m'$, $n = n'$. For this term one may carry out the integration and then solve for the coefficient. The final result is

$$A_{mn} = \frac{\epsilon_m \epsilon_n}{ab} \int_0^b \int_0^a F(x, y) \cos \frac{m\pi x}{a} \cos \frac{n\pi y}{b} \, dx \, dy. \tag{B–12}$$

The Neumann factor ϵ_i has the value $\epsilon_i = 1$ if $i = 0$ but $\epsilon_i = 2$ if $i \geq 1$.

Of all possible modes of the waveguide, often only a few contribute to the ultimate acoustic field. First, $F(x, y)$ may have properties that make the integral in Eq. B–12 vanish for certain values of m, n. This means that the modes denoted by these values of m, n are not excited by the source. Second, not all the modes that are excited by the source share in the subsequent propagation. Propagation in a given mode can occur only if the source frequency exceeds the cutoff frequency of the mode. If the source frequency is below cutoff, the mode is evanescent: its amplitude decays exponentially because its wave number β_{mn} is imaginary. The field downstream of the source may therefore consist of just a few modes.

3. Example

Suppose the source produces a pressure that is independent of x but varies with y, as shown in Fig. 12.2, i.e.,

$$F(x, y) = F(y) = \begin{cases} F_0, & 0 < y < b/2, \\ -F_0, & b/2 < y < b \end{cases}. \tag{B–13}$$

When this expression for F is used in Eq. B–12, the integration over x yields

$$\int_0^a \cos \frac{m\pi x}{a} \, dx = \begin{cases} 0, & m \neq 0, \\ a, & m = 0. \end{cases}$$

Therefore, the only "x mode" that is excited is the one for which $m = 0$. This is not a surprising result, since the source excitation does not depend on x. The expression for A_{mn} now becomes

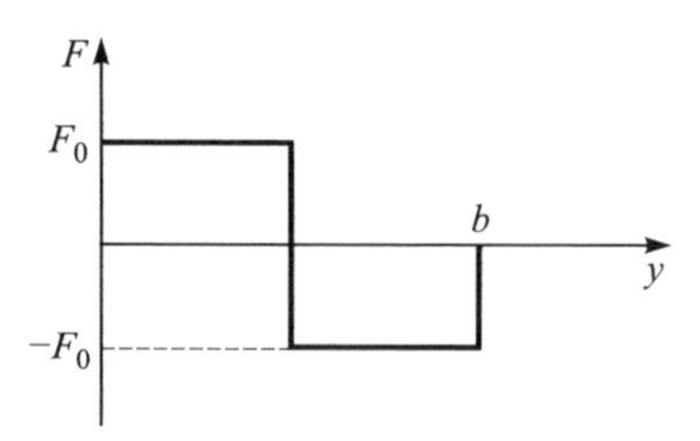

Figure 12.2 Source amplitude at the beginning of a waveguide.

$$A_{0n} = \frac{\epsilon_n}{b} \int_0^b F(y) \cos \frac{n\pi y}{b}\, dy.$$

For $n = 0$ the integral vanishes identically. For other values of n ($\epsilon_n = 2$) we have

$$\begin{aligned} A_{0n} &= \frac{2F_0}{b}\left[\int_0^{b/2} \cos \frac{n\pi y}{b}\, dy - \int_{b/2}^b \cos \frac{n\pi y}{b}\, dy\right] \\ &= \frac{4F_0}{n\pi} \sin \frac{n\pi}{2}, \qquad n \geq 1. \end{aligned} \tag{B–14}$$

The coefficients vanish for even values of n and alternate their signs for odd values of n. In particular, the fact that $A_{00} = 0$ means that the plane-wave mode is not excited.

Now consider the mode selection process from a physical standpoint. The sketches in Fig. 12.3 help show why the even-numbered modes are not needed to represent the source function and why every other odd-numbered mode has a negative coefficient. A unique feature of the excitation function $F(y)$ is its anti-symmetry about the midpoint $y = b/2$. Since the $n = 0$ mode does not have this property, it cannot contribute to the representation of $F(y)$. For the same reason, all the other even modes fail to contribute. The burden of

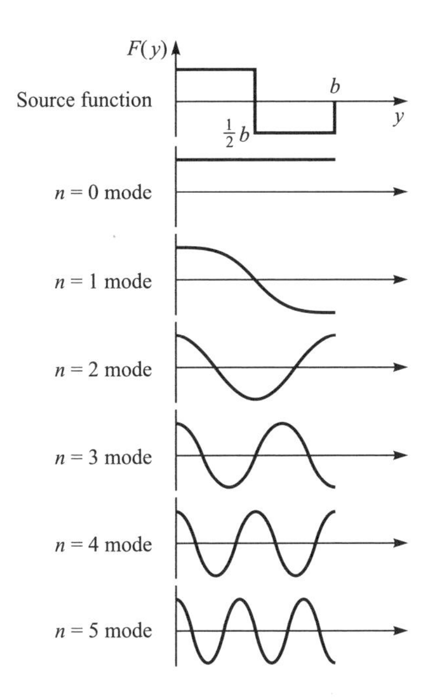

Figure 12.3 What modes are needed to build up the source function?

representing $F(y)$ therefore falls solely on the (anti-symmetric) shoulders of the odd modes. However, if all the odd modes were added just as they are drawn in the figure, it is clear that the sum of them would be a curve too high at $y = 0$, too low at $y = b$. Moreover, the curve would not make the rapid change required at $y = b/2$. If every other odd mode is first inverted, however, the curve will have the required shape. Qualitative reasoning therefore confirms and explains the main features of the analytical results.

What effect does cutoff have on the modes that are excited? Since for this problem the only nontrivial value of m is $m = 0$, Eqs. B–7, B–6, and B–8 reduce to

$$f_{0n}^c = \frac{c_0 n}{2b}, \tag{B–15a}$$

$$\beta_{0n} = \frac{\omega}{c_0}\sqrt{1 - (f_{0n}^c/f)^2}, \tag{B–15b}$$

$$c_{0n}^{\text{ph}} = c_0/\sqrt{1 - (f_{0n}^c/f)^2}, \tag{B–15c}$$

respectively. If, for example, the source frequency lies between f_{03}^c and f_{05}^c, the wave number is real for the first and third modes but imaginary for all higher modes. Propagation therefore occurs in the first and third modes (at speeds c_{01}^{ph} and c_{03}^{ph}, respectively) but not in the higher modes. Consequently, the acoustic field downstream ($z > 0$) is considerably different from that at the source ($z = 0$).

Some practical conclusions may be drawn from this example. It was found that anti-symmetry of the excitation function $F(y)$ about the midpoint prevents the even modes from being generated. Conversely, it may be seen that if $F(y)$ were symmetric about the midpoint, only the even modes would be generated.[3] Sometimes one wishes to generate only the plane-wave mode ($m = 0$, $n = 0$). Ideally, this can be done by making F a pure constant (see Eq. B–12), for instance, by using as a source a plane piston that completely fills the entrance to the guide. A rectangular piston may, however, not be available or in any case may be impractical. A horn driver with a round throat may have to be used. Such a source will generate many higher modes in addition to the plane-wave mode. This does not mean, however, that one is forced to accept the presence of the higher modes. If the horn driver mouth is mounted symmetrically in the entrance plane of the waveguide, all modes for which either m or n is odd will be automatically excluded. The remaining higher modes (*both* m and n even) may be eliminated by choosing the drive frequency low enough that they are cut off.

In other applications one may wish to study one or more of the higher modes in the absence of the plane-wave mode. The plane-wave mode may be difficult to avoid, however. It is not subject to cutoff and may therefore be

[3]In the general case, where F is a function of x as well as y, the symmetry properties of F in the x direction may be used to determine whether the odd or even values of m are selected.

prevented only by using a source that is perfectly antisymmetric about the midline $x = 0$ (or $y = 0$) of the entrance plane. However, even a small amount of residual symmetry will cause the plane-wave mode to be excited. As an example, the first nonplanar mode ($m = 0$, $n = 1$) may be produced by mounting two identical horn drivers in the entrance plane of the waveguide and driving them push-pull, as indicated in Fig. 12.4. This arrangement is ideal in theory, but if the two drivers are not perfectly positioned or their outputs are not identical, some amount of the plane-wave mode will be generated. Once generated, the plane wave has no difficulty propagating downstream.

One other practical consideration should be mentioned. Irregularities in the waveguide, such as an imperfect junction between two lengths of guide or a hole that has been drilled in a sidewall to house a microphone, can cause scattering of sound. The scattered sound may be represented in terms of modes of the waveguide just as the source sound is. Undesired higher order modes may sneak into the wave field in this way.

4. Pressure Release Walls

A Styrofoam trough filled with water, either topped with styrofoam or left open to the air, is an example of a waveguide having pressure release bounding surfaces. The analysis of this waveguide is left as an exercise for the reader. What is the fate of the plane-wave mode in this case?

5. Phase and Group Velocity

The reader should review Sec. 6D at this point for a discussion of phase and group velocity for rectangular waveguides. The general definition of phase velocity is

$$c^{\mathrm{ph}} = \frac{\omega}{\beta}. \tag{B–16}$$

For a rectangular waveguide, Eqs. B–8 and B–9 give the specific expressions. The discussion in Sec. 6D makes clear that phase velocity is simply the trace

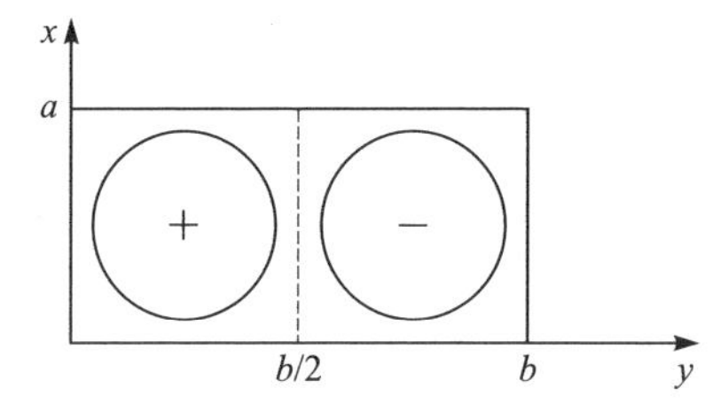

Figure 12.4 Waveguide excitation designed to generate the 0, 1 mode.

speed c_{trace} for a ray that bounces back and forth zigzag fashion between the walls of the waveguide. In terms of the angle of incidence θ_i the ray makes with the normal to the waveguide wall, the phase velocity is

$$c^{ph} = \frac{c_0}{\sin\theta_i}, \tag{B–17}$$

where

$$\sin\theta_i = \sqrt{1-\cos^2\theta_i} = \sqrt{1-(f^c/f)^2}. \tag{B–18}$$

The group velocity, on the other hand, is defined by

$$c^{gr} \equiv \frac{d\omega}{d\beta}. \tag{B–19}$$

Some authors characterize the group velocity as the speed with which a small wave packet travels down the guide, others as the speed with which the energy is transmitted through the guide. A very simple interpretation is given in Sec. 6D: the group velocity is simply the axial component c_{axial} of the propagation velocity, i.e.,

$$c^{gr} = c_{axial} = c_0 \sin\theta_i = c_0\sqrt{1-(f^c/f)^2}. \tag{B–20}$$

Notice that the relation between c^{gr} and c^{ph} is

$$c^{gr}c^{ph} = c_0^2. \tag{6D–15}$$

At cutoff the ray bounces back and forth between the waveguide walls at an angle of incidence $\theta_i = 0$. Since the propagation velocity then has no axial component, the group velocity is zero. At the same time the phase velocity is infinite because the wavefront, being parallel to the waveguide wall, takes zero time to trace out the entire wall.

C. CYLINDRICAL WAVES IN WAVEGUIDES

Two kinds of waveguides are well known for their ability to support cylindrical wave fields. The first kind is a tube or duct of circular cross section. In this guide the traveling-wave component is represented by the same function used for the rectangular waveguide, namely $\exp j(\omega t - \beta z)$, where z is the distance along the axis of the tube. The standing-wave component, a function of r and θ, is cylindrical. The second kind of waveguide for cylindrical waves is made up of two parallel reflecting planes, such as the top and bottom of a body of water, and a source is located in the fluid between the two planes. In this case progressive waves travel outward in the radial direction and are described by the product $e^{j\omega t}H_m^{(2)}(\beta r)$; the angular function $\Theta(\theta)$ acts as a directivity factor. The standing-wave field, set up between the two parallel planes, is described by the $\mathcal{Z}$ function, which consists of sine and cosine functions.

1. Cylindrical Tube

The formal analysis for a circular waveguide (radius a) with rigid walls closely parallels that for the rectangular waveguide. In place of Eq. B–3 we start with

$$p = \begin{Bmatrix} J_m(k_r r) \\ N_m(k_r r) \end{Bmatrix} \begin{Bmatrix} \cos m\theta \\ \sin m\theta \end{Bmatrix} \begin{Bmatrix} e^{j\beta z} \\ e^{-j\beta z} \end{Bmatrix} \begin{Bmatrix} e^{j\omega t} \\ e^{-j\omega t} \end{Bmatrix}, \tag{C–1}$$

where (see Eq. 11A–9)

$$\beta^2 = (\omega/c_0)^2 - k_r^2. \tag{C–2}$$

Sometimes complex exponentials $\exp \pm jm\theta$ are used in place of $\cos m\theta$ and $\sin m\theta$ to emphasize the traveling-wave nature of so-called spinning modes. Except for annular ducts, one discards the Neumann function and chooses $\exp j(\omega t - \beta z)$ to represent propagation in the positive z direction. The boundary condition at the wall, $u^{(r)}(k_r a) = 0$, is satisfied by choosing

$$k_r = \alpha'_{mn}/a, \tag{C–3}$$

where α'_{mn} is the nth zero of J'_m. The expression for the pressure of the (mn)th mode is therefore

$$p_{mn} = J_m\left(\frac{\alpha'_{mn} r}{a}\right)[A_{mn} \cos m\theta + B_{mn} \sin m\theta]\, e^{j(\omega t - \beta_{mn} z)}. \tag{C–4}$$

The full solution is of course a sum over all values of m and n.

Use of Eqs. C–2 and C–3 yields formulas for the wave number,

$$\beta_{mn} = \sqrt{(\omega/c_0)^2 - (\alpha'_{mn}/a)^2}, \tag{C–5}$$

the phase velocity,

$$c_{mn}^{\text{ph}} = \frac{\omega}{\beta_{mn}} = \frac{c_0}{\sqrt{1 - (\alpha'_{mn} c_0 / 2\pi f a)^2}}, \tag{C–6}$$

and the cutoff frequency,

$$f_{mn}^{c} = \frac{\alpha'_{mn} c_0}{2\pi a}. \tag{C–7}$$

The modal coefficients A_{mn} and B_{mn} are determined by the source excitation. The procedure is similar to that used for a rectangular waveguide. Morse (Ref. 8) shows how to make the calculation for axisymmetric source functions ($m = 0$). Dyer (Ref. 4) treats the more general case and considers both pressure and particle velocity sources. For application to fan noise in circular ducts, see, for example, Refs. 1, 5, and 6.

An interesting problem is described in Ref. 7. A flexible membrane stretched over a cross section of a tube acts as a partial obstruction to an incident plane wave traveling down the tube. The presence of the membrane, which is set into its own modal vibration, leads to reflected and transmitted fields consisting of various modes of the tube.

2. Parallel Planes

Figure 12.5 shows a sound source S located in the region between two parallel planes. Because the sound is confined between the two planes, it can ultimately spread in only two dimensions. This implies that the traveling waves are cylindrical. The Bessel and Neumann functions in Eq. C–1 are replaced by Hankel functions (of argument βr to indicate r is the direction of propagation). At the same time, for the $\mathcal{Z}$ function, the complex exponentials are replaced with sine and cosine functions (of argument $k_z z$).

As an example, let the parallel planes be rigid. In this case the pressure associated with the (mn)th mode may be written

$$p_{mn} = A_{mn} \cos m(\theta - \theta_{0mn}) \cos \frac{n\pi z}{D} H_m^{(2)}(\beta_{mn} r)\, e^{j\omega t}, \tag{C–8}$$

where D is the distance between the planes, and the wave number is given by

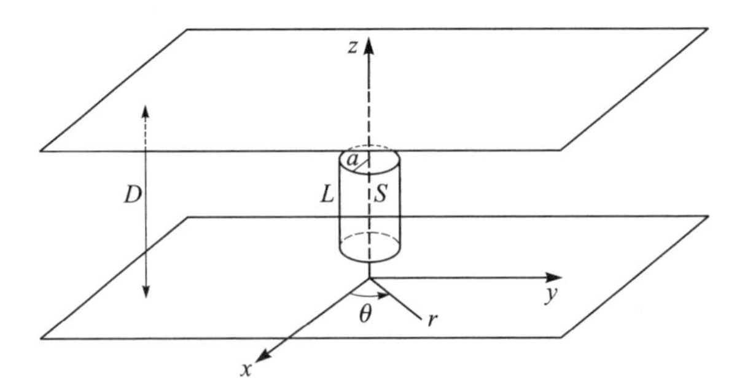

Figure 12.5 Cylindrically spreading sound in a parallel-plane waveguide.

$$\beta_{mn} = \sqrt{(\omega/c_0)^2 - (n\pi/D)^2} = (\omega/c_0)\sqrt{1 - (f_n^c/f)^2}. \tag{C–9}$$

The formula for the cutoff frequency is $f_n^c = nc_0/2D$. (How does the Hankel function behave for frequencies below cutoff?) The total solution is the sum of all the modes,

$$p = \sum_{m,n} p_{mn}. \tag{C–10}$$

The values of the constants A_{mn} and θ_{0mn} are set by conditions at the source. Consider, for example, uniform radial pulsation of the cylinder. Since the sound generated by such a source cannot depend on θ, only the $m = 0$ modes can be excited, i.e., $A_{mn} = 0$ if $m \neq 0$. If, moreover, the cylinder extends from the top of the guide to the bottom, that is, $L = D$, the field cannot depend on z, in which case only the 0, 0 mode (simple radial wave) can be excited. The result is

$$p = A_{00} H_0^{(2)}(kr)\, e^{j\omega t},$$

where $k = \omega/c_0$. Note the absence of any cutoff restriction. The coefficient A_{00} is found by matching the solution to the motion prescribed at the surface of the cylinder $r = a$. The wave field in the region $0 \leq z \leq D$ is the same as that generated by uniform pulsation of an infinitely long cylinder in an open medium; see Eqs. 11C–15 to 11C–17. (Would this be true if one or both of the planes were pressure release surfaces?)

If the source does not fill the guide completely, other modes are excited. For example, let the source condition be

$$u^{(r)}(a, \theta, z; t) = u_0 f(z) e^{j\omega t}. \tag{C–11}$$

Although m is still zero, the pressure and (radial) particle velocity are given by more complicated expressions:

$$p = \sum_n A_n \cos \frac{n\pi z}{D}\, H_0^{(2)}(\beta_n r)\, e^{j\omega t} \tag{C–12}$$

and (see Eq. A–3)

$$u^{(r)} = \frac{1}{jk\rho_0 c_0} \sum_n \beta_n A_n \cos \frac{n\pi z}{D}\, H_1^{(2)}(\beta_n r)\, e^{j\omega t}. \tag{C–13}$$

Requiring Eq. C–13 to satisfy Eq. C–11 yields

$$jk\rho_0 c_0 u_0 f(z) = \sum_n A_n \beta_n H_1^{(2)}(\beta_n a) \cos\frac{n\pi z}{D}, \tag{C–14}$$

from which the coefficients A_n may be obtained by the usual methods of Fourier series,

$$A_n = \frac{\epsilon_n jk\rho_0 c_0 u_0}{D\beta_n H_1^{(2)}(\beta_n a)} \int_0^D f(z) \cos\frac{n\pi z}{D}\, dz. \tag{C–15}$$

Suppose the source frequency is well above the cutoff frequency for a given mode. How does the modal amplitude vary as the source frequency approaches f_n^{c}?

An elementary acoustical model of the ocean is a fluid bounded by two parallel planes, the bottom one rigid, the top pressure release. In this case the $\mathcal{Z}$ eigenfunctions are qualitatively different. The pressure must vanish at the top and be an extremum at the bottom. A uniformly pulsating cylinder therefore excites many modes even if it extends from bottom to top. See Problem 12–13. See Ref. 2 for a detailed treatment of the oceanic waveguide.

REFERENCES

1. V. Boleter and R. C. Chanaud, "Propagation of fan noise in cylindrical ducts," *J. Acoust. Soc. Am.* **49**, 627–638 (1971).
2. H. Medwin and C. S. Clay, *Fundamentals of Acoustical Oceanography* (Academic Press, San Diego, 1998), Chap. 11.
3. P. E. Doak, "Elements of sound propagation," in E. J. Richards and D. J. Mead, Eds., *Noise and Acoustic Fatigue in Aeronautics* (John Wiley and Sons, New York, 1968), Sec. 2.1.
4. I. Dyer, "Measurement of noise sources in ducts," *J. Acoust. Soc. Am.* **30**, 833–841 (1958).
5. C. L. Morfey, "Rotating pressure patterns in ducts: Their generation and transmission," *J. Sound Vib.* **1**, 60–87 (1964).
6. C. L. Morfey, "Sound transmission and generation in ducts with flow," *J. Sound Vib.* **14**, 37–55 (1971).
7. P. M. Morse, "Waves and targets: Generalizations and specifics," *Am. J. Phys.* **53**, 25–40 (1985).
8. P. M. Morse, *Vibration and Sound*, 2nd ed. (McGraw-Hill Book Co., New York, 1948), Secs. 26 and 31. Reprinted by Acoustical Society of America, New York, 1981.
9. P. M. Morse and K. U. Ingard, *Theoretical Acoustics* (McGraw-Hill Book Co., New York, 1968), Secs. 9.1–9.4. Reprinted by Princeton University Press, Princeton, NJ, 1986.

PROBLEMS

12–1. A rigid-wall rectangular waveguide of height a and width b is excited by a time-harmonic source located at the center of the waveguide, in particular,

$$p(x, y, 0; t) = \delta\left(\frac{x}{a} - \frac{1}{2}\right)\delta\left(\frac{y}{b} - \frac{1}{2}\right) F_0 e^{j\omega t},$$

where δ is the delta function.

(a) Solve for the pressure field in the waveguide.

(b) What modes are not excited? Why are they not excited? Give a geometrical or physical explanation.

(c) The dimensions are $a = 20$ mm, $b = 50$ mm and the medium is air. What frequency restriction allows nonplanar modes to be avoided?

12–2. Obtain the expression for the pressure coefficients A_{mn} for a rectangular, rigid-wall waveguide (see Eq. B–10) for a source specified in terms of the particle velocity w (component in the z direction), namely,

$$w(x, y, 0; t) = w_0 F(x, y)\, e^{j\omega t}.$$

(a) Find the expression for A_{mn}.

(b) Qualitatively, how does A_{mn} behave as the frequency is decreased toward the mode cutoff frequency?

(c) From part (b), draw a brief conclusion about mode response.

12–3. This problem is a variation on the rectangular waveguide example analyzed in Sec. B. You are given the same source excitation, Eq. B–13, but the top and bottom surfaces of the waveguide, $x = 0$ and $x = a$, are pressure release rather than rigid. The other two surfaces, $y = 0$ and $y = b$, remain rigid. Find the expressions for β_{mn}, f^c_{mn}, p_{mn}, and the modal coefficients A_{mn}. Why does the same source excite more modes in this case?

12–4. Consider the propagation of sound down a rectangular water trough of depth d and width w. The sides and bottom of the trough are rigid. The top is open to the air. Assume a time-harmonic pressure source at the plane $z = 0$.

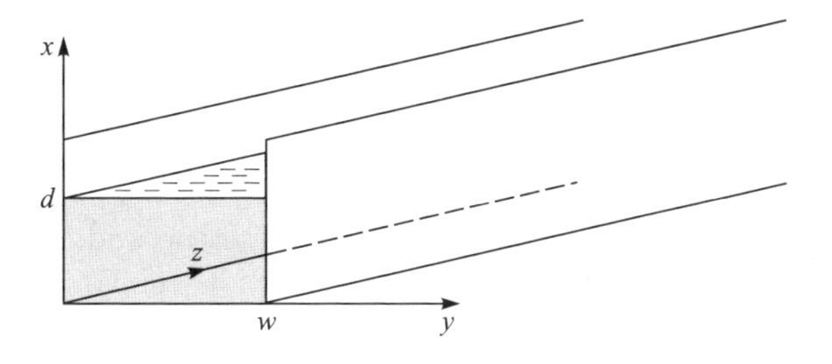

(a) Find the expression for the acoustic pressure of the (mn)th mode p_{mn}, where m is an index associated with the x dependence and n an index associated with the y dependence.

(b) Find the expression for the cutoff frequency f_{mn}^c.

(c) The trough contains seawater ($c_0 = 1500$ m/s), and the source frequency is 3 kHz. Only the lowest mode (i.e., the mode with the lowest cutoff frequency) is desired, and its phase velocity is to be 3000 m/s. What values of w and d should be used?

12–5. A horn driver at one end of a long circular pipe excites all waveguide modes of the pipe. At the other end is an ideal termination that prevents reflections. The pipe i.d. is 5 cm and the medium is air. List the modes with cutoff frequencies f_{mn}^c below 10 kHz, beginning with $f_{01}^c = 0$ Hz. In each case give the value of f_{mn}^c and the expression for the mode function p_{mn}. Arrange your answers in a table, ordered by increasing values of f_{mn}^c.

12–6. A flexible membrane is attached to two diverging straight rails to form the pie-shaped waveguide shown in the sketch. The boundary conditions are that $\eta = 0$ at $\theta = 0, \theta_0$. A source at $r = r_0$ drives the membrane as follows:

$$\eta(r_0, \theta; t) = F(\theta)e^{j\omega t},$$

where the function $F(\theta)$ (unspecified in this problem) gives the dependence of the driving amplitude on θ.

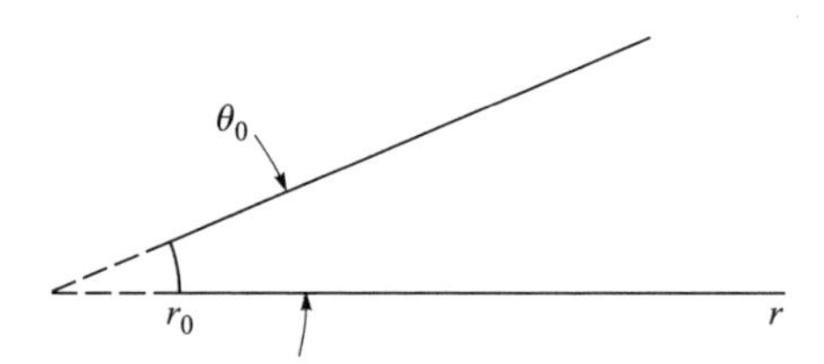

(a) Find the mode functions $\eta_n(r, \theta; t)$; identify the index n.

Next let $\theta_0 = 22\frac{1}{2}^\circ$.

(b) What are the mode functions for this case?

(c) Find the amplitudes of the modes for the case $F(\theta) = 1$.

(d) Find the general expression for the phase velocity. By taking the asymptotic limit, show that $c^{\text{ph}} \sim c_0$ for large values of kr.

12–7. A pie-shaped waveguide has rigid walls. The top and bottom walls, at $z = 0, d$ (see sketch), are parallel. The side walls are flat but diverge at an angle θ_0. The source at $r = r_0$, described by $u_{\text{source}}^{(r)} = u_0 f(\theta, z)e^{j\omega t}$, sends waves down the guide in the radial direction. In this particular waveguide, $\theta_0 = 18^\circ$ ($\pi/10$ rad).

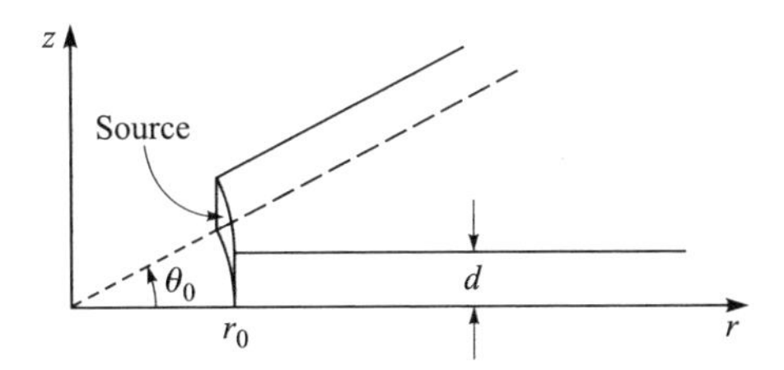

(a) Find the expression for the mode function P_{mn} (r, θ, z) for the waveguide, where $p_{mn} = P_{mn}e^{j\omega t}$. What (nontrivial) values of m, n are allowed? It is not necessary to solve for the modal coefficients A_{mn}.

(b) Find the expression for the cutoff frequencies for the modes. On what indices do the cutoff frequencies depend?

(c) What is the lowest cutoff frequency, and with what family of modes is it associated?

(d) What restrictions must be put on the source shading function $f(\theta, z)$ to excite just the lowest mode of the family identified in part (c)?

12–8. A rigid-wall circular tube of radius a has a rigid longitudinal septum that extends halfway across the tube; see the sketch. Consider axial propagation generated by a time-harmonic source $p = f(r, \theta)e^{j\omega t}$ at $z = 0$.

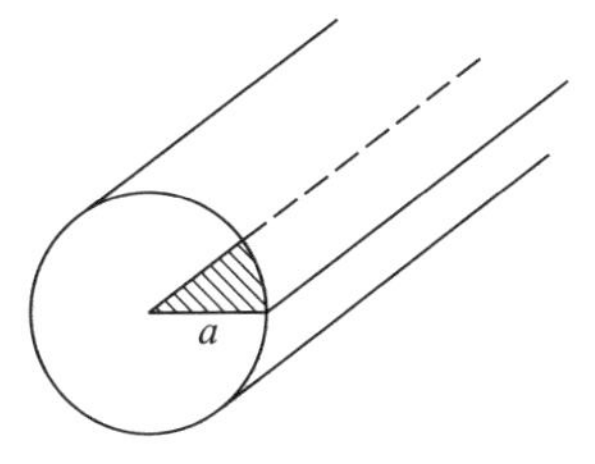

(a) Find the expression for the mode functions p_{mn}. Give all particulars. For example, what are the restrictions, if any, on the index m?

(b) Find the expression for the cutoff frequency f_{mn}^c.

(c) Not counting the plane-wave mode, what is the lowest cutoff frequency?

12–9. Consider the waveguide described by the sketch in the previous problem. In this case, however, the fluid in the pipe is (fresh) water, and the septum is made of Styrofoam (pressure release material).

(a) Find the expression for the mode functions p_{mn}. Give the appropriate values for the indices.

(b) The tube radius is 5 cm. An attempt is made to send sound of frequency 5000 Hz down the waveguide. Will the attempt be successful? If so, what is (are) the phase speed(s)? If not, why not?

12–10. A horizontal oil pipeline of diameter 1 m (i.d.) is half filled with oil (see the sketch). The pipeline wall is rigid. The sound speed in the oil is 400π m/s, while that in the mixture of gaseous vapor and air above the oil is 100π m/s. The impedance of the oil is much, much greater than the impedance of the vapor-air mixture. The problem thus has two waveguides, one formed by the oil, the other formed by the vapor-air gas. The pressure in each fluid may be expressed as

$$\text{Oil:} \quad p_{mn}^{\text{oil}} = F_{mn}(r, \theta)e^{j(\omega t - \beta_{mn}^{\text{oil}} z)},$$

$$\text{Vapor-air:} \quad p_{m'n'}^{\text{vap}} = G_{m'n'}(r, \theta)e^{j(\omega t - \beta_{m'n'}^{\text{vap}} z)}.$$

To keep the problem simple, assume that the sound fields in the two waveguides are not coupled. That is, assume that once a signal gets started in one waveguide, it propagates independently of whatever sound field exists in the other waveguide and does not "leak" into the other waveguide. This assumption may be justified to some extent by the very large difference in the impedances of the two fluids.

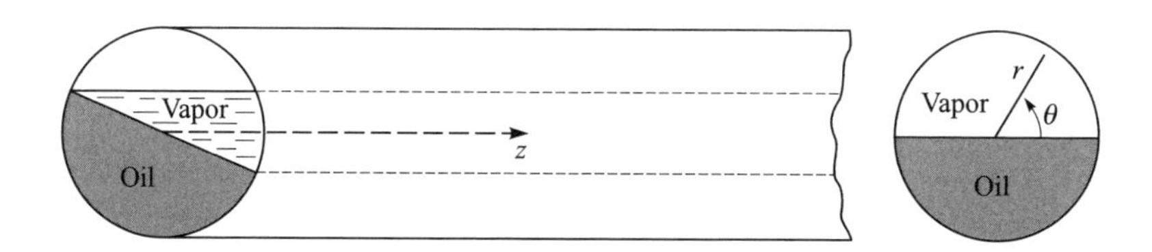

(a) Find the functions F_{mn} and $G_{m'n'}$ that apply to this problem and that satisfy the boundary conditions.

(b) Find the expression for the cutoff frequencies f_{mn}^{oil} and $f_{m'n'}^{\text{vap}}$ for the modes of both waveguides.

(c) The source frequency is 300 Hz. What modes can propagate in each waveguide?

12–11. Waves travel down a liquid-filled pipe of semicircular cross section, radius a. The walls of the pipe are Styrofoam (pressure release material). A source of angular frequency ω generates sound at the pipe entrance.

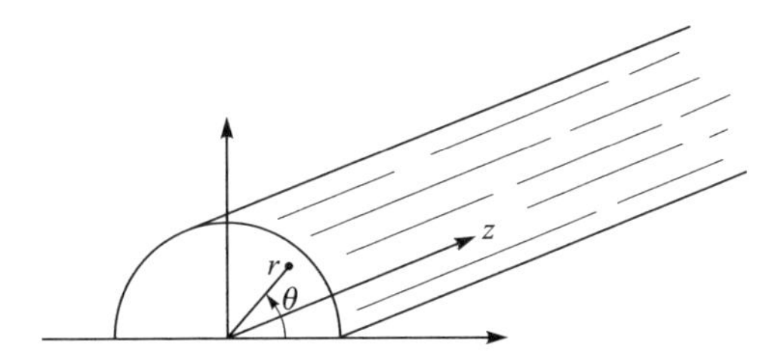

(a) Find the expression for the mode functions p_{mn}, where m is the azimuthal index.

(b) Determine the formula for the cutoff frequencies f_{mn}^c. What is the lowest cutoff frequency (nontrivial modes only)?

Suppose the source is a semicircular piston that just fits the pipe and vibrates uniformly with velocity $u^{(z)} = u_0 e^{j\omega t}$.

(c) Modes with what value(s) of m are excited by this source? If any modes are missing, explain why.

12–12. A source radiates sound into a semi-infinite, gas-filled, cylindrical pipe of radius a. The axial particle velocity prescribed by the source is

$$u^{(z)}(r, \theta, 0, t) = u_0 F(r/a) e^{j\omega t},$$

where F shows how the velocity amplitude varies over the surface of the source.

(a) Find the expression for the pressure as an infinite sum of all the appropriate modes of the pipe. Use the notation A_{mn} (or A_n, if appropriate) for the mode coefficient.

(b) Obtain the expression for the coefficients in terms of an integral involving $F(r/a)$.

(c) What effect does the source frequency have on the magnitude of the coefficients?

12–13. Waves are radiated into the ocean by uniform pulsation of a vertical cylinder of radius a. The cylinder extends from the ocean surface to the bottom, at depth D. Use the coordinate system shown. Take the water surface ($z = 0$) to be a pressure release boundary, the bottom ($z = D$) a rigid boundary.

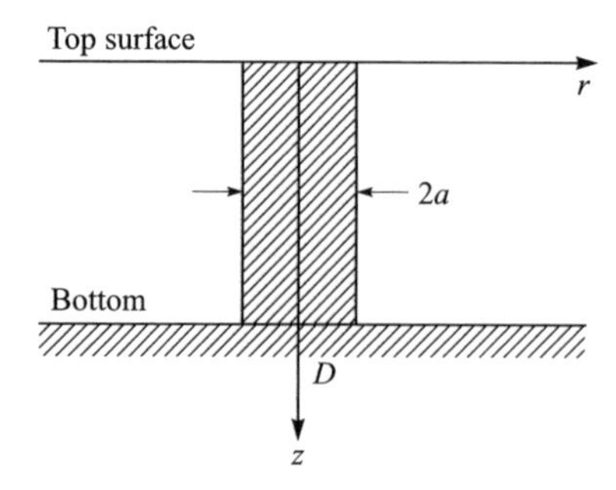

(a) Give the solution (in terms of the pressure p) as an infinite sum of the modes, identified by the index n, appropriate for this problem. The mode amplitude coefficients A_n are determined in part (c).

(b) Find the expression for the radial particle velocity $u_n^{(r)}$ for each mode. Sketch the mode shape as a function of z for the modes having the three lowest cutoff frequencies. For example, if the z dependence of $u^{(r)}$ for one of the modes were $\sin 2\pi z/D$, the profile would be that sketched below.

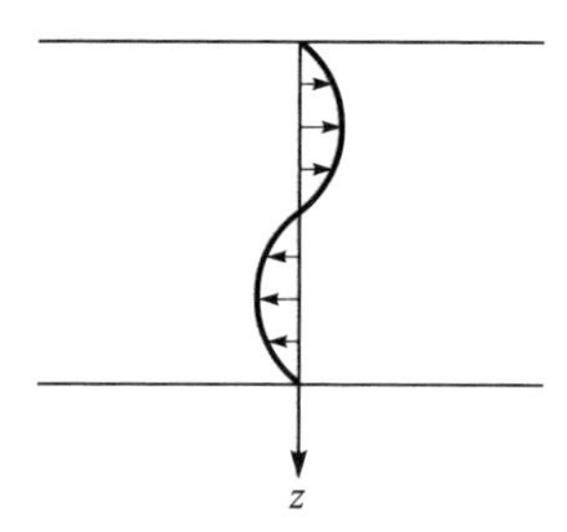

(c) The actual source condition is that the radial particle velocity at the surface of the cylinder is $u^{(r)} = u_0 e^{j\omega t}$. Use this information to determine the coefficient A_n. Then give the final complete solution.

(d) Consider a case in which $D = 3.75$ m, $c_0 = 1500$ m/s, $a = 4$ cm, $f = 200$ Hz, and $u_0 = 2$ cm/s. Find the SPL (rms pressure re 1 μPa) at range $r = 20$ m and depth $z = D/2$.

12–14. The shallow water near the ocean shore is an oceanic waveguide bounded above by the air surface and below by the sloping bottom. Model the waveguide as a wedge of angle θ_0. Idealize the air surface as a smooth pressure release plane, the bottom as a smooth rigid plane, and the shore as an infinite straight line, as shown in the sketch. For simplicity, let $\theta_0 = 5.625^\circ (\pi/32$ rad). Assume that the surf near the shore acts as a uniform source of noise along the entire shoreline.

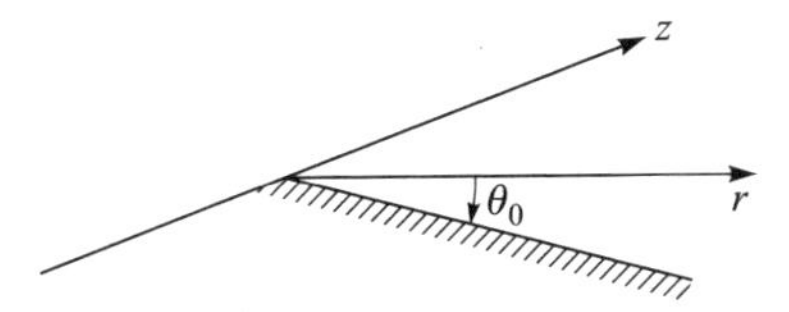

(a) Find the mode solutions $p_m(r, \theta; t)$ appropriate for the problem. Consider a single-frequency component (angular frequency ω) of the noise, and assume that the noise source is located close enough to the wedge apex that you need be concerned only with sound that travels outward into the deeper ocean. Determine whether a cutoff frequency is associated with each mode (give the cutoff frequency if one exists).

(b) Next assume that the shore extends only a distance W in the z direction, that is, that the water is restricted to an estuary of width W. Model the estuary as a wedge bounded by rigid vertical planes at $z = 0$ and $z = W$. Find the mode solutions $p_{mn}(r, \theta, z; t)$ for this case and again answer the question about cutoff frequency.

13

RADIATION FROM A BAFFLED PISTON

Up to now we have considered sources of very simple configuration, namely vibrating planes, spheres, and cylinders. Most practical sources of sound—loudspeakers, sonars, motor housings, and so on—have more complicated shapes. The piston vibrating in a baffle is a source of sufficient complexity to serve as a reasonable model for many practical sound radiators, for example, a loudspeaker mounted in a large cabinet. Yet at the same time the baffled piston is simple enough to be treated in detail analytically.

What is a baffled piston? Consider a surface of very large area, usually but not necessarily a plane. All of the surface is fixed except for a section, ordinarily in the center, that vibrates normal to the surface. The moving section is the piston, the fixed section the baffle. The purpose of the baffle is to restrict the sound field to the forward hemisphere (more generally, to decouple the sound fields in the forward and backward hemispheres). As shown by Example 10.2 (Sec. 10D.3), without the baffle the piston would tend to act as a dipole, which radiates poorly. The baffle is most often designed to be rigid, but pressure release baffles, and even baffles of intermediate impedance, are sometimes used.

Our attention in this chapter is on the classical problem of radiation from a uniformly excited circular piston mounted in a plane rigid baffle. A very large literature exists on piston radiation; only the surface is scratched in this chapter. More complicated piston problems that have been considered include those in which (1) the piston is noncircular, e.g., the vibrating surface is rectangular (Ref. 11) or elliptical; (2) the excitation is nonuniform (Ref. 3), e.g., the amplitude or phase or both vary over the face of the piston (this topic is treated briefly in Sec. F); and (3) the baffle is incomplete or curved (Ref. 8). It should be pointed out that although we treat the piston here as an acoustical transmitter, it works equally well as a receiver. For example, the (farfield) directional characteristics of a given piston are the same whether used as a transmitter or receiver.

In Sec. A an integral expression for the pressure field radiated by a piston in a rigid plane baffle is derived. The piston shape is allowed to be arbitrary, but the excitation is taken to be uniform and time harmonic. As an example, a ring piston is considered. Most of the remainder of the chapter is devoted to radiation by a full circular piston. The farfield is treated in Sec. B, the field along the axis in Sec. C, and radiation impedance of the piston in Sec. D. Transient radiation is taken up in Sec. E. The chapter ends (Sec. F) with a brief look at radiation from nonuniformly driven pistons.

A. GENERAL SOLUTION: THE RAYLEIGH INTEGRAL

Our general model for the pressure field produced by vibration of a piston mounted in a rigid baffle is a solution called the Rayleigh integral. The shape of the piston is arbitrary, and u_p, the normal component of piston velocity, is an arbitrary function of time. The Rayleigh integral may be obtained as a special case of Eq. 10D–11. Two caveats are needed, however. The first caveat is mentioned in the discussion following Eq. 10D–11: that equation is derived on the assumption that the simple sources on the distributed area radiate into full space, whereas in the case of the baffled piston, the baffle restricts the radiation to the forward hemisphere. See Fig. 13.1. Consequently, the factor 4π in the denominator of Eq. 10D–11 must be replaced by 2π. Another way to justify the change to 2π is to recognize that restricting the radiation to a half space effectively doubles the strength of the simple sources. The second caveat is about the source strength function $\dot{q}_a$. Since $\dot{q}_a$ is the volume acceleration per unit area, it is simply the piston acceleration $\dot{u}_p$. The final result is

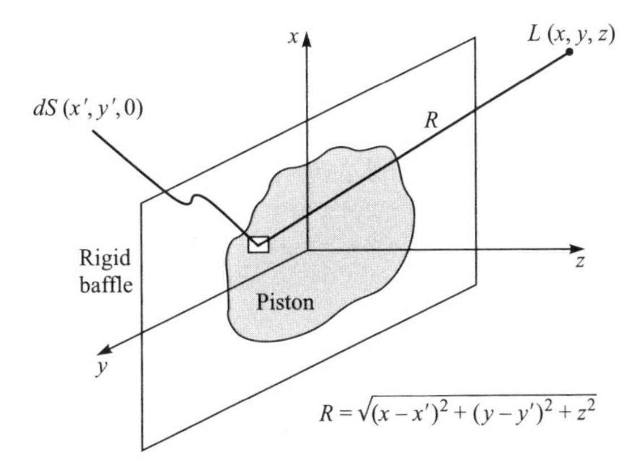

Figure 13.1 Radiation from a baffled piston of arbitrary shape.

$$p(x, y, z; t) = \rho_0 \int_S \frac{\dot{u}_p(x', y'; t - R/c_0)}{2\pi R} \, dS, \tag{A–1}$$

which is the Rayleigh integral (Ref. 10). Here x', y' are the coordinates of the source point on the piston, and R is the distance to the field point of interest $L(x, y, z)$:

$$R = \sqrt{(x - x')^2 + (y - y')^2 + z^2}.$$

Note the use of primed coordinates to designate a source point and unprimed coordinates to designate a field point.

Although Eq. A–1 is an exact solution, the integration is generally difficult to carry out. Except for the few problems that can be done analytically, one must perform the integration numerically.

1. Time-Harmonic Piston Vibration

Most of the rest of this chapter is devoted to piston vibration that is time harmonic:

$$u_p = u_0 e^{j\omega t}, \tag{A–2}$$

where it is assumed that u_0 is a constant, i.e., the piston is uniformly driven. For this case Eq. A–1 becomes

$$p(x, y, z; t) = \frac{jk\rho_0 c_0 u_0 e^{j\omega t}}{2\pi} \int_S \frac{e^{-jkR}}{R} dS. \tag{A–3}$$

Before we apply Eq. A–3 to solve specific piston radiation problems, some general remarks are in order. First, if the piston velocity were to vary in magnitude over the surface of the piston, u_0 would have to appear inside the integral (phase variation would also be handled in this way). In Sec. F a problem in which u_0 varies radially from the center of a circular piston is considered. Second, if the piston velocity were not normal to the surface dS, the arithmetic product $u_0 \, dS$ would have to be replaced by the dot product $\mathbf{u}_0 \cdot d\mathbf{S}$; only the normal component of particle velocity produces sound radiation.

2. Example: Ring Piston

A very simple radiator that may be used to illustrate the use of Eq. A–3 is the ring piston. The ring piston is also useful as a stepping stone toward our main problem of interest, the full circular piston. The ring shown in Fig. 13.2 has average radius a and (small) thickness w. The ring is imbedded in a rigid plane

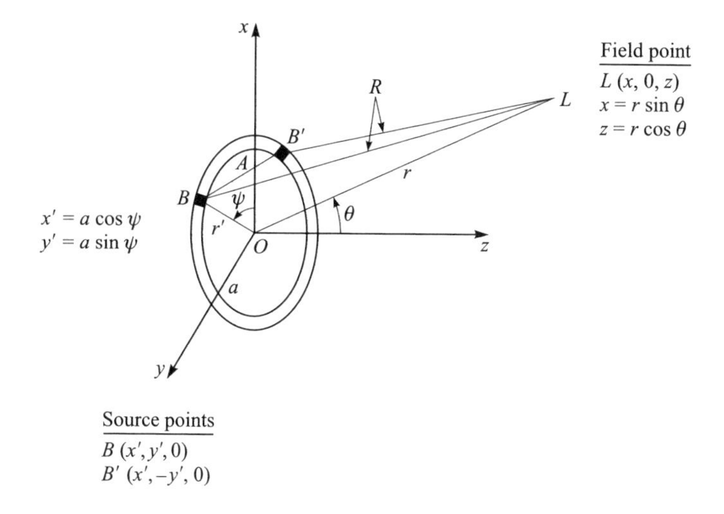

Figure 13.2 Calculation of the farfield pressure generated by a ring piston.

baffle. Because of the symmetry of the ring, the field point L may, without loss of generality, be located in the x, z plane. Morever, L is more conveniently described in terms of the polar coordinates r,θ defined in Fig. 13.2. The center of the ring is the origin O of the coordinate system. Let $\mathbf{r}$ represent the vector from O to L:

$$\mathbf{r} = \mathbf{i}x + \mathbf{j}0 + \mathbf{k}z = \mathbf{i}r\sin\theta + \mathbf{k}r\cos\theta.$$

Similarly, let $\mathbf{r}'$ represent the vector to the radiating element B of the ring:

$$\mathbf{r}' = \mathbf{i}x' + \mathbf{j}y' + \mathbf{k}0 = \mathbf{i}a\cos\psi + \mathbf{j}a\sin\psi.$$

The distance R between source and field points is therefore

$$R \equiv |\mathbf{r} - \mathbf{r}'| = \sqrt{(r\sin\theta - a\cos\psi)^2 + a^2\sin^2\psi + r^2\cos^2\theta}$$
$$= \sqrt{r^2 + a^2 - 2ra\sin\theta\cos\psi}. \qquad \text{(A–4)}$$

Because of symmetry, a second radiating element B', at the other end of the dashed line segment BAB' (parallel to the y axis), is at the same distance R from L. The two elements have combined area

$$dS = 2aw\,d\psi.$$

The two-dimensional integral in Eq. A–3 therefore reduces to a single integral over the angle ψ, from 0 to π:

$$p = \frac{jka\, w\rho_0 c_0 u_0 e^{j\omega t}}{\pi} \int_0^{\pi} \frac{e^{-jkR}}{R}\, d\psi. \qquad \text{(A–5)}$$

Equation A–5 is now evaluated for two special but important cases.

First, the field point L is assumed to be in the farfield in the sense that $r \gg a$. Expansion of Eq. A–4 for this case yields

$$R = r[1 - (a/r)\sin\theta\cos\psi + O(a/r)^2].$$

Stopping with the first two terms, we obtain

$$R = r - a\sin\theta\cos\psi. \qquad \text{(A–6)}$$

Physically, Eq. A–6 represents the relation between R and r when L is far enough away that (1) R may be approximated by AL (the distance from L to the center of the dotted line BAB') and (2) AL and OL are parallel. Given these conditions, one may easily derive Eq. A–6 directly.

The expression for R in Eq. A–6 is now substituted in the exponential function in Eq. A–5. It is not, however, necessary to use Eq. A–6 for the amplitude factor 1/R. Here the simpler approximation $R \doteq r$ is sufficient (because L is in the farfield, signals emitted from all points on the ring have very nearly the same amplitude when they arrive at L). The farfield expression for the pressure is therefore

$$p = \frac{jkaw\rho_0 c_0 u_0}{\pi r} e^{j(\omega t - kr)} \int_0^{\pi} e^{j(ka\sin\theta)\cos\psi}\, d\psi.$$

The integral is recognized as $\pi J_0(ka\sin\theta)$ (see Eq. 11A–27b), and we have

$$p = \frac{jkaw\rho_0 c_0 u_0}{r} J_0(ka\sin\theta) e^{j(\omega t - kr)}. \qquad \text{(A–7)}$$

The farfield radiation is thus a spherically spreading wave of amplitude proportional to the excitation amplitude u_0, the ring area $2\pi aw$, the frequency f, and the directional function $J_0(ka\sin\theta)$.

This is a convenient point to introduce a quantity to describe the directional characteristics of a source. The *amplitude directivity factor D* is defined as the pressure at any angle θ relative to that (at the same range r) at $\theta = 0$[1]:

[1]This is the appropriate definition of D for sources for which the peak radiation occurs at $\theta = 0$. If the peak radiation occurs at another angle $\theta_{\max}$, the directivity factor is usually defined as the ratio $p(\theta)/p(\theta_{\max})$. Notice that the definition given here is appropriate for axisymmetric radiation. When the radiation is a function of two angles, D is defined to depend on both.

$$D(\theta) = \frac{P(r, \theta)}{P(r, 0)}, \tag{A–8}$$

where P is defined by $p = P(r, \theta)e^{j\omega t}$. When Eq. A–8 is evaluated for the ring piston, the result is

$$D(\theta) = J_0(ka \sin \theta). \tag{A–9}$$

It is clear that the radiation is strongest on the z axis ($\theta = 0$) and has nulls at angles given by $ka \sin \theta = \alpha_{0n}$ (recall that α_{0n} is the nth zero of J_0). In between the nulls are secondary radiation maxima, of monotonically decreasing prominence, at angles given by $ka \sin \theta = \alpha'_{0n}$ ($= \alpha_{1n}$). The number of nulls and secondary maxima is determined by the size of ka ($= 2\pi a/\lambda$). For example, if $ka < \alpha_{01} = 2.405$, J_0 is limited to its first (or major) lobe, and no nulls or secondary maxima occur. The fact that the major lobe of the radiation is spread over the entire hemisphere, $0 < \theta < 90°$, means that the radiation is not very directional. On the other hand, if ka is very large, the major lobe is confined to the much narrower sector $0 < \theta < \sin^{-1} \alpha_{01}/ka$, i.e., the radiation is highly directional. This result is typical of many directional sources: in order for the radiation to be highly directive, the source must usually be large compared to a wavelength. A side effect, however, is that the high-ka piston has many secondary maxima, called minor lobes, in its radiation pattern [if ka is large, $J_0(ka \sin \theta)$ goes through many oscillations as θ varies from 0° to 90°].

For our second calculation of the field of the ring piston, we specialize to the z axis, the so-called acoustic axis of the piston. No restriction to the farfield is made. Since L is now equidistant from all radiating points on the ring ($R = \sqrt{r^2 + a^2}$), the integration in Eq. A–5 is trivial. We obtain

$$p = jkaw\rho_0 c_0 u_0 \frac{e^{j(\omega t - k\sqrt{r^2+a^2})}}{\sqrt{r^2 + a^2}}. \tag{A–10}$$

This seemingly uninteresting result, which shows that the axial amplitude decreases monotonically with range r, becomes more interesting when compared with the result for a full piston.

3. Circular Piston (Disk)

The full circular piston (Fig. 13.3) is much more commonly encountered in practice than the ring piston. We may, however, make use of the ring piston calculations by regarding the full piston as made up of infinitesimal vibrating ring elements. The ring radius a thus becomes the variable σ, and the thickness w becomes $d\sigma$. See Fig. 13.3. Integrating over Eq. A–5 (or starting directly from Eq. A–3), we obtain

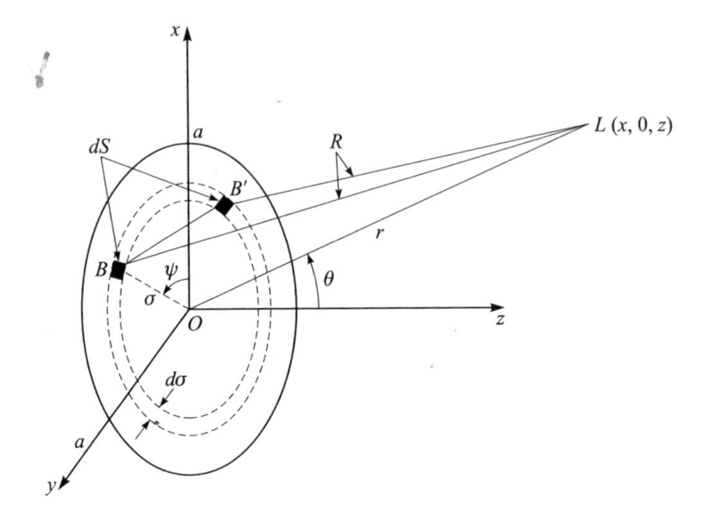

Figure 13.3 Radiation from a circular piston.

$$p(r, \theta; t) = \frac{jk\rho_0 c_0 u_0}{\pi} e^{j\omega t} \int_0^{\pi} d\psi \int_0^a \sigma \frac{e^{-jkR}}{R} d\sigma, \tag{A–11}$$

where R is now given by Eq. A–4, with a replaced by σ:

$$R = \sqrt{r^2 + \sigma^2 - 2r\sigma \sin\theta \cos\psi}. \tag{A–12}$$

Equation A–11 is the exact general expression for the radiation from a uniform circular piston. In the following special but important cases the integral may be evaluated in closed form and analytical expressions for the pressure obtained:

1. Field point L in the farfield
2. Field point L on the axis
3. Field point L on the piston face

These three cases are taken up in Secs. B, C, and D, respectively.

B. FARFIELD RADIATION

The calculation of the farfield pressure due to a full circular piston is easily carried out by starting with the farfield expression for the ring piston, Eq. A–7. Replacing a and w by σ and $d\sigma$, respectively, and integrating over σ yield

$$p = \frac{jk\rho_0 c_0 u_0}{r} e^{j(\omega t - kr)} \int_0^a \sigma J_0(k\sigma \sin\theta)\, d\sigma. \tag{B–1}$$

Let $k\sigma \sin\theta = \mu$. The integral becomes

$$\frac{1}{k^2 \sin^2\theta} \int_0^{ka \sin\theta} \mu J_0(\mu)\, d\mu = \frac{a}{k \sin\theta} J_1(ka \sin\theta),$$

where Eq. 11A–24c has been used. Substitution in Eq. B–1 yields

$$p(r, \theta; t) = \frac{ja\rho_0 c_0 u_0}{r} \frac{J_1(ka \sin\theta)}{\sin\theta} e^{j(\omega t - kr)}. \tag{B–2}$$

To prepare for calculation of the directivity of the circular piston, first evaluate Eq. B–2 at $\theta = 0$. The result is

$$p(r, 0; t) = \frac{ja\rho_0 c_0 u_0}{r} \frac{ka}{2} e^{j(\omega t - kr)}.$$

When this expression and Eq. B–2 are used to evaluate Eq. A–8, the piston directivity function is obtained:

$$D(\theta) = \frac{2J_1(ka \sin\theta)}{ka \sin\theta}, \tag{B–3}$$

which is tabulated in Appendix E.[2] A Maclaurin series expansion of the directivity function written in terms of the argument $A = ka \sin\theta$ is

$$D(A) = 1 - \frac{1}{8}A^2 + \frac{1}{192}A^4 + \cdots = \sum_{n=0}^{\infty} (-1)^n \frac{(A/2)^{2n}}{n!(n+1)!}. \tag{B–4}$$

A practical analysis of the piston farfield is now presented.

1. Rayleigh Distance

Combining various factors in the amplitude of the farfield pressure leads to an interesting and useful rearrangement of Eq. B–2. First, note that $\rho_0 c_0 u_0$ is a pressure amplitude, which we may call P_0. Keep in mind, however, that P_0 is not the true value of the pressure amplitude at the face of the piston. Rather, it is a fictitious quantity that represents the pressure amplitude that *would* exist at the face if the piston diameter were infinite. Another quantity of interest in

[2]From Ref. 4, Appendix, Table IV, which also includes a tabulation of $[2J_1(A)/A]^2$ (some authors prefer to define directivity in terms of intensity rather than pressure amplitude).

piston radiation problems is the so-called Rayleigh distance, defined in general by

$$R_0 \equiv S/\lambda, \tag{B–5}$$

where S is the piston area, whatever its shape, and λ is the wavelength. For the circular piston, $S/\lambda = \pi a^2/\lambda$, the Rayleigh distance is

$$R_0 = ka^2/2. \tag{B–6}$$

In terms of $D(\theta)$, P_0, and R_0, Eq. B–2 takes on the simpler form

$$p = j\frac{R_0 P_0}{r} D(\theta) e^{j(\omega t - kr)}. \tag{B–7}$$

An interesting interpretation of the Rayleigh distance develops if we consider the pressure radiated by a pulsating sphere of radius R_0 and pressure amplitude P_0 at its surface:

$$p = \frac{P_0 R_0}{r} e^{j[\omega t - k(r - R_0)]}.$$

Comparison with Eq. B–7 shows that on axis [$D(\theta) = 1$] the piston signal seems to be coming from a spherical source of radius R_0 (except for certain phase differences, which are discussed below). The *starting* amplitude P_0 at R_0 is the (fictitious) piston face pressure found by using the plane-wave impedance relation. This interpretation mildly suggests that piston radiation starts out as a collimated plane-wave beam, where the amplitude is P_0 from the face out to $r = R_0$, beyond which point the beam spreads spherically, as shown in Fig. 13.4. Thus R_0 roughly marks the end of the nearfield (which we have not yet investigated) and the beginning of the farfield. The model is a gross oversim-

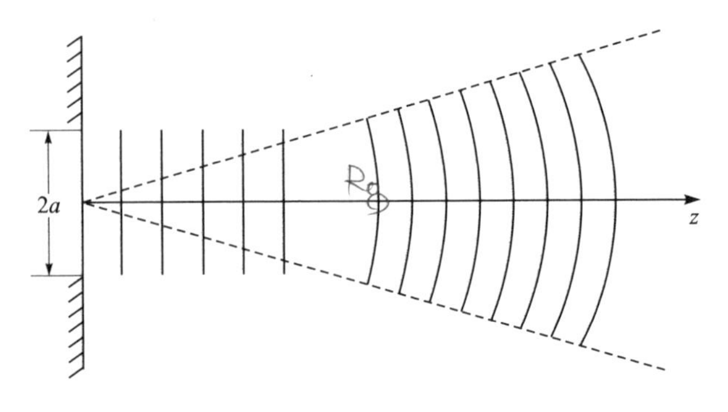

Figure 13.4 Simple geometrical model of the radiation field of a circular piston.

plification and must not be taken too literally (see, for example, Ref. 12), but it is sometimes a useful interpretation of the radiation field.

Now return to the phase differences that distinguish farfield piston radiation from radiation due to a pulsating sphere. The phase factor e^{jkR_0} in Eq. B–7 is readily explained by the fact that the radiation begins at $r = R_0$ for the sphere but at $r = 0$ for the piston. The multiplicative factor j in Eq. B–2 or B–7 is more subtle. The 90° phase shift implied by this factor is characteristic of piston radiation. A physical interpretation emerges if R_0 is expressed as $\omega a^2/2c_0$ and P_0 as $\rho_0 c_0 u_0$ in Eq. B–7:

$$p = j\omega \frac{\rho_0 u_0 a^2}{2r} D(\theta) e^{j(\omega t - kr)} = \frac{\partial}{\partial t} \frac{\rho_0 u_0 a^2}{2r} D(\theta)\, e^{j(\omega t - kr)}. \tag{B–8}$$

Both forms affirm the proportionality of the farfield pressure to the piston acceleration, not the piston velocity (see Eq. A–1). The derivative form is revisited in Sec. E below.

2. Size of *ka*

We see from Eq. B–4 that for $ka \ll 1$ (i.e., the piston circumference much less than the wavelength) the directivity factor reduces to

$$D(\theta) = 1 - \frac{(ka \sin\theta)^2}{8} \cdots \doteq 1.$$

The radiation is thus omnidirectional in the hemisphere $0 < \theta < \pi$. Figure 13.5*a* shows the beam pattern (plot of $|D(\theta)|$) for this case. The piston is so small compared with the wavelength of the sound it is emitting that it acts as a point source set atop a reflecting plane. Equation B–6 shows that a very small value of ka generally means that the nearfield is very short. The farfield is established almost immediately. If ka is large, however, that is, if the piston is many wavelengths in circumference, $J_1(ka \sin\theta)$ passes through zero several

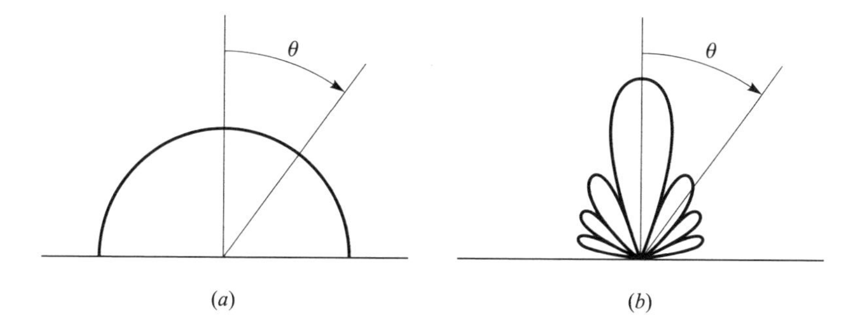

Figure 13.5 Beam patterns: $|D(\theta)|$. (*a*) $ka \ll 1$; (*b*) $ka \gg 1$.

times as θ increases from 0 to $\pi/2$. In this case the directivity pattern $D(\theta)$ has several nulls (one for each zero of J_1), as shown in Fig. 13.5*b*. Note that the minor lobes, also called sidelobes, decrease monotonically with increasing order. Moreover, since $D(\theta)$ changes sign when a null is traversed, the radiation in one lobe is 180° out of phase with the radiation in the next lobe.

Example 13.1. First null and piston diameter. Find the piston diameter for which the first null occurs at exactly 90°. Answer: Since $\sin\theta = 1$ in this case, ka must be the first zero of J_1. Solving for the diameter, we find

$$2a = 3.832\,\lambda/\pi = 1.220\,\lambda.$$

3. First Null, Minor-Lobe Suppression, Beamwidth, and Phase

As indicated in Example 13.1, the first null occurs at an angle θ_1 given by $ka\sin\theta_1 = \alpha_{11} = 3.832$:

$$\theta_1 = \sin^{-1}\frac{3.832}{ka} = \sin^{-1}\frac{0.610\lambda}{a}. \tag{B–9}$$

Unless $a \geq 0.610\lambda$, therefore, the directivity pattern has no null.

Minor-lobe suppression is a term used by transducer engineers to mean the difference between the sound pressure level at the major lobe and that at the most prominent sidelobe. It is a rough measure of how well a transducer emits sound in a desired direction rather than in some unwanted direction. In the case of the circular piston the most prominent minor lobe is the first one. To identify the minor-lobe peaks, set the derivative $D'(\theta)$ equal to zero. As Problem 11–2 shows, the roots are the zeros of $J_2(ka\sin\theta)$. In particular, the first minor-lobe peak is defined by $ka\sin\theta = \alpha_{21} = 5.136$, for which value $|D(\theta)| = 0.132$. The minor-lobe suppression is therefore

$$|20\log_{10}(0.132)| = 17.6 \text{ dB}.$$

A quantitative measure of the directivity is the half-power beamwidth, which is defined as the angle between the half-power points on the major lobe ($I = I_{\text{axis}}/2$, where I_{axis} is the intensity on the axis). The half-power angle θ_{HP} is therefore defined by

$$D(\theta_{\text{HP}}) = 1/\sqrt{2}. \tag{B–10}$$

For the circular piston θ_{HP} is given by

$$\frac{2J_1(ka\sin\theta_{\text{HP}})}{ka\sin\theta_{\text{HP}}} = \frac{1}{\sqrt{2}},$$

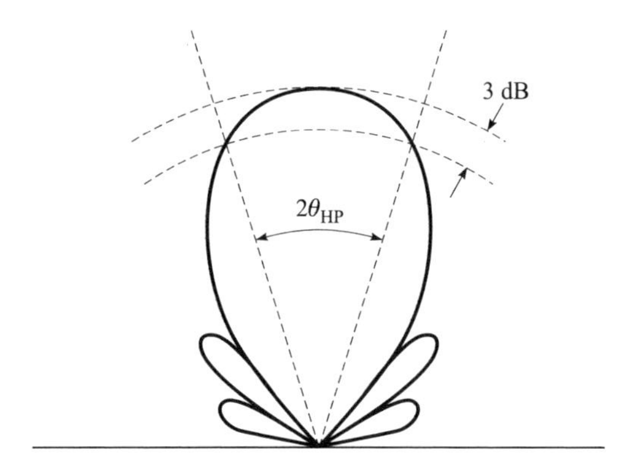

Figure 13.6 Half-power beamwidth.

the solution of which is $ka \sin\theta_{HP} = 1.616$, i.e., $\theta_{HP} = \sin^{-1}(1.616/ka)$. The half-power beamwidth $2\theta_{HP}$ (see Fig. 13.6) is therefore

$$2\theta_{HP} = 2\sin^{-1}(1.616/ka). \tag{B–11}$$

For narrow beams (large values of ka), $\sin\theta_{HP} \doteq \theta_{HP}$, in which case

$$2\theta_{HP} = 3.232/ka. \tag{B–12}$$

It should be noted that other definitions of beamwidth are in common use. For example, the 6-dB beamwidth is the angle between the 6-dB down points on the major lobe.

4. Intensity, Power, and Source Level

Finally a few comments are made about intensity, power, and source level. In the farfield, where the pressure and particle velocity are in phase, the intensity is proportional to p^2_{rms}. If a fixed amount of acoustic power is available, the intensity at a given distance from the piston may be raised by making the source more directional. In other words, the intensity at a certain angle may be enhanced at the expense of the intensity at other angles.

To calculate the total power radiated W, one could in principle integrate the farfield value of $p^2_{\mathrm{rms}}/\rho_0 c_0$ over a hemisphere surrounding the source. It turns out to be easier, however, to calculate the power emitted at the piston face:

$$W = S u^2_{\mathrm{rms}} \mathrm{Re}(Z_p),$$

where S is the area of the piston, $u^2_{\mathrm{rms}} = u_0^2/2$ is the mean square of the piston velocity, and Z_p is the radiation impedance of the piston, that is, the impedance presented to the piston by the medium. The calculation of Z_p is carried out in Sec. D below; the real part is (see Eq. D–5)

$$\mathrm{Re}(Z_p) = \rho_0 c_0 R_1(2ka),$$

where $R_1(2ka) = 1 - [2J_1(2ka)]/(2ka)$. The power may thus be expressed as

$$W = (\pi/2)a^2 u_0^2 \rho_0 c_0 R_1(2ka) \tag{B–13a}$$

$$= \frac{\pi a^2}{2\rho_0 c_0} P_0^2 R_1(2ka). \tag{B–13b}$$

Source level (SL), a term used to characterize the relative strength of a source, is the sound pressure level measured on axis in the farfield but extrapolated by spherical spreading back to unit distance, e.g., 1 m, from the center of the source. In the case of a circular piston, the rms pressure on axis in the farfield is, from Eq. B–7,

$$p_{\mathrm{rms}} = \frac{P_0 R_0}{\sqrt{2} r}.$$

If r and R_0 are measured in meters, the farfield SPL may be expressed as

$$\mathrm{SPL}(r) = 20 \log_{10} \frac{P_0 R_0 \ (1\ \mathrm{m})}{\sqrt{2} r\, p_{\mathrm{ref}}\ (1\ \mathrm{m})}$$

$$= 20 \log_{10} \frac{P_0 R_0}{\sqrt{2}\ p_{\mathrm{ref}}\ (1\ \mathrm{m})} - 20 \log_{10} \frac{r}{1\ \mathrm{m}}. \tag{B–14}$$

In this example the first term on the right-hand side is the source level, since it equals the SPL when $r = 1$ m. One application is the design of a source for which a specified SPL is desired at a given distance r. See, for example, Problem 13–11. Note that the SL is usually not the actual SPL 1 m from the source, since 1 m may be a point in the nearfield.

C. PRESSURE FIELD ON THE AXIS

Figure 13.7 shows the geometry for evaluating Eq. A–3 when the field point L is on the axis. Since the element of area is now a ring ($dS = 2\pi\sigma\, d\sigma$), Eq. A–3 becomes

$$p = jk\rho_0 c_0 u_0 e^{j\omega t} \int_0^a \frac{e^{-jkR}}{R}\ \sigma\, d\sigma.$$

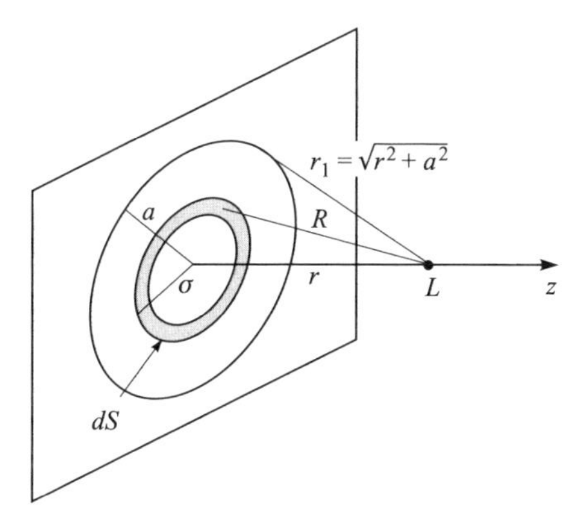

Figure 13.7 Calculation of the axial pressure field.

The expression for R is now given by $R^2 = r^2 + \sigma^2$, the differential form of which, $\sigma\, d\sigma = R\, dR$, may be used to change the integral from one over σ to one over R. The integration is then very simple:

$$p = jkP_0 e^{j\omega t} \int_r^{r_1} e^{-jkR}\, dR$$

$$= P_0\left[e^{j(\omega t - kr)} - e^{j(\omega t - kr_1)}\right], \qquad \text{(C–1)}$$

where, as noted in Fig. 13.7, $r_1 = \sqrt{r^2 + a^2}$. On the axis the pressure seems to be the sum of two plane-wave signals of the same amplitude, one coming directly from the center of the piston, the other from the rim, or edge, of the piston. The latter is often called the "edge wave." Note that the edge wave is phase inverted relative to the direct wave.

Interpreting the axial field as made up of two plane waves does make some sense physically. Consider the case in which the piston radius a becomes very large. The edge wave is delayed a long time. Suppose the piston vibration is a finite train of cycles, that is, a pulse. The edge pulse may be delayed as long as we choose simply by making the piston radius large enough. In the limit as $a \to \infty$, the edge pulse never arrives at L. The solution is then just an ordinary plane wave

$$p = P_0 e^{j(\omega t - kr)},$$

which is precisely what one would expect from the vibration of an infinite plane. The phase inversion of the edge wave is due to the reaction of the medium to the abrupt termination of the vibrating source at $r = a$. That is, if the piston were not terminated at $r = a$, we would observe no second signal. Note also that because point L is on the axis, it is equidistant from all points on the edge. All waves generated at (or scattered from) the edge thus arrive at L in phase; in fact, they add up to form a wave equal in amplitude to the direct wave. What happens if the observation point L is moved off the axis? Although the calculation for this case is much more difficult, it is clear that the arrivals from the edge no longer have the same phase. As a result, the edge wave signal is degraded. In general, the further off axis L is moved, the more the degradation. More on this topic is given in Chap. 14.

1. Transition to the Farfield

All this is very well, one may say, but how do two plane waves combine to produce a spherically spreading wave, which we know must exist in the farfield? The "$1/r$ decay" is produced by the tendency of the direct and edge waves to cancel as r becomes large. Let the phase difference due to the different travel paths r and r_1 be designated 2ξ:

$$2\xi = k(r_1 - r) = k(\sqrt{r^2 + a^2} - r). \tag{C–2}$$

Equation C–1 may then be expressed as

$$p = P_0(1 - e^{-j2\xi})e^{j(\omega t - kr)}.$$

For $r \gg a$, $\sqrt{r^2 + a^2}$ may be represented by the first two terms of its binomial expansion, namely, $r + a^2/2r$, and 2ξ becomes

$$2\xi \doteq ka^2/2r = R_0/r.$$

When 2ξ becomes small, we have $1 - e^{-j2\xi} = 1 - (1 - j2\xi \cdots) \doteq j2\xi = jR_0/r$, and the expression for the axial pressure reduces to

$$p = j\frac{P_0 R_0}{r} e^{j(\omega t - kr)}, \tag{C–3}$$

in agreement with Eq. B–7. Beyond the Rayleigh distance, therefore, the progressive narrowing of the difference between the path lengths of the two signals manifests itself as the $1/r$ decrease characteristic of a spherical wave. Incidentally, although the series expansion we have used implies that the farfield formula is valid only for $r \gg R_0$, in fact the $1/r$ dependence is well established by about $r = 2R_0$, as the next analysis shows.

2. Nearfield Structure

Although passage to the farfield limit is interesting, the axial solution is more important for the glimpse it provides us of the remarkable behavior of the nearfield. By using the phase difference defined by Eq. C–2, we may rewrite Eq. C–1 as follows:

$$\begin{aligned} p &= P_0 e^{j(\omega t - kr - \xi)}\left[e^{j\xi} - e^{-j\xi}\right] = j2P_0 \sin \xi e^{j(\omega t - kr - \xi)} \\ &= j2P_0 \sin (k/2)\left(\sqrt{r^2 + a^2} - r\right) e^{j(\omega t - kr - \xi)}. \end{aligned} \tag{C–4}$$

Figure 13.8 shows how the pressure amplitude varies along the axis. In graph (*a*) the normalized pressure amplitude $|p|/P_0$ is plotted against r/R_0 for a piston of radius $a = 5\lambda$. Graph (*b*) is the same data, but the ordinate is relative sound pressure level and the abscissa is logarithmic. Graph (*c*) shows the effect of doubling the piston radius. All three sketches demonstrate that the axial pressure amplitude in the nearfield goes through a series of peaks and nulls, not at all what one would expect if the nearfield were really the collimated plane-wave beam implied by Fig. 13.4. The pressure amplitude at the peaks is $2P_0$, that is, 6 dB higher than the level of the fictitious pressure $P_0 = \rho_0 c_0 u_0$ at the piston face. The number of peaks and nulls in the nearfield depends on the value of a/λ. If a/λ is an integer n, the number of nulls is n, and the first null occurs at the piston itself. If a/λ is not an integer, the number of nulls is equal to the integer just smaller than a/λ, and the first null is in front of the piston (the proof is left to the reader). For the axial field to have a null, therefore, the piston radius must exceed a wavelength.

Several asymptotes are shown on plots (*b*) and (*c*). First, note the farfield asymptote. Its position relative to the exact curve seems to confirm the often made statement "Spherical spreading begins at the Rayleigh distance." However, from an analysis of the entire nearfield (not just the axial part), Zemanek (Ref. 12) concluded that in some sense the farfield begins much earlier, at about $R_0/4$. The other two asymptotes are model curves sometimes used to represent the nearfield as a plane-wave region. Curve MM represents the function (Ref. 5)

$$|P| = \frac{P_0}{\sqrt{1 + (r/R_0)^2}},$$

while curve M represents the function (Ref. 6)

$$|P| = \frac{2P_0}{\sqrt{1 + (2r/R_0)^2}}.$$

3. Intensity

The expression for the axial intensity is given by Greenspan (Ref. 3),

$$I_{\text{axial}} = \frac{P_0^2}{\rho_0 c_0}\left(1 + \frac{r}{r_1}\right) \sin^2\left(k\,\frac{r_1 - r}{2}\right). \tag{C–5}$$

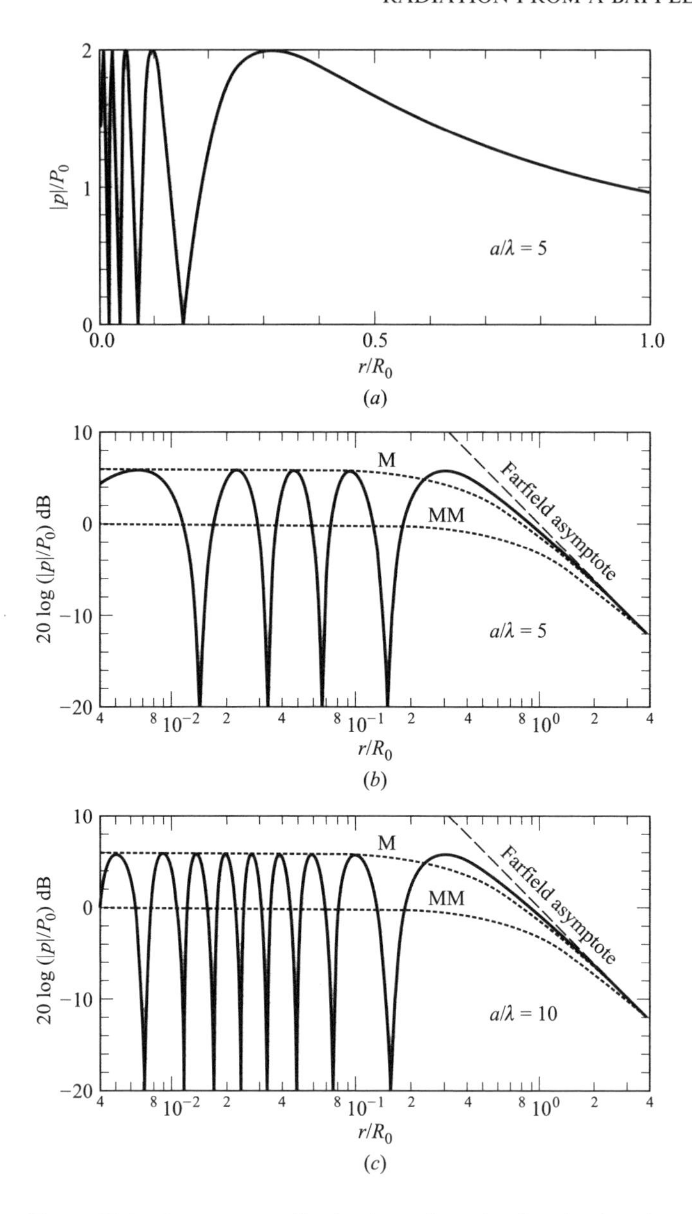

Figure 13.8 Pressure amplitude along the axis of a circular piston.

Note that many textbooks erroneously give the intensity as $|p|^2/2\rho_0 c_0$, where $|p|$ is found from Eq. C–4. The pressure and particle velocity are in general not in phase in the nearfield. They do, however, get in phase in the farfield (in this case meaning both $r > R_0$ and $r \gg a$) for which region both the correct and incorrect formulas reduce to

$$I_{\text{axial}} = \frac{P_0^2 R_0^2}{2\rho_0 c_0 r^2} \qquad \text{(farfield).} \tag{C–6}$$

Notice that Eq. C–6 is implied by Eq. B–7.

D. PRESSURE ON THE FACE OF THE PISTON

In this section the radiation impedance of the piston is calculated. Let $p = Pe^{j\omega t}$ and let the average pressure amplitude on the face of the piston be denoted P_{av}. Since the piston velocity amplitude is u_0, the radiation impedance is

$$Z_p = \frac{P_{\text{av}}}{u_0} = \frac{F/\pi a^2}{u_0}, \tag{D–1}$$

where F is the total force (complex amplitude) exerted by the sound field on the piston face. To find F, we need to integrate the local pressure over the face of the piston,

$$F = \int P(A)\, dS', \tag{D–2}$$

where A is an arbitrary point on the face (see Fig. 13.9). In turn, to find $P(A)$, we must evaluate the integral in Eq. A–3 for the case in which the field point L is at A. Calculating the force F thus requires two integrations.

In computing F by means of Eq. D–2, we count the force on dS' due to the pressure at dS as well as the force on dS due to the pressure at dS'. In other words the two integrations give us twice what we shall get if we arrange the

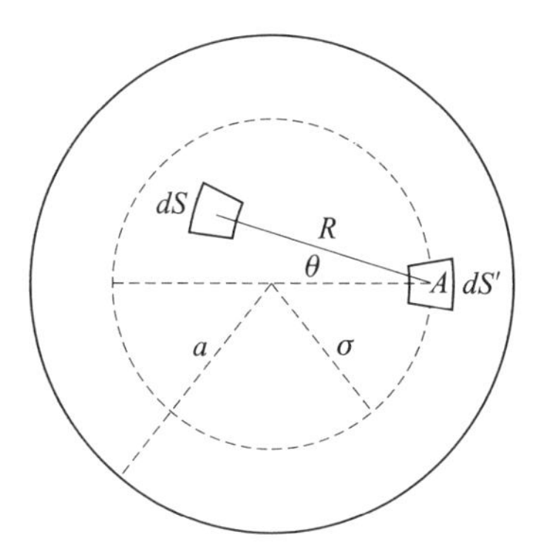

Figure 13.9 Geometry for calculating the pressure on the piston face. (From Ref. 4, Fig. 7.10.)

limits of integration to count the force of one element on the other only once. We so arrange the integration and then multiply the final result by 2.[3]

First compute $P(A)$—the incremental area is $dS = R\,dR\,d\theta$, and the limits on R are 0 and $2\sigma\cos\theta$:

$$P(A) = \frac{jkP_0}{2\pi}\int_{-\pi/2}^{\pi/2} d\theta \underbrace{\int_0^{2\sigma\cos\theta} \frac{e^{-jkR}}{R}\,R\,dR}_{(1-e^{-jk2\sigma\cos\theta})/jk}$$

$$= \frac{P_0}{2\pi}\left[\pi - 2\int_0^{\pi/2} e^{-jk2\sigma\cos\theta}\,d\theta\right]$$

$$= \frac{P_0}{2}\left[1 - \frac{2}{\pi}\int_0^{\pi/2} \cos(2k\sigma\cos\theta)\,d\theta + j\frac{2}{\pi}\int_0^{\pi/2} \sin(2k\sigma\cos\theta)\,d\theta\right].$$

The first integral (with its multiplier $2/\pi$) is $J_0(2k\sigma)$, as can readily be shown by manipulating Eq. 11A–27a. The second integral (again with the $2/\pi$ factor) is the zero-order Struve function $K_0(2k\sigma)$ (see Ref. 1).

$$\therefore\ P(A) = (P_0/2)[1 - J_0(2k\sigma) + jK_0(2k\sigma)]. \tag{D–3}$$

The force is found by substituting Eq. D–3 in Eq. D–2 and remembering to multiply by 2, as indicated previously. Here $dS' = \sigma\,d\sigma\,d\beta$, where β is the polar angle of A with respect to the center. The limits on σ are 0, a while the limits on β are $0, 2\pi$. The result is

$$F = P_0\int_0^{2\pi} d\beta \int_0^a [1 - J_0(2k\sigma) + jK_0(2k\sigma)]\,d\sigma$$

$$= 2\pi P_0\int_0^a [\sigma - \sigma J_0(2k\sigma) + j\sigma K_0(2k\sigma)]\,d\sigma.$$

Use of the recursion relation Eq. 11A–24c, which applies to Struve as well as to Bessel functions (Ref. 1), allows one to complete the integration and obtain

$$F = \pi a^2 P_0\left[1 - \frac{2J_1(2ka)}{2ka} + j\frac{2K_1(2ka)}{(2ka)}\right], \tag{D–4}$$

where K_1 is the Struve function of order 1. Substitution of Eq. D–4 in Eq. D–1 yields the expression for the radiation impedance of the piston

[3]This ingenious approach was put forward by Lord Rayleigh (Ref. 10), who first made the calculation of the force on the piston face.

$$Z_p = \rho_0 c_0 \left[\underbrace{1 - \frac{2J_1(2ka)}{2ka}}_{R_1(2ka)} + j \underbrace{\frac{2K_1(2ka)}{(2ka)}}_{X_1(2ka)} \right]. \tag{D–5}$$

Morse's notation (Ref. 7) is θ_0 is in place of R_1 and $M(2ka)$ in place of X_1. Values of θ_0 and M are tabulated in Table VIII, p. 447, of Morse's book, but notice that Morse uses $-i$ whereas $+j$ is used here.

The functions R_1 and X_1, frequently called piston functions, are tabulated in Appendix E. They are sketched in Fig. 13.10. Their "low frequency" ($2ka \ll 1$) and "high frequency" ($2ka \gg 1$) behaviors are of interest. The Maclaurin expansions are

$$R_1 = \frac{(ka)^2}{1 \cdot 2} - \frac{(ka)^4}{1 \cdot 2^2 \cdot 3} + \frac{(ka)^6}{1 \cdot 2^2 \cdot 3^2 \cdot 4} \cdots, \tag{D–6}$$

$$X_1 = \frac{4}{\pi} \left[\frac{2ka}{3} - \frac{(2ka)^3}{3^2 \cdot 5} + \frac{(2ka)^5}{3^2 \cdot 5^2 \cdot 7} \cdots \right]. \tag{D–7}$$

The first term in each of these series is used in Chap. 4 to approximate the impedance seen by a wave in a tube when the wave reaches an open end (the end is assumed flanged); see Sec. 4C.3. At the other end of the scale, for large values of $2ka$ (i.e., piston large compared with the wavelength), $R_1 \doteq 1$ and $X_1 \doteq 0$; Eq. D–5 then reduces to

$$Z_p = \rho_0 c_0.$$

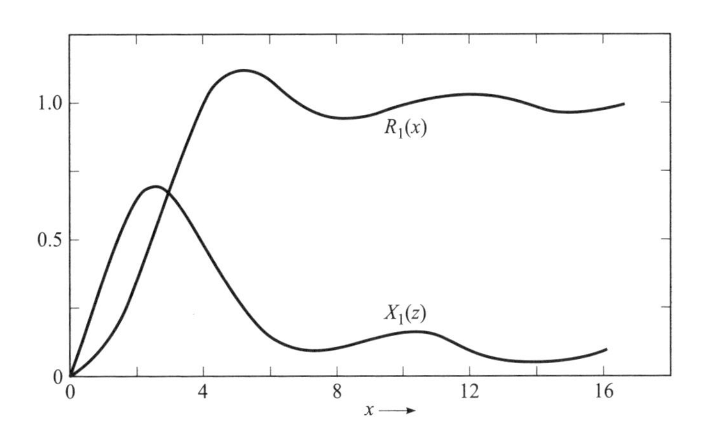

Figure 13.10 Real and imaginary parts of piston radiation impedance. (From Ref. 4, Fig. 7.11.)

The quantity $\rho_0 c_0 u_0$ characterized earlier as P_0 is therefore the true (average) pressure at the face of the piston in the high frequency limit. What is the minimum frequency at which $R_1 = 1$ is a good approximation? At its first maximum ($2ka = \alpha_{21} = 5.1356$), R_1 has a value 1.132, which is 1.08 dB above the high frequency asymptote. If we take $20 \log_{10} R_1 = \pm 1.08$ dB as a criterion, the lower limiting value of the argument is $2ka = 3.36$. This translates into lower limiting frequency $f_{\text{lower}} = 183/2a$ (a in meters) for air, $f_{\text{lower}} = 792/2a$ for water. For example, for a piston of diameter $2a = 0.1$ m, the radiation resistance would be within ± 1.08 dB of its high frequency value for frequencies above about 1800 Hz in air, 8000 Hz in water. Note, however, that in the lower portion of this range the reactive component X_1 is not negligible.

E. TRANSIENT RADIATION FROM A PISTON

If the piston face has a velocity different from $u_0 e^{j\omega t}$, it is still possible to compute the radiated signal. We revert to the Rayleigh integral Eq. A–1, here simplified by suppressing the dependence on x', y' (the coordinates on the piston):

$$p(\mathbf{r}; t) = \frac{\rho_0}{2\pi} \int_S \frac{\dot{u}_p(t - R/c_0)}{R} \, dS. \qquad \text{(E–1)}$$

The integral is evaluated first for a field point on the axis and then for an arbitrary point in the farfield.

1. Signal on the Axis

The calculation of the axial signal is particularly simple. The geometry is shown in Fig. 13.7. Since in this case $dS = 2\pi\sigma \, d\sigma = 2\pi R \, dR$, Eq. E–1 becomes

$$p = \rho_0 \int_r^{r_1} \dot{u}_p(t - R/c_0) \, dR.$$

A perfect differential is achieved by making use of the relation between the time and space derivatives,

$$\dot{u}_p\left(t - \frac{R}{c_0}\right) = -c_0 \frac{\partial}{\partial R} u_p\left(t - \frac{R}{c_0}\right).$$

We obtain $p = -\rho_0 c_0 \int_r^{r_1} (\partial u_p / \partial R) \, dR$, which may be integrated immediately:

$$p = \rho_0 c_0 \left[u_p(t - r/c_0) - u_p(t - r_1/c_0)\right]. \qquad \text{(E–2)}$$

This result reinforces the conclusion reached earlier that the signal along the axis seems to be two plane waves, the first one originating at the center of the piston, the second one, a phase-inverted replica of the first, coming from the edge of the piston. An interesting result is obtained when the farfield limit, $r \gg a$, is taken. First, as shown in Sec. C, $r_1 = \sqrt{r^2 + a^2}$ may be approximated by $r_1 = r + a^2/2r$. The expression for the second term in Eq. E–2 is then

$$u_p\left(t - \frac{r_1}{c_0}\right) = u_p\left[t - \frac{r}{c_0} - \frac{a^2}{2rc_0}\right],$$

or, on expansion by Taylor series,

$$u_p\left(t - \frac{r_1}{c_0}\right) = u_p\left(t - \frac{r}{c_0}\right) - \frac{a^2}{2rc_0}\, u_p'\left(t - \frac{r}{c_0}\right) + \cdots. \tag{E–3}$$

It is sufficient (given $r \gg a$) to stop with the second term of the expansion. Note also that $u_p' = \dot{u}_p$. The result of substituting Eq. E–3 into E–2 is

$$p = \frac{\rho_0 a^2}{2r}\, \dot{u}_p\left(t - \frac{r}{c_0}\right). \tag{E–4}$$

This equation is a generalization of Eq. C–3 and reinforces interpretations previously given for time-harmonic piston radiation: In the farfield (1) the axial disturbance is a spherical wave and (2) the pressure is proportional to the *acceleration*, not the velocity, of the piston.

2. Farfield

The geometry for the analysis when L is not restricted to the axis is shown in Fig. 13.11. The element of radiating area dS, each point of which is equidistant from L, is in general an arc. When L is in the farfield, however, the arc may be approximated by the rectangular strip $dS = 2\sqrt{a^2 - x^2}\,dx$ shown in Fig. 13.11. The distance from the strip to L is $R \doteq r - x \sin\theta$. This value is used for R in the argument of u_p in Eq. E–1, but the simpler approximation $R \doteq r$ is justified in the denominator of the integrand. Equation E–1 thus reduces to

$$p = \frac{\rho_0}{\pi r}\int_{-a}^{a} \dot{u}_p\left(t - \frac{r - x\sin\theta}{c_0}\right)\sqrt{a^2 - x^2}\,dx. \tag{E–5}$$

Morse (Refs. 7, 8) has given one simplification of this integral and has evaluated it for a piston having a step function in displacement. See Pierce (Ref. 9) for other references and a solution not limited to the farfield.

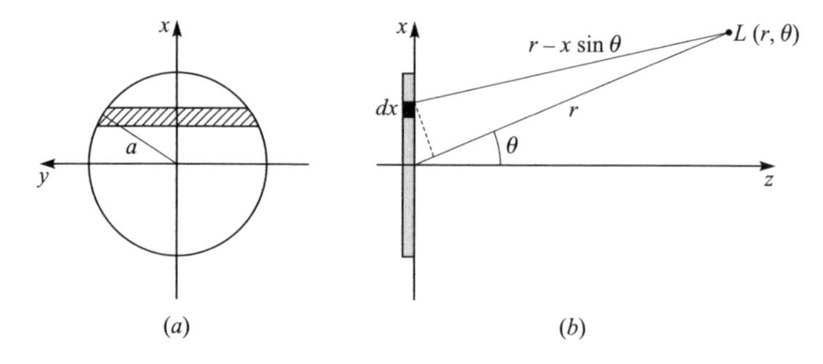

Figure 13.11 Strip method of evaluating the Rayleigh integral when L is in the farfield: (*a*) front view; (*b*) side view.

Our method of evaluating Eq. E–5 leads to a generalization, in operator form, of Eq. B–2. Introduction of the notation $t' = t - r/c_0$, $\mu = x/a$, $b = (a/c_0)\sin\theta$, and $D_t = \partial/\partial t$ allows Eq. E–5 to be written

$$p = \frac{\rho_0 a^2}{\pi r} D_t \int_{-1}^{1} u_p(t' + b\mu)\sqrt{1-\mu^2}\, d\mu. \tag{E–6}$$

The piston velocity is now expanded in Taylor series about t':

$$u_p(t' + b\mu) = u_p(t') + b\mu D_t u_p(t') + \frac{(b\mu)^2}{2!} D_t^2 u_p(t') + \cdots .$$

When this series is introduced in Eq. E–6, the integral may be evaluated term by term. When the odd and even properties of each integrand are considered, E–6 may be reduced to

$$p = \frac{2\rho_0 a^2}{\pi r}\left[D_t \sum_{n=0}^{\infty} \frac{(bD_t)^{2n}}{(2n)!} \int_0^1 \mu^{2n}\sqrt{1-\mu^2}\, d\mu \right] u_p(t'). \tag{E–7}$$

The integral is known to be equal to $[\pi(2n)!]/[2^{2n+2}\, n!\,(n+1)!]$ (Ref. 2). Substitution in Eq. E–7 and replacement of $(bD)^{2n}$ by $(-1)^n (jbD_t)^{2n}$ (in order to create a series that is more recognizable) give

$$p = \frac{\rho_0 a^2}{2r} \sum_{n=0}^{\infty} (-1)^n \frac{(jbD_t/2)^{2n}}{n!(n+1)!}\, \dot{u}_p(t').$$

Comparison with Eq. B–4 shows the series here to be the piston directivity function but with argument jbD_t. The final result is

$$p = \frac{\rho_0 a^2}{2r}\left[\frac{2J_1[j(a/c_0)\sin\theta D_t]}{j(a/c_0)\sin\theta D_t}\right]\dot{u}_p\left(t - \frac{r}{c_0}\right). \tag{E–8}$$

The "directivity function" in Eq. E–8 (the quantity in large brackets) is to be regarded as an operator. Two special cases are noted here. First, if the piston velocity is time harmonic ($u_p = u_0 e^{j\omega t}$), D_t becomes $j\omega$, and Eq. E–8 reduces to Eq. B–2. Second, for small values of the angle θ, Eq. E–8 becomes

$$p = \frac{\rho_0 a^2}{2r}\left[\dot{u}_p(t') + \frac{1}{8}\left(\frac{a}{c_0}\sin\theta\right)^2 \dddot{u}_p(t') + \frac{1}{192}\left(\frac{a}{c_0}\sin\theta\right)^4 u_p^{(v)}(t') + \cdots\right],$$

which reproduces Eq. E–4 on the axis and shows that off-axis contributions are strongly dependent on higher derivatives of the piston velocity function u_p.

F. NONUNIFORM PISTON

The nonuniform, or shaded, piston is of interest because (1) some radiators are restricted (at the edge, for instance) so that they cannot vibrate uniformly and (2) shading gives the designer an extra degree of freedom in achieving a desired radiation field. In Sec. A.1 we noted that radiation from a nonuniform but time-harmonic piston may be calculated by retaining the piston velocity amplitude inside the integral in Eq. A–3. We consider here a class of circular nonuniform pistons described by

$$u_0 = (n+1)u_0^{\text{av}}\left(1 - \frac{\sigma^2}{a^2}\right)^n, \tag{F–1}$$

where σ and a are defined in Fig. 13.3 and u_0^{av} is the average velocity amplitude (as one may easily show, $\pi a^2 u_0^{\text{av}}$ is equal to $\int_0^a u_0 2\pi\sigma\, d\sigma$). For example, $n = 0$ gives the velocity amplitude for a uniform piston, $n = 1$ that of a disk simply supported at its edge ($\sigma = a$), and $n = 2$ that of a disk clamped at its edge. Greenspan (Ref. 3) has provided a thorough description of the radiation from pistons shaded according to Eq. F–1. He has given results for the farfield, the axial field, and the reaction force on the face of the piston. We are content here to consider just the farfield.

The farfield pressure may be obtained easily by starting with Eq. A–7, the pressure due to a ring piston. After replacing a in Eq. A–7 by σ, w by $d\sigma$, and u_0 by the expression in Eq. F–1, we integrate over σ to obtain the pressure due to all the rings making up the actual piston,

$$p = j\frac{ka^2\rho_0 c_0 u_0^{\text{av}}}{2r}e^{j(\omega t - kr)}2(n+1)\int_0^1 (1-\mu^2)^n \mu J_0(A\mu)\, d\mu, \tag{F–2}$$

where $\mu = \sigma/a$ and $A = ka \sin\theta$. The integral is shown in Problem 11–5 to be equal to

$$\frac{2^n n! J_{n+1}(x)}{x^{n+1}}.$$

Equation F–2 therefore reduces to

$$p = j\frac{ka^2 \rho_0 c_0 u_0^{\text{av}}}{2r} e^{j(\omega t - kr)} \frac{(n+1)! 2^{n+1} J_{n+1}(ka \sin\theta)}{(ka \sin\theta)^{n+1}}.$$

The limit of $J_{n+1}(x)/x^{n+1}$ as $x \to 0$ is $[2^{n+1}(n+1)]^{-1}$. It is thus convenient to express the final result as

$$p = j\frac{R_0 P_0^{\text{av}}}{r} D(\theta) e^{j(\omega t - kr)}, \tag{F–3}$$

where $R_0 = ka^2/2$ is the usual Rayleigh distance, $P_0^{\text{av}} = \rho_0 c_0 u_0^{\text{av}}$ is the (fictitious) average face pressure, and

$$D(\theta) = \frac{2^{n+1}(n+1)! J_{n+1}(ka \sin\theta)}{(ka \sin\theta)^{n+1}} \tag{F–4}$$

is the directivity function.

The expression for the farfield pressure, Eq. F–3, has the same form as that for the uniform piston. The only substantive difference is in the directivity function, which is strongly dependent on the index n. For example, the directivity of the simply supported disk ($n = 1$) is

$$D_{\text{SSD}} = \frac{8 J_2(ka \sin\theta)}{(ka \sin\theta)^2}, \tag{F–5}$$

while that for the clamped disk ($n = 2$) is

$$D_{\text{CD}} = \frac{48 J_3(ka \sin\theta)}{(ka \sin\theta)^3}. \tag{F–6}$$

The sidelobe suppression for the simply supported disk is better than that for the uniform piston. The improvement is even more for the clamped disk; see Problem 13–16.

REFERENCES

1. M. Abramowitz and I. A. Stegun (Eds.), *Handbook of Mathematical Functions*, National Bureau of Standards, Applied Mathematics Series 55 (U.S. Government Printing Office, Washington, DC, 1964).
2. D. Bierens de Haan, *Nouvelles Tables d'Intégrales Définies* (Amsterdam, 1867, reprinted by Hafner Publishing Co., New York), p. 33, Table 7, No. 3.
3. M. Greenspan, "Piston radiator: Some extensions of the theory," *J. Acoust. Soc. Am.* **65**, 608–621 (1979).
4. L. E. Kinsler and A. R. Frey, *Fundamentals of Acoustics*, 2nd ed. (John Wiley and Sons, New York, 1962), Chap. 7.
5. R. H. Mellen and M. B. Moffett, "A model for parametric sonar radiator design," *U. S. Navy J. Underwater Acoust.*, **22**, 105–116 (1972).
6. H. M. Merklinger, "High intensity effects in the non-linear acoustic parametric end-fire array," Ph.D. thesis, University of Birmingham (1971).
7. P. M. Morse, *Vibration and Sound*, 2nd ed. (McGraw-Hill Book Co., New York, 1948), Sec. 28. Reprinted by Acoustical Society of America, New York, 1981.
8. P. M. Morse and K. U. Ingard, *Theoretical Acoustics* (McGraw-Hill Book Co., New York, 1968), Secs. 7.2, 7.4. Reprinted by Princeton University Press, Princeton, NJ, 1986.
9. A. D. Pierce, *Acoustics: An Introduction to Its Physical Principles and Applications* (McGraw-Hill, New York, 1981), pp. 227–231. Reprinted by Acoustical Society of America, New York, 1989.
10. Lord Rayleigh, *Theory of Sound*, 2nd ed., Vol. 2 (Macmillan, London, 1896), Secs. 278, 302. Reprinted by Dover, New York, 1945.
11. H. Stenzel, "Die akustische Strahlung der rechteckigen Kolbenmembran," *Acustica* **2**, 263–281 (1952). English translation by R. D. Wallace, "The Acoustical Radiation of the Rectangular Piston Membrane," Tech. Rep. ARL-TR-73-58, Applied Research Laboratories, The University of Texas at Austin (16 July 1973).
12. J. Zemanek, "Beam behavior within the nearfield of a vibrating piston," *J. Acoust. Soc. Am.* **49**, 181–191 (1971).

PROBLEMS

13–1. Two concentric ring pistons of radii a_1 and a_2, where $a_2 > a_1$, are mounted in a rigid baffle and radiate at an angular frequency ω. The volume velocity amplitude $Q_i = 2\pi a_i w_i u_{0i}$ is the same for both pistons.

(a) Give the expression for the farfield radiation (not restricted to the axis) when the two pistons are driven in phase.

The rest of the problem is devoted to the case in which (i) the two pistons are driven 180° out of phase ($Q_2 = -Q_1$) and (ii) the radii are small enough that $ka_1 \ll 1$ and $ka_2 \ll 1$.

(b) Give the expression for the farfield radiation, taking advantage of the fact that the argument of each Bessel function is small.

(c) Find the directivity $D(\theta)$.

(d) Given the answer to part (c), find an elementary source that has a similar directivity function.

13–2. The width w of a ring piston is very small compared to the ring radius a (see sketch). Find the farfield pressure p_{ring} and the directivity D_{ring} by considering the ring to be two concentric full pistons, of radii a and $a + w$, superposed but driven 180° out of phase. Their excitations cancel except in the outer section $a < \sigma < a + w$ of the larger piston. If $p(a + w)$ and $p(a)$ represent the pressures due to the larger and smaller pistons, respectively, their difference represents the ring piston pressure:

$$p_{\text{ring}} = p(a+w) - p(a) = \frac{p(a+w) - p(a)}{w} w \doteq \frac{\partial p}{\partial a} w.$$

Use the last expression, with p given by the farfield formula for the full piston, to find p_{ring}. Then obtain D_{ring}.

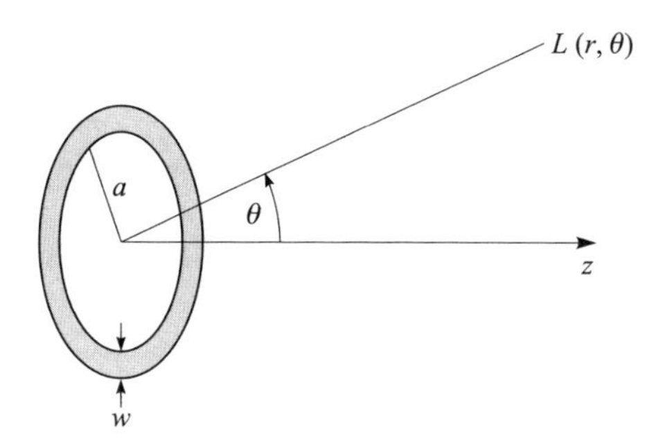

13–3. Compare the radiation pattern of the circular ring with that of the ordinary circular piston. In particular, consider

(a) Half-power beamwidth $2\theta_{\text{HP}}$.

(b) Sidelobe suppression.

Be quantitative in your comparisons.

13–4. Using the approach outlined in Problem 13–2, find the pressure field along the axis of a ring piston. Do not make the farfield assumption.

(a) Find the expression for p_{ring} on the axis.

(b) Check your expression against the farfield result (obtained in Problem 13–2). From this comparison deduce the meaning of *farfield* for the ring piston.

(c) Let A stand for the pressure amplitude P at $r = 0$. Sketch P/A as a function of r/a. Does the axial field near the piston have a structure of peaks and nulls like that for the full circular piston?

13–5. Use the strip method (Fig. 13.11) to derive Eq. B–7, the expression for the farfield pressure due to a circular piston driven at angular frequency ω. The strip method is described in Sec. E.2

13–6. A baffled loudspeaker (model it as a circular piston) is to be used to transmit sound in air in the frequency range 1–4 kHz. The desired half-power beamwidth $2\theta_{HP}$ is 60° for 1-kHz signals.

(a) Find the required radius a of the loudspeaker for 1-kHz sound.

Assume that a mechanical filter in the loudspeaker diaphragm allows you to vary the effective radius a as a function of frequency f.

(b) How must a vary with f in order that the beamwidth be the same for all frequencies in the 1–4-kHz band?

(c) Given the variation of a with f found in (b), how does the farfield pressure amplitude vary with f? Assume that the piston velocity u_0 is the same for all frequencies.

13–7. Ground-level sonic boom measurements are often made with a condenser microphone M mounted flush in a large panel of plywood (see sketch). The panel is laid on the ground. The system thus constitutes a baffled piston used as a receiver. Since the sensitivity of the measuring system should be independent of the angle of incidence θ (sonic booms are incident at a variety of angles), the receiver should be omnidirectional in the upper hemisphere. If the diameter of the microphone is $\frac{1}{2}$ in., what is the maximum frequency at which the receiver is omnidirectional to a tolerance of 3 dB?

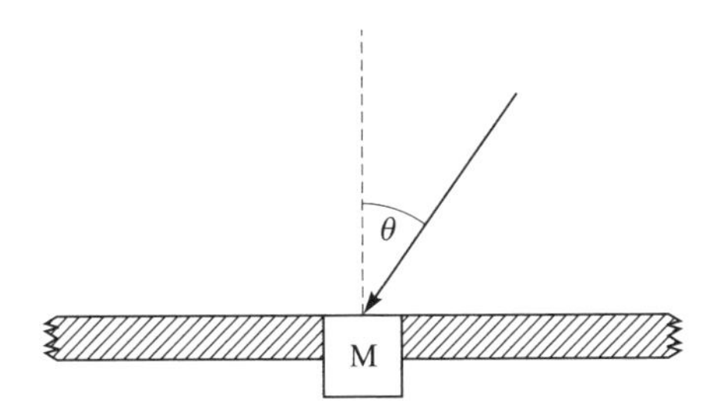

13–8. A baffled piston in the form of a narrow rectangle (length L, width dx) vibrates with a uniform velocity $u_0e^{j\omega t}$ in the z direction. The field point Q is in the farfield in the y, z plane. Find the directivity function $D(\theta)$ for this case and show that the expression for the pressure at Q is

$$p(r, \theta) = j\frac{P_0R_0}{r}D(\theta)e^{j(\omega t - kr)},$$

where $R_0 = S/\lambda$ is the Rayleigh distance and S is the area of the piston.

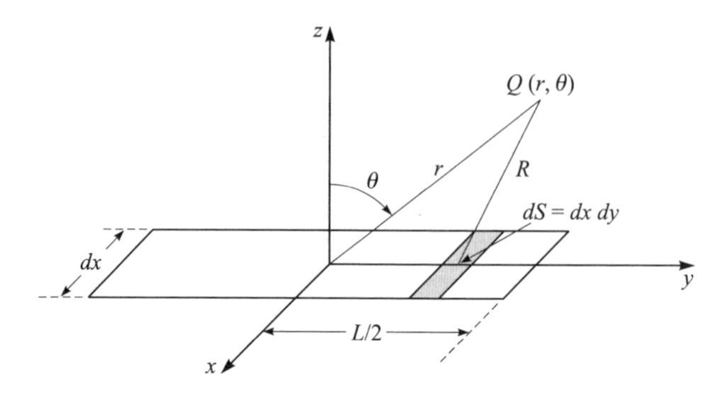

13–9. Consider the radiation from a baffled circular piston in which the piston is fixed and the baffle vibrates with velocity $u_0 \exp(j\omega t)$. The field may be represented as the superposition of two fields: one due to the vibration of an infinite plane (a baffle without any piston), the other due to radiation from an ordinary baffled piston (baffle fixed, piston moving). By proper phasing of the two sources, one may achieve the true boundary condition for the problem posed (baffle moving, piston fixed).

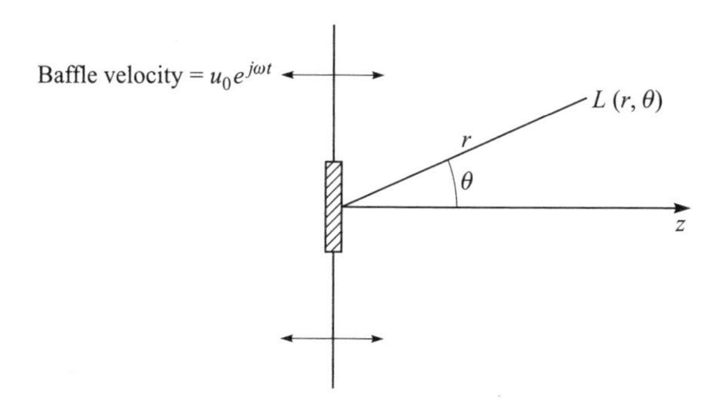

(a) Find the expression for the pressure in the region $r > R_0$, where R_0 is the Rayleigh distance in the conventional baffled piston problem. Notice that since $z = r\cos\theta$, the answer may be expressed in the form $p(r, \theta)$, $p(z, \theta)$, or $p(r, z, \theta)$, as convenient.

(b) Find the expression for the axial field (no restriction on z). Does the axial field have a succession of peaks and nulls like that for the ordinary piston? If not, why not?

13–10. A circular baffled piston generates 10-kHz sound in seawater. At a distance $r = 1$ m from the piston the rms pressure is measured as a function of the angle θ. At $\theta = 0$, i.e., on the piston axis, the sound pressure level is 180 dB re 1 μPa. At $\theta = 37.58°$ a null is recorded. The measurement is repeated at $r = 2$ m, and exactly the same directivity pattern is observed.

(a) What is the SPL on axis for the second set of measurements ($r = 2$ m)?

(b) Find the radius a of the piston.

(c) Find the piston velocity amplitude u_0.

(d) Find the SPL at the center of the piston face.

13–11. As a transducer engineer you are asked to design a transducer to operate at 15 kHz in seawater at 13°C ($c_0 = 1500$ m/s). The design goal is to achieve a sound pressure level of 200 dB (re 1 μPa) at a point 500 m distant from the source. Electrical power available is 1000 watts, and you have the capability of making transducers that are 50% efficient (conversion of electrical power to acoustic power) at the design frequency.

(a) Can you achieve the design SPL with an omnidirectional transducer such as a pulsating sphere?

(b) If not, can you design a circular piston transducer that will meet the goal? By "design" is meant a selection of the piston radius a. It is suggested that you start out by assuming that the design distance, $r = 500$ m, is a point in the farfield. Once you have picked a, also find the beamwidth $2\theta_{HP}$. Finally, check your design by calculating the Rayleigh distance R_0 to be sure the design distance ($r = 500$ m) is indeed in the farfield.

(c) On the basis of the calculations in (b), and other calculations if you feel they are necessary, do you conclude that the design goal can be met with a circular piston transducer (of any radius a) under the prescribed conditions (frequency, power, and efficiency)?

13–12. A circular loudspeaker of diameter 25 cm is mounted in a rigid baffle and driven in air at a frequency of 3430 Hz. Model the speaker as a flat piston.

(a) How much acoustic power is radiated by the loudspeaker when the SPL on axis is 100 dB at a distance of 10 m?

(b) Find the half-power beamwidth $2\theta_{HP}$ (in the farfield).

(c) Find the most distant axial null. Do other nulls exist?

13–13. The electrical signal to a baffled, circular piston causes the piston displacement ξ_p to be one cycle of a sine wave, i.e.,

$$\xi_p = \begin{cases} \xi_0 \sin \omega t, & 0 \le \omega t \le 2\pi, \\ 0, & \text{otherwise.} \end{cases}$$

Predict the axial pressure signal at a distance r in the farfield. Illustrate your answer with a sketch of the time waveform; specify the amplitude.

13–14. A baffled, circular piston is excited by a random electrical signal. Assume that the piston velocity u_p is white noise (amplitude spectrum flat with frequency). What is the shape of the (amplitude) spectrum of the axial pressure signal in the farfield? Be quantitative. For example, what is the relation of the SPL at frequency $2f$ to that at f?

13–15. A traveling wave is a one-sided pulse shown in the sketch. The pulse is defined by $p = A \sin \omega t'$, $0 < \omega t' < \pi$, where τ is the duration, A is the amplitude, and t' is the time starting when the front of the pulse reaches the listener. See the sketch. Investigate the possibility of using a baffled piston to produce such a pulse. In particular, let the pulse be the axial farfield signal generated by the piston.

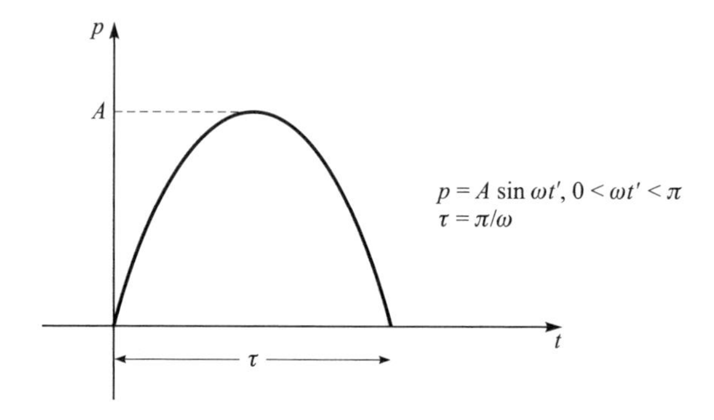

(a) Design the piston excitation $u_p(t)$ to produce the pulse shown in the sketch. For definiteness, let d be the distance from the piston to the listener so that $t' = t - d/c_0$.

(b) Sketch the required piston velocity $u_p(t)$ found in (a). Then comment on the practicality of producing the required motion with a real piston.

(c) Generalize your results by answering the following question: What can you say about the feasibility of producing any one-sided pulse with a real baffled piston (on axis in the farfield)?

13–16. Find the sidelobe suppression for the nonuniform pistons described as the "simply supported disk" ($n = 1$) and "clamped disk" ($n = 2$) in Sec. F.

13–17. A 1.48-MHz ultrasonics transducer has a radius $a = 10$ mm and is driven in water for which $c_0 = 1480$ m/s and $\rho_0 = 1000$ kg/m^3.

(a) Show that the wavelength is $\lambda = 1$ mm and that the Rayleigh distance is $R_0 = 100\pi$ mm.

(b) Find the number of nulls on the axis.

(c) How many nulls does the farfield directivity pattern have? If you have to estimate, explain your procedure.

(d) The SPL on axis at $r = 4R_0$ is 134 dB re 1 μPa. Find the velocity amplitude u_0 at the piston face.

(e) Now let the piston be driven at the same amplitude as in part (d) but for only two cycles of the 1.48-MHz sine wave. Find the SPL on axis at $r = 49.5$ mm.

14

DIFFRACTION

A. INTRODUCTION

Put somewhat unscientifically, diffraction is what surprises you when you expect ray theory to work. The terms *diffraction* and *scattering* are often used interchangeably.[1] Sound is diffracted by obstacles, by discontinuities at a source, and even by continuous changes in the acoustical medium (e.g., sound leaks into a shadow zone by diffraction). In many cases acoustical communication is totally dependent on diffraction. One person can carry on a conversation with another person who is out of sight around the corner of a building, even though no direct or reflected path for the sound exists. The sound striking the corner of the building scatters in all directions, and some of it arrives at the ears of the person around the corner. In other cases the direct and scattered sound coexist and interfere. The interference may produce maxima and minima, even nulls, in the sound field. The complicated directivity pattern of a vibrating piston ($ka \gg 1$) is a good example. In this case diffraction is due to abrupt termination of the radiation at the rim of the piston.

Diffraction may be treated in a variety of ways. For example, the scattering of a plane wave by a sphere may be calculated by expanding the incident plane wave as an infinite series of spherical waves, which are better suited to satisfy boundary conditions on the surface of the sphere. See, for example, Refs. 6 and 7. Similarly, diffraction by a cylindrical object may be treated by expanding the incident wave as an infinite series of cylindrical waves.

[1]Scattering may be a broader term, however. For example, scattering from a rough surface may not involve diffraction.

Because of space limitations we concentrate on a method of calculation based on the Helmholtz-Kirchhoff integral (see, for example, Refs. 2, 3, 8, and 9). In this method one calculates the pressure at any given point by specifying the wave field on a convenient surface (not necessarily spherical) surrounding the point. After deriving the Helmholtz-Kirchhoff integral in the frequency domain, we transform to the time domain. The integral is then used to solve diffraction problems involving apertures and disks.

B. HELMHOLTZ-KIRCHHOFF INTEGRAL THEOREM

1. Derivation

The derivation is carried out in the frequency domain. Let $\bar{p}(x, y, z; \omega)$ be the Fourier transform of the acoustic pressure p, i.e.,

$$\bar{p}(x, y, z; \omega) = \mathcal{F}[p(x, y, z; t)] = \int_{-\infty}^{\infty} p(x, y, z; t) e^{-j\omega t}\, dt.$$

To take the Fourier transform of the wave equation,

$$\nabla^2 p - \frac{1}{c_0^2} p_{tt} = 0,$$

multiply by $e^{-j\omega t}\, dt$ and integrate over time from $-\infty$ to $+\infty$. The result is the Helmholtz equation,

$$\nabla^2 \bar{p} + k^2 \bar{p} = 0, \qquad \text{(B–1)}$$

where $k^2 = \omega^2/c_0^2$.

Next employ Green's theorem. Let S be a closed surface surrounding volume V, as in Fig. 14.1. In this chapter, following Ref. 3, we take the normal $\mathbf{n}$ to S to be inward pointing. Let U and U_1 be two functions having continuous first and second partial derivatives within and on S. Green's theorem states that

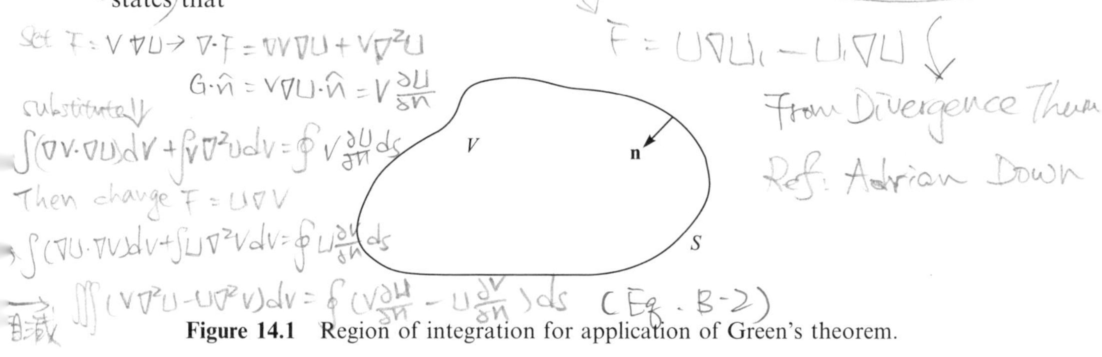

Figure 14.1 Region of integration for application of Green's theorem.

$$\int_V (U\,\nabla^2 U_1 - U_1\,\nabla^2 U)\,dV = \int_S \left(U_1 \frac{\partial U}{\partial n} - U \frac{\partial U_1}{\partial n} \right) dS, \tag{B–2}$$

where $\partial U/\partial n$ means the gradient of U in the direction normal (inward) to S at the point in question. Next let U and U_1 be functions that satisfy the Helmholtz equation; in particular, take $U = \overline{p}$ and $U_1 = \overline{p}_1$, where $\overline{p}$ and $\overline{p}_1$ satisfy Eq. B–1. In this case the volume integral (left-hand side of Eq. B–2) vanishes because, by application of Eq. B–1,

$$\begin{aligned} U\,\nabla^2 U_1 - U_1\,\nabla^2 U &= \overline{p}\,\nabla^2 \overline{p}_1 - \overline{p}_1\,\nabla^2 \overline{p} \\ &= \overline{p}(-k^2 \overline{p}_1) - \overline{p}_1(-k^2 \overline{p}) \\ &= 0, \end{aligned}$$

and Eq. B–2 reduces to

$$\int_S \left(\overline{p}_1 \frac{\partial \overline{p}}{\partial n} - \overline{p} \frac{\partial \overline{p}_1}{\partial n} \right) dS = 0. \tag{B–3}$$

Up to this point $\overline{p}$ and $\overline{p}_1$ may be any solutions of the Helmholtz equation. We now take $\overline{p}_1$ to be the specific solution representing the field from a point source located at an interior point Q, i.e.,

$$\overline{p}_1 = A \frac{e^{-jks}}{s},$$

where s is the distance from point Q to point x,y,z on S. See Fig. 14.2. This function has the requisite continuity properties, *except* at the point Q itself. To avoid the singularity at Q, we redraw the surface S to exclude Q. Let Q be surrounded by a small sphere S' of radius ϵ, as shown in Fig. 14.2 The surface of integration then becomes $S + S'$, and Eq. B–3 becomes

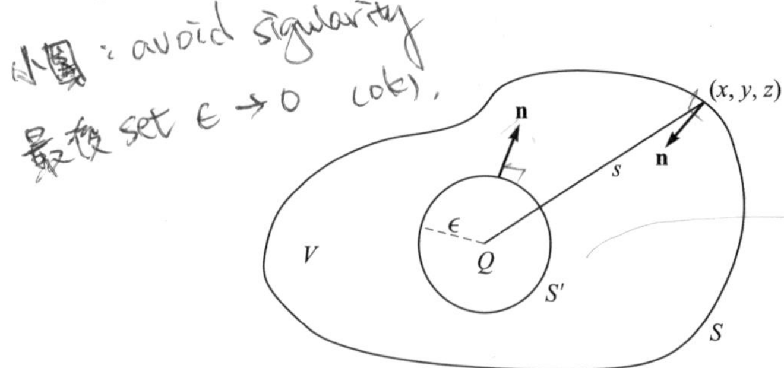

Figure 14.2 Modified region of integration.

$$\int_{S+S'} \left\{ \frac{e^{-jks}}{s} \frac{\partial \overline{p}}{\partial n} - \overline{p} \frac{\partial}{\partial n} \left(\frac{e^{-jks}}{s} \right) \right\} dS = 0,$$

or, if the integral is broken up into parts over S and S',

$$\int_{S'} \left\{ \frac{e^{-jks}}{s} \frac{\partial \overline{p}}{\partial n} - \overline{p} \frac{\partial}{\partial n} \left(\frac{e^{-jks}}{s} \right) \right\} dS' = -\int_{S} \left\{ \frac{e^{-jks}}{s} \frac{\partial \overline{p}}{\partial n} - \overline{p} \frac{\partial}{\partial n} \left(\frac{e^{-jks}}{s} \right) \right\} dS. \quad \text{(B–4)}$$

The integral over S' may easily be evaluated. On S' we have

$$s = \epsilon, \qquad \frac{\partial}{\partial n} = \frac{\partial}{\partial s},$$

and

$$dS' = \epsilon^2 \, d\Omega,$$

where Ω is the solid angle. Given these substitutions, the left-hand side of Eq. B–4 becomes

$$\int_0^{4\pi} \left\{ \frac{e^{-jk\epsilon}}{\epsilon} \frac{\partial \overline{p}}{\partial s} + \overline{p} \frac{e^{-jk\epsilon}}{\epsilon} \left(jk + \frac{1}{\epsilon} \right) \right\} \epsilon^2 \, d\Omega.$$

Now take the limit as $\epsilon \to 0$. Only the last term of the integrand survives, and the integral reduces to $\int_0^{4\pi} \overline{p}(Q) \, d\Omega = 4\pi \overline{p}(Q)$. Solved for $\overline{p}$, B–4 becomes

$$\overline{p}(Q, \omega) = \frac{1}{4\pi} \int_S \left\{ -\frac{e^{-jks}}{s} \frac{\partial \overline{p}}{\partial n} + \overline{p} \frac{\partial}{\partial n} \left(\frac{e^{-jks}}{s} \right) \right\} dS. \quad \text{(B–5)}$$

This is the Helmholtz-Kirchhoff integral solution of the wave equation. To compute the pressure at any point Q, all we need is data for $\overline{p}$ and its normal derivative on any surface S that surrounds Q.

2. Time Domain Version

A time domain version of the Helmholtz-Kirchhoff integral may be found by taking the inverse Fourier transform of Eq. B–5. It is first assumed that inverse transformation and integration over the surface S are operations that commute. Let the integrand in Eq. B–5 be designated $\overline{B}(\omega)$. With its second term expanded, $\overline{B}$ becomes

$$\overline{B}(\omega) = -\frac{e^{-j\omega s/c_0}}{s}\frac{\partial \overline{p}}{\partial n} + \left\{\frac{\partial}{\partial n}\left(\frac{1}{s}\right) - \frac{j\omega}{c_0 s}\frac{\partial s}{\partial n}\right\} e^{-j\omega s/c_0}\, \overline{p}. \tag{B–6}$$

The inverse transform of Eq. B–6 may be taken by making use of the time shift property of Fourier transforms, which is as follows: given $\mathcal{F}(f(t)) = \overline{f}(\omega)$, then

$$\mathcal{F}(f(t-\tau)) = e^{-j\omega\tau}\overline{f}(\omega).$$

In Eq. B–6 s/c_0 plays the role of τ. The inverse transform of $\overline{B}(\omega)$ is thus

$$B(t) = -\frac{1}{s}\left[\frac{\partial p}{\partial n}\right] + \frac{\partial}{\partial n}\left(\frac{1}{s}\right)[p] - \frac{1}{c_0 s}\frac{\partial s}{\partial n}\left[\frac{\partial p}{\partial t}\right],$$

where the square brackets here mean that the quantity bracketed is to be evaluated at the retarded time $t - s/c_0$, for example,

$$[p] \equiv p(x, y, z; t - s/c_0).$$

Note that because $B(t)$ is a function that is to be integrated over the surface S, the coordinates x, y, z identify a point on S. Having transformed the integrand in Eq. B–5, we arrive finally at the time domain version of the Helmholtz-Kirchhoff integral,

$$p(Q, t) = \frac{1}{4\pi}\int_S \left\{-\frac{1}{s}\left[\frac{\partial p}{\partial n}\right] + \frac{\partial}{\partial n}\left(\frac{1}{s}\right)[p] - \frac{1}{c_0 s}\frac{\partial s}{\partial n}\left[\frac{\partial p}{\partial t}\right]\right\} dS. \tag{B–7}$$

A practical difficulty must now be faced. Although Eqs. B–5 and B–7 are exact, evaluation of the integral requires knowledge of p and its normal derivative at all points on the surface S. Here's the rub. In many cases S may be chosen so that part of it is so far away that one can safely assume p and $\partial p/\partial n$ vanish over that part. However, on other, usually nearby, parts of S one frequently does not know both p and $\partial p/\partial n$ completely and thus has to guess values that seem to be reasonable. The exact solution then becomes an approximate one.

C. CIRCULAR APERTURE

Our first application is the diffraction of a plane wave by a circular aperture. The plane wave is normally incident on a rigid plate that contains the aperture. If two approximations are made, the Helmholtz-Kirchhoff integral may easily be evaluated for observation points Q on the axis of the aperture. The calculation is carried out here. Also treated briefly in this section is the case in which the incident wave is spherical.

1. Plane Wave Normally Incident on a Circular Aperture

a. Analysis

For this problem S is conveniently chosen to be the surface of the hemispherical dome indicated by the dashed line in Fig. 14.3. If the radius of C is made large enough, no signals from the aperture can reach C in finite time. Consequently, both p and $\partial p/\partial n$ are zero on C, and integration over C yields nothing. Next we shall take the integral over B, the back side of the plate, to be zero. Here is where the first approximation is made. Because the plate is rigid, the first term in the integrand of Eq. B–7 vanishes. But for the other two terms to vanish as well, we must assume $p = 0$ on B. In fact, however, the presence of an acoustic field to the right of the aperture implies that in general p is not zero on B. On the other hand it seems reasonable that most of the radiation proceeding through the aperture will travel in the forward direction at angles less than 90°. (What does this imply about frequency?) Accordingly, we assume that the value of p on the back side of the plate is small enough that the integral over B may be neglected. This step reduces the integral over S to an integral over A, the surface of the aperture itself.

The geometry for computing the integral over the aperture is shown in Fig. 14.4. The surface element dS is an annulus of radius r and thickness dr, i.e., $dS = 2\pi r\, dr$. It is convenient, however, to shift the integration so that it is over s instead of r. The relation $r^2 = s^2 - s_0^2$ leads to

$$dS = 2\pi s\, ds.$$

To obtain $\partial s/\partial n$, first notice that s, being the distance from Q to the annulus, may be expressed as $s = x' \cos\theta + r \sin\theta$ (see Fig. 14.4*b*), where x' is opposite in direction to x, and $\cos\theta = s_0/s$:

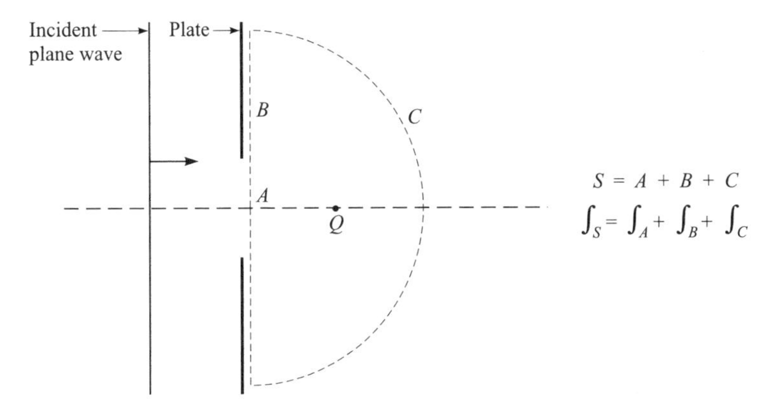

Figure 14.3 Surface integration for diffraction by a circular aperture.

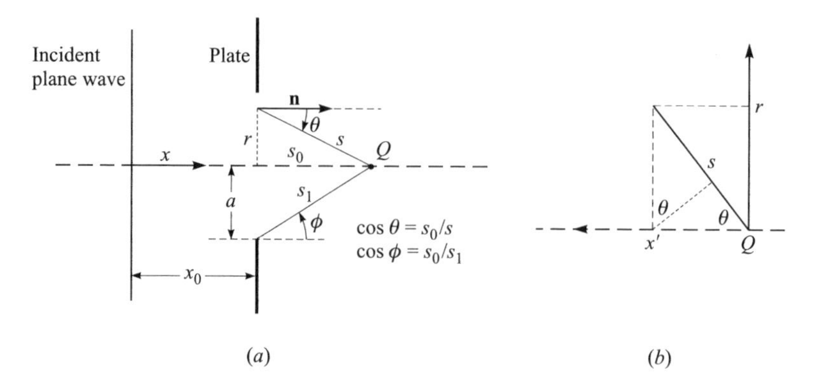

Figure 14.4 Details for the integration over the aperture.

$$\frac{\partial s}{\partial n} = \frac{\partial s}{\partial x} = -\frac{\partial s}{\partial x'} = -\cos\theta = -\frac{s_0}{s}.$$

Equation B–7 therefore becomes

$$p(Q, t) = \frac{1}{2}\int_{s_0}^{s_1}\left\{-\left[\frac{\partial p}{\partial n}\right] + \frac{s_0}{sc_0}\left[\frac{\partial p}{\partial t}\right] + \frac{s_0}{s^2}[p]\right\} ds. \qquad \text{(C–1)}$$

Expressions for the pressure and its derivatives in the aperture must next be found. Let the incident plane wave be represented by $p_{\text{inc}} = f(t - x/c_0)$, or, when it arrives at the aperture,

$$p_{\text{inc}} = f(t - x_0/c_0) = f(\tau), \qquad \text{(C–2)}$$

where x_0 is the distance of the aperture from the initial position of the plane wave (an arbitrary reference); τ thus represents time beginning with the arrival of the incident wave in the aperture. We assume (and this is the second approximation) that the pressure field in the aperture is the same as it would be in a completely free field, that is, if the plate were not present. This assumption, which implies a neglect of any scattered field in the aperture itself, leads to the following expressions for $[p]$, $[\partial p/\partial t]$, and $[\partial p/\partial n]$:

$$[p] = f\left(\tau - \frac{s}{c_0}\right) \qquad \text{(C–3a)}$$

$$\left[\frac{\partial p}{\partial t}\right] = f'\left(\tau - \frac{s}{c_0}\right) = -c_0\frac{\partial f}{\partial s}, \qquad \text{(C–3b)}$$

$$\left[\frac{\partial p}{\partial n}\right] = \left[\frac{\partial p}{\partial x}\right]_{x=x_0} = \left[-\frac{1}{c_0} f'\left(t - \frac{x}{c_0}\right)\right]_{x-x_0} = -\frac{1}{c_0} f'\left(\tau - \frac{s}{c_0}\right) = \frac{\partial f}{\partial s}, \quad \text{(C–3c)}$$

where f written without its argument means $f(\tau - s/c_0)$. When these expressions are substituted in Eq. C–2, a perfect differential is obtained,

$$p(Q, t) = \frac{1}{2}\int_{s_0}^{s_1} \left\{-\frac{\partial f}{\partial s} - s_0 \frac{\partial}{\partial s}\left(\frac{f}{s}\right)\right\} ds \quad \text{(C–4)}$$

$$= \frac{1}{2}\int_{s_1}^{s_0} \frac{\partial}{\partial s}\left\{\left(1 + \frac{s_0}{s}\right) f\right\} ds$$

$$= f\left(\tau - \frac{s_0}{c_0}\right) - \frac{1 + \cos\phi}{2} f\left(\tau - \frac{s_1}{c_0}\right), \quad \text{(C–5)}$$

where $\cos\phi = s_0/s_1$. The first term represents the incident wave, which has traveled straight through the aperture. The second term represents the scattered wave. It seems to be coming from the rim and is a phase-inverted replica of the incident wave, multiplied by an amplitude factor $(1 + \cos\phi)/2$, which is called the "obliquity factor." Figure 14.5 shows an example in which the incident wave is a triangular positive pulse.

b. Discussion

Far from the aperture the obliquity factor approaches unity. The result is then in perfect agreement with the calculation made for transient radiation from a circular piston. Indeed, the scattered wave is an "edge wave" like that produced by a vibrating piston (see Secs. 13C and 13E). However, although they are similar, the piston and aperture problems are not identical. The velocity of the piston provides an exact boundary condition over its face. In the aperture problem only the form of the incident wave is known. We have had to convert information about the incident wave to reasonable approximations about the field in the aperture.

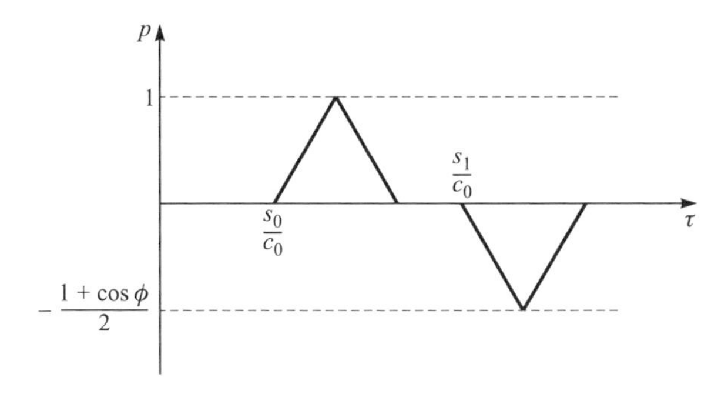

Figure 14.5 Direct and edge wave for a plane pulse incident on a circular aperture.

To test the approximations made about the aperture field, let $s_0 = 0$, i.e., let Q be at the center of the aperture. For this case Eq. C–5 reduces to

$$p(0, t) = f(\tau) - \frac{1}{2} f\left(\tau - \frac{a}{c_0}\right),$$

which is not consistent with our assumption that the pressure in the aperture should be simply $f(\tau)$. The inconsistency is typical of the method and is presumably due to the approximations used to evaluate the Helmholtz-Kirchhoff integral. The method gives good results for high frequencies, far from the diffracting object, but inconsistent results close to the object.

Given the physical interpretation of the terms in Eq. C–5, what signal might one expect to observe if Q is off the axis but still "in the beam," i.e., exposed to the direct wave?

2. Spherical Wave Incident on a Circular Aperture

The aperture diffraction problem may also be solved when the incident wave is spherical. Because the analysis is more involved, however, the actual calculation is not given here, only the final result; see Ref. 2 for a derivation. The geometry of the problem is shown by Fig. 14.6. The source is located at point A on the aperture axis. The pressure of the incident spherical wave, when it fills the aperture (arc of radius r), is $p = f(\tau)$, where $\tau = t - r/c_0$ is the time beginning when the incident wave reaches point B. Evaluation of Eq. B–7 yields

$$p(Q, t) = \frac{r}{r + s_0} \left\{ f\left(\tau - \frac{s_0}{c_0}\right) - \frac{1}{2}(1 + \cos\phi) f\left(\tau - \frac{s_1}{c_0}\right) \right\}. \qquad \text{(C–6)}$$

In Ref. 1 the calculation is carried out for an absorptive medium.

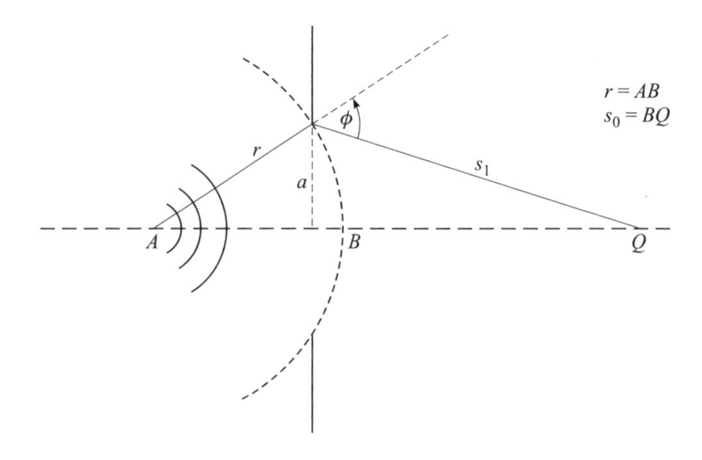

Figure 14.6 Diffraction of a spherical wave by a circular aperture.

Figure 14.7 shows an experimental verification of Eq. C–6. The measurements were made in air with a spark source, which produced an N wave of about 20 μs duration, and a laboratory-constructed, wideband condenser microphone. For (*a*) the diameter of the aperture was 146 mm, and the distances r and s_1 (see Fig. 14.6) were such that the incident and diffracted N waves are resolved. Just as predicted, the diffracted signal is an inverted replica of the incident signal. For (*b*) the aperture was much smaller; its diameter was 38.1 mm. In this case, for the values of r and s_1 used, the delay between the two signals was so small (2.5 μs) that they are not resolved. The resulting waveform resembles the derivative of the incident waveform, a result that may be predicted from Eq. C–6 for the case in which both source and receiver are far from the aperture.[2]

Much more detailed calculations and measurements of diffraction of a spherical N wave by a circular aperture were carried out by Lockwood (Ref. 4). The results are presented in Fig. 14.8, which shows how the edge wave deteriorates as Q (the microphone) is moved off the axis. Although the scattered signals making up the edge wave are perfectly correlated on axis, they become increasingly more poorly correlated as Q moves off axis.

D. REFLECTION BY A RIGID DISK

The Helmholtz-Kirchhoff integral is also useful for computing the signal produced when a plane wave is reflected by a disk. Here a rigid disk is considered. In Fig. 14.9 the disk A is centered in the plane $x = 0$, the observation point Q lies on the axis of the disk, and the head of the incident wave I, which is

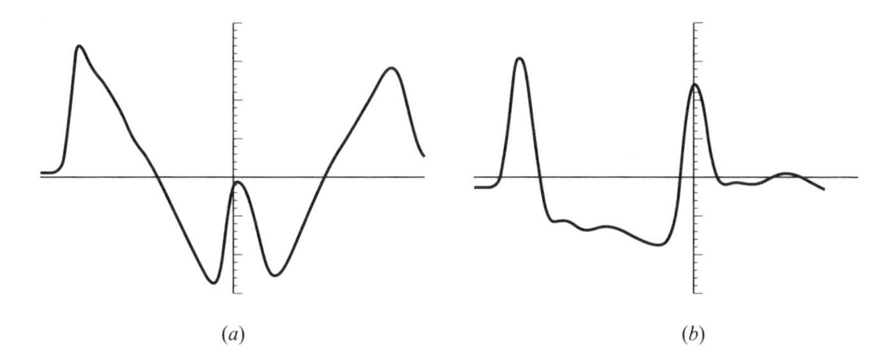

Figure 14.7 Diffraction of a spherical N wave by a circular aperture (time scale 5 μs/div): (*a*) large aperture; (*b*) small aperture.

[2]The small difference in amplitude of the two signals cannot be attributed to the obliquity factor (which was 0.99). Rather, the difference is due to the finite area of the active element of the microphone, which discriminates slightly against the diffracted signal.

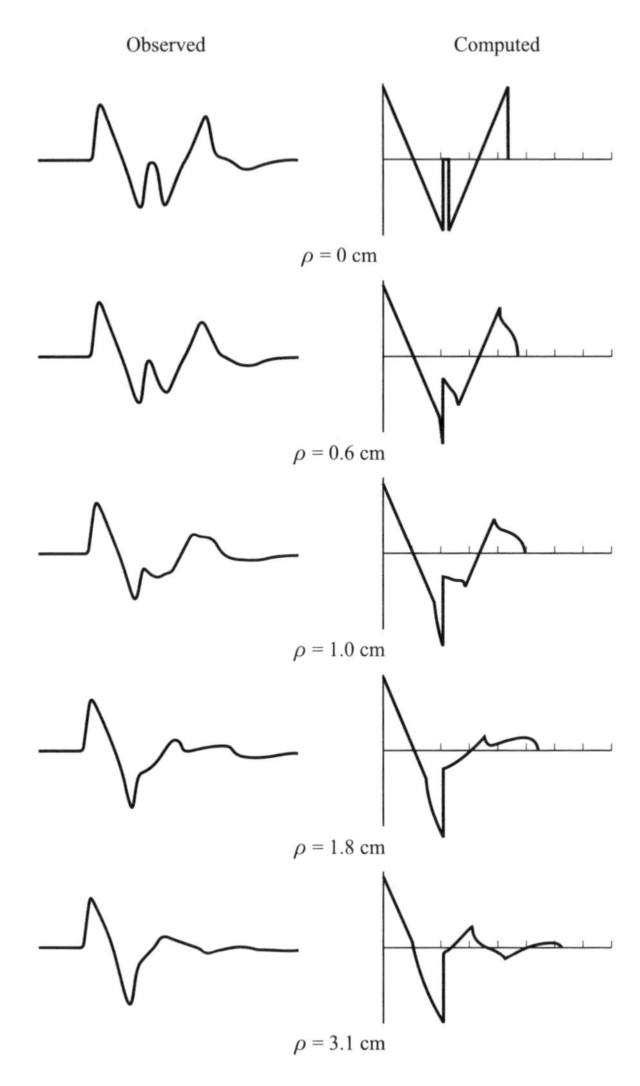

Figure 14.8 Observed and computed waveforms for diffraction of a spherical N wave (duration 24 μs) by a circular aperture (see Fig. 14.6). Source-to-aperture distance is 171 cm, aperture-to-receiver distance is 35 cm, a is 7.25 cm, and ρ is the distance off axis.

traveling in the negative x direction, is at $x = x_0$ at time $t = 0$. If the time waveform of the pressure at $x = x_0$ is $f(t)$, the incident wave is represented by the expression

$$p_{\text{inc}} = f\left(t + \frac{x - x_0}{c_0}\right). \tag{D–1}$$

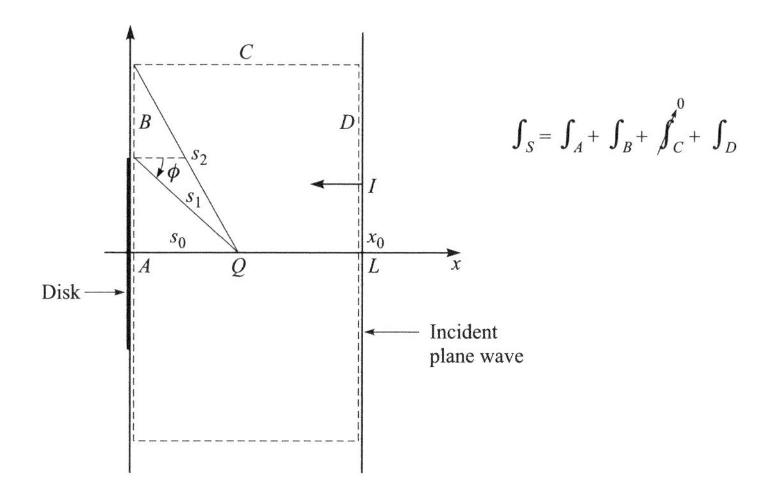

Figure 14.9 Reflection and diffraction of a plane wave by a rigid disk.

The surface S is conveniently divided into elements A (the disk itself), B, C, and D. The cylindrical element C is taken to be so remote that signals from it do not arrive at Q in finite time. The integral over C may therefore be neglected.

First perform the integration over A, the disk itself. Because the geometry is exactly the same as in the aperture problem, Eq. C–1 is applicable here. At A the expression for the incident wave is

$$p_{\text{inc}} = f\left(t + \frac{x - x_0}{c_0}\right)\bigg|_{x=0} = f\left(t - \frac{x_0}{c_0}\right) = f(\tau), \tag{D–2}$$

where $\tau = t - x_0/c_0$ represents time starting when the incident wave arrives at the disk. The pressure at the disk is not p_{inc}, however. Because the disk is rigid, the normal gradient of the pressure vanishes there, and pressure doubling occurs. The requisite quantities for integration over A are therefore

$$[p] = 2f\left(\tau - \frac{s}{c_0}\right), \tag{D–3a}$$

$$\left[\frac{\partial p}{\partial t}\right] = 2f'\left(\tau - \frac{s}{c_0}\right) = -2c_0\frac{\partial f}{\partial s}, \tag{D–3b}$$

$$\left[\frac{\partial p}{\partial n}\right] = 0. \tag{D–3c}$$

Again, f written without its argument means $f(\tau - s/c_0)$.[3] Substitution of Eqs. D–3a,b,c in Eq. C–1 leads to

$$p_A(Q, t) = s_0 \int_{s_0}^{s_1} -\frac{\partial}{\partial s}\left(\frac{f}{s}\right) ds$$
$$= f\left(\tau - \frac{s_0}{c_0}\right) - \cos\phi f\left(\tau - \frac{s_1}{c_0}\right), \qquad \text{(D–4)}$$

where $\cos\phi = s_0/s_1$ and the subscript A denotes the contribution due to integration over A.

With a simple change in limits (from s_0, s_1 to s_1, s_2, respectively, where eventually $s_2 \to \infty$), Eq. C–1 may also be used to carry out the integration over B. The appropriate data for B are as follows:

$$[p] = f; \quad \left[\frac{\partial p}{\partial t}\right] = -c_0 \frac{\partial f}{\partial s}; \quad \left[\frac{\partial p}{\partial n}\right] = -\frac{\partial f}{\partial s}.$$

The first two of these expressions are the same as those for surface A in the aperture problem, namely, Eqs. C–3a and C–3b, respectively. The last expression is the same as Eq. C–3c, but with a minus sign. The change in sign occurs because in this case the incident wave is backward traveling. Integration over surface B yields

$$p_B(Q, t) = \frac{1}{2}\int_{s_1}^{s_2} \left\{\frac{\partial f}{\partial s} - s_0 \frac{\partial}{\partial s}\left(\frac{f}{s}\right)\right\} ds$$
$$= \frac{1}{2}\left(1 - \frac{s_0}{s}\right) f\left(\tau - \frac{s}{c_0}\right)\Bigg|_{s_1}^{s_2}.$$

In the limit as $s_2 \to \infty$ the signal $f(t - s_2/c_0)$ is delayed indefinitely and thus does not contribute to the pressure observed at Q.

$$\therefore\ p_B(Q, t) = -\tfrac{1}{2}(1 - \cos\phi) f(\tau - s_1/c_0). \qquad \text{(D–5)}$$

Integration over surface D yields the incident signal at Q, that is,

$$p_D(Q, t) = f\left(t - \frac{x_0 - s_0}{c_0}\right) = f\left(\tau + \frac{s_0}{c_0}\right). \qquad \text{(D–6)}$$

[3]Notice that although Eq. D–3c is rigorously correct, Eqs. D–3a,b are not. The true pressure on the surface of the disk must include, besides the doubled incident signal, the signals scattered from the edge of the disk. Having no prior information about the scattered signals on the face of the disk, we assume they are small and ignore them.

Although the result is obvious, the calculation has instructional value and is left as an exercise.

Adding Eqs. D–4, D–5, and D–6, we obtain the final result,

$$p(Q, t) = f(\tau + s_0/c_0) + f(\tau - s_0/c_0) - \tfrac{1}{2}(1 + \cos\phi) f(\tau - s_1/c_0). \quad \text{(D–7)}$$

The first term represents the incident wave, the second term the reflected wave, and the third term the wave diffracted from the edge of the disk. For example, let the incident wave be a triangular pulse of amplitude P_0. The waveform of the signal at Q is that shown in Fig. 14.10; I, R, and D indicate the incident, reflected, and diffracted signals, respectively. We have assumed that Q is far enough away from the disk that $\frac{1}{2}(1 + \cos\phi) = 1$.

Two interesting limiting cases of disk reflection are now considered; in both it is assumed that the obliquity factor is unity.[4] First, let the point Q approach the disk. As it does, the incident and reflected pulses approach each other; in the limit as $s_0 \to 0$, they superpose and form a pulse of amplitude $2P_0$, as shown in Fig. 14.11. The diffracted wave is now delayed only by the travel distance a over the face of the disk. Next, let the radius a become very small. The delay of the diffracted signal decreases until it is no longer resolved from the combined incident-reflected signal. In the limit as $a \to 0$, the three signals superpose exactly, and the resultant pulse is simply the incident wave by itself. The reflected and diffracted signals have in effect canceled each other.

From the preceding discussion, one may understand the effect of diffraction on the transient response of a microphone. Most microphones have a pressure-sensitive element centered in a circular housing; let a be the housing radius. Because the impedance of the sensing element is generally large compared to Z_0

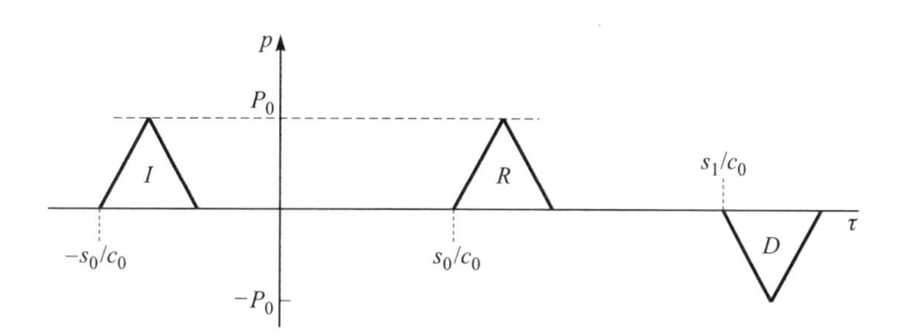

Figure 14.10 Time waveforms when the wave incident on the disk is a triangular pulse. Observation point Q is in front of the disk.

[4]As partial justification for this assumption, we cite Lockwood and Wright's solution of the disk reflection problem (Ref. 5). Their method was based on use of a Green function. Their expression for the axial pressure is the same as Eq. D–7 except that the obliquity factor is replaced by 1. That is, the diffracted wave is predicted to have the same amplitude as the incident wave regardless of the distance between Q and the disk.

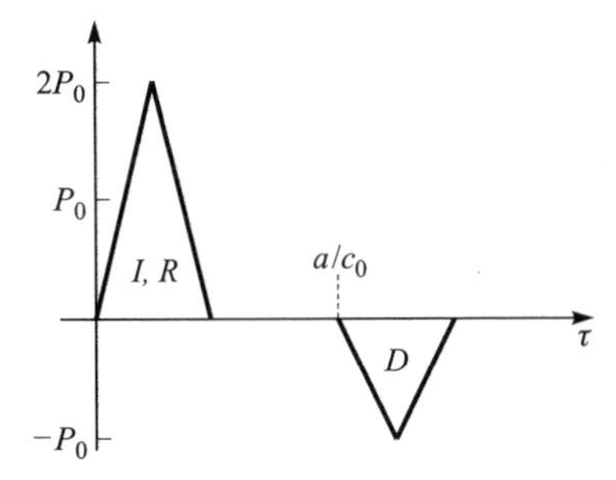

Figure 14.11 Observation point Q on the disk.

for air, the microphone face may be taken to be a rigid disk. If the incident transient wave is slowly varying and of long duration, the delay of the diffracted signal is unimportant: the reflected and diffracted signals effectively cancel. In this case the microphone acts as a point probe. If the transient is short, however, say of order a/c_0, the partial overlay of the diffracted signal on the incident-reflected signal may severely modify the shape of the received signal. In such a case it may be desirable to mount the microphone in a baffle, i.e., enlarge the disk, so as to postpone sufficiently the arrival of the diffracted wave. Figure 14.12 shows the response of two condenser microphones to an incident N wave produced by an electric spark (Ref. 5). The response in (*a*) is from a laboratory-constructed microphone mounted in a baffle 9.9 cm in diameter. The pressure-doubled N comes first followed by the delayed and smoothed upside-down N.[5] The response of an unbaffled $\frac{1}{4}$-in. commercial microphone is shown in (*b*). Here the response is much more complicated because the diffracted signal interferes with the incident-reflected signal. The reason the computed peak and trough values in (*b*) greatly exceed the measured values is that the limited bandwidth of the commercial microphone was not taken into account in computing the predicted response.

E. BABINET'S PRINCIPLE

Babinet's principle is that complementary diffracting objects have complementary diffraction patterns. As a demonstration (not a proof), consider diffraction by a circular aperture in an infinite plate and by a disk of the same radius. The disk is the geometrical complement of the aperture. Refer to Fig. 14.3, in particular, the plane made up of surfaces A and B, and Eq. C–4, the integral over the opening (surface A). If the opening were infinite, we could break the integral up into two parts as follows:

[5] Because the pressure-sensitive element of the microphone was finite in area, not a point, the diffracted wave was integrated as it traveled across the element. The integration caused a smoothing of the diffracted signal, making it appear rounded rather than sharp.

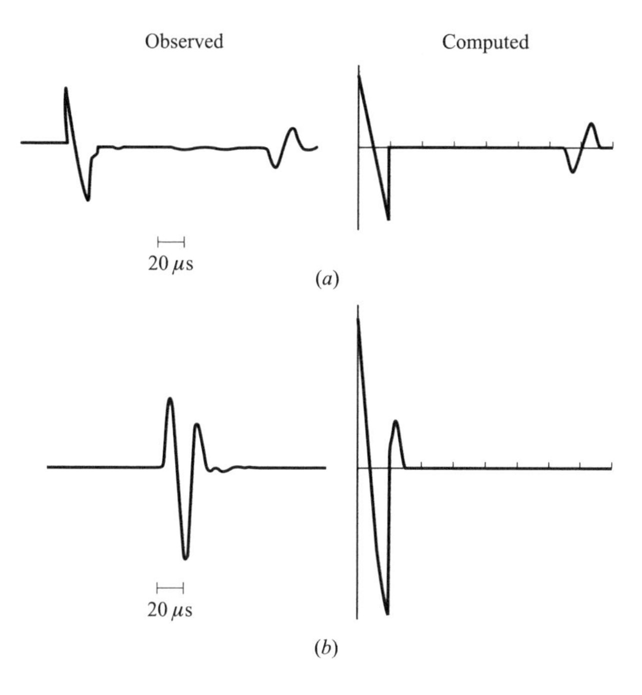

Figure 14.12 Microphone response to an N wave normally incident on the microphone face. (*a*) Laboratory-constructed microphone mounted at the center of a large baffle (active area diameter 0.20 cm; baffle diameter 9.9 cm). (*b*) Commercial $\frac{1}{4}$-in. microphone with no baffle other than its own housing (active area diameter 0.32 cm; baffle diameter 0.60 cm). "Active area" refers to the pressure-sensitive element of the microphone. (From Ref. 5.)

$$\frac{1}{2}\int_{s_0}^{\infty} \{\cdot\}\, ds = \frac{1}{2}\int_{s_0}^{s_1} \{\cdot\} ds + \frac{1}{2}\int_{s_1}^{\infty} \{\cdot\} ds, \tag{E–1}$$

where the curly brackets represent the integrand in Eq. C–4. The left-hand side of Eq. E–1 is simply the free-field pressure,

$$\frac{1}{2}\int_{s_0}^{\infty} \{\cdot\} ds = f\left(\tau - \frac{s_0}{c_0}\right) = p_{\text{free field}}. \tag{E–2}$$

The first term on the right-hand side of Eq. E–1 is of course the pressure found in the circular aperture problem, p_{aperture}, given by Eq. C–5. The second term on the right-hand side is the pressure one would calculate if surface *B* were the opening and surface *A* were closed. In other words, the second term represents the pressure due to diffraction around a disk, p_{disk}. Equation E–1 may therefore be expressed as

$$p_{\text{free field}} = p_{\text{aperture}} + p_{\text{disk}}, \tag{E–3}$$

from which, knowing $p_{\text{free field}}$ and p_{aperture}, we may calculate p_{disk}:

$$p_{\text{disk}} = p_{\text{free field}} - p_{\text{aperture}} = \frac{1 + \cos\phi}{2} f\left(\tau - \frac{s_1}{c_0}\right). \tag{E–4}$$

As an example, let the incident signal be a triangular pulse. Equation E–3 for this case may be represented by the sketches in Fig. 14.13. The figure illustrates Babinet's principle very clearly. The aperture and disk are complementary diffracting objects, and their diffraction patterns, represented by the shaded waveforms in Fig. 14.13, are complements of each other.

The result for the disk is interesting in its own right. Surprise may be the initial reaction to the prediction of a very strong signal on the axis behind the disk, where ray theory analysis would lead one to expect the shadow to be deepest. Sometimes described in optics as the "bright spot behind the penny," the axial signal owes its strength to the unique position of the axis, which is equidistant from all points on the rim of the disk. The signals scattered from the rim all arrive in phase, that is, perfectly correlated. Off axis the correlation is poor, and the shadowing effect of the disk is better realized. It may also be seen that the high degree of correlation on the axis may be destroyed by tilting the disk so that it faces the incoming radiation at an angle. A disk with a microphone mounted axially behind it could therefore be used as a direction finder.

An experimental demonstration of the "bright spot" behind the disk is shown in Fig. 14.14. Produced by an electric spark, an N wave was normally incident from a distance of 830 mm on a 146-mm-diameter disk in air. The sound field was recorded on the axis 720 mm behind the disk.[6] The upper waveform is for the disk removed, the lower waveform for the disk in place. The diffracted wave is seen to be an almost perfect replica of the incident wave.

Finally, attention is called to the similarity between the field behind the disk and the radiation produced by the moving baffle (fixed piston) in Problem 13–9.

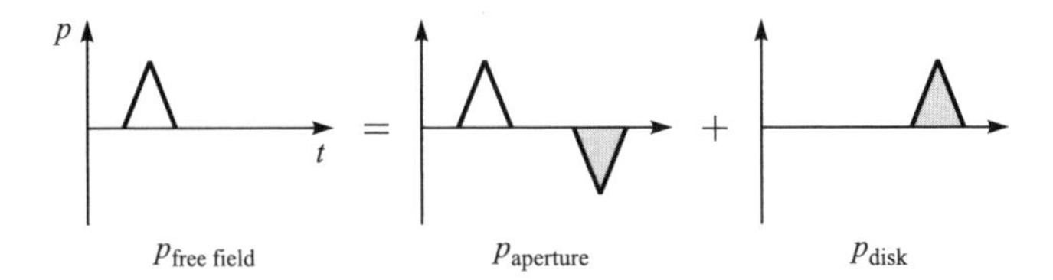

Figure 14.13 An illustration of Babinet's principle. The diffracted signals are shaded.

[6]The experimental apparatus is the same as that described for the aperture measurements shown in Fig. 14.7. In fact, the diameters of the disk and the large aperture (Fig. 14.7*a*) are the same. Because the distances r and s_1 (Fig. 14.6) are different, however, the diffracted wave delays are not quite the same.

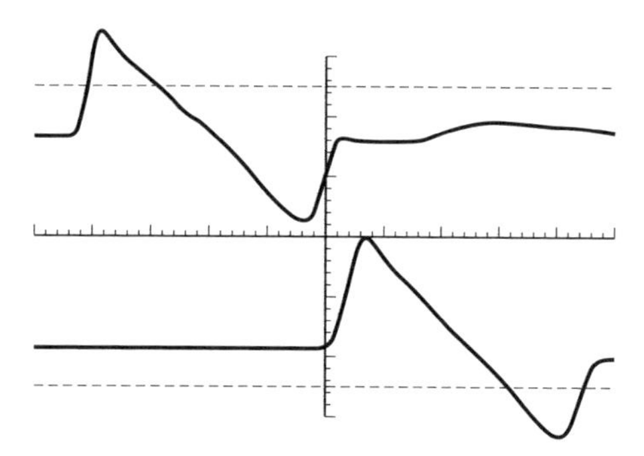

Figure 14.14 Diffraction by a disk, microphone on axis behind the disk. Upper trace: disk not present. Lower trace: disk in place.

REFERENCES

1. M. O. Anderson, "The Propagation of a Spherical N Wave in an Absorbing Medium and Its Diffraction by a Circular Aperture," Tech. Rep. ARL–TR–74–25, Applied Research Laboratories, University of Texas at Austin, August 1974 (AD 787 878).
2. C. L. Andrews, *Optics of the Electromagnetic Spectrum* (Prentice-Hall, Englewood Cliffs, NJ, 1960), Appendix II, pp. 483–487.
3. M. Born and E. Wolf, *Principles of Optics*, 6th ed. (Pergamon Press, New York, 1980), pp. 375–382.
4. J. C. Lockwood, "Two Problems in High-Intensity Sound. I. Finite-Amplitude Sound Propagation in the Farfield of Nonuniform Sources. II. The Diffraction of N waves by a Circular Aperture in a Plane Baffle," Tech. Rep. ARL–TR–71–26, University of Texas at Austin, July 1971 (AD 740 498).
5. J. C. Lockwood and W. M. Wright, "Effect of edge diffraction on microphone response to transients," *J. Acoust. Soc. Am.* **51**, 106(A) (1972).
6. P. M. Morse, *Vibration and Sound*, 2nd ed. (McGraw-Hill Book Co., New York, 1948). Reprinted by Acoustical Society of America, New York, 1981.
7. P. M. Morse and K. U. Ingard, *Theoretical Acoustics* (McGraw-Hill Book Co., New York, 1968). Reprinted by Princeton University Press, Princeton, NJ, 1986.
8. B. Rossi, *Optics* (Addison-Wesley, Reading, MA, 1957).
9. H. A. Wright, "Time-domain analysis of broad-band refraction and diffraction," *J. Acoust. Soc. Am.* **46**, 661–666 (1969).

PROBLEMS

14–1. A plane wave is normally incident on a rigid disk. By evaluating the Helmholtz-Kirchhoff integral, compute the signal at an axial point Q behind the disk. Let the incident wave be a triangular pulse. Sketch the time waveform of the signal $p(Q, t)$. Compare it with the signal that would be received at Q if the disk were not in place.

14–2. A normally incident plane sound wave is diffracted by an annular hole, of outer radius b and inner radius a, in a plate (see sketch).

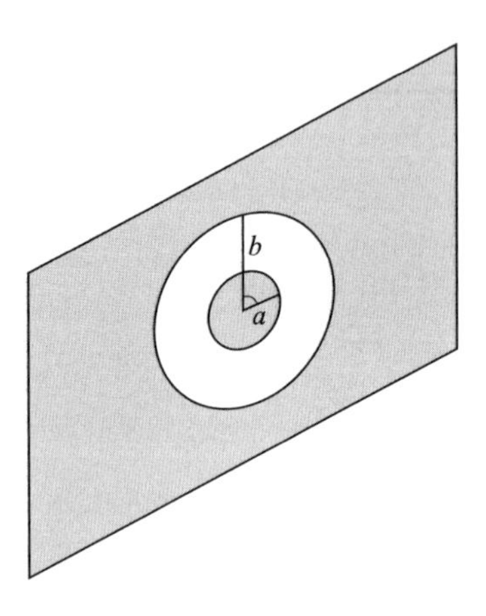

(a) Compute the axial signal behind the plate.

(b) Sketch the time waveform of the axial signal for a triangular pulse incident wave. Identify appropriate parts of the signal.

(c) Determine whether your answer agrees with known results in the following limiting cases:

(i) $a \to 0$

(ii) $b \to \infty$

14–3. A plane wave is normally incident on an infinite rigid plate in which there is a semicircular hole of radius a. Find the pressure at any point Q on the axis of the hole behind the plate. (The problem is the same as that analyzed in Sec. C.1 except that the hole is a semicircle rather than a full circle).

(a) Obtain the expression for $p(Q, t)$.

(b) Illustrate your answer with a sketch, say for the case in which the incident pressure disturbance is a triangular pulse.

(c) Use your solution to find the pressure behind a semi-infinite rigid wall (barrier) at a point even with the top of the wall.

14–4. Solve the problem of reflection from a pressure release disk, that is, a disk on which $p = 0$. The listening point Q is on the axis of the disk and is on the same side of the disk as the incident plane wave (see Fig. 14.7). Illustrate and interpret your answer with a sketch, say for a triangular incident pulse.

14–5. A disk is made of a material for which the plane-wave reflection coefficient is R, where R is a real number. A plane wave is normally incident on the disk.

(a) Compute the signal on the axis in front of the disk, that is, on the incident-wave side of the disk.

(b) To check your answer, assume the incident wave is a triangular pulse and sketch the time waveforms for the following limiting cases:

(i) $R = 1$

(ii) $R = -1$

(iii) $R = 0$

For case (iii) assume, for example, that the disk is a perfect absorber, that is, that all the incident sound is absorbed, none reflected.

Discuss and interpret your results.

14–6. Consider the disk reflection problem of Sec. 14D for a disk half of which is rigid ($R = 1$, $0 < \psi < \pi$), the other half pressure release ($R = -1$, $\pi < \psi < 2\pi$), where ψ is the angular coordinate in the plane of the disk. The incident wave $f(t)$ is normally incident on the disk. Find $p(Q)$, where the listening point Q is on the axis in front of the disk, as in Fig. 14.9.

(a) Begin by guessing the answer (no penalty for a wrong guess).

(b) Now use the Helmholtz-Kirchhoff integral to find $p(Q, t)$. No need to repeat calculations that are in the text or that you worked out in other homework problems; simply cite the appropriate material and proceed.

(c) Interpret your answer to part (b) and comment on whether it agrees with your guess in part (a).

(d) Qualitatively compare $p(Q, t)$ for this disk to the field you would expect for a disk having a uniform impedance equal to $\rho_0 c_0$ ($R = 0$), for the following two cases:

(i) Let Q be an axial point, as in the rest of the problem.

(ii) Let Q be a point off axis.

14–7. A zone plate is a plate having a sequence of concentric ring apertures like the one shown in the sketch. It is used to obtain amplification of a sinusoidal signal of given frequency at a specific point Q on the axis behind the zone plate. Amplification occurs when the inner and outer radii of each ring are chosen so that all the diffracted signals (and the transmitted signal if, as in the sketch, the central element is a hole rather than a disk) arrive in phase at Q. Design a zone plate to provide 12 dB of gain at a receiver located a distance d behind the zone plate. Assume that the obliquity factor is 1. The plate is designed to operate in a given medium (sound speed c_0) at a given frequency f (wavelength $\lambda = c_0/f$).

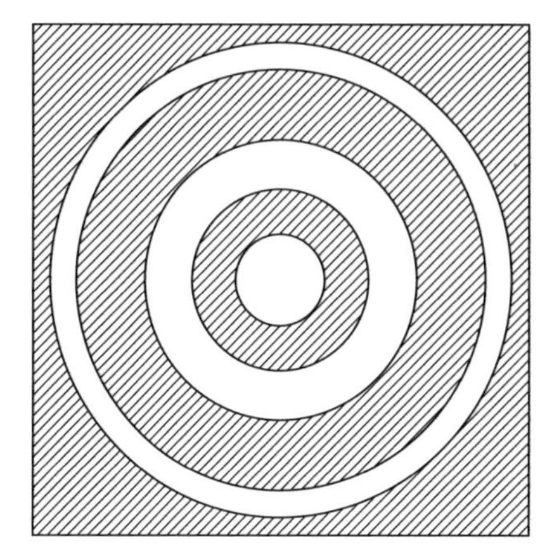

(a) Determine how many rings are necessary to achieve the desired gain.

(b) Derive formulas for the hole and ring radii a_n in terms of d and λ. Let a_1 denote the smallest radius, whether it be for a hole or a disk.

14–8. A cork disk of radius a floats on the surface of fresh water. A plane wave [time waveform $f(t)$] from below is normally incident on the water surface. At point Q on the axis of the cork disk, a distance s_0 from the disk, is a hydrophone.

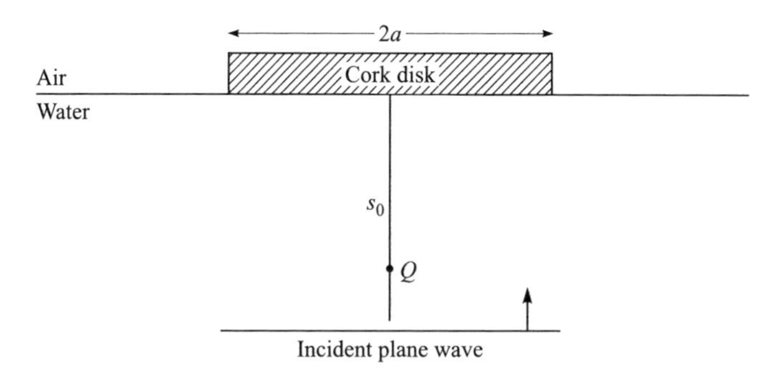

(a) Find the expression for the sequence of signals recorded by the hydrophone.

(b) Illustrate your answer with a sketch. Assume that the waveform of the incident signal is a triangular pulse. Be sure to identify each signal in the sequence.

You may assume that the cork is sufficiently lossy that you do not need to consider propagation back and forth inside the cork itself.

14–9. The cork disk in the previous problem is replaced by a rigid disk.

(a) Find the expression for the sequence of signals recorded by the hydrophone at Q.

(b) Illustrate your answer with a sketch of the time waveform at at the hydrophone. Assume the incident signal to be a triangular pulse. Identify the various signals.

(c) Use physical arguments to explain the features of the sequence of pulses found in part (b).

14–10. Given the result derived in Sec. 14D for reflection by a rigid disk (Eq. D–7 and Fig. 14.10), use Babinet's principle to solve the problem of reflection by a rigid plate having a circular hole in it. The observation point Q is on the axis of the hole, in front of the plate.

14–11. The hole in a rigid infinite plate consists of two semicircular sections, of radii a_1 (the smaller) and a_2 (the larger); see the sketch. A plane pressure disturbance of time waveform $f(t)$ is normally incident on the plate as shown.

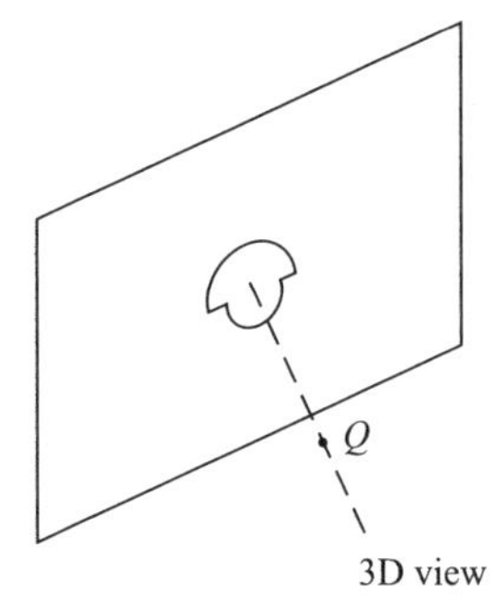

3D view

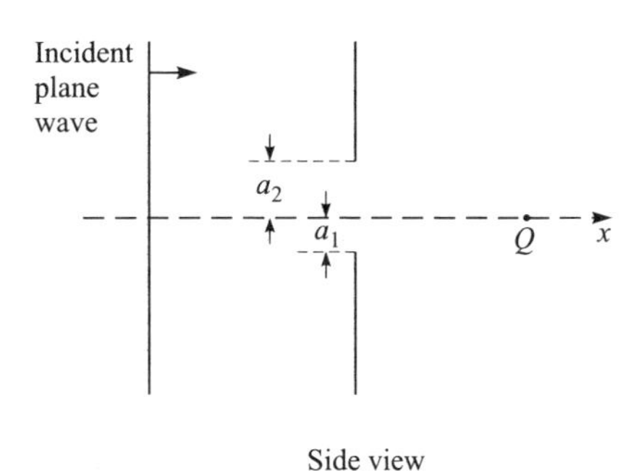

Side view

(a) For the signal at Q:

(i) Find the time waveform of the pressure signal at point Q on the axis of the hole.

(ii) Illustrate your answer with a time waveform sketch for the case in which the incident signal is a single cycle of a sine wave of period T. Assume the geometry is such that the incident wave is resolved in time from the diffracted wave(s) but that if multiple diffracted waves are present, they overlap.

(b) Modify the result from part (a) so that it applies to a hole made up of:

(i) Three arcs of $2\pi/3$ rad each, of radii a_1, a_2, and a_3, respectively.

(ii) N equal arcs of $2\pi/N$ radians each, of radii, $a_1, a_2, \ldots, a_N$, respectively.

(c) Qualitatively apply your results to estimate the signal transmitted through a jagged aperture, i.e., an aperture having a radius that varies randomly about a mean value a_0. Illustrate with a sketch of the time waveform you expect for the transmitted signal.

15

ARRAYS

A. DIRECTIVITY: NOMENCLATURE AND DEFINITIONS

In this chapter consideration is restricted to time-harmonic radiation in the farfield. In particular, this means

$$p = P(r, \theta)e^{j\omega t}, \qquad I = \frac{PP^*}{2\rho_0 c_0} = \frac{\hat{p}^2}{\rho_0 c_0},$$

where P^* is the complex conjugate of P and $\hat{p}$ stands (in this chapter) for rms pressure. The symbol $D(\theta)$ is used throughout the text to mean the *amplitude* directivity function,

$$D(\theta) = \frac{P(r, \theta)}{P(r, \theta_{\max})}, \tag{A–1}$$

where $\theta_{\max}$ is the angle at which the radiation pattern has its maximum.[1] Frequently the maximum is on the axis of the source, i.e., $\theta_{\max} = 0$. The *intensity* directivity factor, for which we use the symbol D_i, is

$$D_i = \frac{I_{\max}}{I_{\text{ave}}}, \tag{A–2}$$

[1] In case the response is a function of two angles, say θ and ψ, the directivity is expressed as a function $D(\theta, \psi)$.

where I_{max} is the intensity at point r at the maximum ($\theta = \theta_{max}$) and I_{ave} is the average intensity if the total power radiated were uniformly spread over a spherical surface of radius r, i.e., over area $4\pi r^2$. Notice that D_i is not equal to $|D(\theta)|^2$. If for simplicity we assume that $\theta_{max} = 0$, Eq. A–2 becomes

$$D_i = \frac{4\pi r^2 P(r, 0)P^*(r, 0)}{2\rho_0 c_0 \int I\, dS} \tag{A–3a}$$

$$= \frac{4\pi r^2 P(r, 0)P^*(r, 0)}{\int P(r, \theta)P^*(r, \theta)\, dS}, \tag{A–3b}$$

or, in terms of the increment of solid angle $d\Omega = dS/r^2$ and the directivity $D(\theta)$,

$$D_i = \frac{4\pi}{\int_\Omega |D(\theta)|^2 d\Omega}. \tag{A–4a}$$

An alternative form in terms of a surface integral is

$$D_i = \frac{4\pi r^2}{\int_s |D(\theta)|^2\, dS}. \tag{A–4b}$$

Finally, the *directivity index* DI, in decibels, is

$$\mathrm{DI} = 10 \log_{10} D_i. \tag{A–5}$$

The usefulness of the quantities D_i and DI are illustrated by an example. Suppose a specified sound pressure level SPL_{req} is required at a given distance r from a source. The source is yet to be designed, but the available acoustical power is known to be W. First consider whether a simple omnidirectional source (a monopole) will do. The intensity produced by such a source is

$$I_{omni} = \frac{W}{4\pi r^2},$$

the rms pressure $\hat{p}_{omni}$ is

$$\hat{p}_{omni} = \sqrt{\rho_0 c_0 I_{omni}} = \sqrt{\frac{\rho_0 c_0 W}{4\pi r^2}},$$

and the sound pressure level is

$$\mathrm{SPL}_{omni} = 20 \log_{10} \frac{\hat{p}_{omni}}{p_{ref}}.$$

This level is related to the required level as follows:

$$\begin{aligned}
\mathrm{SPL}_{\mathrm{req}} &= 10\log_{10}\left(\frac{\hat{p}_{\mathrm{req}}}{p_{\mathrm{ref}}}\right)^2 \\
&= 20\log_{10}\frac{\hat{p}_{\mathrm{omni}}}{p_{\mathrm{ref}}} + 10\log_{10}\left(\frac{\hat{p}_{\mathrm{req}}}{\hat{p}_{\mathrm{omni}}}\right)^2 \\
&= \mathrm{SPL}_{\mathrm{omni}} + 10\log_{10} D_i \\
&= \mathrm{SPL}_{\mathrm{omni}} + \mathrm{DI}. \qquad \text{(A–6)}
\end{aligned}$$

Thus if an omnidirectional source is insufficient, a source having a DI that makes up the difference should be designed. The directivity index is a measure of the degree to which the available acoustical power may be concentrated in a given direction.

Example 15.1. Monopole next to a rigid plane. To illustrate the computation of D, D_i, and DI, consider the radiation from a small, time-harmonic monopole placed a short distance h above a rigid plane (Fig. 15.1). If the frequency is assumed low enough that $kh \ll 1$, the radiation is omnidirectional in the hemisphere above the plane ($\theta < \pi/2$). Since no sound exists in the space below the plane ($\pi/2 < \theta < \pi$), the directivity function is

$$D = \begin{cases} 1, & 0 < \theta < \pi/2, \\ 0, & \pi/2 < \theta < \pi. \end{cases}$$

Consequently, the intensity directivity factor D_i is

$$D_i = \frac{4}{\displaystyle\int_0^{2\pi} (1)\, d\Omega + \int_{2\pi}^{4\pi} (0)\, d\Omega} = \frac{4\pi}{2\pi} = 2, \qquad \text{(A–7a)}$$

and the directivity index is

$$DI = 10\log_{10}(2) = 3 \text{ dB}. \qquad \text{(A–7b)}$$

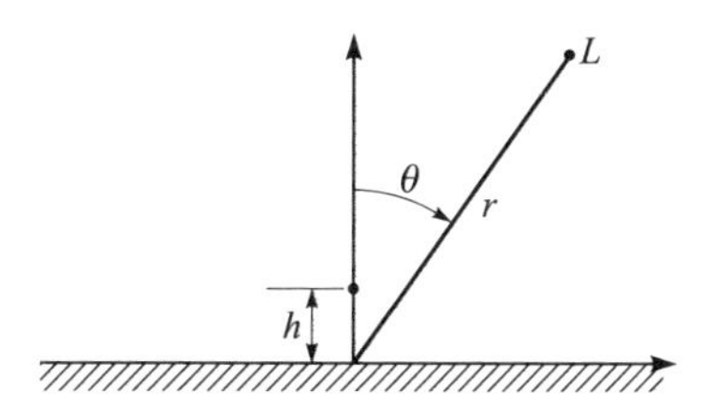

Figure 15.1 Radiation produced by a monopole above a rigid plane.

In other words, the SPL in the upper hemisphere is 3 dB higher than it would be if the plane were not present.[2]

As a more involved example, compute D_i and DI for a circular piston in a rigid baffle. Two approaches are presented. First, the calculation is done as an application of Eq. A–4a. As can be seen from Fig. 15.2, the increment of solid angle is $d\Omega = 2\pi \sin\theta \, d\theta$. Substituting the expression for the piston directivity (Eq. 13B–3) in Eq. A–4a, we obtain

$$\frac{1}{D_i} = \frac{1}{4\pi} \int_0^{\pi/2} \left[\frac{2J_1(ka \sin\theta)}{ka \sin\theta} \right]^2 2\pi \sin\theta \, d\theta.$$

The integration, which is not trivial, yields

$$\frac{1}{D_i} = \frac{1}{k^2a^2} \left[1 - \frac{2J_1(2ka)}{2ka} \right] = \frac{R_1(2ka)}{k^2a^2},$$

where $R_1(2ka)$ is the real part of the (normalized) piston radiation impedance (see Sec. 13D), or

$$D_i = k^2a^2 / R_1(2ka). \tag{A–8}$$

An alternative derivation of Eq. A–8 is based on results available from Chap. 13. Start with Eq. A–3a and recognize that $\int I \, dS$ is the piston power W, which is given by Eq. 13B–13b. Equation 13B–7 shows that $P(r, 0)P^*(r, 0) = P_0^2 R_0^2 / r^2$. Substitution in Eq. A–3a leads to Eq. A–8.

To analyze Eq. A–8, let us consider two special but important cases.

1. If the piston is very small compared to a wavelength ($ka \ll 1$), the asymptotic expression for $R_1(2ka)$ is $k^2a^2/2$, and one obtains

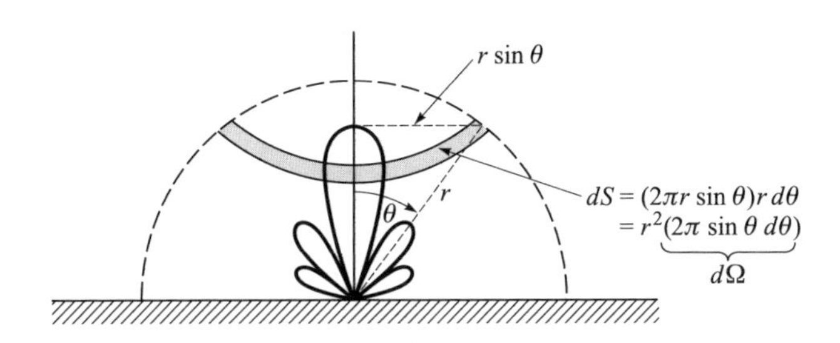

Figure 15.2 Radiation from a circular piston.

[2]Why is the answer not 6 dB? After all, the wave reflected from a rigid surface has the same amplitude as the incident wave. *Hint*: What is being held constant as the plane is brought up next to the source, the source pressure amplitude or the power output of the source?

$$D_i = 2, \qquad \mathrm{DI} = 3\ \mathrm{dB}.$$

Not surprisingly, this is a repeat of the results for a small monopole very close to a rigid plane.

2. If the piston is very large compared to a wavelength ($ka \gg 1$), then $R_1(2ka) \doteq 1$, in which case

$$D_i = k^2a^2, \tag{A–9a}$$

$$\mathrm{DI} = 20 \log_{10} ka. \tag{A–9b}$$

The way to obtain a high directivity index, therefore, is to use either a very large piston or a very high frequency.

B. ARRAY OF TWO POINT SOURCES

We now calculate the farfield radiation from a pair of point sources, each radiating as follows:

$$p = \frac{A}{r} e^{j(\omega t - kr)}. \tag{B–1}$$

Because the point of observation is in the farfield, the rays from sources 1 and 2 (Fig. 15.3) take parallel paths to the field point. Let the distance from the center of the array be r and the distance between sources be d. Ignoring the difference between r and $r \pm (d/2) \sin\theta$ in the amplitude factor but taking account of it in the phase factors, we obtain

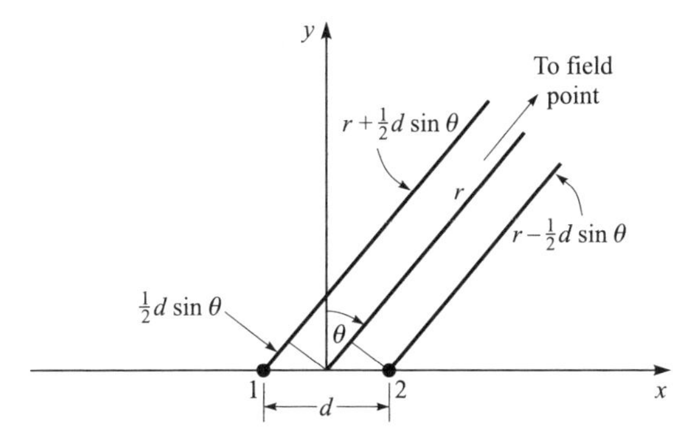

Figure 15.3 Radiation from a two-source array.

$$\begin{aligned} p &= p_2 + p_1 \\ &= \frac{A}{r} e^{j(\omega t - kr)} [e^{j\phi} + e^{-j\phi}] \\ &= 2\frac{A}{r} e^{j(\omega t - kr)} \cos\phi, \end{aligned} \tag{B–2}$$

where $\phi = (kd/2)\sin\theta$ is half the difference in phase between signals from the two sources. The directivity for the two-source array is, from Eqs. A–1 and B–2,

$$D(\theta) = \cos\phi = \cos\left(\frac{kd}{2}\sin\theta\right). \tag{B–3}$$

The shape of the beam pattern defined by Eq. B–3 is determined by the size of $kd/2 = \pi d/\lambda$. First let $kd/2 \ll 1$ so that $D(\theta) \doteq 1$. The two sources are so close together that they practically coincide. Pattern (*a*) in Fig. 15.4, omnidirectional radiation, results. As $kd/2$ is increased from near zero, $D(\theta)$ drops most rapidly near $\theta = \pm 90°$, and the pattern begins to pinch in at its equator. The pinch becomes a null [pattern (*b*)] when $kd/2 = \pi/2$, i.e., when $d = \lambda/2$, and a figure-eight pattern results. The radiation is nil in the end-fire direction ($\theta = 90°$), because the two signals emitted in that direction are exactly 180° out of phase. If $kd/2$ is increased further, the beam narrows, but the null moves to lower angles and sidelobes appear [pattern (*c*)]. Sidelobes are the penalty paid for a narrower beam. When $kd/2$ reaches the value π ($d = \lambda$), the sidelobes become as prominent as the main lobes at $\theta = 0$, 180° [pattern (*d*)]. The beam pattern is then said to have secondary major lobes. How does the pattern change as $kd/2$ increases beyond π?

Next consider the introduction of phase shift at the source. Let the difference in phase between the two sources be 2ψ. As can easily be shown, the directivity then becomes

$$D(\theta) = \cos\left(\frac{kd}{2}\sin\theta - \psi\right). \tag{B–4}$$

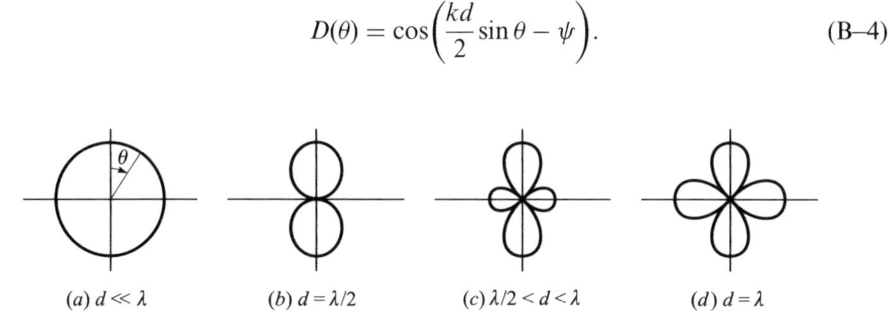

Figure 15.4 Radiation patterns for an array of two-point sources for various values of separation distance d.

The major lobe now occurs at $\theta = \theta_{max} = \sin^{-1}(2\psi/kd)$ (provided $\psi \leq kd/2$); see Fig. 15.5. If the frequency is varied, θ_{max} can be kept constant by varying ψ linearly with frequency. Notice that the dependence of the major-lobe angle θ_{max} on ψ provides a means of steering the beam electrically instead of mechanically. The two elements 1 and 2 can of course be receivers instead of transmitters. Then the variation of the phase angle ψ allows a steering of the receive "beam."

A special case occurs when $\psi = \pi/2$, for then the two sources are exactly out of phase with each other. Their contributions therefore cancel in the normal direction $\theta = 0$. The directivity function becomes

$$D(\theta) = \cos\left(\frac{kd}{2}\sin\theta - \frac{\pi}{2}\right) = \sin\left(\frac{kd}{2}\sin\theta\right), \tag{B–5}$$

which not only indicates the null at $\theta = 0$ but also shows a maximum at $\theta = \pi/2$ (provided $kd/2 \leq \pi/2$, i.e., $d \leq \lambda/2$).[3] See Fig. 15.6. The pair of sources now constitutes an *end-fire* array (the term *broadside array* applies to the array we considered first, $\psi = 0$). That is, maximum radiation is along the line connecting the two sources. When the sources are separated by $\lambda/2$, the signal from one source reaches the other source at exactly the right time to perfectly reinforce the emission of the other source.

Return for a moment to the simple broadside array ($\psi = 0$). Equation B–3, written in terms of ϕ, may be reexpressed as follows:

$$D(\theta) = \cos\phi \times \frac{2\sin\phi}{2\sin\phi} = \frac{\sin 2\phi}{2\sin\phi}. \tag{B–6}$$

The reason for writing the directivity in this fashion becomes apparent when we consider arrays of more than two sources.

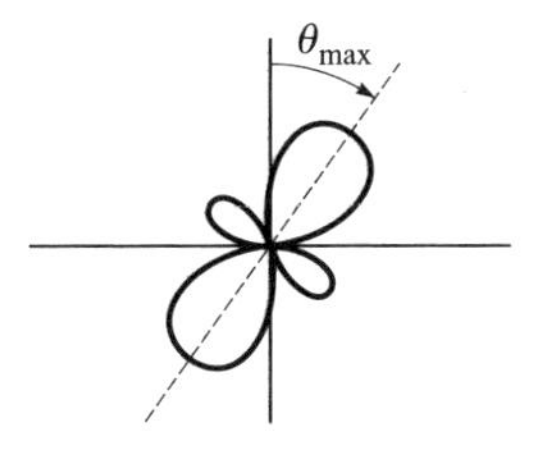

Figure 15.5 Effect of a phase shift between the two sources.

[3] Strictly speaking, the expression should be normalized so that $D = 1$ at its maximum.

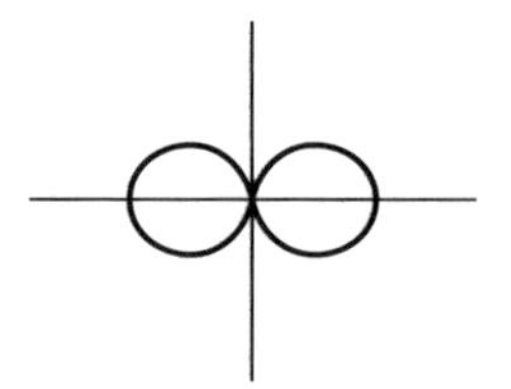

Figure 15.6 End-fire array ($\psi = \pi/2$).

C. ARRAY OF N POINT SOURCES

Only *unshaded* arrays of equally spaced sources are considered here.[4] See Fig. 15.7. The pressure p_0 due to the source designated zero is[5]

$$p_0 = \frac{A}{r} e^{j(\omega t - kr)}, \tag{C–1}$$

that due to the source designated p_{-1} is

$$p_{-1} = \frac{A}{r} e^{j[\omega t - k(r + d \sin\theta)]} = p_0 e^{-j2\phi},$$

that due to the source designated p_{-2} is

$$p_{-2} = p_0 e^{-j4\phi},$$

and so on. The pressure p due to all N sources is the finite sum

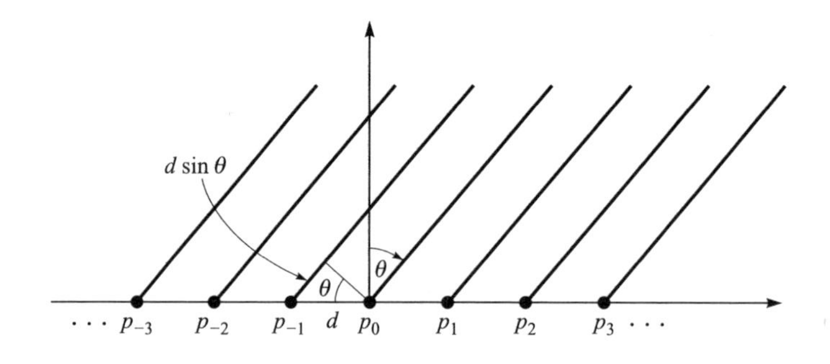

Figure 15.7 An array of N point sources.

[4]In an unshaded array, amplitude is the same for all elements. If the amplitudes are different, the array is said to be *shaded.* See, for example, Refs. 1 and 3.
[5]Note that p_0 used here does not mean ambient pressure.

$$p = p_0[\cdots + e^{-j4\phi} + e^{-j2\phi} + 1 + e^{j2\phi} + e^{j4\phi} + \cdots].$$

Factor out the final term $e^{j(N-1)\phi}$:

$$p = p_0 e^{j(N-1)\phi}[1 + e^{-2j\phi} + e^{-j4\phi} + \cdots + e^{-j2(N-1)\phi}].$$

The term in brackets is a geometric progression of the form $1 + x + x^2 + \cdots + x^{N-1}$, which sums to $(1 - x^N)/(1 - x)$.

$$\therefore\ p = p_0 \frac{e^{jN\phi}}{e^{j\phi}} \frac{1 - e^{-j2N\phi}}{1 - e^{-j2\phi}} = p_0 \frac{e^{jN\phi} - e^{-jN\phi}}{e^{j\phi} - e^{-j\phi}} = p_0 \frac{\sin N\phi}{\sin\phi}. \tag{C–2}$$

Since on axis ($\theta = 0$, $\phi = 0$) the pressure is $p = Np_0$, the directivity function is

$$D(\phi) = \frac{\sin N\phi}{N\sin\phi}, \tag{C–3}$$

which is a generalization of Eq. B–6.

The derivation that has been given here is for an odd number of sources. If N is even, the center of the array is midway between the central pair of sources. Although the bookkeeping is a little different, the same end result, Eq. C–3, is obtained.

Example 15.2. Six-element array. To illustrate the dependence of the directivity function on d/λ, we make a rough graphical calculation of Eq. C–3 for the case $N = 6$. The numerator (amplitude 1) and denominator (amplitude 6) are sketched as a function of ϕ in Fig. 15.8*a*. Division of the $\sin 6\phi$ curve by the $6\sin\phi$ curve results in the curve $D(\phi)$ sketched in Fig. 15.8*b*. Although the directivity is shown for values of ϕ out to $7\pi/6$, of physical interest is only that part of the curve that corresponds to $0 \le \theta \le \pi/2$. Since $\phi = (kd/2)\sin\theta$, the maximum value ϕ can have is $\phi_{max} = \pi d/\lambda$. For example, if $d = \lambda$, $\phi_{max} = \pi$, and the entire curve $0 \le \phi \le \pi$ in Fig. 15.8*b* is used for $D(\theta)$.[6] If $d = \lambda/2$, however, $\phi_{max} = \pi/2$, in which case $D(\theta)$ is made up of only the $0 \le \phi \le \pi/2$ portion of the $D(\phi)$ curve. Some observations based on this example are as follows:

1. To avoid any part of the secondary major lobe, one must put $\phi_{max} = 5\pi/6$, that is, the spacing between sources must be no grcater than $d = 5\lambda/6$.
2. The minor lobes decrease in amplitude until the value $\phi = \pi/2$ is reached. A monotonically decreasing envelope of the lobes may therefore be achieved by choosing $d \le \lambda/2$.

[6] Note that since plots of the directivity function are always $|D(\theta)|$, not $D(\theta)$ itself, the $D(\theta)$ curve in Fig. 15.8*b* will be rectified when plotted.

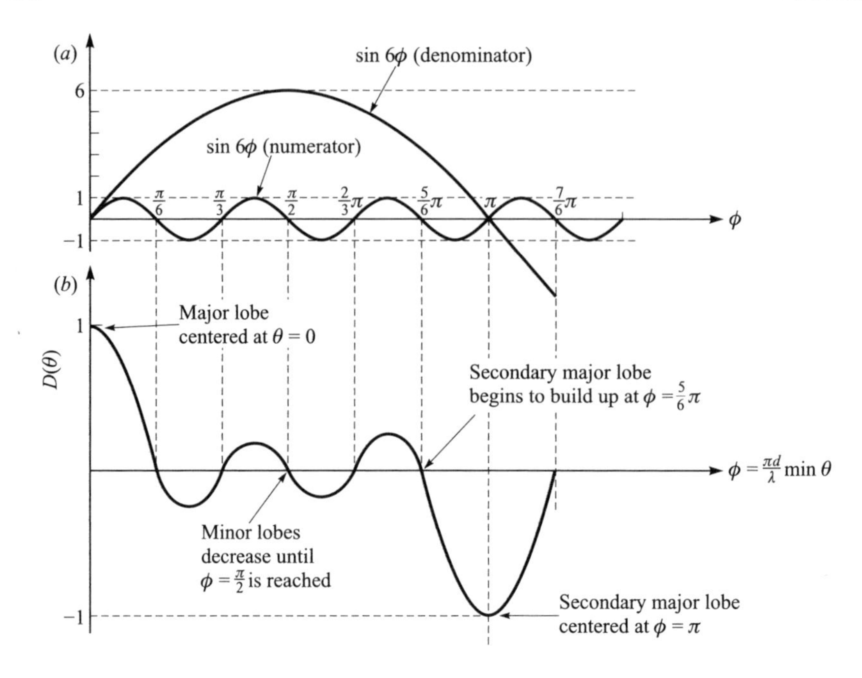

Figure 15.8 Graphical calculation of $D(\phi)$ for $N = 6$.

3. A pattern without any minor lobes may be obtained by setting $\phi_{\mathrm{max}} = \pi/6$, that is, $d = \lambda/6$.

It is particularly useful to generalize observation 1 from this example. If secondary major lobes are to be avoided, the spacing must be somewhat less than a wavelength. In particular, one must have

$$d < \frac{N-1}{N}\lambda, \tag{C–4}$$

or since the length L of the array is $L = (N-1)d$,

$$L < \frac{(N-1)^2}{N}\lambda \doteq (N-1)\lambda. \tag{C–5}$$

D. CONTINUOUS LINE ARRAY

A continuous line array of length L may be considered the limit of the array of N point sources as the number N becomes very large and the spacing d becomes very small. When d is replaced by $L/(N-1)$ in the expression for ϕ,

$$\phi = \frac{kd}{2}\sin\theta = \frac{kL}{2(N-1)}\sin\theta,$$

Eq. C–3 becomes

$$D(\theta) = \sin\left\{\frac{N}{N-1}\frac{kL}{2}\sin\theta\right\} \Big/ N\sin\left\{\frac{kL}{2(N-1)}\sin\theta\right\}.$$

In the limit as $N \to \infty$, the outer sine function in the denominator is replaced by its argument, and the result is

$$D(\theta) = \frac{\sin(\frac{1}{2}kL\sin\theta)}{\frac{1}{2}kL\sin\theta}. \tag{D–1}$$

The directivity is simply a $\sin x/x$ function, shown in Fig. 15.9.

The same result may also be obtained by integrating over a line source the same way we did in the case of the circular piston. In fact, Eq. D–1 is the directivity of a rectangular piston of length L and infinitesimal width dW (Problem 13–8). If the width W is finite (Fig. 15.10), another $(\sin x)/x$ pattern is obtained for the directivity in the x, z plane (see sketch). The complete directivity for the rectangular piston is

$$D(\theta, \gamma) = \frac{\sin(\frac{1}{2}kL\sin\theta)}{\frac{1}{2}kL\sin\theta}\,\frac{\sin(\frac{1}{2}kW\sin\gamma)}{\frac{1}{2}kW\sin\gamma}. \tag{D–2}$$

The sidelobe suppression for a continuous line array (or a rectangular piston) is a little over 13 dB.

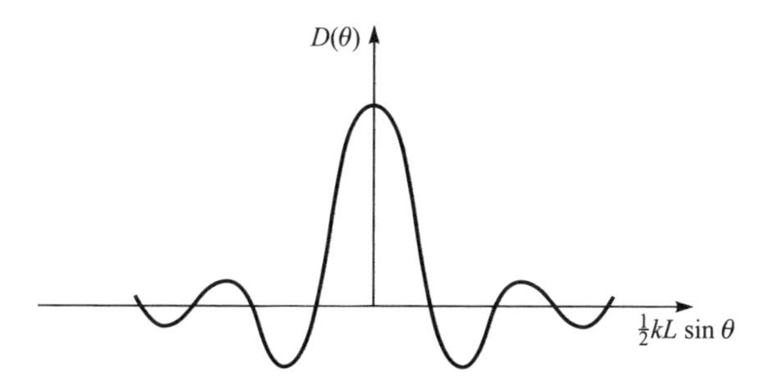

Figure 15.9 Directivity function of a continuous line source.

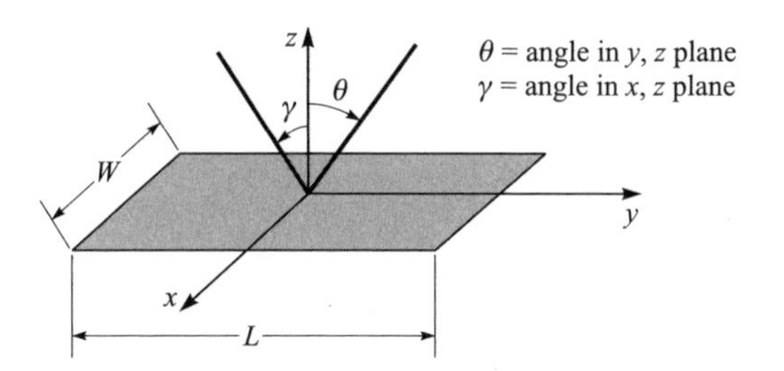

Figure 15.10 Rectangular piston.

E. ARRAY OF DIRECTIONAL SOURCES: PRODUCT THEOREM

If the elements in the array are directive, each with the same directivity D_e (the subscript e denotes element), the only change in the derivation in Sec. C is that p_0 is no longer given by Eq. C–1 but by

$$p_0 = \frac{A}{r} D_e(\theta) e^{j(\omega t - kr)}.$$

When this replacement for p_0 is made in Eq. C–2, the new version of Eq. C–3 is the overall directivity:

$$D(\theta) = D_e(\theta) D_a(\phi), \tag{E–1}$$

where $D_a = (\sin N\phi)/(N \sin \phi)$ is appropriately called the array directivity. In other words the overall directivity is simply the product of the element directivity and the array directivity.

REFERENCES

1. V. M. Albers, *Underwater Acoustic Handbook II* (Pennsylvania State University Press, University Park, PA, 1965), pp. 23–24, 27, 180–205.
2. Y. L. Luke, *Integrals of Bessel Functions* (McGraw-Hill Book Co., New York, 1963), p. 299.
3. D. G. Tucker and B. K. Gazey, *Applied Underwater Acoustics* (Pergamon Press, New York, 1966), Chap. 6.

PROBLEMS

15–1. Measurements made in a free field show that the sound produced by a whistle is omnidirectional and that the SPL achieved at 1 m is 111 dB. If the whistle is placed in a two-dimensional corner (intersection of two planes) of a large rectangular room having hard surfaces, what SPL (again at 1 m) can be expected? What SPL is expected if the whistle is placed in a three-dimensional corner (intersection of three planes)? For both cases state your assumptions.

15–2. Find the intensity directivity factor D_i and the directivity index DI for a ring piston. You should find the following integral useful (Ref. 2):

$$\int_0^{\pi/2} J_\mu(z\sin t)J_\nu(z\sin t)\sin t\tan^{2\mu} t\,dt = 2^{-\mu}\left(\tfrac{1}{2}\right)_{-\mu}(2z)^{-1}\int_0^{2z} t^\mu J_\nu(t)\,dt,$$

where the notation $(a)_k$ means $\Gamma(a+k)/\Gamma(a)$.

(a) Find an expression for D_i (or $1/D_i$) in terms of $\int_0^{2ka} J_0(t)\,dt$.

(b) Although the integral over J_0 may be evaluated in closed form to give a final answer in terms of Struve functions, it is suggested that for the purposes of this problem you simply express J_0 by its Maclaurin series expansion and then integrate term by term. Give the expression for $1/D_i$ in terms of the integrated series.

(c) Using the result from part (b), find D_i and DI for the following cases:

(i) $ka \ll 0$

(ii) $ka = 1$

(iii) $ka = 2$

15–3. Consider your two ears as an array composed of two point receivers. The directivity of the array is $D(\theta)$.

(a) Measure (nondestructively) the distance d between your own ears.

(b) At low enough frequency the array behaves as an omnidirectional receiver. Find the highest frequency at which, over the entire range $0 \le \theta \le 90°$, $20\log_{10}|D(\theta)|$ departs from 0 dB by no more than 3 dB. At what angle (at this frequency) does $D(\theta)$ depart most from unity?

(c) Find the frequency and angle at which the first null sets in.

(d) Find the frequency at which the first pair of sidelobes becomes as prominent as the original major lobes.

(Modeling the two ears as a pair of simple point receivers is of course a gross oversimplification. For one thing, the diffracting effect of the head is not considered. However, the calculations in this problem should provide a rough estimate of one of the physical attributes that contribute to binaural hearing.)

15–4. A three-transducer array is shown in the sketch. The two outer transducers are in phase and of relative source strength 1. The central transducer is 180° out of phase with its two neighbors and is of relative source strength 2. Let the farfield signal due to the central transducer by itself be denoted

$$p = \frac{2A}{r} e^{j(\omega t - kr)}.$$

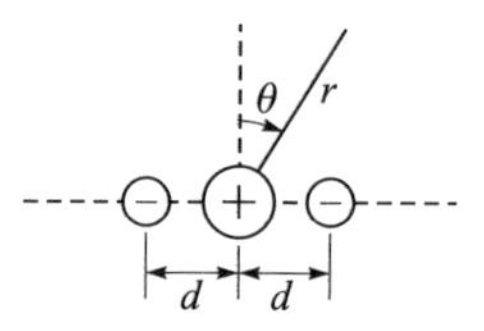

(a) Find the expression for the farfield pressure of the array. Identify the directivity function $D(\theta)$.

(b) The spacing of the transducers is $d = 0.075$ m, the medium is seawater ($c_0 = 1500$ m/s), and the frequency is 10 kHz. However, because the transducers behave somewhat nonlinearly, each also emits a small amount of second-harmonic distortion (20 kHz), still in the same phase and amplitude relation originally described. Give $D(\theta)$ and sketch $|D(\theta)|$ (polar plot) for both the 10-kHz main signal and the second-harmonic component.

15–5. Transmission of a narrow beam of 10-kHz sound in air is desired. The half-power beamwidth is required to be $2\theta_{HP} = 10°$. Consider the following different ways of producing the required beam:

(a) Two point sources

(b) Five point sources

(c) Continuous line array

(d) Circular piston

After making calculations, in a summary table compare the four choices as to (i) size (overall length, or, in the case of the piston, the diameter) and (ii) prominence of the sidelobes. As a measure of sidelobe prominence, use the sidelobe suppression figure in decibels, i.e., $-20 \log_{10} |D(\theta_s)|$, where θ_s is the angle at which the most prominent sidelobe has a maximum. In particular, does the array have secondary major lobes?

15–6. An array is composed of two dipoles with their axes aligned along the element line, as shown in the sketch. Since the pressure due to this array vanishes at $\theta = 0$, the conventional definition of directivity function $D(\theta) = P(r, \theta)/P(r, 0)$ cannot be used. However, the expression for the farfield pressure signal may still be put in the form

$$p(r, \theta, t) = \frac{A}{r} G(\theta) e^{j(\omega t - kr)}.$$

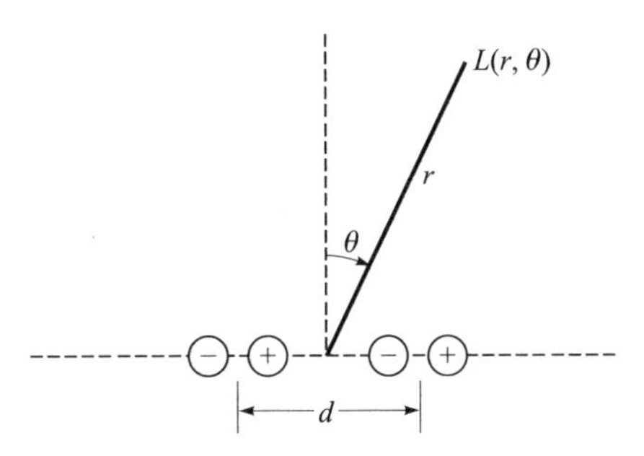

(a) Find $G(\theta)$.

(b) Determine the angles at which $G(\theta)$ has extrema (without further information one cannot determine whether a given extremum defines a major or a minor lobe).

(c) Develop expressions for the directivity factor

$$D(\theta) = G(\theta)/|G(\theta_{\text{max}})|$$

for the following two cases:

(i) $d = \lambda/2$

(ii) $d = \lambda$

In each case locate the major and minor lobes and give a rough sketch of the directivity pattern.

15–7. Let $D_1(\theta)$ stand for the directivity of an ordinary circular piston. The sidelobe suppression of the first sidelobe is 17.6 dB. As a transducer designer, you are faced with the problem of decreasing the level of this sidelobe, i.e., increasing the sidelobe suppression. One idea that occurs to you is to use an array of two circular pistons and arrange the array so that $D_2(\theta)$, the combined directivity of the two sources in the plane containing the two sources, has a null at the angle of the first piston sidelobe. What distance d between piston centers, in terms of piston radius a, should be used? Remember that to be practical, the pistons cannot overlap. Would problems with secondary major lobes arise?

APPENDIX A

ELASTIC CONSTANTS, VELOCITY OF SOUND, AND CHARACTERISTIC IMPEDANCE

Elastic Constants, Velocity of Sound, and Characteristic Impedance

Solid	Density, ρ_0 (kg/m^3)	Young's Modulus, Y (GPa)	Shear Modulus, G (GPa)	Bulk Modulus, B (GPa)	Poisson's Ratio, σ	Velocity (m/s)		Characteristic Impedance, $\rho_0 c_0$ (MKS Mrayls)	
						Bar $\sqrt{\frac{Y}{\rho_0}}$	Bulk $\sqrt{\frac{Y}{\rho_0}\frac{1-\sigma}{(1+\sigma)(1-2\sigma)}}$	Bar	Bulk
Aluminum	2,700	71	24	75	0.33	5,150	6,300	13.9	17.0
Brass	8,500	104	38	136	0.37	3,500	4,700	29.8	40.0
Copper	8,900	122	44	160	0.35	3,700	5,000	33.0	44.5
Iron (cast)	7,700	105	44	86	0.28	3,700	4,350	28.5	33.5
Lead	11,300	16.5	5.5	42	0.44	1,200	2,050	13.6	23.2
Nickel	8,800	210	80	190	0.31	4,900	5,850	43.0	51.5
Silver	10,500	78	28	105	0.37	2,700	3,700	28.4	39.0
Steel	7,700	195	83	170	0.28	5,050	6,100	39.0	47.0
Glass (Pyrex)	2,300	62	25	39	0.24	5,200	5,600	12.0	12.9
Quartz (X-cut)	2,650	79	39	33	0.33	5,400	5,750	14.5	15.3
Lucite	1,200	4	1.4	6.5	0.4	1,800	2,650	2.15	3.2
Concrete	2,600	—	—	—	—	—	3,100	—	8.0
Ice	920	—	—	—	—	—	3,200	—	2.95
Cork	240	—	—	—	—	—	500	—	0.12
Oak	720	—	—	—	—	—	4,000	—	2.9
Pine	450	—	—	—	—	—	3,500	—	1.57
Rubber (hard)	1,100	23	1	5	0.4	1,450	2,400	1.6	2.64
Rubber (soft)	950	0.005	—	1	0.5	70	1,050	0.065	1.0
Rubber (rho-c)	1,000	—	—	2.4	—	—	1,550	—	1.55

Liquids

Liquid	Temperature, t (°C)	Density, ρ_0 (kg/m^3)	Bulk Modulus, B_T (GPa)	Ratio of Specific Heats, γ	Velocity c_0 (m/s)	Characteristic Impedance, $\rho_0 c_0$ (MKS Grayls)	Coefficient of Viscosity, η (N–s/m)
Water (fresh)	20	998	2.18	1.004	1,481	1.48	0.001
Water (sea)	13	1,026	2.28	1.01	1,500	1.54	0.001
Alcohol (ethyl)	20	790	—	—	1,150	0.91	0.0012
Caster oil	20	950	—	—	1,540	1.45	0.96
Mercury	20	13,600	25.3	1.13	1,450	19.7	0.0016
Turpentine	20	870	1.07	1.27	1,250	1.11	0.0015
Glycerin	20	1,260	—	—	1,980	2.5	1.2

Gases (pressure 1.013 × 10^{5}N/m^2)

Gas	Temperature, t (°C)	Density, ρ_0 (kg/m^3)	Ratio of Specific Heats, γ	Velocity, c_0 (m/s)	Characteristic Impedance, $\rho_0 c_0$ (MKS rayls)	Coefficient of Viscosity, η (N-s/m^2)
Air	0	1.293	1.402	331.6	428	0.000017
Air	20	1.21	1.402	343	415	0.0000181
Oxygen	0	1.43	1.40	317.2	453	0.0002
CO_2 (low freq.)	0	1.98	1.304	258	512	0.0000145
CO_2 (high freq.)	0	1.98	1.40	268.6	532	0.0000145
Hydrogen	0	0.09	1.41	1269.5	114	0.0000088
Steam	100	0.6	1.324	404.8	242	0.000013

Source: Kinsler and Frey, *Fundamentals of Acoustics*, 2nd ed., pp. 502–503.

APPENDIX B

ABSORPTION FORMULAS FOR THE ATMOSPHERE AND OCEAN

1. ATMOSPHERE

Absorption of sound in the atmosphere is the subject of two standards, one by the American National Standards Institute (Ref. 1) and one by the International Standards Organization (Ref. 6). The following formulas, taken from Ref. 2, are (with minor variation) used in both standards.

The absorption coefficient for the atmosphere has three terms, one for nitrogen relaxation (subscript N), one for oxygen relaxation (subscript O), and one for all other processes (principally viscosity and thermal conduction):

$$\alpha = \frac{B_1 f_{r,\mathrm{N}} f^2}{f^2 + f_{r,\mathrm{N}}^2} + \frac{B_2 f_{r,\mathrm{O}} f^2}{f^2 + f_{r,\mathrm{O}}^2} + B_3 \frac{p_s}{p_{s0}} f^2 \qquad \frac{\text{nepers}}{\text{m}}, \tag{B–1}$$

where f is frequency in hertz, $f_{r,\mathrm{N}}$ and $f_{r,\mathrm{O}}$ are the nitrogen and oxygen relaxation frequencies, respectively, p_s is local atmospheric pressure,[1] p_s0 is the reference atmospheric pressure (1 atm), and B_1, B_2, and B_3 are functions of temperature. In particular, the expressions for the relaxation frequencies are

$$f_{r,\mathrm{N}} = \frac{p_s}{p_{s0}} \left(\frac{T_0}{T}\right)^{1/2} (9 + 280 h e^{-4.17[(T_0/T)^{1/3} - 1]}), \tag{B–2a}$$

[1]Lowercase p here means total pressure, not acoustic pressure.

$$f_{r,\mathrm{O}} = \frac{p_s}{p_{s0}}\left(24 + 4.04 \times 10^4 h \frac{0.02 + h}{0.391 + h}\right), \tag{B–2b}$$

where T is absolute temperature of the atmosphere in kelvins, $T_0 = 293.15$ K is the reference value of T (20°C), and h is the absolute humidity (molar concentration of water vapor) in percent, which is related to the relative humidity h_r by

$$h = h_r \frac{p_{\mathrm{sat}}/p_{s0}}{p_s/p_{s0}} = p_{s0}\left(\frac{h_r}{p_s}\right)\left(\frac{p_{\mathrm{sat}}}{p_{s0}}\right)\%, \tag{B–3}$$

The accepted expression for the saturation vapor pressure p_{sat} is given in the ANSI standard (Refs. 1 and 2) but is very complicated. Very nearly as accurate is the following simpler alternative (Ref. 2), which is given in the ISO standard (Ref. 6):

$$\log_{10}[p_{\mathrm{sat}}/p_{s0}] = -6.8346(T_{01}/T)^{1.261} + 4.6151. \tag{B–4}$$

The coefficients B_i, which have units of nepers/meter · hertz, are

$$B_1 = 1.068(T/T_0)^{-5/2} e^{-3352/T}, \tag{B–5a}$$

$$B_2 = 0.01275(T/T_0)^{-5/2} e^{-2239.1/T}, \tag{B–5a}$$

$$B_3 = 1.84 \times 10^{-11}\sqrt{T/T_0}(p_{s0}/p_s). \tag{B–5c}$$

If pressure is expressed in atmospheres, the reference pressure has value unity, and the formulas take on somewhat simpler form.

In Ref. 2 the absorption coefficient, frequency, and relative humidity are scaled by atmospheric pressure. In terms of the scaled frequencies $F = f/p_s$, $F_{r,\mathrm{N}} = f_{r,\mathrm{N}}/p_s$, and $F_{r,\mathrm{O}} = f_{r,\mathrm{O}}/p_s$, Eq. B–1 becomes

$$\frac{\alpha}{p_s} = f^2\left\{\left(\frac{T}{T_0}\right)^{-5/2}\left[\frac{0.1068 e^{-3352/T}}{F_{r,\mathrm{N}} + F^2/F_{r,\mathrm{N}}} + \frac{0.01275 e^{-2239.1/T}}{F_{r,\mathrm{O}} + F^2/F_{r,\mathrm{O}}}\right] + 1.84 \times 10^{-11} p_{s0}\left(\frac{T}{T_0}\right)^{1/2}\right\} \frac{\text{nepers}}{\text{m} \cdot \text{atm}}. \tag{B–6}$$

This formula is particularly useful for computing α at higher altitudes, where p_s may differ appreciably from p_{s0}.

Equation B–6 also makes it clear that, for a fixed temperature, a single graph of absorption vs. frequency suffices for all atmospheric pressures. Curves of $\bar{\alpha}/p_s$ (dB/100 m · atm) as a function of f/p_s for a temperature of 20°C are shown in Fig. B.1 for several values of relative humidity (actually

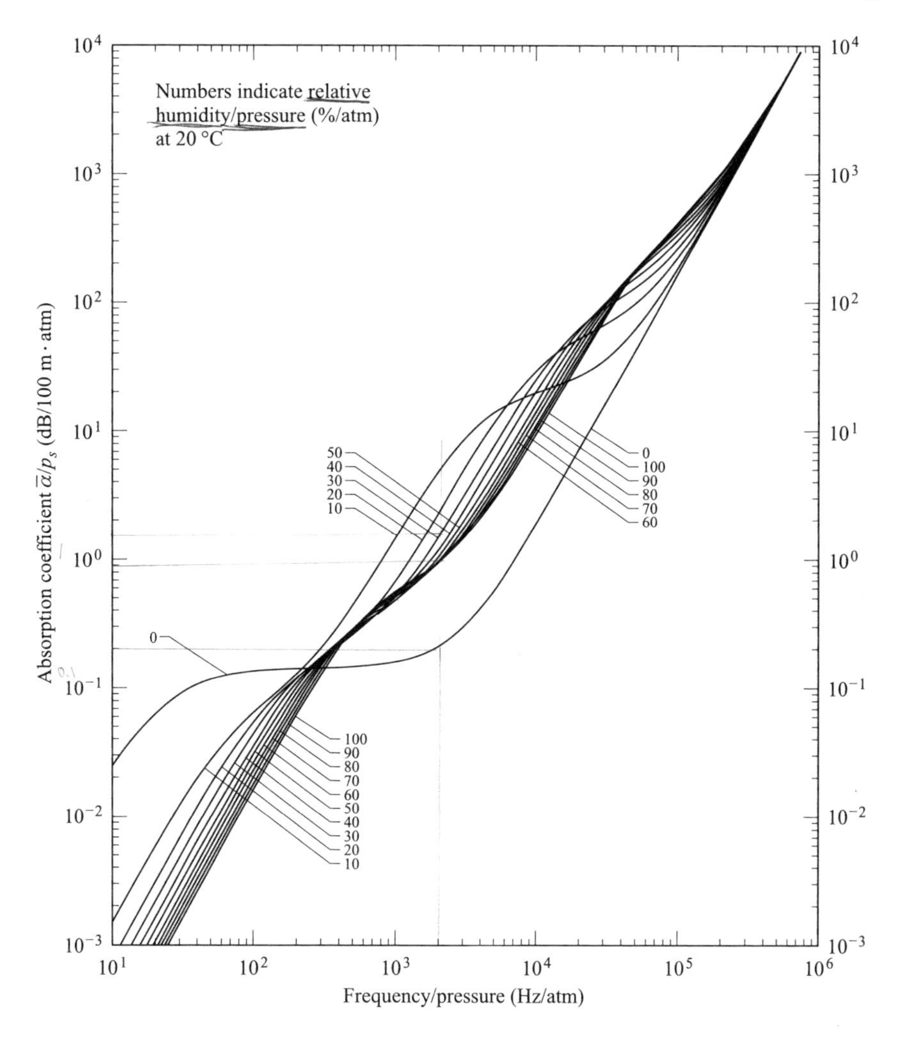

Figure B.1 Atmospheric absorption coefficient, SI units, at 20°C. (From Ref. 2.)

h_r/p_s). Curves in terms of English units (dB/kft · atm) are given in Fig. B.2. The relation between α and $\bar{\alpha}$ is given by Eq. 9A–7.

The figures may be read to an accuracy of about two significant digits. As an example, find the absorption coefficient for an 8-kHz wave under the following conditions: atmospheric pressure 0.8 atm, relative humidity 40%, and temperature 20°C. To find $\bar{\alpha}$ from Fig. B.1, use the values $f/p_s = 8000/0.8 = 10{,}000$ Hz/atm and $h_r/p_s = 40/0.8 = 50\%$ atm, for which reading the graph yields $\bar{\alpha}/p_s = 16$ dB/100 m · atm, or $\bar{\alpha} = 13$ dB/100 m. The value by direct calculation, by either Eq. B–6 or Eq. B–1, is $\bar{\alpha} = 12.7$ dB/100 m. A byproduct of direct calculation is information about

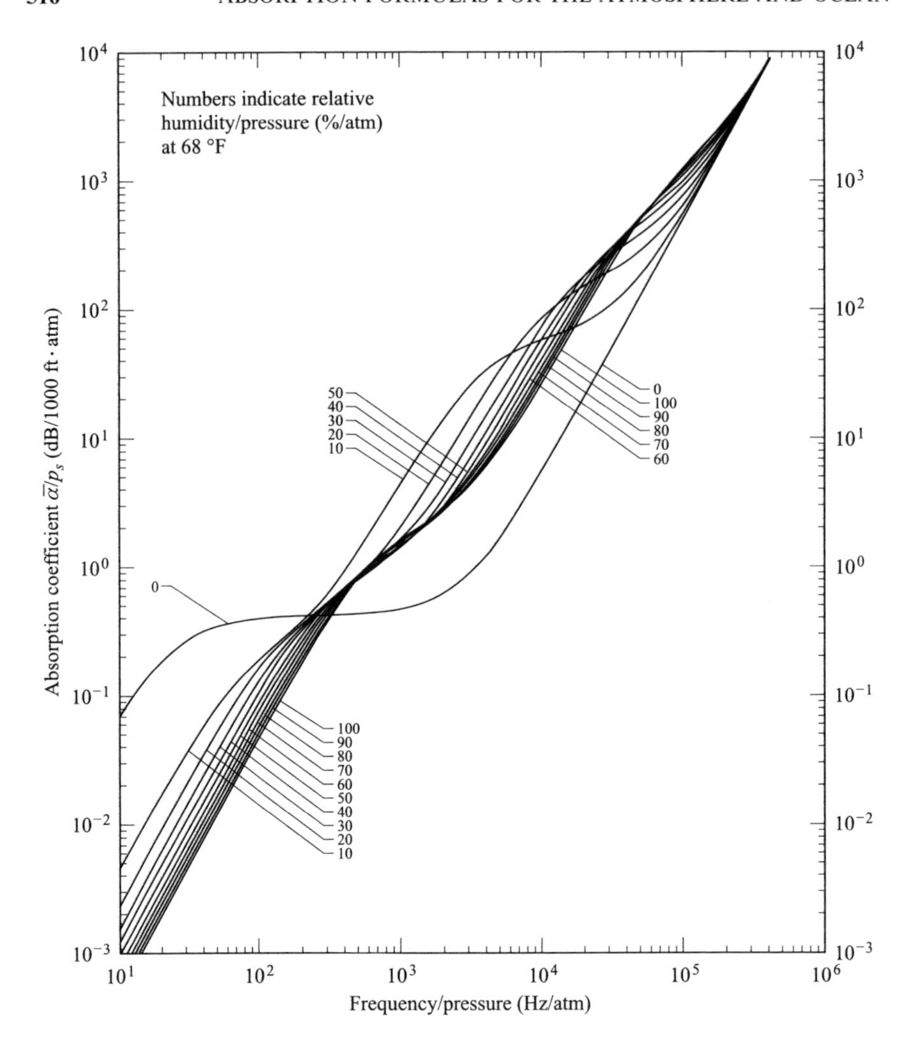

Figure B.2 Atmospheric absorption coefficient, English units, at 20°C. (From Ref. 2.)

individual contributions to the absorption: of the 12.7 dB/100 m, about 88% is due to oxygen relaxation, 2% to nitrogen relaxation, and 10% to thermoviscous effects.

2. OCEAN

Although no standard has been written for sound absorption in the ocean, the topic has been thoroughly investigated. The formulas given below are from Ref. 5.

The formula for absorption in the ocean, like that for the atmosphere, has three main terms, two for relaxation processes and one associated with pure water (principally viscosity). Boric acid is responsible for the low frequency relaxation process, magnesium sulfate for the one at high frequency. Temperature, salinity, and depth play roles analogous to those played by temperature, relative humidity, and atmospheric pressure for the atmosphere. The expression is

$$\bar{\alpha} = \frac{A_1 P_1 f_1 f^2}{f^2 + f_1^2} + \frac{A_2 P_2 f_2 f^2}{f^2 + f_2^2} + A_3 P_3 f^2 \qquad \text{dB/km}. \tag{B–7}$$

The first term is associated with boric acid, the second with magnesium sulfate, and the third with pure water. All frequencies are in kilohertz, the P_i factors are depth functions, and the proportionality factors A_i are functions of various properties, such as pH, temperature, salinity, and sound speed.

Expressions for the various quantities in Eq. B–7 are as follows. First, the relaxation frequencies are

$$f_1 = 2.8(S/35)^{0.5} 10^{[4-1245/(t+273)]} \text{ kHz}, \tag{B–8a}$$

$$f_2 = \frac{8.17 \times 10^{[8-1990/(t+273)]}}{1 + 0.0018(S - 35)} \text{ kHz}, \tag{B–8b}$$

where t is temperature in degrees Celsius, S is salinity in parts per thousand (‰). The (dimensionless) depth functions are

$$P_1 = 1, \tag{B–9a}$$

$$P_2 = 1 - 1.37 \times 10^{-4} D + 6.2 \times 10^{-9} D^2, \tag{B–9b}$$

$$P_3 = 1 - 3.83 \times 10^{-5} D + 4.9 \times 10^{-10} D^2, \tag{B–9c}$$

where D is depth in meters. The proportionality factors A_i have units decibels/kilometer · kilohertz and are given by

$$A_1 = \frac{8.86}{c} \times 10^{(0.78p\text{H}-5)}, \tag{B–10a}$$

$$A_2 = \frac{21.44S}{c}(1 + 0.025t), \tag{B–10b}$$

$$\text{if } t \le 20°\text{C}: \quad A_3 = 4.937 \times 10^{-4} - 2.59 \times 10^{-5} t + 9.11 \times 10^{-7} t^2 - 1.50 \times 10^{-8} t^3, \tag{B–10c}$$

$$\text{if } t > 20°\text{C}: \quad A_3 = 3.964 \times 10^{-4} - 1.146 \times 10^{-5} t + 1.45 \times 10^{-7} t^2 - 6.5 \times 10^{-10} t^3, \tag{B–10d}$$

where the sound speed c is calculated in meters per second from

$$c = 1412 + 3.21t + 1.19S + 0.0167D. \tag{B–11}$$

Here the symbol t denotes temperature in °C; the symbol T is used in Ref. 5.

REFERENCES

1. ANSI S1.26-1995, "American National Standard method for calculation of the absorption of sound by the atmosphere" (Acoustical Society of America, New York, 1995).
2. H. E. Bass, L. C. Sutherland, A. J. Zuckerwar, D. T. Blackstock, and D. M. Hester, "Atmospheric absorption in air: Further developments," *J. Acoust. Soc. Am.* **97**, 680–683 (1995); "Erratum," *J. Acoust. Soc. Am.* **99**, 1259 (1996).
3. L. B. Evans, H. E. Bass, and L. C. Sutherland, "Atmospheric absorption of sound: Theoretical predictions," *J. Acoust. Soc. Am.* **51**, 1565–1575 (1972).
4. R. E. François and G. R. Garrison, "Sound absorption based on ocean measurements: Part I: Pure water and magnesium sulfate contributions," *J. Acoust. Soc. Am.* **72**, 896–907 (1982).
5. R. E. François and G. R. Garrison, "Sound absorption based on ocean measurements: Part II. Boric acid contribution and equation for total absorption," *J. Acoust. Soc. Am.* **72**, 1879–1890 (1982).
6. ISO 9613-1:1993, "Acoustics—Attenuation of sound during propagation outdoors—Part 1: Calculation of the absorption of sound by the atmosphere" (International Organization for Standardization, Geneva, Switzerland, 1993).

APPENDIX C

ABSORPTION DUE TO TUBE WALL BOUNDARY-LAYER EFFECTS

Absorption of low frequency sound in tubes and ducts is largely due to viscous shear friction and heat transfer at the walls of the tube or duct, not to the mainstream thermoviscous effects described in Sec. 9B. Shear friction occurs in the acoustical viscous boundary layer, a very thin region where the axial particle velocity rapidly decreases from its mainstream value to zero at the wall. Heat transfer takes place in a similarly thin layer that marks the transition from adiabatic conditions in the mainstream to isothermal conditions at the wall.[1] In the analysis presented here, only the viscous boundary layer is treated in detail. After the effect of viscous shear is analyzed, that due to heat transfer to (from) the wall is simply stated. The two effects superpose in a single dissipation term from which total boundary-layer absorption and dispersion may be calculated.

The traditional calculation of absorption due to tube wall effects is rigorous but quite complicated. Since the acoustic field is actually not constant over the tube cross section, the plane-wave model is not applicable. The true two-dimensionality (or three-dimensionality) of the sound field must be taken into account. The rigorous approach was used by Kirchhoff (Ref. 2) to investigate thermoviscous effects on sound propagation in a circular tube. After obtaining expressions that constitute an exact (but extremely complicated) dispersion relation, he found approximate formulas for the absorption α and phase velocity c^{ph}. Thorough reexamination of the problem by the rigorous method led Weston (Ref. 4) to a very useful classification of the results, which depend markedly on

[1] For conditions at the wall to be isothermal, the wall material must have high thermal conductivity and high heat capacity so that the wall acts as an infinite heat sink (or source). Most common tube walls qualify.

the relation of the dimensionless boundary thickness $k\delta^{\mathrm{BL}}$ to the dimensionless tube radius ka, where $k = \omega/c_0$ is the wave number, δ^{BL} is the dimensional boundary-layer thickness, and a is the tube radius. Weston's *wide tube* category, defined by Eq. 9D–6, covers a very broad range of gases, tube sizes, and frequencies. Kirchhoff's expressions for α and c^{ph} apply to the wide tube. In physical terms a wide tube is one for which the radius is much, much larger than the boundary-layer thickness but not so large that mainstream dissipation is important.[2]

The wide tube is also assumed here, but a much simpler analysis is given. The wave motion in a wide tube remains planar in the mainstream, which occupies almost the entire cross section of the tube; the viscous friction and heat transfer are concentrated in two very thin overlapping regions (boundary layers) adjacent to the tube wall. This description suggests that a plane-wave (one-dimensional) model of the propagation may succeed if we account properly for the effects the two boundary layers pass on to the fluid in the mainstream. The first to develop this idea was Chester (Ref. 1). Viewing the viscous boundary layer as an agent that feeds mass into (or withdraws mass from) the mainstream, Chester developed a modified mass conservation (continuity) equation. In our approach the viscous boundary layer allows the mainstream to feel the drag force exerted by the wall; we therefore modify the momentum equation. Both approaches lead to the same expression for the absorption coefficient.

Our procedure is as follows. The boundary layer and mainstream are considered separately. First the boundary layer flow is found, in particular, the velocity profile across the layer. The drag shear force (drag) per unit wall area exerted by the wall on the fluid is then calculated. Next the mainstream is considered. A one-dimensional momentum equation, much like that derived in Sec. 1C.3.b, is developed, but including the drag force (found from the boundary-layer calculation) exerted on the fluid by the tube wall. The new momentum equation is combined with the usual continuity and state equations to produce a plane-wave equation that includes the effect of viscous wall drag. Finally, the modification to include the effect of the thermal boundary layer is stated.

1. VISCOUS BOUNDARY LAYER

For a wide tube, flow in the viscous boundary layer may be modeled as oscillatory motion of an incompressible fluid over a flat plate. Such a simple description is justified by the extreme thinness of the layer. First, to a fluid particle in the boundary layer, the major changes are across the layer, not parallel to it. The curvature of the bounding surface is thus not noticed; the

[2]As shown in Sec. 9D, the boundary-layer absorption coefficient is inversely proportional to tube radius.

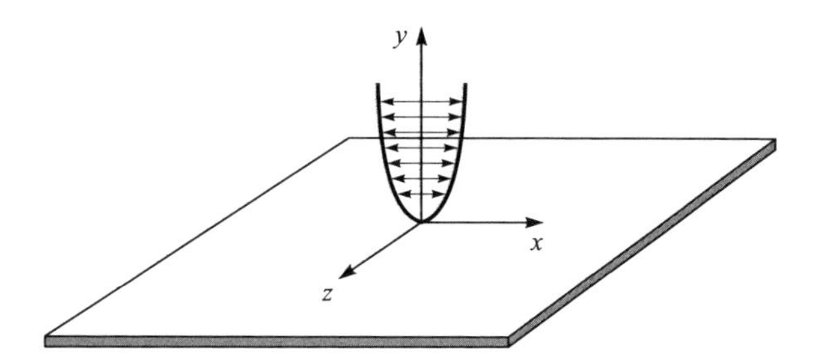

Figure C.1. Oscillatory flow over a flat plate.

surface seems flat. Second, because $\delta_{\text{visc}}^{\text{BL}} \ll \lambda$, where λ is the acoustic wavelength, sound appears to traverse the boundary layer almost instantaneously, i.e., the fluid in the layer appears incompressible.

The geometry for the flat plate model is shown in Fig. C.1. The linearized version of Eq. 2A–19, the momentum equation for a viscous fluid, is

$$\rho_0 \mathbf{u}_t + \nabla p = (\lambda + 2\mu)\nabla(\nabla \cdot \mathbf{u}) - \mu \nabla \times \nabla \times \mathbf{u}. \tag{C–1}$$

Because $D\rho/Dt = 0$ for an incompressible fluid, the continuity equation, Eq. 2A–5, reduces to $\nabla \cdot \mathbf{u} = 0$. The $(\lambda + 2\mu)$ term in Eq. C–1 therefore drops out, and $\nabla \times \nabla \times \mathbf{u}$ may be replaced by $-\nabla^2 \mathbf{u}$ (see Eq. 2A–20). The flow is assumed to be in the x-direction (parallel to the axis of the duct) and thus has no z-component. The two remaining components of Eq. C–1 are

$$\rho_0 u_t + p_x = \mu \nabla^2 u, \tag{C–2a}$$

$$\rho_0 v_t + p_y = \mu \nabla^2 v, \tag{C–2b}$$

where u and v are the x and y components of the particle velocity, respectively. Further simplifications occur because the flow (1) is in the x-direction only, i.e., $v = 0$, and (2) varies only in the y direction. Equations C–2 reduce to

$$\rho_0 u_t + p_x = \mu u_{yy}, \tag{C–3a}$$

$$p_y = 0. \tag{C–3b}$$

Equation C–3b shows that the pressure, though time-dependent, does not vary across the boundary layer. We therefore replace p with $p^{(m)}(t)$, where the superscript (m) denotes value in the mainstream, and rewrite Eq. C–3a as

$$\rho_0 u_t - \mu u_{yy} = -p_x^{(m)}(t). \tag{C–4}$$

The pressure-gradient term $p_x^{(m)}(t)$ therefore plays the role of a forcing function: it causes the fluid in the boundary layer to move.[3] Note that in this model the entire mass of fluid above the plate sloshes back and forth parallel to the plate. The boundary layer is so thin compared to λ that the flow does not recognize the very gradual (in terms of boundary-layer thicknesses) variation of $p^{(m)}$ with x.

a. Boundary-Layer Profile and Thickness for Time-Harmonic Flow

To prepare Eq. C–4 for solution, first rewrite it as

$$u_{yy} - \nu^{-1} u_t = \mu^{-1} p_x^{(m)}(t), \qquad \text{(C–5)}$$

where $\nu = \mu/\rho_0$ is the kinematic viscosity. If the forcing function is time harmonic,

$$p_x^{(m)}(t) = P_x^{(m)} e^{j\omega t},$$

where $P_x^{(m)}$ is the amplitude, the particle velocity must have a similar form, namely, $u = U(y)e^{j\omega t}$. Substitution in Eq. C–5 yields

$$U_{yy} - j\omega\nu^{-1} U = \mu^{-1} P_x^{(m)}, \qquad \text{(C–6)}$$

for which the solution that remains bounded as $y \to \infty$ and vanishes at $y = 0$ is

$$U = -\frac{P_x^{(m)}}{j\omega\rho_0}\left(1 - e^{-\sqrt{j\omega/\nu}\,y}\right),$$

where $\sqrt{j}$ means the "positive" root, $(1+j)/\sqrt{2}$. Expansion yields

$$u = -\frac{P_x^{(m)}}{j\omega\rho_0}\left[e^{j\omega t} - e^{-\sqrt{\omega/2\nu}\,y}\, e^{j(\omega t - \sqrt{\omega/2\nu}\,y)}\right]. \qquad \text{(C–7)}$$

The first term represents the oscillatory mainstream flow. The second term represents the shear wave set up as a reaction to cancel the mainstream velocity at the surface of the plate $y = 0$. Notice that $\sqrt{\omega/2\nu}$ represents the (very large) attenuation of the shear wave and $\sqrt{2\nu\omega}$ its phase velocity. Let the boundary-layer thickness for the oscillating flow be defined as the distance over which the shear-wave amplitude decays by $1/e$:

[3]Notice that the fluid acceleration is proportional to the negative of the pressure gradient. This is to be expected. The pressure must be higher to the left ($p_x^{(m)} < 0$) if the fluid is to be accelerated to the right.

$$\begin{aligned}\delta_{\text{visc}}^{BL} &= \sqrt{2\nu/\omega} \\ &= 2.19/\sqrt{f}\ \text{mm}, \quad \text{(for air at 20°C)}.\end{aligned} \tag{C–8}$$

For example, for air the boundary-layer thickness is 0.219 mm at 100 Hz, 0.0693 mm at 1 kHz, 0.0219 mm at 10 kHz, and so on. Except for capillary tubes and very low frequencies, therefore, the boundary layer is but a tiny fraction of the tube radius.

b. Drag Force on the Fluid (Arbitrary Time Dependence)

The tangential or drag force per unit area the plate exerts on the fluid is given by

$$f_{\text{drag}} = -\mu u_y|_{y=0}. \tag{C–9}$$

Following Lamb (Ref. 3), we could use Eq. (C–9) to compute the drag, but our result would then be restricted to time-harmonic motion. A more general approach, suggested by Chester's analysis (Ref. 1), is to solve Eq. C–5 for an arbitrary forcing function $p_x^{(m)}(t)$. Laplace transforms are used. Let an overbar indicate the Laplace transform of a quantity, for example,

$$\overline{u}(y, s) = L[u(y, t)] \equiv \int_0^\infty u(y, t)e^{-st}dt.$$

Similarly, $\overline{p}_x^{(m)}(s)$ stands for the Laplace transform of $p_x^{(m)}(t)$. Given a fluid at rest at time $t = 0$, i.e., $u(y, 0+) = 0$, the Laplace transform of Eq. C–5 is

$$\overline{u}_{yy} - \frac{s}{\nu}\overline{u} = \frac{1}{\mu}\overline{p}_x^{(m)}(s). \tag{C–10}$$

The bounded solution that vanishes at $y = 0$ is

$$\overline{u}(y, s) = -\frac{\overline{p}_x^{(m)}(s)}{\rho_0 s}\left(1 - e^{-\sqrt{s/\nu y}}\right). \tag{C–11}$$

Substitution in Eq. C–9 yields the drag per unit area (in the transform domain):

$$\overline{f}_{\text{drag}} = \sqrt{\nu/s}.$$

The inverse transform may be found by convolution [note that $L^{-1}(1/\sqrt{s}) = 1/\sqrt{\pi t}$]:

$$f_{\text{drag}} = \sqrt{\frac{\nu}{\pi}} \int_0^t \frac{p_x^{(m)}(x, t - \tau)}{\sqrt{\tau}} d\tau, \tag{C–12}$$

where we have finally admitted that the mainstream pressure varies with x as well as t. Now that the drag force is known, it may be included among the forces in the momentum equation for the mainstream, which is now derived.

2. QUASI-PLANE-WAVE EQUATION

Reconsider the derivation of the momentum equation given in Sec. 1C.3.b. The only modification needed here is to include the effect of the viscous drag exerted by the wall. Refer to Fig. 1.21. However, to emphasize the fact that the cross section need not be circular, let its shape be arbitrary (though constant) and have area S and perimeter C.[4] Acting on the lateral surface area $C\Delta x$ of the control volume (CV), is the drag force $C\Delta x f_{\mathrm{drag}}$. The momentum equation for the CV is therefore

$$\frac{\partial}{\partial t}[(\rho u)^{(m)}S\Delta x] = (\rho u^2)^{(m)}S|_x - (\rho u^2)^{(m)}S|_{x+\Delta x} + p^{(m)}S|_x - p^{(m)}S|_{x+\Delta x} + C\Delta x f_{\mathrm{drag}},$$

where the force Sp_0 due to the ambient pressure p_0 has been omitted because it is the same at both ends of the CV. From now on the superscript (m) is omitted, but all field variables are understood to be those in the mainstream. Divide by S and take the limit as $\Delta x \to 0$. The result is

$$\rho\frac{Du}{Dt} + p_x = \frac{4}{\mathrm{HD}}\sqrt{\frac{\nu}{\pi}}\int_0^t \frac{p_x(x, t-\tau)}{\sqrt{\tau}}\,d\tau, \tag{C–13}$$

where $\mathrm{HD} = 4S/C$ is the hydraulic diameter (for a circular tube of radius a, $\mathrm{HD} = 2a$). In this way we preserve the one-dimensional nature of the momentum balance while still taking account of the wall drag.

When Eq. C–13 is linearized and combined with the linear versions of the continuity and state equations, the following wave equation is obtained:

$$u_{xx} - \frac{1}{c_0^2}u_{tt} = \frac{4}{\mathrm{HD}}\sqrt{\frac{\nu}{\pi}}\left[\int_0^t u_{xx}(x, t-\tau)\frac{d\tau}{\sqrt{\tau}} - \frac{p_x(x,0)}{\rho_0 c_0^2\sqrt{t}}\right]. \tag{C–14}$$

In steady state ($t \to \infty$) the equation becomes

$$u_{xx} - \frac{1}{c_0^2}u_{tt} = \frac{4}{\mathrm{HD}}\sqrt{\frac{\nu}{\pi}}\int_0^\infty u_{xx}(x, t-\tau)\frac{d\tau}{\sqrt{\tau}}. \tag{C–15}$$

[4]Strictly speaking, S and C should be values for just the fluid in the mainstream, i.e., not include the fluid in the boundary layer. For a circular cross section, for example, S should be $\pi(a - y_{\mathrm{visc}}^{\mathrm{BL}})^2$, not πa^2. For a wide tube, however, the error in using $S = \pi a^2$ is negligible.

3. EFFECT OF THE THERMAL BOUNDARY LAYER

A similar analysis may be carried out to account for heat transfer to and from the tube wall. A temperature profile in the thermal boundary layer is obtained under the condition that the temperature remain constant at the wall surface. The profile is used to calculate the wall heat transfer, which in turn is included in the derivation of the one-dimensional energy equation for the mainstream (see Sec. 9B.2 and Fig. 9.5). Linearization and combination with the linearized versions of the continuity equation and the momentum equation derived above (Eq. C–13) yields the quasi-plane-wave equation that includes both viscous and thermal wall effects. The effect of this procedure is to replace $\sqrt{\nu}$ in Eq. C–15 as follows:

$$\sqrt{\nu} \quad \rightarrow \quad \sqrt{\nu}\left(1 + \frac{\gamma - 1}{\sqrt{\mathrm{Pr}}}\right).$$

Equation 9D–3 is then obtained.

REFERENCES

1. W. Chester, "Resonant oscillations in closed tubes," *J. Fluid Mech.* **18**, 44–64 (1964).
2. G. Kirchhoff, "Über den Einfluss der Wärmeleitung in einem Gase auf die Schallbewegung," *Ann. Phys. Chem.* **134**, 177–193 (1868). English translation: "On the influence of thermal conduction in a gas on sound propagation," in *Physical Acoustics*, R. B. Lindsay, ed., Vol. **4** in Benchmark Papers in Acoustics series (Dowden, Hutchinson, and Ross, Stroudsberg, Pa., 1974), pp. 7–19.
3. H. Lamb, *Hydrodynamics*, 6th ed. (Dover Publications, Inc., New York, 1945), Art. 328.
4. D. E. Weston, "The theory of the propagation of plane sound waves in tubes," *Proc. Phys. Soc. (Lond.)* **B66**, 695–709 (1953).

APPENDIX D

SOLUTION OF LEGENDRE'S EQUATION BY POWER SERIES

Legendre's equation is Eq. 10B–8, repeated here for convenience,

$$(1 - z^2)P_{zz} - 2zPz + CP = 0, \tag{D–1}$$

where C is the separation constant identified in Eq. 10B–6.

The domain is $-1 \leq z \leq 1$ (corresponding to all points on the surface of a unit sphere; the relation $z = \cos\theta$ gives the projection on the z axis of a given point on the sphere). A solution bounded everywhere in the domain is sought.

A power series method is used here. We assume a solution in the form

$$P = a_0 + a_1 z + a_2 z^2 + a_3 z^2 + \cdots \tag{D–2}$$

and substitute it in Eq. D–1,

$$(1 - z^2)[2a_2 + 6a_3 z + 12a_4 z^2 + \cdots] - 2z[a_1 + 2a_2 z + 3a_3 z^2 + \cdots] \\ + C[a_0 + a_1 z + a_2 z^2 + \cdots] = 0.$$

Like-order terms are now gathered,

$$[2a_2 + Ca_0] + [6a_3 - (2 - C)a_1]z + [12a_4 - (6 - C)a_2]z^2 \\ +[20a_5 - (12 - C)a_3]z^3 + \cdots = 0.$$

Since this series must be zero for arbitrary values of z, each of the coefficients must vanish separately. This condition leads to the following set of relations among the a_i:

$$a_2 = -\frac{1}{2!}Ca_0,$$
$$a_3 = \frac{1}{3!}(2 - C)a_1 = \frac{1}{3!}(2 \cdot 1 - C)a_1,$$
$$a_4 = \frac{1}{12}(6 - C)a_2 = -\frac{1}{4!}(3 \cdot 2 - C)a_0,$$
$$a_5 = \frac{1}{20}(12 - C)a_3 = \frac{1}{5!}(2 \cdot 1 - C)(3 \cdot 4 - C)a_1,$$

etc. Substitution of these results in Eq. D–2 yields

$$P = a_0\left[1 - \frac{C}{2!}z^2 - \frac{C(6 - C)}{4!} - \frac{C(6 - c)(20 - C)}{6!}z^6 \cdots\right]$$
$$+a_1\left[z + \frac{2 - C}{3!}z^3 + \frac{(2 - C)(12 - C)}{5!}z^5 + \frac{(2 - C)(12 - C)(30 - C)}{7!}z^7 + \cdots\right], \tag{D–3}$$

i.e., an even and an odd series. This series satisfies Legendre's equation and, as expected for the general solution of a second-order differential equation, contains two arbitrary constants, a_0 and a_1.

As it stands, however, Eq. D–3 does not satisfy the requirement that the solution be bounded everywhere in the domain $-1 \le z \le 1$. In particular, the series diverges at the poles, $z = \pm 1$. Boundedness can, however, be achieved for special values of the separation constant C. We have two alternatives. First, let $a_1 = 0$ and choose C to be one of these values:

$$C = 0, 6, 20, 42, \ldots, n(n + 1), \quad \text{for } n \text{ even}.$$

Second, let $a_0 = 0$ and choose C to be one of these values:

$$C = 2, 12, 30, 56, \ldots, n(n + 1), \quad \text{for } n \text{ odd}.$$

In either case, one series is eliminated and the other is truncated, and P becomes a polynomial, in fact a family of polynomials, each identified by the index n. The polynomials so defined satisfy the boundedness requirement and are called the *Legendre polynomials*. Having chosen one of the two constants, a_0 or a_1, to be zero, we could pick the other one in any way desirable. However, it is traditional to pick it in such a way that $P_n = 1$ at $z = 1$. Under this normalization, the Legendre polynomials are as follows: $P_0 = 1$, $P_1 = z$, $P_2 = (3z^2 - 1)/2$, $P_3 = (5z^3 - 3z)/2$, etc. See Eqs. 10B–13.

APPENDIX E

DIRECTIVITY AND IMPEDANCE FUNCTIONS FOR A CIRCULAR PISTON

Directivity and Impedance Functions for a Circular Piston

x	Directivity Functions ($x = ka \sin\theta$)		Impedance Functions ($x = 2ka$)	
	Pressure	Intensity	Resistance	Reactance
	$\frac{2J_1(x)}{x}$	$\left[\frac{2J_1(x)}{x}\right]^2$	$R_1(x)$	$X_1(x)$
0.0	1.0000	1.0000	0.0000	0.0000
0.2	0.9950	0.9900	0.0050	0.0874
0.4	0.9802	0.9608	0.0198	0.1680
0.6	0.9557	0.9134	0.0443	0.2486
0.8	0.9221	0.8503	0.0779	0.3253
1.0	0.8801	0.7746	0.1199	0.3969
1.2	0.8305	0.6897	0.1695	0.4624
1.4	0.7743	0.5995	0.2257	0.5207
1.6	0.7124	0.5075	0.2876	0.5713
1.8	0.6461	0.4174	0.3539	0.6134
2.0	0.5767	0.3326	0.4233	0.6468
2.2	0.5054	0.2554	0.4946	0.6711
2.4	0.4335	0.1879	0.5665	0.6862
2.6	0.3622	0.1326	0.6378	0.6925
2.8	0.2927	0.0857	0.7073	0.6903
3.0	0.2260	0.0511	0.7740	0.6800
3.2	0.1633	0.0267	0.8367	0.6623
3.4	0.1054	0.0111	0.8946	0.6381
3.6	0.0530	0.0028	0.9470	0.6081
3.8	+0.0068	0.00005	0.9932	0.5733

(continued)

x	Directivity Functions ($x = ka\sin\theta$)		Impedance Functions ($x = 2ka$)	
	Pressure	Intensity	Resistance	Reactance
	$\frac{2J_1(x)}{x}$	$\left[\frac{2J_1(x)}{x}\right]^2$	$R_1(x)$	$X_1(x)$
4.0	−0.0330	0.0011	1.0330	0.5349
4.5	−0.1027	0.0104	1.1027	0.4293
5.0	−0.1310	0.0172	0.1310	0.3232
5.5	−0.1242	0.0154	1.1242	0.2299
6.0	−0.0922	0.0085	1.0922	0.1594
6.5	−0.0473	0.0022	1.0473	0.1159
7.0	−0.0013	0.00000	1.0013	0.0989
7.5	+0.0361	0.0013	0.9639	0.1036
8.0	0.0587	0.0034	0.9413	0.1219
8.5	0.0643	0.0041	0.9357	0.1457
9.0	0.0545	0.0030	0.9455	0.1663
9.5	0.0339	0.0011	0.9661	0.1782
10.0	+0.0087	0.00008	0.9913	0.1784
10.5	−0.0150	0.0002	1.0150	0.1668
11.0	−0.0321	0.0010	1.0321	0.1464
11.5	−0.0397	0.0016	1.0397	0.1216
12.0	−0.0372	0.0014	1.0372	0.0973
12.5	−0.0265	0.0007	1.0265	0.0779
13.0	−0.0108	0.0001	1.0108	0.0662
13.5	+0.0056	0.00003	0.9944	0.0631
14.0	0.0191	0.0004	0.9809	0.0676
14.5	0.0267	0.0007	0.9733	0.0770
15.0	0.0273	0.0007	0.9727	0.0880
15.5	0.0216	0.0005	0.9794	0.0973
16.0	10.0113	0.0001	0.9887	0.1021

Source: *Fundamentals of Acoustics* by L. E. Kinsler and A. R. Frey, 2nd ed. (John Wiley and Sons, 1962), p. 506.

INDEX

Taylor series:

$$f(x) = f(a) + \frac{f'(a)}{1!}(x-a) + \frac{f''(a)}{2!}(x-a)^2 + \cdots$$